THE WORLDWIDE
HISTORY OF
TELECOMMUNICATIONS

THE WORLDWIDE HISTORY OF TELECOMMUNICATIONS

ANTON A. HUURDEMAN

WILEY-INTERSCIENCE

A JOHN WILEY & SONS, INC., PUBLICATION

Cover: Inauguration of the New York–Chicago telephone line by A. Graham Bell on October 18, 1892. (Scanned with permission of the ITU from Catherine Bertho Lavenir, *Great Discoveries: Telecommunications,* International Telecommunication Union, Geneva, 1990, p. 39.)

For general information on our other products and services please contact our Customer Care Department within the U.S. at 877-762-2974, outside the U.S. at 317-572-3993 or fax 317-572-4002.

Wiley also publishes its books in a variety of electronic formats. Some content that appears in print, however, may not be available in electronic format.

Library of Congress Cataloging-in-Publication Data:

Huurdeman, Anton A.
    The worldwide history of telecommunications / Anton A. Huurdeman.
        p.   cm.
    "A Wiley-Interscience publication."
    Includes index.
    ISBN 0-471-20505-2 (cloth : alk. paper)
    1. Telecommunication—History.   I. Title.
 TK5102.2 .H88   2003
 384—dc21

                                                    2002027240

Printed in the United States of America

10 9 8 7 6 5 4 3 2 1

CONTENTS

FOREWORD   xv

PREFACE   xvii

ACKNOWLEDGMENTS   xix

PART I   INTRODUCTION AND PERIOD BEFORE 1800   1

1   Introduction   3

1.1 Definition of Telecommunications, 3
1.2 Telecommunications Tree, 7
1.3 Major Creators of Telecommunications, 11

2   Evolution of Telecommunications Up to 1800   14

2.1 Evolution of Telecommunications Prior to 1750, 14
2.2 Evolution of Telecommunications from 1750 to 1800, 16

3   Optical Telegraphy   18

3.1 Tachygraphe of Claude Chappe, 18
3.2 Optical Telegraph of Claude Chappe, 20
3.3 Beginning of Optical Telegraphy, 24

PART II   PERIOD FROM 1800 TO 1850   27

4   Evolution of Telecommunications from 1800 to 1850   29

v

5 Optical Telegraph Systems Worldwide   **34**

    5.1 Optical Telegraph Systems in France, 34
        5.1.1 Chappe Systems, 34
        5.1.2 Other Optical Telegraph Systems in France, 37
    5.2 Optical Telegraphy Outside France, 45

6 Electrical Telegraphy   **48**

    6.1 Evolution Leading to Electrical Telegraphy, 48
    6.2 Electrical Telegraphy in the United States, 55
        6.2.1 Morse Telegraph, 55
        6.2.2 Washington–Baltimore Electrical Telegraph Line, 59
        6.2.3 Pioneering Telegraph Companies, 61
        6.2.4 House Direct Printing Telegraph Systems, 65
    6.3 Electrical Telegraphy in Canada, 66
    6.4 Electrical Telegraphy in Great Britain, 66
        6.4.1 Electrical Telegraphs of Cooke and Wheatstone, 66
        6.4.2 Electrochemical Telegraph of Bain, 72
    6.5 Electrical Telegraphy in France, 72
    6.6 Electrical Telegraphy in Germany, 74
        6.6.1 Railway Telegraph Lines in Germany, 74
        6.6.2 German Electrical Telegraph Equipment for Public Use, 76
    6.7 Electrical Telegraphy in Austria, 83

PART III   PERIOD FROM 1850 TO 1900   **85**

7 Evolution of Telecommunications from 1850 to 1900   **87**

8 Electrical Telegraph Systems Worldwide   **91**

    8.1 Telegraph Transmission Technology, 91
        8.1.1 Open-Wire Lines, 91
        8.1.2 Underground Cable, 94
        8.1.3 Submarine Cable, 95
    8.2 Electrical Telegraph Lines in the United States, 98
        8.2.1 Western Union, 98
        8.2.2 The Pony Express, 98
        8.2.3 First Transcontinental Telegraph Line, 99
        8.2.4 Collins Overland Telegraph Line and the Purchase of
            Alaska, 100
        8.2.5 The Hughes Direct Letter Printing Telegraph, 103
    8.3 Electrical Telegraph Lines in Canada, 104
    8.4 Electrical Telegraph Lines in Great Britain, 106
    8.5 Summary of National Electrical Telegraph Achievements, 107
    8.6 Major Terrestrial Telegraph Lines, 119
        8.6.1 Australian Overland Telegraph Line, 119
        8.6.2 Indo-European Telegraph Line, 124

8.6.3 Great Northern Telegraph Line, 128

8.6.4 Central American Telegraph Line, 128

8.7 Submarine Telegraph Cables, 129

8.7.1 European Submarine Cables, 129

8.7.2 Transatlantic Telegraph Cables, 130

8.7.3 Submarine Telegraph Cables Connecting Europe
Worldwide, 135

8.7.4 Inter-American Submarine Telegraph Cables, 138

8.8 Worldwide Electrical Telegraph Network, 139

8.9 Morse, the Father of Electrical Telegraphy, 141

8.10 Morse Codes, 143

8.11 Morse Telegraphers, 145

9 Image Telegraphy   147

9.1 Facsimile Device of Bain, 147

9.2 Image Telegraph of Bakewell, 148

9.3 Pantelegraph of Caselli, 149

9.4 Autographic Telegraph of Bernhard Meyer, 151

9.5 Telautograph of Elisha Gray, 151

10 Telephony   153

10.1 Evolution Leading to Telephony, 153

10.2 The Telephone of Alexander Graham Bell, 156

10.2.1   Alexander Graham Bell, the Father of Telephony, 156

10.2.2   Early Days of Bell in Great Britain, 159

10.2.3   Bell's Telephone Experiments in the United States, 159

10.2.4   Bell's Telephone: "It DOES Speak", 163

10.2.5   Bell Telephone Company, 165

10.2.6   Bell's Honeymoon Trip to Europe, 167

10.2.7   Telephone Developments in Sweden, 174

10.2.8   Biggest Patent Battle on Telecommunications, 176

10.2.9   Battle of David Against Goliath, 178

10.2.10 Pioneers Leave the Telephone Business, 179

10.3 Companies with Common Bell Roots, 180

10.4 Worldwide Introduction of Telephony, 181

10.5 International Telephony, 181

10.6 The Art of Telephone Sets, 185

11 Telephone Switching   188

11.1 Manual Switching, 188

11.2 Evolution Leading to Automatic Switching, 192

11.3 Strowger System, 194

11.3.1 Strowger's First Operating Exchange, 194

11.3.2 Strowger's Up-and-Around Switch, 195

12   Radio Transmission   199

    12.1 Evolution Leading to Radio Transmission, 199
    12.2 Experiments of Heinrich Hertz, 201
    12.3 Radio Transmission from Theory to Practice, 204
    12.4 The Radio Invented by Marconi, 207
    12.5 Radios of Marconi's Competitors, 212

13   International Cooperation   217

PART IV   PERIOD FROM 1900 TO 1950   223

14   Evolution of Telecommunications from 1900 to 1950   225

15   Worldwide Telephone Penetration   229

    15.1 Worldwide Telephone Statistics, 229
    15.2 Telephone Penetration in the United States, 231
    15.3 Telephone Penetration Outside the United States, 234

16   Electromechanical Telephone Switching   237

    16.1 Worldwide Introduction of the Strowger System, 237
        16.1.1 Strowger System in the United States, 237
        16.1.2 Strowger System in Canada, 238
        16.1.3 Strowger System in Japan, 240
        16.1.4 Strowger System in Germany, 241
        16.1.5 Strowger System in Great Britain, 244
        16.1.6 Strowger System in Austria, 246
        16.1.7 Strowger System in Sweden, 246
    16.2 Automatic or Semiautomatic Switching?, 247
    16.3 Electromechanical Indirect-Control Systems, 250
        16.3.1 Automanual and All-Relay Systems, 251
        16.3.2 Lorimer System, 252
        16.3.3 Panel System, 255
        16.3.4 Rotary System, 258
        16.3.5 Uniselector System in France, 260
        16.3.6 LME 500-Point System, 261
        16.3.7 Hasler Hs 31 System, 262
        16.3.8 Automatic Switching Systems in the USSR, 264
    16.4 Crossbar Switching, 264
    16.5 Private Switching, 266

17   High-Frequency Radio Transmission   269

    17.1 Evolution of Radio Technology, 269
        17.1.1 Spark Radio Transmitters, 269
        17.1.2 Squenched Spark Radio Transmitter, 271
        17.1.3 Poulsen Convertor Arc Radio Transmitter, 274

17.1.4 Frequency Alternator Radio Transmitter, 277
17.1.5 Electronic Radio Equipment, 279
17.1.6 Shortwave Transmission, 280
17.2 Maritime Radio, 281
17.3 Mobile Radio, 285
17.4 Intercontinental Radiotelephony, 287
17.5 RCA and C&W Created to Beat Marconi, 289
17.5.1 Radio Corporation of America, 289
17.5.2 Cable & Wireless, 290

18 Phototelegraphy   **294**

18.1 Kopiertelegraph of Gustav Grzanna, 294
18.2 Telautograph of Arthur Korn, 294
18.3 Telegraphoscope of Edouard Belin, 295
18.4 Siemens–Karolus–Telefunken Picture Transmission System, 296
18.5 Facsimile Machines of AT&T and Western Union, 297
18.6 Photograph Transmission Equipment in Japan, 298

19 Teleprinters   **300**

19.1 Teleprinter Development in the United States, 300
19.2 Teleprinter Development in Great Britain, 303
19.3 Teleprinter Development in Germany, 306
19.4 Teleprinter Development in Japan, 307

20 Copper-Line Transmission   **308**

20.1 Telegraphy Transmission on Copper Lines, 308
20.2 Telephony Transmission on Copper Lines, 314
20.3 Phantom Circuits, 316
20.4 Pupin Coils, 317
20.5 Krarup Cable, 321
20.6 Telephone Amplifiers, 322
20.7 Analog Multiplexing, 324
20.8 Digital Multiplexing, 327
20.9 Coaxial Cable, 331

21 Radio-Relay Transmission   **337**

21.1 Evolution Leading to Radio-Relay Transmission, 337
21.2 World's First Radio-Relay Link, 342
21.3 Initial Radio-Relay Systems, 343

22 Cryptography   **350**

22.1 Manual Coding, 351
22.2 Automatic Coding, 352

23 International Cooperation   **357**

PART V   PERIOD FROM 1950 TO 2000   361

24 Evolution of Telecommunications from 1950 to 2000   363

   24.1 The Semiconductor Era, 364
   24.2 Digitalization, 366
   24.3 New Telecommunications Networks, 367

25 Radio-Relay Networks   369

   25.1 Technological Development of Radio-Relay Systems, 369
         25.1.1 All-Solid-State Radio-Relay Systems, 370
         25.1.2 Digital Radio-Relay Systems, 371
         25.1.3 Radio-Relay Systems for the Synchronous Digital
                 Hierarchy, 374
         25.1.4 Transhorizon Radio-Relay Systems, 375
   25.2 Radio-Relay Systems Worldwide, 376
         25.2.1 Radio-Relay Systems in North America, 376
         25.2.2 Radio-Relay Systems in Latin America, 376
         25.2.3 Radio-Relay Systems in Europe, 379
         25.2.4 Radio-Relay Systems in Asia, 381
         25.2.5 Radio-Relay Systems in Australia, 382
         25.2.6 Radio-Relay Systems in Africa, 383
   25.3 Wireless Access Systems, 386
   25.4 Radio-Relay Towers and Aesthetics, 391

26 Coaxial Cable Transmission   397

   26.1 Terrestrial Coaxial Cable, 397
   26.2 Submarine Coaxial Cable, 399
         26.2.1 Transatlantic Coaxial Telephone Cables, 399
         26.2.2 Worldwide Submarine Coaxial Telephone Cables, 404

27 Satellite Transmission   407

   27.1 Evolution Leading to Satellite Transmission, 407
         27.1.1 Rocketry Pioneers, 408
         27.1.2 Passive Satellites, 410
         27.1.3 Postwar Rocket Development in the United States, 410
         27.1.4 Postwar Rocket Development in the USSR, 411
         27.1.5 Sputnik, the First Satellite, 412
         27.1.6 First Communication Satellites, 413
   27.2 First Synchronous Communication Satellites, 419
   27.3 Satellite Launching, 421
   27.4 Satellite Transmission Systems, 426
         27.4.1 Global Satellite Systems, 427
         27.4.2 Regional Satellite Systems, 428
         27.4.3 Domestic Satellite Systems, 431
         27.4.4 Mobile Satellite Systems, 433

27.4.5 Global Mobile Personal Communication by Satellite, 435
27.4.6 Multimedia Satellite Systems, 439

28 Optical Fiber Transmission   **445**

28.1 Evolution Leading to Optical Fiber Transmission, 445
28.2 Terrestrial Optical Fiber Cable Systems, 456
28.3 Submarine Optical Fiber Cable Systems, 459
    28.3.1 Transatlantic Optical Fiber Cables, 460
    28.3.2 SEA–ME–WE Cable System, 461
    28.3.3 Caribbean ARCOS Network, 463
    28.3.4 Global Submarine Optical Fiber Cable Systems, 463
    28.3.5 African Cable Network Africa ONE, 466
    28.3.6 Various Submarine Cable Systems, 467
    28.3.7 Repeaterless Submarine Cable Systems, 467
28.4 Fiber-in-the-Loop Systems, 471
    28.4.1 Worldwide Testing of FITL Solutions, 472
    28.4.2 Delay of FITL Deployment, 475

29 Electronic Switching   **480**

29.1 Continuation of Deployment of the Prewar Switching
    Systems, 480
    29.1.1 Crossbar Switching, 480
    29.1.2 Siemens Rotary Switch, 480
    29.1.3 End of the Strowger Switch, 482
29.2 Implementation of Automatic Telephone Switching, 483
    29.2.1 National Automatic Switching, 483
    29.2.2 International Automatic Switching, 484
29.3 Electronic Switching Systems, 485
    29.3.1 Evolution toward Electronic Switching, 485
    29.3.2 Preliminary Electronic Switching Systems, 489
    29.3.3 Commercial Electronic Switching Systems, 494
29.4 Digital Switching Systems, 495
29.5 Data Switching, 500
29.6 Integrated Services Digital Network, 505
29.7 Broadband Switching, 506
29.8 Private Switching, 507

30 Telex   **510**

30.1 Continuation of Teleprinter Deployment, 510
30.2 Telex Service, 510
30.3 Teletex, 512
30.4 Termination of Telex Services, 512

31 Telefax   **515**

31.1 Technological Development of Telefax, 515
31.2 Worldwide Telefax Penetration, 517

32 Cellular Radio   **519**

32.1 Evolution of Cellular Radio, 519
32.2 Analog Cellular Radio, 521
    32.2.1 Analog Cellular Radio in Japan, 522
    32.2.2 Analog Cellular Radio in Scandinavia, 523
    32.2.3 Analog Cellular Radio in North America, 523
    32.2.4 Analog Cellular Radio in West Europe, 524
32.3 Digital Cellular Radio, 524
    32.3.1 Global System for Mobile Communication, 528
    32.3.2 D-AMPS System, 532
    32.3.3 Personal Digital Cellular System, 534
32.4 Personal Communications Network, 535
    32.4.1 CT1–CT3 Systems, 536
    32.4.2 Japanese Personal Handyphone System, 537
    32.4.3 Digital European Cordless Telecommunications, 537
    32.4.4 Personal Access Communications System, 539
32.5 International Mobile Telecommunication System, 540

33 Telephony and Deregulation   **546**

33.1 Telecommunications Deregulation and Liberalization, 546
33.2 Telephony and Deregulation in the Americas, 551
    33.2.1   Telephony and Deregulation in the United States, 551
    33.2.2   Telephony and Deregulation in Canada, 555
    33.2.3   Telephony and Deregulation in Mexico, 556
    33.2.4   Telephony and Deregulation in Central America, 557
    33.2.5   Telephony and Deregulation in the Caribbean, 558
    33.2.6   Telephony and Deregulation in Brazil, 558
    33.2.7   Telephony and Deregulation in Chile, 559
    33.2.8   Telephony and Deregulation in Argentina, 561
    33.2.9   Telephony and Deregulation in Peru, 561
    33.2.10 Telephony and Deregulation in Venezuela, 561
    33.2.11 Telephony and Deregulation in Colombia, 561
    33.2.12 Telephony and Deregulation in Ecuador, 562
    33.2.13 Telephony and Deregulation in Bolivia, 563
    33.2.14 Telephony and Deregulation in Uruguay, 563
    33.2.15 Telephony and Deregulation in Paraguay, 563
33.3 Telephony and Deregulation in Africa, 563
    33.3.1   Telephony and Deregulation in North Africa, 565
    33.3.2   Telephony and Deregulation in South Africa, 566
    33.3.3   Telephony and Deregulation in Sub-Saharan Africa, 566
33.4 Telephony and Deregulation in Asia, 567
    33.4.1   Telephony and Deregulation in India, 568
    33.4.2   Telephony and Deregulation in China, 569
    33.4.3   Telephony and Deregulation in Japan, 571
    33.4.4   Telephony and Deregulation in Other Asian
        Countries, 573
33.5 Telephony and Deregulation in Europe, 574

33.5.1   Telephony and Deregulation in the European Union, 575
33.5.2   Telephony and Deregulation in Eastern Europe, 575
33.6 Telephony and Deregulation in Oceania, 577

34   Multimedia   **580**

34.1 Evolution Leading to Multimedia, 580
34.2 Computers and Communications, 581
34.3 Global Information Infrastructure, 581
34.4 Internet, 583
34.5 Global Village, 589
34.6 Multimedia Services, 590

35   International Cooperation   **597**

APPENDICES

**A   Chronology of the Major Events in the Two Centuries of
     Telecommunications**   **601**

**B   Worldwide Statistics of Population, Internet Users, Cellular Phones, and
     Main Telephones**   **607**

C   Glossary   **613**

INDEX   **621**

FOREWORD

Communication—the exchange of information—is essential both for the social life of mankind and the organization of nature. Since human communication is restricted to a relatively small spatial environment, due to physiological and physical conditions, human endeavor has always been directed at the enhancement of natural communication possibilities: achieving *telecommunications*. The history of the development of telecommunications is therefore not only of technical interest but also of general cultural importance.

The beginning of modern telecommunications in the nineteenth century is marked by the discovery of electromagnetism, which initiated new effective methods for long-distance information transmission. Further progress soon resulted from the application of electromagnetic wave propagation in telecommunications systems. In this period the development was stimulated additionally by economic, political, and military requirements. In the twentieth century the introduction of electronics and semiconductor physics led to more rapid and dramatic technical progress, followed by widespread dissemination of telecommunications. Today, telecommunications governs nearly all economic, social, and scientific domains of life with ever-increasing intensity.

This book presents the fascinating story of the technical development of telecommunications. It shows the impressive scientific and technical efforts and the achievements of many ingenious inventors, discoverers, physicists, and engineers during the long journey from telegraphy and telephony—via radio, fiber, and satellite transmission—to mobile radio, Internet, and multimedia services.

This representation offers a concise overview of the field based on the large professional experience and competence of the author. A special feature of the book is the detailed documentation of the worldwide development of telecommunications, covering various countries, thus filling a gap in the relevant technical literature.

The author treats the vast and complex matter in a well-structured and comprehensive form that avoids tedious theoretical detail. The text is enriched by many instructive graphics and photos, together with a lot of historical and technical data and observations. The wide range of original material utilized by the author is cited extensively at the end of each chapter. Historically and technically interested readers—not only those with a scientific background but also persons in the fields of economics, politics, and sociology—will find the book to be an invaluable guide to the basic ideas and most current aspects of global communications and its sources.

The author's opus deserves broad attention.

PROF. DR. PHIL. NAT. DR.-ING. E.H. DIETRICH WOLF

Institute of Applied Physics
Johann Wolfgang Goethe-Universität
Frankfurt am Main, Germany
March 2002

PREFACE

This book has been written to present a comprehensive overview of the worldwide development of telecommunications in a single volume. Ample information on the evolution of various domains of telecommunications in specific countries is preserved in numerous specialized books, magazines, and other publications in libraries of universities and museums, but a single book in the English language covering the entire field of worldwide telecommunications does not exist. To fill this gap, I have collected, evaluated, interpreted, and cross-checked almost a hundred books and even more journals over the last 15 years. Based on my experience and knowledge of telecommunications, I have condensed their contents into a chronological story of the worldwide development of telecommunications. In the interest of truly worldwide coverage, I give information on telecommunications events that took place in over 100 countries and include statistics for over 200 countries. Writing the worldwide history of telecommunications necessarily means using information already published by many experts in their fields. George P. Oslin spent 35 years researching telecommunications documents and interviewing the pioneers or their descendants before, at the age of 92, submitting the manuscript for his fascinating book *The Story of Telecommunications*. Oslin's book, which he relates to his nation's history, was a great inspiration and a valuable resource for me to write this book related to worldwide history. Instead of interviewing pioneers, I have endeavored to find the best published sources available for each subject covered in my book, and thus I could reduce the manuscript preparation time to about five years.

Numerous statements in the book are the result of combined information found in two or more sources and occasionally, cross-checked with a third or even a fourth source. Making reference to all those sources within the text would have a negative effect on readability, so I have cited the references at the end of each chapter. I have mentioned the source directly in the text only in the few cases where a larger portion was based on a single source.

I cover telecommunications starting with optical telegraphy at the end of the eighteenth century; followed in the nineteenth century by electrical telegraphy of coded signals, images, and written text transmitted via open wires and terrestrial and submarine cables; followed by telephony and telephone switching and by radio transmission. Then follows the entire range of new technologies developed in the twentieth century: intercontinental radio, mobile radio, radio-relay, cryptography, satellites, coaxial and optical fiber, terrestrial and submarine cable networks, telex, telefax, electronic switching, cellular radio, and the convergence into multimedia of most of those technologies via a global information infrastructure. To enable adequate coverage of this wide range of technologies within the scope of a single volume, I have excluded the domains of radio and television broadcasting, navigation, telemetry, and computers. This book covers the history from the end of the eighteenth century up to the end of the twentieth century. Some events that happened between January 2001 and December 2002 which were relevant to the status described in the text are covered in footnotes.

I took special care to present the evolution and development of telecommunications as a human achievement attained thanks to the perseverance of many ingenious pioneers who had the vision and capability to turn the discoveries and inventions of contemporary scientists into new and useful applications. Wherever appropriate and available, I include personal information about the major pioneers and protagonists. Applications of the new telecommunication devices have been introduced by several newly founded companies, manufacturers, and service providers, which grew, merged, and still exist or have disappeared. Those companies made telecommunications happen and, again, wherever appropriate and available, I include briefly the relevant industrial history.

For those readers who are not familiar with telecommunications technologies and want to obtain a better understanding of the subject matter, I have included Technology Boxes, which give a concise description of the underlying technologies in more technical language. Numerous footnotes give additional related details.

I have written this book without sponsorship or obligation to any company. Moreover, I have endeavored to present the history in an objective way with balanced coverage of significant events in several countries without overemphasizing the achievements of particular companies or specific countries.

Telecommunications is a nonpolluting employment-generating industry which plays an ever-increasing role in our human relations. It is an indispensable tool for economic growth and better distribution of wealth. My hope is that this book will contribute to the preservation and greater awareness of the worldwide heritage of telecommunications and to responsible future applications.

ANTON A. HUURDEMAN

Todtnauberg, Germany
April 2002

ACKNOWLEDGMENTS

I have written this book in the English language, although my mother tongue is Dutch and I have been living in Germany since 1958. I am very grateful, therefore, to an English friend, Peter Jones, for proofreading, removing Dutch and German influences, and very conscientiously safeguarding the *Queen's English*. I am also grateful to my previous department director, Dipl.-Ing. Gerd Lupke, for proofreading to safeguard the historical truth and for supplying interesting photographs from his private archive. My thanks also go to Prof. Dr. Dietrich Wolf for his valuable suggestions on manuscript improvements and, especially, for writing the foreword.

It was my objective to provide wide international coverage with historical photos, adequately balanced among various companies and museums. Unfortunately, the international response was not sufficient to meet that objective. Personal contacts proved to be important in obtaining the right information and historical photos. I am very obliged, therefore, to those persons who supplied me generously with numerous photos that are reproduced in the book. I thank especially Dr. Helmut Gold, Dieter Herwig, and Jürgen Küster of the Museum für Kommunikation Frankfurt, Germany; Dr. Lanfredo Castellitte of the Musei Civici Como, Italy; Gertrud Braune, Karin Rokita, Dr. Marie Schlund, and Dr. Lothar Schön, of Siemens AG, Munich, Germany; and Gerhard Schränkler of Alcatel SEL, Stuttgart, Germany.

I thank Dr. Julie Lancashire of Artech House Books, London, for permission to reuse drawings and parts of the text that I prepared originally for the book *Radio-Relay Systems*, published by Artech House Books in 1995, and for the book *Guide to Telecommunications Transmission Systems*, published in 1997. Text from those two books has been reused for all Technology Boxes except Boxes 11.1 and 16.2 to 16.5, which are based on information given in great detail in the two volumes of *100 Years of Telephone Switching*, by R. J. Chapuis and A. E. Joel Jr., published by North-Holland Publishing Company, New York, in 1982 and 1990.

Finally, I thank Mrs. Marie-José Urena of the ITU, Geneva, for granting permission to use information from the ITU *Indicators Updates*, and to scan photos from the ITU publications *From Semaphore to Satellite, Great Discoveries: Telecommunications*, and *ITU News*. ITU made this permission subject to an acknowledging statement indicating that:

1. The texts/figures extracted from ITU material have been reproduced with the prior authorization of the Union as copyright holder.
2. The sole responsibility for selecting extracts for reproduction lies with the beneficiary of this authorization alone and can in no way be attributed to the ITU.
3. The complete volume(s) of the ITU material, from which the texts/figures reproduced are extracted, can be obtained from:

   International Telecommunication Union

   Sales and Marketing Division

   Place des Nations–CH-1211 Geneva 20 (Switzerland)

   Telephone: +41 22 730 61 41 (English)
   +41 22 730 61 42 (French)
   +41 22 730 61 43 (Spanish)

   Telex: 421 000 uit ch / Fax: +41 22 730 51 94

   E-mail: *sales@itu.int | http://www.itu.int/publications*

Upon request of the ITU and Artech House Books, I have given in the caption of each figure covered by a publication permission the complete and exact source. Figures without such a credit line in the caption are either from photos taken from my own archive or from drawings that I made especially for this book.

PART I

INTRODUCTION AND PERIOD BEFORE 1800

1

INTRODUCTION

1.1 DEFINITION OF TELECOMMUNICATIONS

Telecommunication is a technology that eliminates distance between continents, between countries, between persons. To contact another person by telephone, only the distance between one's actual location and the next telephone needs to be covered. This distance can be mere centimeters in the industrialized world and kilometers in the developing world. For centuries, messages were transported by messengers, or couriers, who either walked or were transported by horse, coach, or boat, and when fire, smoke, or sound signals were sent they simply confirmed prearranged messages. With telecommunications a message does not need a messenger. Telecommunications eliminated a master-to-servant relationship: replacing the service of a messenger by mechanical telegraph in 1794, by copper wires in 1837, by electromagnetic waves in 1896, and by optical fiber in 1973. Telecommunications enormously reduces the time required to transport messages, accelerates business transactions, and improves human relationships.

The word *communications*, derived from the Latin word *communicatio*, the social process of information exchange, covers the human need for direct contact and mutual understanding. The word *telecommunication*, adding *tele* (= distance), was created by Edouard Estaunié (1862–1942)[1] in 1904 in his book *Traité pratique de télécommunication electrique (télégraphie–téléphonie)* (Figure 1.1), in which he defined telecommunication as "information exchange by means of electrical signals." Estaunié thus limited telecommunications explicitly to "electrical signals." In the

---

[1] Director of the Ecôle Supérieure des Postes et Télégraphes de France, author of various books in which he criticized the prevailing social conditions, and member of the Académie Française.

The Worldwide History of Telecommunications, By Anton A. Huurdeman
ISBN 0-471-20505-2   Copyright © 2003 John Wiley & Sons, Inc.

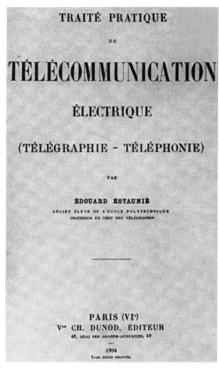

Figure 1.1 Facsimile of the title page of the book in which the word *telecommunication* was created. (Scanned from Catherine Bertho, *Histoire des télécommunications en France*, 1984, p. 13.)

preface to his book, he modestly apologized for the invention of the new word, stating: "I have been forced to add a new word to a glossary that is already too long in the opinion of many electricians. I hope they will forgive me. Words are born in new sciences like plants in spring. We must resign ourselves to this, and the harm is not so great after all, because the summer that follows will take care of killing off the poor shoots." Fortunately, the word *telecommunication* did not belong to the "poor shoots" and has already survived a hundred summers. Telecommunications became more complex, and new definitions were created, as summarized in Technology Box 1.1.

In order to telecommunicate, local, regional, national, and international telecommunication networks are required. Figure 1.2 shows the basic configuration of the classical telecommunication networks. In local telecommunication networks, also called *access networks*, individual telecommunication users (the telecommunication originators as well as the telecommunication recipients) are all connected with one or more *local switches* (also called *local exchanges* or *central offices*). Telecommunication users such as the subscribers of public networks are connected by their local exchange—primarily by means of a single cable pair but previously also by open wire, at distant or isolated locations by radio, and currently, increasingly, by broadband optical fiber or wireless systems. In regional and national telecommunication

TECHNOLOGY BOX 1.1

Definitions of Telecommunications

The *International Telecommunication Union* (ITU) officially recognized the term *telecommunications* in 1932 and defined it as: "any telegraph or telephone communication of signs, signals, writings, images and sound of any nature, by wire, radio, or other system or processes of electric or visual (semaphore) signaling." Currently, the ITU defines telecommunications as "any transmission, emission, or reception of signs, signals, writings, images, and sounds; or intelligence of any nature by wire, radio, visual, or other electromagnetic systems."

In this definition the ITU postulates transmission as a basic function of telecommunications. The word *transmission*, from the Latin *trans mettere* for transfer or transport in the figurative sense, however, quite confusingly, is used for many purposes. It was used in the industrial revolution to represent a *transmission system* for the transmission by mechanical means of power from a central steam engine to the various production machines in a factory. In electrical power technology, *high-tension transmission line and ht-transmission grid* are well-known names for high-voltage overhead electricity distribution lines.

In the book *Transmission Systems for Communications* published by members of the technical staff of Bell Labs in 1954, which used to be *the bible of transmission*, the primary function of a transmission system is described as being "to provide circuits having the capability of accepting information-bearing electrical signals at a point and delivering related signals bearing the same information to a distant point."

In my book *Guide to Telecommunications Transmission Systems*, published in 1997, transmission within the context of telecommunications is defined concisely as the "technology of information transport."

In the context of telecommunications, a transmission system transports information between the source of a signal and a recipient. Transmission thus stands for the *tele* part of the word telecommunications and as such is the basis of all telecommunication systems. Transmission equipment serves to combine, send, amplify, receive, and separate electrical signals in such a way that long-distance communication is made possible.

In terms of technology, telecommunications transmission systems are divided into *line transmission* and *radio transmission* systems:

- *Line transmission* is the technology of sending and receiving electrical signals by means of copper wire, and nowadays, increasingly by means of optical fiber, on overhead lines, by underground cable, and by submarine cables.
- *Radio transmission* in the context of telecommunications stands for the technology of information transmission on electromagnetic waves by means of high-frequency radio and mobile radio, including cellular radio systems, radio relay, and satellites.

Source: Adapted from A. A. Huurdeman, *Guide to Telecommunications Transmission Systems*, Artech House, Norwood, MA, 1997; with permission of Artech House Books.

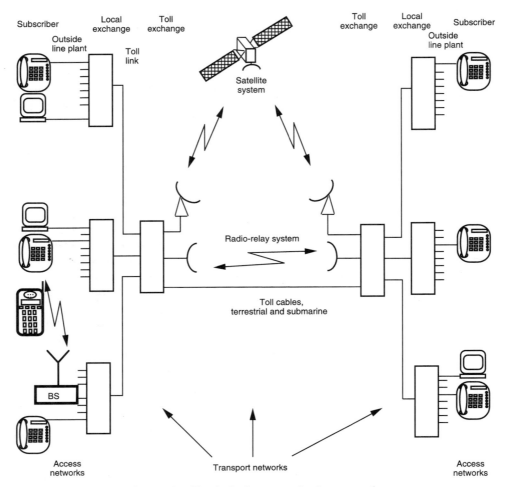

Figure 1.2 Classical telecommunication networks.

networks, a number of local exchanges are connected via transmission links in *transport networks* with a *tandem exchange* (also called a *toll* or *trunk exchange*); all the tandem exchanges of a region or a nation are also interconnected by transmission links. A transmission link can consist of copper wire or optical fiber cable, radio relay, or satellite.

In international telecommunication networks, telecommunication users are connected via their local exchange and one or more tandem exchanges with international exchanges in their country. International exchanges worldwide are interconnected by transmission links either directly or by means of one or more other international exchanges. Currently, this vast network of hierarchically arranged circuit-switching exchanges is being complemented by a new network based on packet switching, using the Internet Protocol.

1.2 TELECOMMUNICATIONS TREE

The evolution of telecommunications from optical telegraphy (also called semaphore communication) to multimedia in its various stages can be visualized in a telecommunications tree (Figure 1.3). Indeed, a tree appears to be very appropriate to illustrate this evolution. On the one hand, it enables us to show the formation of a new technology as soon as the necessary prerequisites are given, and on the other, it demonstrates the complementary function of the various domains that together make telecommunications happen. The trunk of the tree represents the technological prerequisites for successive unfolding of the various telecommunication domains into the branches of the tree. The leaves of the branches represent evolution within the separate telecommunication domains. In the following summary of the evolution, the evolution stages that appear in the tree are italicized.

The bases of telecommunications, and thus the roots of the tree, are *science* and *industrialization*. They made telecommunications possible, and on the other hand, cannot now exist without telecommunications.

Optical telegraphy became possible, and thus telecommunications could germinate, once the *telescope* was available and basic mechanical constructions could be made with sufficient accuracy. *Optical telegraph lines* were constructed within a number of countries for exclusive communications within those countries. Optical telegraphy is the only telecommunications domain so far that disappeared completely and was replaced by a better technology (electrical telegraphy); thus this branch is only a historical relic and therefore has been sawn off.

The theory of *electromagnetism* and the development of *precision mechanics* nourished the growth of an electrical telegraphy branch. *Electrical telegraphy* started with code-writing telegraphs and needle telegraphs, which were replaced by direct text-writing *teleprinters*, which are still used in the international *telex* network but are giving way increasingly to *telefax*, which developed over a 100-year period from *image telegraphy* and *photo telegraphy*. Telegraphy, especially radio-telegraphy, could easily be intercepted, so that *cryptography* became widely used to increase the privacy of telegraphic messages.

Understanding the basic *laws of electricity* and the discovery of *gutta-percha* began the evolution of *copper-line transmission* systems on open wire, copper cable, and coaxial cable. *Carrier-frequency systems* were used to increase the number of channels per physical circuit. *Digitalization* substantially improved the quality and reduced the cost of transmission and switching.

The basic *theory of sound* developed by Helmholtz supported the evolution from telegraphy to *telephony*. *National telephone networks* were built in most countries. Direct dialing in *international telephone networks* became possible worldwide when submarine telephone cables and satellite systems were installed.

The early *automation* of industrial processes enabled the replacement of manual switchboards by automatic *switching* devices. In switching, quite unnoticed by the general public, a tremendous evolution happened in a 100-year period, from *electromechanical switching* by means of *crossbar* and *electronic switching* to *digital switching* with integrated services digital network (*ISDN*) functions.

The discovery of *electromagnetic radiation* and the subsequent development of devices for generating and detecting such waves led to the development of *radio-*

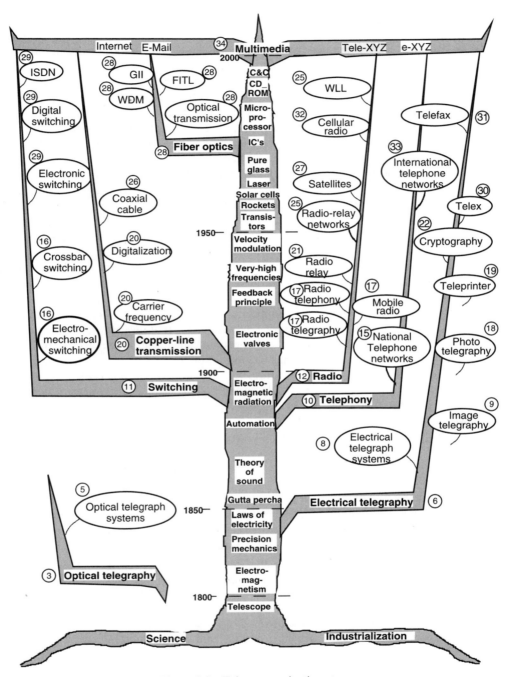

Figure 1.3 Telecommunications tree.

telegraphy. The creation of *electronic tubes* (diodes and triodes) started the electronic era, which enabled the evolution from radio-telegraphy to *radio-telephony* and *mobile radio.* Marine radio was the first mobile radio application, followed by vehicle-mounted private mobile radio and since the 1970s, by infrastructure-sharing trunk systems.

The *feedback* principle applied in electronic circuitry facilitated the generation of high frequencies and thus the development of medium- and shortwave radio transmission and a new technology of circuit combination: carrier frequency, or multiplexing. *Carrier frequency* equipment enabled transmission on a single medium (copper-wire pair, coaxial cable pair, radio-relay, satellite, or optical fiber pair) of thousands of telephone channels. With analog and later also with digital multiplex equipment, installation of national and international *coaxial cable networks* all over the world began in the 1960s.

The development of *very high frequency generators* in 1920 and *velocity-modulated electronic tubes* in the early 1930s made *radio-relay* transmission possible, whereby relay stations, suitably located within lines of sight, receive, amplify, and retransmit radio signals over hundreds and even thousands of kilometers. *Radio-relay networks* were installed beginning in the 1950s, mainly for the distribution of television channels but also as a standby or instead of coaxial cable systems, especially in difficult regions where laying cable would be more expensive. Currently, radio-relay systems are used increasingly for direct access of single subscribers to the public telephone network with wireless-local-loop (*WLL*) systems. This replaces the expensive "last-mile" cable connection between a telephone subscriber and the nearest telephone exchange.

Rockets, transistors, and *solar cells* were the ingredients for the *satellite* branch. Here an evolution is going on of the complementary operation of fixed global and international regional satellite networks, national domestic networks, and global mobile personal communication by satellite networks for person-to-person communication.

The *laser* and extremely *pure glass* enabled the *fiber optics* branch to grow. Long-distance systems with digital signal regenerating repeaters with optoelectronic components are evolving by means of optical amplifiers to regeneration-free soliton transmission systems. For new subscriber access networks, optical fiber cable with fiber-in-the-loop (*FITL*) systems is used increasingly instead of copper cable. Currently, with wavelength-division multiplexing (*WDM*), a number of composite data streams, each with a capacity of 2.5 to 40 Gbps, are transmitted on a single optical fiber pair. Optical fiber cables using WDM are currently being installed between the continents as a major contribution to a global information infrastructure (GII).

ICs (integrated circuits) and *microprocessors* were the nourishment for the cellular radio branch. *Cellular radio* is currently the quickest-growing domain of telecommunications. Here a rapid evolution took place from vehicle-bound analog cellular radio, via vehicle-bound and handheld digital cellular radio, handheld cordless systems, and currently to personal communications networks with person-to-person communication under a single worldwide personal telephone number independent of home, office, or leisure-time location. By year-end 2000, in addition to 987 million fixed telephone lines, some 740 million mobile phones were in use worldwide, and in 36 countries there are more mobile than fixed telephones.

The convergence of communications and computers (*C&C*) and the application of *CD-ROMs* for high-volume data storage is currently leading to *multimedia* services, such as the *Internet* for worldwide interactive information exchange, *tele* working/ medicine/banking/learning/shopping/booking/travel scheduling/entertaining services, and the almost costless *e-mail*.

A global information infrastructure with satellites and optical fiber cable spans the globe. As soon as this infrastructure has been completed, with sufficient connections to communities still unserved, the objective postulated by the International Telecommunication Union (ITU) can be met: "that everybody on this planet can obtain the right answer to her/his questions in a matter of seconds, at affordable cost."

Within two centuries, telecommunications experienced tremendous progress. Especially in the last 100 years, with the application of electronics, transistors, microprocessors, satellites, and optoelectronics, telecommunications became the decisive technology for global human development. This development is best demonstrated by the example of transatlantic submarine cable transmission:

- *1866*. The first transatlantic telegraph cable installed and operated by private enterprise transmitted one Morse-coded telegraph channel with a speed of about 5 words per minute.
- *1956*. The first transatlantic telephone cable, TAT-1, co-owned by the U.S. AT&T, the Deutsche Bundespost, the French France Telecom, the U.K. General Post Office, and other administrations, operated 36 telephone channels on two separate cables.
- *2000*. The state-of-the art transatlantic fiber optic cable, Flag Atlantic-1, owned by the private company Flag Telecom, has 12 fibers each with a capacity of 40 WDM 10-Gbps channels, thus a total of 4.8 Tbps, which is equivalent to 58,060,800 telephone channels.

In another recent comparison it was stated that if automobile technology had progressed at the same pace as telecommunications, a Rolls-Royce would cost less than $2 and get 40,000 miles to the gallon (equal to 17,000 km/L).

Despite its age, the telecommunications tree will continue to grow during a still unpredictable future. Some leaves will drop, as already indicated for the leaves that represent image telegraph, photo telegraphy, and teleprinter, which are no longer used. In the near future other leaves will disappear, such as those representing electromechanical switching, crossbar switching, and telex. Complete branches will probably disappear in the first quarter of the twenty-first century, such as electrical telegraphy and copper-line transmission. New leaves will grow. The first new leaf will probably represent an entirely new range of combined optical transmission-switching systems. Another leaf might represent wireless broadband links in metropolitan areas provided by "subspace" flying base stations located in unmanned balloons and airplanes circling in the stratosphere.

The chronological development of telecommunications for the period 1790–2000 is shown in Figure 1.4. A more detailed chronological summary of the major telecommunications events for this period is given in Appendix A.

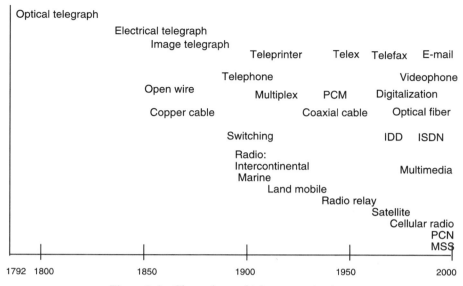

Figure 1.4 Chronology of telecommunications.

1.3 MAJOR CREATORS OF TELECOMMUNICATIONS

Telecommunications development has been the result of timely use of newly discovered technical features by ingenious pioneers who had the vision to create new applications. Those persons, in their time, however, usually faced strong opposition and needed to put forth substantial effort to obtain recognition and acceptance of their invention. Most of them experienced the fate of any discoverer, described very appropriately by the French physicist Dominique François Jean Arago (1786–1853): "Those who discover a new fact in the sciences of observation must expect, first, to have its correctness denied,—next its importance and utility contested,—and afterwards will come the chapter of priority,—then, passages, obscure, insignificant, and previously unnoticed, will be brought forward in crowds as affording evident proofs of the discovery not being new." In this introduction, a brief homage is given to the major pioneers who created telecommunications.

Claude Chappe (1763–1805)   Claude Chappe began the era of telecommunications with the successful operation of his optical telegraph between Paris and Lille on August 15, 1794. People accused him, however, of having copied what they claimed to be their idea. Chappe took these attacks so seriously that he became depressed, and he committed suicide at the age of 42.

Samuel Finley Breese Morse (1791–1872)   The electrical telegraph had many "fathers" and they all developed unique solutions, so that a dozen different electrical telegraph systems operated simultaneously in various countries. In the worldwide competition for the best technology, the writing telegraph of Morse proved its superiority and found worldwide use. Morse became an internationally respected telecommunications expert. To celebrate his eightieth birthday in 1871, a bronze statue of Morse with his 1844 telegraph instrument was placed in Central Park in New York. He died one year later.

Alexander Graham Bell (1847–1922)   The telephone era begun in 1876 in the United States with the operation of a telephone line across a 2-mile stretch between Boston and Cambridge, Massachusetts, with telephone apparatus produced by Bell. For Bell a 10-year patent battle started that in the end, legally gave him the honor and satisfaction of being the inventor of the telephone. Bell then became the most successful of all telecommunications pioneers and gained international prestige. With his wife, Mabel Bell, he lived a prosperous life—the last years in Nova Scotia, where he died at the age of 74.

Heinrich Rudolf Hertz (1857–1894)   Heinrich Hertz laid the basis for radio transmission with successful experiments in 1887–1889 that proved the existence of electromagnetic radiation and its similarity to the behavior of light. Hertz very soon gained substantial international appreciation. Unfortunately, he became ill and died at the age of 37.

Guglielmo Marconi (1874–1937)   It was Marconi who two years after Hertz died began the radio era. Marconi succeeded in transmitting a radio signal over a few kilometers at Bologna in 1896. He successfully combined technical ingenuity with commercial aptitude. Famous and wealthy, at the age of 53 Marconi turned to private life and Italian politics. A heart attack stopped his life at the age of 63. All radio transmitters worldwide observed 2 minutes of silence.

Almon Brown Strowger (1839–1902)   A funeral director in Kansas City, Missouri, Almon B. Strowger, with his "girl-less, cuss-less, out-of-order-less, and wait-less telephone exchange" started a development that has resulted in today's gigantic worldwide telephone network, which is interconnected through thousands of automatic telephone exchanges. For health reasons, Strowger retired at the age of 57 to Florida, where he died in Greenwood at the age of 62.

Michael Idvorsky Pupin (1858–1935)   Michael Idvorsky Pupin was born the son of a "free and independent farmer" in Idvor, Serbia–Croatia. He emigrated to the United States, where he developed the Pupin coil, which made him a millionaire. In 1923 he published his autobiography, *From Immigrant to Inventor*, for which he obtained a Pulitzer Prize. Pupin died in New York at the age of 76.

Alec H. Reeves (1902–1971)   Alec H. Reeves conceived the idea of digitizing speech and patented his pulse-code-modulation (PCM) procedure, but at a time when the prevailing technology prevented its economical realization. Thirty years later, when his ideas could be realized, the importance of his fundamental invention was recognized by the award to Reeves in 1965 of the Ballantine Medal of the Franklin Institute, by the City of Columbus Gold Medal in 1966 and in 1969 by the inclusion of PCM on the 1-shilling postage stamp in the United Kingdom.

Remarkably, eight of the nine creators above are honored for achievements in the eighteenth and nineteenth centuries. They conceived devices that could be realized with the help of only a few persons. To move the arms of his optical telegraph, Chappe needed a mechanical motion device, which he obtained from the experienced clockmaker Abraham-Louis Breguet. To construct an electromagnetic writing device on his easel, Morse availed himself of the technical skill of his student, Alfred Lewis Vail. Bell had his electrician, Watson, and Almon B. Strowger, fortunately, had a technically talented nephew, Walter S. Strowger. Marconi experimented with his radio with the assistance of his brother. Production of the first Pupin coil needed the idea rather than much technical skill. However, to realize in the twentieth century Reeves's idea of PCM, even once the transistor was available, some 10 to 20 engineers at Bell Telephone Laboratories had to undertake years of research work.

In contrast to these essentially one-person inventions of the eighteenth and nineteenth centuries, teamwork was required for the big telecommunications achievements of the twentieth century, and even then, success was not always guaranteed. Development of the first electronic telephone exchange in the 1950s, the No. 1 ESS of AT&T, took seven years and absorbed $100 million. Before the world's first commercial cellular radio system, conceived by the engineers of Bell Labs in 1946, could be put into operation in Japan in 1979, 100 Japanese engineers and technicians required a 12-year period of development. ITT spent a record $1 billion in the 1980s for the development of their digital switching system (System 12), and then abandoned telecommunications. Iridium, the first global mobile personal communication satellite system, was conceived in 1987. Over 1000 engineers, technicians, and mathematicians, mainly in the United States but also in Europe and Asia, with great skill and energy, worked out elaborate designs for components and systems for software, management plans, and logistics at a cost of $3.4 billion before the system could be put into operation on November 1, 1998. By then, unfortunately, they were too late. The unexpectedly rapid worldwide penetration of cellular radio made the Iridium system superfluous.

2

EVOLUTION OF TELECOMMUNICATIONS UP TO 1800

2.1 EVOLUTION OF TELECOMMUNICATIONS PRIOR TO 1750

In its definition of telecommunications in 1932, the ITU expressively postulated "visual (semaphore) signaling" as a means of telecommunications, which implies that telecommunications started with the optical telegraph developed by Claude Chappe during the French Revolution. Visual signaling was not invented by Chappe, but his system was the first that found systematic use in various countries over about half a century and then gradually became replaced by telegraphy with electrical signals.

The Greeks, Persians, and Romans used smoke and fire signals for transmission of predefined information about singular occurrences. For his attack on Troy, Agamemnon erected a 500-km line of beacons in 1084 B.C. After 10 years without being used, the news of the fall of Troy was suddenly transmitted one night and then the beacons became obsolete. The Persian King Darius I (550–485 B.C.) had a fire-telegraph network throughout Persia, enabling him to obtain timely information about any planned rebellion or attack from outside.

In addition to using smoke-and-fire signals, the Romans raised and lowered wooden beams on a platform of special towers placed in a straight line of sight in various areas throughout their empire, up to Hadrian's Wall in northern England. In the Middle Ages, smoke and fire signals were employed between Crusader-built towns and forts in Palestine and Syria. In Spain by 1340, the Castilian navy had adopted signal telegraphy. The Admiral of Castile, D. Fabrique, made use of different-colored pennants to communicate orders and coded messages to his ships warring against the kingdom of Aragon.

In England, a fire and beacon system was used in the sixteenth century to give

The Worldwide History of Telecommunications, By Anton A. Huurdeman
ISBN 0-471-20505-2   Copyright © 2003 John Wiley & Sons, Inc.

warning of the approach of the Spanish Armada. Many of the hills on which these beacons stood are still remembered by names such as "Beacon Hill."

The North American Indians perfected smoke signaling. Not only did they use a varying number of "puffs," but by throwing substances on a fire, they added information to a signal by changing the colors of the smoke.

All these examples of optical signaling applications were limited in their application, and there was always the danger of misinterpretation. They cannot be considered as telecommunication systems in the sense of the ITU definition. The discovery of the telescope in 1608 by the Dutch optician Hans Lippershey[1] was a vital step forward toward a genuine telecommunication system.

The first documented proposal to use the telescope for the transmission of messages was made in a letter dated March 21, 1651 found in the Bavarian state archives in Wurzburg in 1985. The letter had been written at Trier by a Capuchin monk, astronomer, and discoverer, Anton Maria Schyrleus de Rheita (1604–1660), and was sent together with four telescopes to the archbishop and elector of Mainz, Johann Philipp von Schönborn. The monk was born in Reutte, Tirol (hence the name *de Rheita*), as Johann Burchard von Schyrle. In his letter he proposes placing "fingerlong" black characters of complete words in front of a white cloth so that the words can be read through a telescope located at "an hour's distance" (about 5 km). His telescopes[2] were produced in the workshop of one of his pupils at Augsburg. Some telescopes were exported to England, where they were soon copied and used widely on ships. Schyrleus de Rheita's proposal for optical telegraphy did not find approval; on the contrary, he became a victim of the Inquisition and died in exile at Ravenna in 1660.

The first attempt to work out a reliable optical-mechanical signaling system was made by the British astronomer Robert Hooke (1635–1703). On May 21, 1684 he presented to the Royal Society in London a plan for "optical transmission of one's own thoughts on land and between ships at sea with a combination of a telescope and a signaling frame," later published in the *Philosophical Transactions*. He proposed having boards of different shapes—square, triangular, circular, and others—representing the letters of the alphabet, hung in a large square frame divided into four compartments and shown in the order required from behind a screen. Each board represented a letter according to the compartment in which it was hung. Hooke describes the distance between stations and suggests that the signals could be varied in 10.000 ways. The stations should preferably be high and exposed to the sky. Each intermediate station should have two telescopes and three operators, although two operators should suffice at the terminal stations. At night, lights should be used instead of boards.

Hooke tried his system in 1672 between the garden of Arundel House and a boat moored off the far shore of the Thames half a mile away. The president of the Royal Society objected that Hooke's system would often be hindered by the British weather and discouraged Hooke from making further trials.

---

[1] A description of Lippershey's invention, which enabled a 20- to 30-fold magnification of view, reached Galileo in 1609 and then became widely known through Galileo's publication in 1610 of *Siderius Nuncius*, in which he reported the discovery of the moons of Jupiter with the aid of a telescope.

[2] The telescopes were of his own design, as he described in a book entitled *Oculus Enoch et Eliae Sive Radius Sidereomysticus*, published in Antwerp in 1645.

2.2   EVOLUTION OF TELECOMMUNICATIONS FROM 1750 TO 1800

Limited vision at a distance had always been a major obstacle to the introduction of telecommunications. The telescopes available were very expensive and needed a length of several meters to obtain sufficient magnification, albeit with a very faint view. In 1747, Leonhard Euler (1707–1783) found a way to correct the telescope's chromatic error by sandwiching two lenses together, thus making them achromatic, allowing them to be located closer to each other. A few years later, a Swedish physician, Samuel Klingenstjerna (1698–1765), made detailed studies on the color separation characteristics of different types of glass.

In 1757, an English optician, John Dollond (1706–1761), applied the existing knowledge to the construction of a telescope with substantially improved image quality and resolution. The Dollond telescope, with achromatic lenses became the standard for many years. With a Dollond telescope, visibility on a line of sight between geographical points could be increased to tens of kilometers, and an indispensable tool for optical telegraphy thus became available.

Around 1760, a British–Irish teacher, Richard Lovell Edgwort (1744–1817), constructed a privately operated optical signaling line over 90 km between London and Newmarket, probably based on the Hooke system.

A German scientist from Hanau, Johann Andreas Benignus Bergsträsser (1732–1812), tried all kinds of communications: from fire, smoke, explosions, torches and mirrors, trumpet blasts, and artillery fire, to a gymnastic signaling experiment using Prussian soldiers. He proposed constructing an audio-optical telegraph line between Leipzig and Hamburg using four types of rockets: rockets without detonation, rockets with detonation, illuminating signal rockets, and firework rockets. Moreover, he proposed light signals on clouds and audio signals using ringing bells.[3]

During the American War for Independence from 1775 to 1783, a signaling system was used with flags in the daytime and lanterns at night. In France the first practical application of optical signaling was used by Captain de Courrejolles of the French marines in February 1783 on the west coast of Greece. A British squadron under Admiral Hood had blocked French vessels. Courrejolles quickly erected an optical signaling device on the highest accessible site on the Greek coast, from where he watched the movements of the British and informed his commander on the nearby leading French ship. Thanks to this information advantage, the French were able to defeat the British squadron and reach the harbor safely. However, the French authorities were still not convinced of the usefulness of optical signaling.

It was during the French Revolution, with the creation of a new national republican state, that the merits of a permanently installed communication network were finally recognized. At the height of the revolution, France was surrounded by the allied forces of Britain, Holland, Prussia, Austria, and Spain. Moreover, the French cities of Marseille and Lyon were in revolt and Toulon was held by the British fleet. To master this dangerous situation, an optical telegraph as proposed by Claude Chappe proved to be a very effective instrument. Thus, receiving a message by means

---

[3] In a five-volume work published in 1784 entitled *Sinthematografie*, Bergsträsser reviewed all known means of communications ever devised and even gave an account of signaling using shutters or pivoted arms set at angles and operated through bevel gears, so anticipating the telegraphs that were to be developed in the next few decades.

of an optical telegraph line within minutes instead of within weeks by a messenger was at the time even more impressive than it is for us to send e-mail around the world in seconds instead of a letter in a few days. The news of the creation of Chappe's optical telegraph spread widely throughout the world and encouraged people to construct similar systems in most European countries as well as in India, Australia, Canada, and the United States. Basically, three types of optical telegraphs evolved:

1. *Arms type:* using movable arms whose positions represented coded signals for letters, numerals, phrases, or operating commands
2. *Boards type:* using boards whose raising or lowering made up signals according to the number of boards or partitions visible
3. *Moved-to-fixed type:* using moving elements (spheres, flags, boards, or partitions) by which the signals were formed by the relative positions of the moving elements in relation to fixed flags, boards, or panels

The optical telegraph was the first functional telecommunications device to be used successfully until succeeded by a superior solution: the electrical telegraph. Thus it is that the creator of the optical telegraph, Claude Chappe, deserves to be called the *"father of telecommunications."*

3

OPTICAL TELEGRAPHY

3.1 TACHYGRAPHE OF CLAUDE CHAPPE

On July 14, 1789, the population of Paris claiming "we are the people," attacked the Bastille, liberated the prisoners, and started the French Revolution. On January 21, 1793, King Louis XVI was executed using the newly developed guillotine. The royalty of surrounding countries, in a desperate effort to prevent revolutionary ideas from entering their territories, joined forces against France. On August 23, 1793, the *Convention Nationale* declared the whole of France under a *state of siege* and decided on a military enlistment *en masse*, In this dramatic situation, quick dissemination of information and immediate reactions were essential. Fortunately for France, Abbé Claude Chappe (1763–1805) had just started experiments with what he called a *tachygraphe* (Latin for "rapid writer").

Born at Brûlon, Sarthe, on December 25, 1763, Chappe was the second child of prosperous parents with five sons: Ignace Urbain Jean, Claude, Pierre-François, René, and Abraham. Their uncle was a celebrated astronomer Abbé Jean Chappe d'Auteroche. Claude was trained for the church but was more attracted by science and devoted himself to scientific investigations, including a study of what we now call telecommunications. He first attempted to use electricity for transmission of messages. Due to the limitations of that early stage of electrical development, with poor insulation, low mechanical strength of copper wires, and unreliable sources of electricity, he turned to optical-acoustics and eventually to wholly optical methods. By the age of 20, Claude had already been accepted as a member of the Société Philomatique as an award for articles on his experiments published in the *Journal de Physique*.

He first tried an optical-acoustic system using two large clocks, in 1790. The

The Worldwide History of Telecommunications, By Anton A. Huurdeman
ISBN 0-471-20505-2   Copyright © 2003 John Wiley & Sons, Inc.

Figure 3.1 Tachygraphe of Claude Chappe. (Scanned with the permission of the Museum für Kommunikation, Frankfurt, Germany, from Klaus Beyer et al., *So weit das Auge reicht*, Museum für Kommunikation, Frankfurt, Germany, 1995, p. 35.)

clocks were synchronized and their dials showed agreed-upon signs. When the hand of one dial reached the signal to be sent, two copper pans emitted a sound that could be heard 400 m off. As synchronizing a long line was difficult, with the sound disturbing people and limiting the operating distance, Claude turned to a less elaborate optical solution, his *tachygraphe*. On one tachygraphe a pointer was rotated to a coded signal; on a second tachygraphe, placed within visibility of a telescope, the same signal, if recognized, was repeated. The tachygraphe (Figure 3.1) looked like an upgraded guillotine, an early example of the peaceful use of arms. The first experiments with the tachygraphe were made for local officials over a 15-km distance between Brûlon and Parcé on March 2, 1791. Further experiments were made in Paris. Thanks to the influence of his brother Ignace, who was a member of the legislative assembly, he was allowed to erect a tachygraphe at the Etoile barrier. On two occasions a tachygraphe erected at the Etoile was destroyed by a furious mob which suspected that Chappe was communicating with King Louis XVI, imprisoned in the nearby Temple.

3.2 OPTICAL TELEGRAPH OF CLAUDE CHAPPE

The original tachygraphe was limited in visibility and the number of signal variations. With the assistance of Abraham-Louis Breguet (1747–1823), Chappe constructed a new model using moving arms, which, with minor changes, lasted for over 50 years. In 1792, he submitted details of his new machine to the Convention, successor of the legislative assembly. The Convention referred the matter to the Committee of Public Instruction, which on April 1, 1793, reported most favorably to the Convention and asked for an experiment. In the same month, Chappe's friend, Miot de Mélito (1762–1841), departmental chief in the Ministry of War, convinced Chappe to change the name of his invention from *tachygraphe* to *télégraphe aérien* (from the Greek *tele* = distant and *grapheus* = writer).[1]

Chappe's telegraph (Figure 3.2) consisted of a *regulator*, approximately 4.5 m long and 0.35 m wide, to which two *indicators* were attached, each approximately 2 m long and 0.33 m wide. The indicators were made like a window shutter, with alternating slats and apertures, half the slats being set to the right and half to the left, to lessen wind resistance and increase visibility. The indicators were balanced by thin iron counterweights. Generally, the regulator and indicators were painted black, but to improve visibility where necessary, blue triangles were painted on the white regulator and indicators (blue and white being the colors of the French Revolution).

The positions of the regulator and indicators could be changed via three cranks and wire ropes. The regulator could have four distinguishable positions (horizontal, vertical, right inclined, and left inclined), and each indicator, seven positions (0°, 45°, 90°, 135°, 225°, 270°, and 315°). Altogether, then, $4 \times 7 \times 7 = 196$ different configurations were possible. Inside the station, a miniature version of the apparatus reproduced the movements. Figure 3.3 shows the telegraph of Chappe on top of the Louvre as well as a summary of 77 different configurations of the telegraph. Léon Delaunay, related to the Chappes and a former French consul at Lisbon, drew up the first vocabulary for the telegraph. This vocabulary, derived from diplomatic correspondence, contained 9999 words, phrases, and expressions, each represented by a number. It soon proved to be too slow and inconvenient, as from one to four signals were needed to transmit a group of one to four ciphers.

To increase transmission speed, Chappe introduced a new code in 1795. The horizontal and vertical positions of the regulator were reserved for "assuring" the signals, which were first executed with the regulator oblique, then reported and confirmed in the horizontal and vertical positions. The effective number of "working" signal positions was thereby reduced to 96. Chappe then reserved 92 of the 96 signal positions for sending information. He produced a new vocabulary with three categories each of 92 pages with 92 expressions, for a total of $3 \times 92 \times 92 = 25,392$ different significations. One sign was needed to show the category, another for the page,

---

[1] The word *semaphore* (Greek for "bear a signal"), created in 1801 by a Frenchman named Depillon, is generally used for locally limited indication of a small number of signals by means of arms pivoted directly on a mast.

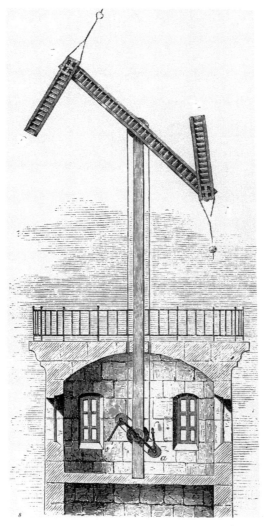

Figure 3.2 Optical telegraph of Chappe. (Scanned with permission of the ITU from Anthony R. Michaelis, *From Semaphore to Satellite*, International Telecommunication Union, Geneva, 1965, p. 18.)

and a third for the number of the word or expression on that page. The first category contained 8464 words; the second category had 8464 expressions or parts of phrases, such as degrees of urgency, incidence of fog, and destination of dispatch; and the third category concerned names of places and phrases used in correspondence.

Once a first message had been given a specific configuration, an operator equipped with a telescope at a second station had to recognize the configuration and bring his telegraph into the same position. As soon as the second telegraph correctly reproduced the configuration of the first, the first would start sending the next signal and an operator at a third station would bring his telegraph into the configuration of the first signal. In this way a signal was repeated from one station to the next.

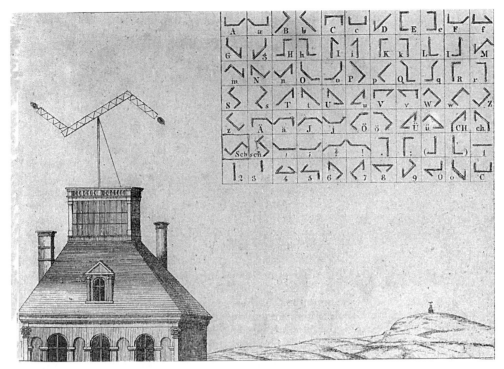

Figure 3.3   Paris terminal. (Courtesy of Museum für Kommunikation, Frankfurt, Germany.)

The coding of the 92 configurations was replaced in 1895 by a codebook with $92 \times 92 = 8464$ words, abbreviations, and complete sentences, each allocated to a group of two configurations. Any information then required two signals, the first indicating the page of the codebook, the second referring to the respective line on that page. Further sophistication eventually resulted in a total of 40,000 codes (kept in the Post Museum in Paris). Coding and decoding were needed only at the terminal stations and at divisional stations (every tenth to fifteenth station). At the repeater stations the operator simply repeated the configuration without knowing the contents. The operators at the intermediate stations needed to know only a few operational codes, such as "error in transmission," "rain [or fog] prevents transmission," and "end of transmission."

The decoding and coding at the divisional stations caused a delay in transmission. However, this was of great value when bad weather interrupted the service on part of the line. Messages thus stopped at a divisional station could be sent on by messenger to a divisional station that did not suffer from bad weather.

The mechanical construction of Chappe's telegraph was made by Abraham-Louis Breguet, the founder of the famous French–Swiss watch manufacturers and the grandfather of Louis Breguet, the constructor of the French electrical telegraph, the great-grandfather of Antoine Breguet, the first manufacturer of Bell telephones in France, and the great-great-grandfather of Louis Breguet, a French aviation pioneer and cofounder of Air France. Abraham-Louis Breguet produced a prototype of

Figure 3.4 World's first telecommunications workshop at 39 Quai de l'Horloge, Paris.

Chappe's telegraph in his watch workshop, which he had opened in 1774 at 51 Quai des Morfondus, Paris, at that time a cosmopolitan location on the Ile de la Cité. For many years, the mechanical parts of Chappe's telegraph were manufactured in this building, which thus can be considered to be the world's first industrial telecommunication equipment production site. The building still exists and is in the possession of a Breguet descendant.[2] In Figure 3.4 it is the four-story building at the right-hand side with the ancient lantern on the wall.

Claude Chappe's brother Ignace Chappe arranged for Claude to demonstrate his improved telegraph to the Convention on March 22, 1793. Claude explained to the Convention that with his telegraph, orders of the Convention could reach French

---

[2] Unfortunately, the building—once called "Maison Breguet"—does not show any signs of the Brequet dynasty and currently accommodates the Librairie du Palais and a papeterie. Due to cadastral modification under Napoleon, the name of the road was changed and the address is now 39 Quai de l'Horloge.

troops outside France within one day. The Convention asked for a trial, which took place on July 12, 1793 over a 35-km line between the park of Saint-Fargeau and Saint-Martin-du-Tertre, with an intermediate station at Ecouen. It took 11 minutes to send the following message: "Danou has arrived here. He announced that the national Convention has just authorized his Committee of General Security to put seals on the papers of the deputies."

3.3   BEGINNING OF OPTICAL TELEGRAPHY

Chappe's experiment was a complete success. The Convention approved adoption of the telegraph as a national utility and instructed the Committee of General Security to map suitable routes. Claude and his younger brother Abraham were appointed as the world's first telegraph engineers. The Chappe brothers were ordered to construct on a turnkey basis at a firm price a telegraph line between Paris and Lille, a town 210 km north of Paris where the Republic's Army of the North was fighting. They were authorized to use any church and castle towers, to remove obstacles, and to cut down trees—a quick *right-of-way* clearance inconceivable nowadays. Local authorities were ordered to contribute personnel and materials. The line started at the Paris terminus, the *Pavillon de Flore* of the Louvre, and ran via the St. Pierre de Montmartre church (Figure 3.5), the trial stations Ecouen and Saint-Martin-du-Tertre, through Clermont, Belloy, Boulogne-la-Grasse, Lihons, and Thélus (east of Arras), to the terminus at St. Catherine's in Lille. The Paris–Lille telegraph line was ready by July 1794. The line had 23 stations at distances varying between 4 and 15 km. Transmission of a signal took 2 minutes (in present terminology, 1 : 120 = 0.008 bps).

The first official message was passed over the Paris–Lille line on August 15, 1794, only one hour after French troops had recaptured the French town of Le Quesnoy (about 200 km north of Paris) from the Austrian troops who had occupied the town since September 12, 1893. The Convention received the good news via the optical telegraph line within one hour of the battle's end and 10 hours before a courier could have arrived. *Thus the era of telecommunications began on August 15, 1794.*

A further triumph came on September 1, 1794, when 20 hours before the courier arrived, the telegraph reported that the town of Condé was restored to the Republic. The position of the telegraph was now assured. The fame of the invention soon spread, and travelers visited the Louvre to observe the mysterious rapid movements of the newly invented telegraph.

The Committee of Public Safety decided on October 3, 1794 to build a second line, from Paris to Landau (near Karlsruhe, Germany), via Metz and Strasbourg, but a shortage of money and changes in priorities delayed implementation. After many interruptions, the Directory, which had replaced the Convention in October 1795, finally, on November 17, 1797, ordered rapid construction of the Paris–Strasbourg line. The planned Peace Congress at Rastatt (starting on December 9, 1797, interrupted on April 23, 1799) suddenly required a line of rapid communication with the French delegates. The 46 stations were erected quickly, and the 480-km line was opened on May 31, 1798. The Strasbourg terminus was located on the tower of the transept of the cathedral. The main station in Metz was on the Palais de Justice, and the Paris terminal was on the tower of St. Sulpice church. It took about 6 minutes to pass a signal between Paris and Strasbourg. A continuation from Strasbourg to

Figure 3.5 Chappe telegraph on the church of St. Pierre. (Scanned with the permission of the Museum für Kommunikation, Frankfurt, Germany, from Klaus Beyer et al., *So weit das Auge reicht*, Museum für Kommunikation, Frankfurt, Germany, 1995, p. 33.)

Huningeu (near the Swiss border north of Basle) with 14 stations (mainly on church towers) was opened on August 16, 1799, but closed again the next year. In an effort to continue the peace talks interrupted in Rastatt, another peace congress was arranged to take place in Lunéville (30 km east of Nancy). Within two weeks, therefore, another offshoot from the Paris–Strasbourg line was constructed in September

1800, with 14 stations, from Vic to Lunéville.[3] The line was closed again after the Peace Treaty of Lunéville was signed on February 8, 1801.

The next very ambitious line went from Paris to Brest. This 870-km line with 55 stations was completed within seven months and opened on August 7, 1798. The line ran from the Ministry of Marine in the Place de la Concorde via Dreux, Avranches, Mont-St. Michel, St. Malo, and St. Brieuc to the Brest terminal.

Claude Chappe, who had always hoped that his invention might be applied to commerce, proposed to Napoleon a pan-European commercial system stretching from Amsterdam to Cadiz (in southern Spain) and even taking in London. He claimed to be able to communicate between Calais and Dover, probably using one or two telegraph ships anchored in the English Channel. He also proposed to relay daily stock exchange and other news. Unfortunately, all these schemes were rejected as being impractical; Napoleon consented only to weekly transmission on the national network of the numbers of the winners in the national lottery.

REFERENCES

Books

Bertho, Catherine, *Telegraphes & telephones de Valmy au microprocesseur*, Hachette, Paris, 1981.

Beyer, Klaus, et al., *So weit das Auge reicht: die Geschichte der optischen Telegrafie*, Museum für Post und Kommunikation, Frankfurt, Germany, 1995.

Brodbeck, Didier, *Journal imaginaire d'Abraham-Louis Breguet*, Editions Scriptar, Lausanne, Switzerland, 1990.

Wilson, Geoffrey, *The Old Telegraphs*, Phillimore & Co., Chichester, West Sussex, England, 1976.

---

[3] Faithful reconstruction of one of the stations of the Paris–Strasbourg line has been built at Haut-Barr, 35 km northwest of Strasbourg. The building was inaugurated by Les Amis de l'Histoire des PTT d'Alsace, May 31, 1968. The last optical telegraph from this station was sent in August 1852. The two-story round-tower station is complete with all mechanical equipment. A former adjacent house for the operators has been rebuilt in original style, but on a slightly larger scale, to serve as a museum of the Chappe telegraph.

PART II

PERIOD FROM 1800 TO 1850

4

EVOLUTION OF TELECOMMUNICATIONS FROM 1800 TO 1850

Optical telegraphy was the only means of telecommunications available for almost the entire first half of the nineteenth century. It was used widely in France, in the countries at war with France, in most other European countries, and eventually, on all continents. Although introduced for military applications, the commercial use of optical telegraphy became popular after the deportation of Napoleon to the isle of St. Helena in 1815. The optical telegraph met the requirements of its era, being the best solution available. A drawback, however, was its dependence on appropriate weather and daylight. During the night, in fog, rain, and snow, as well as on hot days with dusty air or strong winds, the semaphore could not be "online." On the other hand, good visibility made the semaphore an easy military target. Optical telegraphy was an exceedingly costly affair, too. The building of many towers not more than 8 to 15 km apart demanded great capital, and to maintain the staff at each station needed a constant source of money. Optical telegraphy could therefore only be afforded by the state—and then primarily for military or naval messages and for political and police information deemed sufficiently urgent and important. The introduction of electrical telegraphy at the end of this period also meant the end of optical telegraphy in most countries. Simple semaphore signaling, however, was still in use between ships until a few years ago.[1]

The most important invention paving the way toward electrical telecommunications was made in 1799. Alessandro Volta (1745–1827), a self-taught scientist extending the investigations of Luigi Galvani (1737–1798), discovered that "galvanic

---

[1] The railways, too, used semaphore arms for signaling before being replaced by colored electrical lights. In the Spanish language the word *semaphore* has survived as *semáforo*, the name for a traffic light.

The Worldwide History of Telecommunications, By Anton A. Huurdeman
ISBN 0-471-20505-2   Copyright © 2003 John Wiley & Sons, Inc.

Figure 4.1 Two original voltaic piles connected in parallel. (Courtesy of Musei Civici Como.)

electricity" could be generated by placing two different metals in an acidic (electro-lytic) liquid. He constructed an electrolytic cell, which he called the "electro-motor apparatus," later called a *voltaic pile* or *voltaic cell* (Figure 4.1). Each pile consisted of a number of copper and zinc disks separated by coats of acidic water–soaked cloth placed above each other in a jar. Thus the first continuous source of electricity became available. Volta presented this invention officially on March 20, 1800, to the Royal Society of London. In 1801, Napoleon invited Volta to Paris, where Volta repeated his experiments with two voltaic piles at the National Institute in the pres-ence of Napoleon (Figure 4.2), who honored Volta with a gold medal and an annual income.

A British chemist, William Nicholson (1753–1815), learned about the voltaic pile and constructed a similar cell together with Antony Carlile (1768–1840). In 1800 they discovered the electrolytic decomposition of water into its two constituent gases, hydrogen and oxygen. A few attempts were made to use this electrolytic phenome-non as a means of signal transmission. In 1798, the Spanish Franscisco Salvá y Campillo (1751–1828) used the development of hydrogen bubbles on the negative electrode as a signal indicator. A German anatomist, Samuel Thomas von Soem-mering (1755–1830), conceived the same idea and demonstrated it to the Munich Academy of Science in 1809. Soemmering's electrochemical telegraph was also dem-onstrated to Napoleon, who rejected the solution as "une idée germanique." In 1811, von Soemmering, with the assistance of Baron Schilling von Cannstadt, repeated experiments with his electrochemical telegraph using wires, insulated with sealing wax, that were passed through the river Isar.

The 1820s brought a series of discoveries that turned attention away from elec-trochemical signaling toward electromechanical signaling and within three decades

Figure 4.2 Volta explaining the voltaic pile to Napoleon, as painted in 1897 by G. Bertini. (Courtesy of Musei Civici Como.) See insert for a color representation of this figure.

resulted in electrical telegraphy. It started in 1820 when Hans Christian Oerstedt (1777–1851)[2] discovered the electromagnetic field caused by electric current.

Within two months of this discovery, André-Marie Ampère (1775–1836) reported to the Academy of Science in Paris his discovery that an electric current through a wire not only influences a magnetic needle electromagnetically but also has an electrodynamical influence on other electricity-conducting wires. Ampère coined the word *electrodynamics* and proposed to use the deflection of a compass needle when an electric current flows around it, for telegraphy. On October 2, 1820, he proposed an electromagnetic telegraph consisting of 30 magnetic needles each controlled by two conductors. This elaborate device, requiring a 60-wire line between two telegraphs, was never made. In 1822 he constructed the first coil. Ampère became world famous not for this early proposal for the introduction of electrical telegraphy but for his discovery of two basic characteristics of electricity: tension (now expressed in volts) and current, in his honor expressed in amperes.

One month after Ampère reported his discoveries, a German professor of physics, Johann Salomon Christoph Schweigger (1779–1857), presented at the University of Halle the first galvanometer, consisting of a magnetic needle rotating in the middle of a multiturn coil in which each turn adds to the electromagnetic force if a current flows through the coil. Johann Christian Poggendorf (1796–1877), at that time a

---

[2] Oerstedt was a professor of physics at the University of Kiel, then belonging to Denmark, now to Germany. His discovery caused a sensation throughout the scientific world of the time. Until then the fundamental basis of Newtonian science had been built upon the assumption that forces act along straight lines between two points. Oerstedt discovered forces operating in circles, a completely new, unexpected, and inexplicable phenomenon.

student of physics in Berlin, presented independently a similar device, which he baptized the *multiplicator*.

An Italian physicist, Leopoldo Nobili (1784–1835), invented the astatic needle pair in 1825. He made a galvanometer independent of the effect of Earth's magnetism and thus increased its sensitivity by adding a second needle parallel to the first needle on the same axis but at opposite polarization and located outside the coil.

In 1825, a self-educated British physicist, William Sturgeon (1783–1850),[3] constructed the first electromagnet, consisting of a piece of horseshoe-shaped iron with a coil at each end. The two coils consisted of uninsulated copper wire wound spirally around an iron core that was covered with an insulating layer of varnish. He discovered that a current passing through both coils created a magnetic field between the two iron ends.

In 1826, a German physicist, Georg Simon Ohm (1789–1854), demonstrated with a galvanometer in Berlin that the intensity (I) of a current through a resistance (R) is directly proportional to the potential (U) of the electrical source. In 1827, in his book *Die galvanische Kette, mathematisch bearbeitet* (The Galvanic Chain, Elaborated Mathematically), he proposed the basic electrical law $U = IR$, which much later became *Ohm's law*. His book was translated into English, French, and Italian.

In 1828, an American physicist, Joseph Henry (1797–1878), improved Sturgeon's electromagnet by applying various layers of insulated copper wire around each other for the coils, thus substantially increasing the electromagnetic force. He developed practical rules for the construction of electromagnets and constructed the first relay in 1835, both vital prerequisites for the construction of electromagnetic telegraph systems.

In 1830, a British priest and physicist, William Ritchie (1790–1837), inspired by Ampère, demonstrated the transmission of electric signals over a distance of 20 to 30 m using what he called a torsion galvanometer. In the *Journal of the Royal Institution* of October 1830 he prophesied: "We need scarcely despair of seeing the electromagnetic telegraph established for regular communication from one town to another at great distance."

In 1831, another self-educated British scientist, Michael Faraday (1791–1867), starting as a message boy and becoming a famous physicist and member of the Royal Society, discovered that the movement of a magnet relative to a conducting circuit produces an electric current in the circuit, called the *law of electromagnetic induction*. Faraday presented his findings to the Royal Society on November 24, 1831, in a lecture later published under the title "Experimental Researches in Electricity." He revealed the reciprocal nature of the laws of magnetism and predicted the existence of electromagnetic waves, a major achievement for the further development of electromagnetic applications and the development of radio transmission at the end of the nineteenth century.

The accumulation of the aforementioned discoveries together with the optical telegraphy already being used widely in various countries, as well as the need for

---

[3] Sturgeon was born in Lancaster and since 1824 had been a lecturer in science at the Military College, Addiscombe, Surrey. His first electromagnet, with a weight of only 200 g, was able to support up to 4 kg of iron using the current from a single voltaic cell.

rapid signaling for the operation of the newly developing railway routes,[4] created a situation in which electrical telegraphy became not only possible but highly desirable. Almost simultaneously, in 1837, Cooke together with Wheatstone presented an electrical needle telegraph in Great Britain and Morse an electrical writing telegraph in the United States. The two competing electrical telegraph systems, those of Morse and of Cooke and Wheatstone, soon proved their superiority over optical telegraphy. For new telegraph lines, electrical systems were the preferred solution, and existing optical telegraph systems were replaced by electrical systems within a few years.

---

[4]The electrical telegraph became a good solution for communication between the operator of a winding engine located permanently on top of a steep section and the machinist in a locomotive on steep railway slopes. Even more important was its use in the coordination of trains on single-track sections.

5

OPTICAL TELEGRAPH SYSTEMS WORLDWIDE

5.1 OPTICAL TELEGRAPH SYSTEMS IN FRANCE

5.1.1 Chappe Systems

In 1801, all existing telegraph lines were placed under the responsibility of Claude Chappe. In 1802, the Paris–Lille line was extended to Brussels and in 1809 to Antwerp. In 1804, Napoleon ordered the construction of a line from Paris to Milan via Lyon. The line was constructed up to Dijon, then work was interrupted until 1805.

Regretfully, Claude Chappe could not enjoy his success. Various persons accused him of having copied what they claimed to be their idea. The attacks affected Chappe very seriously. He became nervously depressed, and probably aggravated by chronic bladder trouble and an eye disease, committed suicide by throwing himself down a well at the telegraph headquarters on the Rue de Grenelle in Paris on January 23, 1805. He was buried in Vaugirard cemetery, but later reinterred in the famous *Père Lachaise* cemetery with his brother Ignace, who died in 1829. A lead reproduction of the télégraphe was placed on the tomb.

To commemorate 100 years of telegraphy, a handsome bronze statue showing Claude Chappe standing in front of his télégraphe with a telescope in his hand was erected in Paris in 1893. Located at the junction of Boulevard Saint-Germain and the Rue du Bac, close to a metro station, the statue became a favorite rendezvous. Unfortunately, this valuable statue was summarily removed in 1942, melted down, and used for weapons by the German occupation forces. A heavily damaged telegraph (showing the "at rest" sign) on a badly cared for grave (Figure 5.1) is all that is left in Paris as a remembrance of the genius who deserves to be called the "father of telecommunications."

Claude Chappe's brothers Ignace and Pierre-François succeeded Claude as joint

The Worldwide History of Telecommunications, By Anton A. Huurdeman
ISBN 0-471-20505-2   Copyright © 2003 John Wiley & Sons, Inc.

Figure 5.1   Uncared-for grave of the "father of telecommunications."

telegraph administrators. Abraham Chappe was attached to the État Major General of the Grande Armée and was promoted to the rank of colonel by imperial Decree on August 30, 1805. His task was to translate messages for or from the emperor and his immediate staff and keep the emperor informed of troop movements reported by the telegraph.

In 1804, Napoleon prepared for an attack on England, with the Grand Armée encamped at Boulogne, to which the telegraph had been extended via Saint-Omer one year earlier. He ordered Abraham Chappe to devise a means of telegraphing across the English Channel by day and night. For some three years the ends of the arms of the machine on the Louvre had carried lanterns experimentally for night use. Based on these experiments, Abraham, supported by Claude, constructed the "telegraph of Boulogne." To overcome distance and fog, he dispensed with the end arms, made the indicator 5.8 m long and 80 cm wide, and divided it into two separately moving parts. The day-and-night machine was on top of a mast 9.7 m in height. Each moving part, counterbalanced by a 5.8-m-long wooden rod, bore at its end a lantern with a parabolic reflector of diameter 43 cm. A third lantern had a fixed position on top of the mast. Hydrogen, oxygen, and carbonate of lime provided the fuel. Its light was said to be visible 8 leagues (38.6 km) away. The abandonment of the invasion plan ended the experiment.

Work on the Paris–Lyon line was resumed on Napoleon's order in June 19, 1805. Operation from Lyon started two years later. In the meantime, war was renewed with Austria, and Napoleon ordered immediate continuation of the line to Milan. This was to be an achievement, indeed. The line to Milan via Lyon, Chambéry, and Turin was 1100 km long. An extension to Venice added another 320 km. The Mont Cenis Pass at 2082 m was the highest point ever reached by an optical telegraph line. At the end of 1809, operation began from both Milan and Venice. This was just in time to send an order for the execution of Andreas Hofer, the leader of the Tyrolean resistance against Napoleon, by optical telegraph from Paris to Mantua (on the Milan–Venice section) in February 1810.

In September 1821 the Council of Ministers voted for an extension from Lyon via Valence, Orange, and Avignon to Marseille and Toulon. A signal from Paris to Toulon via 100 stations took less then 15 minutes.

On March 13, 1813, when the French army was withdrawing from Germany, Napoleon ordered a branch to be built from Metz to Mainz. The 225-km line with some 18 intermediate stations was completed within two months under Abraham Chappe's direction. The line diverged from Metz, on the roof of the Palais de Justice (Figure 5.2), and ran via Tromborn, Siersberg, Duppenweiler, Humes, Leiterweiler, and Kreuznach. The Mainz terminus was on the Stephen's church. On New Year's Day in 1814, the German army crossed the Rhine, entered Kreuznach, and took the

Figure 5.2   Palais de Justice at Metz with two telegraphs. (Scanned with the permission of the Museum für Kommunikation Frankfurt, Germany, from Klaus Beyer et al., *So weit das Auge reicht*, Museum für Kommunikation Frankfurt, Germany, 1995, p. 47.)

telegraph stations into their possession after the operators had destroyed the telegraph apparatus.

A trunk line linking Paris with Bordeaux and Bayonne was considered in 1820, when war with Spain seemed likely. Begun in 1822, the link, running via Orleans, Tours, Poitiers, and Angoulême, was completed on April 3, 1823. The distance between the stations was relatively short, to overcome the vibration of the air in the flat and sandy region of Les Landes. At Bordeaux, the two machines were on the old tower of St. Michel (built in 1472–1492). Two additional machines were later placed on the same tower: one for the Blaye branch in 1832, and the other for a long cross-country line to Toulouse, Narbonne, Montpellier, and Avignon, which was completed in 1834.

Ignace and Pierre-François Chappe retired in 1823. Count Kerespertz was the succeeding administrator. René and Abraham Chappe became the second and third administrators. In 1824, Ignace brought out his *Histoire de la télégraphy*. In 1840, well after Ignace's death, Abraham Chappe had the book reprinted, including a new preface written by Abraham and citing the advantages of the optical telegraph over those of acoustical systems.

When the revolution of 1830 broke out, Count Kerespertz resigned and Abraham and René Chappe had to retire. They were replaced by a "deputé" named Marchal, soon replaced by an ambitious and very competent administrator, Alphonse Foy. He gave substantial extension to the optical telegraph system and introduced the electrical telegraph. The last important message sent via the Chappe telegraph was the announcement of the capture of Sevastopol on the Black Sea in September 1855. The last Chappe station went out of service in 1856.

The optical telegraph system in France (Figure 5.3) was the world's most developed system. It covered some 5000 km with no fewer than 534 stations, a little more than the total number of optical telegraph stations in the remainder of Europe.

5.1.2   Other Optical Telegraph Systems in France

Shortly after Chappe introduced his optical telegraph, competitive systems appeared which, however, were not successful. A French mathematician Monge (1746–1818) constructed a large and very complex seven-arm optical telegraph enabling the transmission of 823,543 different signals.

Abraham-Louis Breguet returned to Paris in 1795 after two years exile in Switzerland to avoid execution by the revolutionary authorities because of his alleged Jacobin connections. He was very disappointed to find that Chappe had concealed his basic contribution concerning the mechanism that he had constructed for Chappe's telegraph. Still interested in optical telegraphy, however, in 1798, Breguet, together with a friend, Augustin Bétancourt (1758–1826), an engineer of Spanish origin at the time living in Paris (better known for his experiments with steam engines), brought out a simplified telegraph consisting of an arrow turning about a mast. The Breguet–Bétancourt system, with 36 different positions at 10° separation, could produce up to 41,840 signals in a three-sign combination. It was abandoned, however, after a violent dispute with Chappe—including publication of their vigorous letters in French newspapers—and for subsequent lack of government support. A model of the telegraph is in the Technical Museum of Paris.

In 1799, the goverment voted a sum to establish a *vigigraphe* between Paris and Le

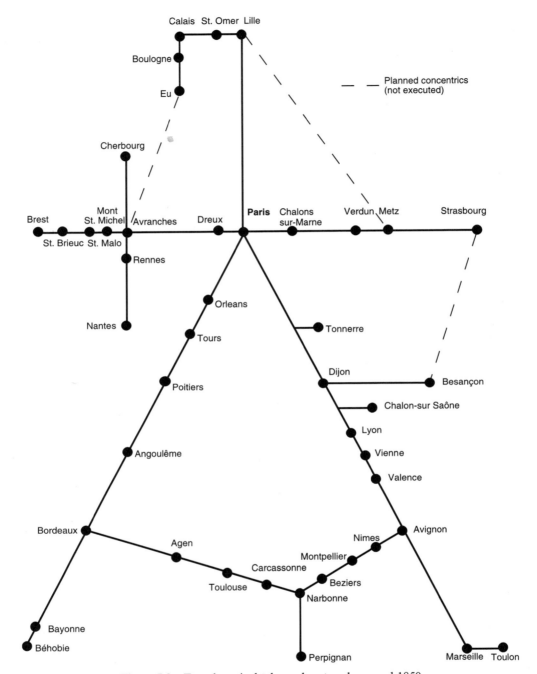

Figure 5.3 French optical telegraph network around 1850.

Havre. The vigigraphe was the invention of a naval engineer named Moncabrier, a harbor official and mathematician. A beam moved up and down one face, and a disk moved up and down the other face of a kind of ladder that had fixed beams at top and bottom. Trials were made over a distance of about 8 km between the church of Saint-Roch in the center of Paris (between Place Vendome and Palais Royal) and a second station in Courbevoie. The vigigraphe had been devised for ship-to-shore communication. In the meantime, however, a former artillery officer named Depillon recommended a simpler solution to the Ministry of Marine, so that the vivigraphe was not used. Depillon proposed a mast with three arms that furnished up to 301 signals. In line with its use for marine signaling, he coined the name *semaphore* (Greek for "bear a sign") for his device in 1801.

The Ministry of Marine adopted Depillon's system for coastal signaling and installed 97 stations along the French coast, 32 stations along the Mediterranean coast, and six stations in Algerian. The main object of the semaphores was to report on movements of the English fleet along the coast and to pass details to the nearest naval base. Depillon's device was the forerunner of similar solutions by Popham, Pasley, Parker, and Watson in England, and by Pistor in Germany. The coastal system in France remained in operation until early in the twentieth century.

In 1809, Vice Admiral de Saint-Haouen proposed another system with some 20 lanterns mounted on a 9-m mast for operation at night and in bad weather. He was authorized in 1822 to construct a line between Paris and Bordeaux with a slightly simplified machine. Twelve stations only were erected, and the system was tested between Paris and Orléans on October 24, 1822. The trial revealed that the de Saint-Haouen system was substantially slower than the Chappe system, and moreover, the Chappe system could be equipped with lanterns for use at night, so the Paris–Bordeaux line was equipped with Chappe telegraph machines in 1823.

The war with Austria gave birth to a rudimentary telegraph in 1809 to link Strasbourg with Vienna. The telegraph consisted of a long pole on which was displayed a white, red, or black flag 12 m in length. The stations were only some 3 km apart and located primarily on church towers. There were more than 200 stations on the route via Durlach, Pforzheim, Göppingen, Landshut, and Schärding. Three to four soldiers were allocated to each station. The signal code was known only to Napoleon; to the Minister of War, Henri Jaquess Guillaume Clarke; and to the Chief of Staff, Alexandre Berthier.

After the July Revolution in 1830, private enterprise also tried to introduce optical telegraph services. Alexandre Ferrier built a line between Paris and Rouen which started operation in July 1833. Another private line between Paris and Brussels was opened in October 1836, informing the people of Brussels by night what had happened during the day in Paris. The telegraph administrator, Alphonse Foy, disliked private services and proposed that the government make telegraphy a state monopoly. The proposal passed the lower house in March 1837, with 212 in favor and 37 against, and a month later the upper house voted, with 86 in favor and 2 against. Thus it became law in France that "anyone who transmits any signal without authorization from one point to another one whether with the aid of mechanical telegraphs or by any other means will be subject to imprisonment for a duration of between one month and one year, and will be liable to a fine of from 1,000 to 10,000 francs (d'or)." This Law on Telegraph Lines, No. 6801, signed by King Louis-Philippe on May 2, 1837, stayed in force with minor modifications until it was

TABLE 5.1 Worldwide Penetration of Optical Telegraphy

System	Major Pioneers	Year	Route	Number of Stations	Remarks
			Algeria		
Chappe	Military	1837	Algiers–Blidah, extended to Orléansville in 1847		Included portable stations
		1848	Algiers–Aumale–Mostaganem–Medéa		
		1853	Oran–Mascara–Sétif–Constantine		
		1854	Constantine–Biskra–Guelma		
			Australia		
Three-arms, 999 codes		1830	Sydney harbor system	3	Served until 1842
		1828	Hobart–Mount Royal (Tasmania)	5	
3 × 2 arms, 999,999 codes	Charles O'Hara Booth	1824	Port Arthur–Hobart (Tasmania)	21	Served until the penal island was closed in 1877
			Channel Islands		
Two arms, 56 signals	P. A. Mulgrave (1778–1847)	1809	Guernsey–Sark–Alderney and Jersey	21	Served until 1815
			Curaçao		
Similar Lipkens		1825	Fort Amsterdam–Eastern Coast	4	Served until 1971; remains found in 1993
			Denmark		
Flaps, 42,221 codes	Lorens Henrich Fisker (1753–1819)	1800	Copenhagen–Nakkehoved; Samsø and Ørby (1807)	10	
		1801	Copenhagen–Kiel	14	
		1849	Østrupgaard–Ravnebjerg–Baaring	3	Served partially until 1862

Description	Inventor	Year	Route	No.	Remarks
Egypt					
Similar Chappe		1823	Cairo–Alexandria	19	160 km
		1835	Alexandria–Cairo–Suez		
England					
Admiralty systems, six shutters, 63 signals	George Murray	1796	London–Deal–Sheerness–Portsmouth	22	
		1805	London–Plymouth	21	Served until 1847
Two arms, 48 signals	Sir Home Riggs Popham (1762–1820)	1824	London–Portsmouth		
Commercial systems, six arms, 9 × 999 codes	B. L. Watson (1801–1865)	1827	Liverpool–Holyhead	5	120 km; served until 1860
		1839	Hull–Spurn Head	12	Served until 1857
		1842	London–North/South Foreland		Stopped by London fire in 1843
		1842	Southampton–Isle of Wight	5	Southampton–London by train
Finland					
5 × 2 boards plus one ball; 2048 signals	C. O. Ramstedt	1854	Kronstadt (Russia)–Helsingfors–Hangö	80	Served until 1855
Germany					
Two cross bars, 44 codes	P. Basilius Sinner (1745–1827)	1801	Monasteries Andechs–Diessen–Seefeld	3	Served until 1803
Commercial systems, three arms, 512 signals	J. L. Schmidt	1837	Hamburg–Hechthausen–Cuxhaven	8	Served until 1847
		1847	Bremen–Bremerhaven–Hechthausen–Hamburg	11	Served until 1852
Prussian military, three pairs of arms, 4096 positions	Carl Ph. H. Pistor (1778–1847)	1834	Berlin–Magdeburg–Coblenz	61	Served until 1852

(Continued)

TABLE 5.1 (*Continued*)

System	Major Pioneers	Year	Route	Number of Stations	Remarks
			India		
Four 1.2- to 1.8-m balls	William Boyce	1821	Calcutta–Chunar	45	650 km; served until 1828; surveyed by George Everest (1790–1866)
			Italy		
Chappe		1810	Venice–Santa Lucia and Venice–Trieste		Served until 1815
			The Netherlands		
Disks and vanes	Jan van Woensel (1740–1816)	1795	Dutch coastline	63	Served until 1813
Chappe	Antoine Lipkens	1810	Antwerp–Rotterdam–Amsterdam	13	Served until 1813
2 × 3 disks, 63 signals		1831	The Hague–Rotterdam–'sHertogenbosch	13	Served until 1839
		1832	Breda to Antwerp and Flushing	9	Served until 1839
			Norway		
2 × 3 flaps, 229 codes	Ole Olhsen	1808	Halden–Oslo–Kristiansand–Stavenger and Bergen up to Trondheim	225	1300 km; served until 1814
			Portugal		
Chappe		1810	Lisbon–Porto	27	Served until 1855

Russia					
Chappe	Modified by Pierre-Jacques Chatau	1824	St. Petersburg–Schlusselburg (now Petrokrepost)		
		1834	St. Petersburg–Kronstadt	8	
		1835	St. Petersburg to Tsarskoe Selo (now Pushkin) and Gatchina		
		1839	St. Petersburg–Warsaw	148	1200 km; served until 1854
South Africa					
Three arms	Henry Hall	1842	Fort Beaufort–Grahamstown–Fort Peddie	10	All stations destroyed in 1846 in the Seventh Kafir War
Spain					
Six shutters	Juan José Larena	1831	Linking the royal palaces at Madrid, Aranjuez, and San Ildefonso		Served until 1839
National system, eight shutters	José Maria Mathé	1846–1850	Madrid to Irun Barcelona and La Jonquera, Valencia, Cadiz, and Badajoz	200	Served until 1855
Sweden					
Ten shutters, 512 signals	Edelcrantz (1754–1821)	1794	Royal castles Stockholm and Drottningholm	3	
		1799	Connecting Stockholm with Gävle, Åland, Artolma, Sandhamm, and Landsort	60	
		1799	Around Karlskrona and around Göteborg	4 7	Served until 1881

(Continued)

43

TABLE 5.1 (*Continued*)

System	Major Pioneers	Year	Route	Number of Stations	Remarks
			United States		
Similar to Chappe	Jonathan Grout, Jr.	1801	Martha's Vineyard (West Chop–Boston	16	Served until 1807
Pasley	John Rowe Parker	1821	Long Island Head–Boston Light, extended in 1825 to Point Allerton		Replaced in 1853 by electrical telegraph
Semaphore	Christopher Colles	1812	New York harbor system		Closed same year
Two arms, five-part dictionary	Jeremiah Thompson	1821/ 1829	Staten Island harbor system		Electrical around 1855
Boards; at night, colored lamps	William C. Briggs	1840	New York–New Jersey–Philadelphia		Electrical in 1846
		1849	San Francisco harbor system		Electrical in 1853

44

replaced on July 26, 1997, to comply with the telecommunications liberalization policy of the European Union.

5.2 OPTICAL TELEGRAPHY OUTSIDE FRANCE

The success of the optical telegraph in France inspired a dozen other countries also to install optical telegraph lines. A detailed description of those lines, unfortunately, is not possible within the scope of this book; those lines are therefore summarized in Table 5.1.

A few optical telegraph stations survived and have been restored. Figure 5.4 shows Chatley Heath Semaphore Tower, which was the fifth station southwest of London of the London–Portsmouth line and was opened to the public in 1989. Figure 5.5 shows the Köln–Flittard tower, the fiftieth tower of the Berlin–Coblenz line. This tower was opened to the public in 1971. Station No. 18 of the same line at Neuwegersleben was restored in 2001. In France the station Haut-Barr of the Paris–

Figure 5.4 Restored optical telegraph station at Chatley Heath. See insert for a color representation of this figure.

Figure 5.5   Restored optical telegraph station at Cologne. (Courtesy of Museum Optischer Telegraph, Köln-Flittard, Germany.)

Strasbourg line was opened as a museum in the 1980s, and a further three stations of that line were restored in the late 1990s. The tower, which still exists on the roof of the Winter Palace in St. Petersburg, was specially erected for the optical telegraph. In India a five-story tower of the Calcutta–Chunar line exists at Nibria, near Mahiari, 8 km west of Howrah. In Australia the Mount Nelson station, 6 km south of Hobar, Tasmania, is now used by the marine board of Hobart. The station in Hobart was restored with its apparatus in working condition and opened as a historical monument in 1940. Some other stations have given a lasting name to their sites; such as Telegraph Hill or Telegraph Bay, in Australia, the United Kingdom, and in the United States; and Telegrafenberg in Germany at Berlin.

REFERENCES

Books

Aschoff, Volker, *Geschichte der Nachrichtentechnik*, Vol. 2, Springer-Verlag, Berlin, 1995.

Beyer, Klaus, et al. *So weit das Auge reicht: die Geschichte der optischen Telegrafie*, Museum für Post und Kommunikation, Frankfurt, Germany, 1995.

Bertho, Catherine, *Histoire des télécommunications en France*, Édition Érès, Toulouse, France, 1984.

Bertho, Catherine, *Telegraphes & telephones de Valmy au microprocesseur*, Hachette, Paris, 1981.

Bertho Lavenir, Catherine, *Great Discoveries: Telecommunications*, Romain Pages Editions, 1991.

Brodbeck, Didier, *Journal imaginaire d'Abraham-Louis Breguet*, Editions Scriptar, Lausanne, Switzerland, 1990.

Carandell, Luis, and Bernardo Riego, *Telefonía: La Gran Evolución*, Lunwerg Editores, Barcelona, Spain 1992.

Michaelis, Anthony R., *From Semaphore to Satellite*, International Telecommunication Union, Geneva, 1965.

Oslin, George P., *The Story of Telecommunications*, Mercer University Press, Macon, GA, 1992.

Wilson, Geoffrey, *The Old Telegraphs*, Phillimore & Co., Chichester, West Sussex, England, 1976.

Articles

Aschoff, Volker, Von Abel Burja bis zum Fächer à la Telegraph, *Archiv für deutsche Postgeschichte*, Vol. 1, 1981, pp. 106–123.

Thewes, Alfons, Eine frühe Beschreibung von optischer Telegrafie, *Archiv für deutsche Postgeschichte*, Vol. 2, 1985, pp. 111–114.

Thewes, Alfons, Die optische Telegrafenverbindung am Ammersee, 1801–1803, *Archiv für deutsche Postgeschichte*, Vol. 1, 1987, pp. 109–114.

6

ELECTRICAL TELEGRAPHY

6.1 EVOLUTION LEADING TO ELECTRICAL TELEGRAPHY

The history of electrical telegraphy is generally considered to have begun on February 17, 1753, when a remarkable letter, signed by a certain C.M. (perhaps standing for Charles Marshall, Renfrew, Scotland; or perhaps for Charles Morrison), was published in the *Scots' Magazine*. Under the heading "An Expeditious Method of Conveying Intelligence," C.M. proposed briefly that "a set of wires equal in number to the letters of the alphabet, be extended horizontally between two given places, parallel to one another and each of them an inch distant from the next to it." The letter then explains in detail how the wires are to be connected to the conductor of an electrostatic machine when it is desired to signal a particular letter. On the receiving side, C.M. explains: "Let a ball be suspended from every wire, and about one sixth to one eight of an inch below the balls, place the letters of the alphabet, marked on bits of paper."

By the middle of the eighteenth century, simple frictional machines producing electrostatic electricity were available and it was known that electrostatic forces would attract small pieces of paper. C.M. thus proposed using the electricity of a frictional machine, channeling it through the appropriate wires, and letting it attract the corresponding pieces of paper with the letter of the alphabet selected on the receiving side. All the principal elements of electrical telegraphy are already present in this early proposal: a source of electricity, its manipulation handling the information to be transmitted, the conducting wires, the mechanism on the receiving end to read the information transmitted, and the high speed of electrical transmission (the speed of light).

The year 1753, however, was hardly the time that practical and economic conditions were ripe for electrical telegraphy. Static electricity was then more often used

The Worldwide History of Telecommunications, By Anton A. Huurdeman
ISBN 0-471-20505-2   Copyright © 2003 John Wiley & Sons, Inc.

to entertain "philosophical" friends of the owner of a frictional machine. For example, it was common then to transmit an electric shock through a circle of 20 to 30 persons, each holding hands with the next; all experienced the shock simultaneously. Evidently, in the year 1753, when Voltaire was discussing philosophy in Potsdam (near Berlin) with King Frederick the Great of Prussia, and when Carolus Linnaeus, the great Swedish botanist, was elected into the fellowship of the Royal Society of London, electrical telegraphy was not really taken seriously.

An early improvement in handling electricity was obtained in 1745 when the German Ewald Jürgen von Kleist (1700–1748), dean of the episcopate Cammin in Pommern, discovered that electricity could be stored in a glass bottle if both the inner and outer surfaces of the bottle were covered with a metallic foil and a metal rod was placed in the middle of the bottle. Von Kleist, who had studied law at the Dutch university at Leyden, informed his friends in Leyden about his discovery. A Dutch physician, Pieter van Muschenbroek (1662–1761), then published the first scientific paper regarding the Kleist bottle, which was then given the name *Leyden jar*.[1]

A Swiss mathematician and physicist, Georges-Louis Lesage (1724–1803), was first to construct an electrostatic telegraph as proposed by "C.M." In 1774, he used 24 pith balls over 24 wires connected with a frictional electricity machine to communicate from room to adjacent room. For use between separate buildings, he proposed to put the 24 wires in special ceramic tubes with 24-hole separating disks at regular distances.

A Spanish engineer, Augustin de Bethencourt y Mollina (1758–1826), later in France named Bétancourt (see Section 5.1.2), in 1787 carried out experiments using Leyden jars and static electricity to send telegraphic messages between Madrid and Aranjuez, a distance of 42 km. Another Spaniard, Franscisco Salvá y Campillo (1751–1828) of Barcelona, proposed a scheme in 1795 to use the discharge of Leyden jars together with multiwire transmission to give electric shocks to operators on the receiving end. Three years later, reportedly, a modification of this scheme, using only a single wire, was actually constructed between Madrid and Aranjuez. Apparently, private messages were sent to the Spanish royal family.

At the end of the eighteenth century, although still in an experimental stage, electrical telegraphy was described in the 1797 edition of the *Encyclopaedia Britannica* in an optimistic and prophetical way: "The capitals of distant nations might be united by chains of posts, and the settling of disputes which at present take up months or years might then be accomplished in as many hours. An establishment of telegraphs might then be made like that of the post; and instead of being an expense, it would produce a revenue." Experiments using Leyden jars and static electricity to send telegraphic messages continued for another few years, until more promising experiments were made, first with electrochemical devices and then, decisively, with electromagnetic devices.

An English merchant, Francis Ronalds (1788–1873), experimented with static electricity to demonstrate the speed of electrical transmission. In 1816, in his garden

---

[1] An experiment with the new source of lectricity was made on a "Guinness Book of Records" scale in 1746 by Abbé Jean-Antoine Nollet (1700–1770), when a shock was passed around a circle formed by 200 Carthusian monks. Together, the monks and their connecting wires formed a line over a mile long. The experiment revealed that electricity could be transmitted over a great distance, and as far as Nollet could observe, it covered that distance instantly.

in the London suburb of Hammersmith, he erected two large wooden frames and suspended between them a wire with a total length of almost 13 km. To one end he connected a frictional electricity machine, to the other a pair of pith balls, which diverged when the line was charged. At the sending station, Ronalds installed a rotating dial with the letters of the alphabet on it, driven by clockwork and running synchronously with a similar dial at the receiving station. The line was charged continuously but was discharged by the operator when the letter desired became visible on his dial. On the receiving end, the other operator would see the pith balls of his dial indicator converge as the same letter came in view. This quite ingenious system worked satisfactorily and deserved serious official consideration, and Ronalds proposed his system to the admiralty. On July 11, 1816 he wrote to Chief Admiral Lord Melville: "Mr Ronalds takes the liberty of soliciting his lordship's attention to a mode of conveying telegraphic intelligence with great rapidity, accuracy and certainty, on all states of the atmosphere, either at night or in the day, at a small expense." On August 5, 1816, Ronalds was informed by the admirality that "telegraphs of any kind are now wholly unnecessary, and that no other than the one now in use will be adopted." The one referred to as "now in use" was the optical telegraph of Lord Murray. The *Encyclopaedia Britannica* changed its above-cited end-of-the-century optimism to new-century pessimism, stating in its 1824 edition: "It has been supposed that electricity might be the means of conveying intelligence, by passing given numbers of sparks through an insulated wire in given spaces of time. A gentleman of the name Ronalds has written a small treatise on the subject; and several persons on the Continent and in England have made experiments on Galvanic or Voltaic telegraphs, by passing the stream through wires in metal pipes to the two extremities ... but there is reason to think that, ingenious as the experiments are, they are not likely ever to become practically useful." In the meantime, fundamental discoveries of electromagnetism, described in Chapter 4, enabled important steps toward electrical telegraphy with the three basic investigations described next.

First Basic Investigation   In 1833, two German professors at the University of Göttingen, Carl Friedrich Gauss (1777–1855), mathematician and astronomer, and Wilhelm Eduard Weber (1804–1891), physicist, working together on an investigation of Earth's magnetism, constructed a huge and very sensitive mirror–galvanometer.[2] They observed the movement of the mirror, and thus the deflection of the magnetic needle, through a telescope placed at a distance of about 5 m from the galvanometer. Instead of using a voltaic cell as a source of electricity, Gauss and Weber made an induction transmitter consisting of a long, heavy permanent-magnetic rod (25 to 50 kg) around which a coil with a winding of some 1000 turns was moved up and down by hand to produce electricity. The output of the transmitter was connected to a polarity switch, which connected either the positive or the negative pole to the outgoing line and thus caused the distant mirror to deflect to either the left or right.

One galvanometer was placed in the astronomical observatory under the responsibility of Weber, a second galvanometer at a distance of 60 m in a magnetic obser-

---

[2] A permanent-magnetic rod with an initial weight of 500 g and a length of 30 cm, later increased to 6 kg and 1.2 m, was placed horizontally in the center of a coil. The coil, initially with 50 (at Weber's laboratory) and 170 (at the laboratory of Gauss) turns and later up to 7000 thin copper-wire turns, had a width of 1.45 m.

Figure 6.1   Gauss and Weber communicating via their electromagnetic telegraph. (Courtesy of the Museum für Kommunikation, Frankfurt, Germany.)

vatory, and a third galvanometer in the physical laboratory of Gauss. The magnetic observatory, a one-floor building 10.7×5.3 m, was constructed entirely of noniron material and placed in the astronomical north–south direction.[3]

To coordinate experiments in the three locations, a galvanic chain was erected connecting the three laboratories, consisting of two separate parallel thin iron wires passing over the roof of the Johannis Church and other buildings in the town center over a distance of 1 km. This line was erected with Weber's participation in March–April 1833. Figure 6.1 is a combination of three original photographs, showing:

- *Left upper corner:* Weber operating the induction transmitter
- *Right lower corner:* Gauss observing the movement of the mirror
- *Main picture:* iron wires passing over the roofs in the town center

During the experiments Gauss and Weber realized that beyond start and stop signals for the experiments or for synchronizing their astronomic clocks, coded signals could also be used for the transmission of additional information, thus for electrical telegraphy. Gauss developed five different telegraph codes for the characters of the alphabet, using combinations of one to six mirror movements to the left or to the right. Although Gauss and Weber were aware of the significance of their electromagnetic telegraph, which according to Gauss could be used "independent from the

---

[3] The magnetic observatory was built on the initiative of Alexander von Humboldt (1769–1859), professor of natural science, who had used similar magnetic observatories in Berlin in 1806–1807 and 1828.

weather and time of the day, even by closed windows, and would enable transmission over many miles as for instance from Petersburg to Odessa,'' they had no incentive to develop the system beyond its scientific achievement. This first electromagnetic telegraph line remained in operation for the transmission of scientific information only until December 14, 1837, when Weber was suddenly dismissed.[4]

Second Basic Investigation   At the age of 18, Carl August von Steinheil (1801–1870)[5] began the study of law in Munich, soon changing to a study of physics, mathematics, and astronomy at the University of Göttingen. In 1835, as professor of mathematics and physics, in his function as a member of the *Royal Bavarian Society of Science*, he visited Gauss and Weber and obtained a detailed introduction to their electromagnetic telegraph. Steinheil immediately recognized the merits of the electromagnetic telegraph for signaling by the newly approaching railways. Back in his laboratory, within one year he developed his own version of an electromagnetic telegraph—the world's first that could write (Figure 6.2). To obtain better operational speed, Steinheil replaced the heavy magnetic rods of Gauss and Weber's telegraph by two magnetic needles each 6 cm long with a width of 0.5 cm. The two needles pivoted above each other, one in the upper and the other in the lower part of a vertically placed coil, with the north and south poles of the two needles facing each other. Each needle had a short arm with a little paint receptacle that had a beak-shaped outlet at its end. Depending on the direction of the current through the coil, one needle would deflect to the right and the other to the left. A paper strip driven by a clock was placed at the left-hand side of the coil so that the needle deflecting to the right would touch the paper and leave a paint dot on the paper. Two small permanent magnets were placed outside the coil near each needle so that the two needles returned immediately to an idle position after each current pulse. The two needles thus produced a number of dots in two rows one above the other.

Steinheil developed a telegraphic code for letters of the alphabet and the numbers 1 to 9 and 0. With this code and his "lean" telegraph, Steinheil obtained a transmission speed of about 40 letters or numbers per minute. For his first telegraph line, in 1836, Steinheil was unsuccessful because he placed two poorly insulated iron wires in underground aqueducts. In the next year, following the example of Gauss and Weber, he erected an overhead line 6 km long, with two wires parallel on a spacing of 1 to 3 m between the *Academy of Science* in Munich and the *Royal Astronomical Observatory* in nearby Bogenhausen. Bavarian King Ludwig I was impressed, but being more inclined to arts than to technology, he did not support further applications.

Steinheil then began with experiments on the railway track between Nuremberg and Fürth, the first railway in Germany, which was constructed in 1835. In his first attempt using the two parallel rails as telegraph lines, he realized that Earth has good electrical conductivity.[6] In 1838, Steinheil erected one single insulated wire on

---

[4] Weber had participated with six other professors in a petition against the change in the constitution and dissolution of the state assembly by the new king Ernst August of Hannover at his accession to the crown.

[5] At the age of 11, Carl Steinheil became seriously ill with typhoid fever, unconcious and near death, said by a physician to be dead. His twin brother had died at the age of 6, but Carl recovered from his illness.

[6] This had been recognized in 1744 by a German philosopher and physicist, Johann Heinrich Winkler (1703–1770), one-time teacher of Johann Wolfgang Goethe, but forgotten in the meantime.

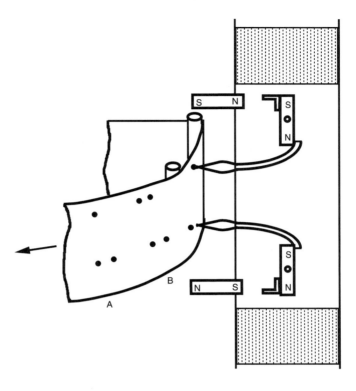

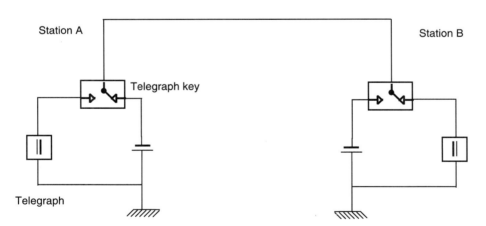

Figure 6.2 Electromagnetic telegraph of Steinheil.

wooden poles parallel to the railway track and used the rails and Earth as return conductors. At the time divided into some 100 autonomous states, territories, and towns, Germany was not, however, mature enough for electrical telegraphy: The first long optical telegraph line (Berlin–Coblenz) had only started to operate three years earlier.

Third Basic Investigation   Baron Pavel Lvovitch Schilling von Cannstadt (1786–1837), born in Reval, Estland, of German parents, was appointed as an attaché to the Russian embassy in Munich, where he assisted Samuel Thomas von Soemmering in experiments with Soemmering's electrochemical telegraph in the period 1809–1811. Back in St. Petersburg, Schilling made various experiments with electrochemical devices, and in 1812 he demonstrated to Tsar Alexander I his version of an electrochemical telegraph. However, the tsar was afraid of revolutionary ideas and forbade further activities and publications on electrical telegraphy. Twenty years later, returning from a two-year expedition through Mongolia, Schilling, probably after reading of Oerstedt's experiments (Chapter 4), changed his basic telegraphy principles from electrochemical to electromagnetic and used the deflection of one to six needles as information code. He demonstrated some parts of his electromagnetic telegraph to Alexander von Humboldt in Berlin in 1832. In the same year he demonstrated his telegraph to Tsar Nicholas I of Russia (brother and successor of Alexander I). Moreover, he demonstrated the world's first needle telegraph with five needles (Figure 6.3) at a congress of physicists in Bonn and at the *Physical Society* in Frankfurt in 1835. George Wilhelm Muncke (1772–1847), professor of physics at Heidelberg University, attended the demonstration in Frankfurt. Valentin Albert, a mechanic in Frankfurt, produced for Muncke a true copy of Schilling's five-needle telegraph, which Muncke then used for his lectures.

In a letter dated September 15, 1836, the British government offered to buy Schilling's new design. This time, however, Nicholas I of Russia also showed interest. In the same year, successful experiments were made and a commission was appointed to advise Nicholas I on the installation of Schilling's telegraph between Kronstadt and his imperial palace, Peterhof. However, on July 25, 1837 Schilling died and the project was canceled.

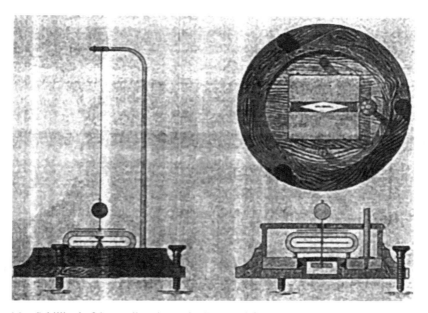

Figure 6.3   Schilling's five-needle telegraph. (Scanned from Ernst Feyerabend, *Der Telegraph von Gauss und Weber*, Reichspostministerium, Berlin, 1933, p. 21.)

Fortunately, William Fothergill Cooke (1806–1879) had attended Muncke's lectures in Heidelberg, and he went back to England with another copy of Schilling's needle telegraph. Together with Charles Wheatstone (1802–1875), he made an improved version for which a patent was applied for on June 12, 1837, two weeks before the inventor of the original idea, Schilling von Cannstadt, died in St. Petersburg. In the same year, Samuel Finley Breese Morse (1791–1872) demonstrated a telegraph at the University of New York in which the electromagnetic force was not used to deflect a needle but to produce a coded written message. Thus the *era of electrical telegraphy started in 1837 almost simultaneously in Great Britain and the United States*.

6.2   ELECTRICAL TELEGRAPHY IN THE UNITED STATES

6.2.1   Morse Telegraph

One hundred kilometers west of the neglected grave of Claude Chappe, in a splendid modern museum, the *Musée Américain Giverny*, opened in 1992, a large painting shows an artist at work in the *Salon Carré* of the Louvre copying 38 masterpieces dispersed throughout the Louvre and reassembled in that painting, called the *Gallery of the Louvre*. The artist painted that work in 1832 to collect Old World treasures which he found worthy of presentation to the New World public. While in Paris he saw Chappe's semaphore in action on top of the Louvre. On October 6, 1832, the artist embarked on the packet ship *Sully* to return to his homeland. On that long voyage back to the New World, he recollected his impressions on art as well as on the highly interesting news on experiments with electricity made at that time by now famous men such as Ampère, Ohm, Faraday, Gauss, Oersted, Steinheil, and others. He discussed his impressions with a fellow passenger, Charles T. Jackson, who had heard about an electromagnet constructed in 1825 by a self-taught British shoemaker–physicist and head of the *Gallery of Practical Science* in Manchester, William Sturgeon. Jackson explained that electricity would pass through miles of wire as fast as lightning. The artist exclaimed: "If this be so, and the presence of electricity can be made visible in any desired part of the circuit, I see no reason why intelligence might not be instantaneously transmitted by electricity to any distance." Thus an idea came to his mind which five years later resulted in the beginning of electrical telecommunications in the United States. The artist, of course, was Samuel Finley Breese Morse (1791–1872), one of the fathers, and certainly the most successful father, of electrical telegraphy.

Samuel Finley Breese Morse was born April 27, 1791, in Charlestown, Massachusetts (now part of Boston). His father, Jedediah Morse, was a Congregational pastor and author of *Geography Made Easy*, the first book on geography printed in the United States, in 1784. His mother, Elizabeth Ann Breese, was a daughter of the man who founded Shrewsbury, New Jersey. Morse devoted his first 41 years primarily to art, although he combined this with a great interest in electricity. As a student at Yale from 1808 to 1810, he attended lectures on electricity and spent a vacation assisting with electrical experiments. In 1810 he went to England to study art at the *Royal Academy* in London. He had to return home four years later, as his father could no longer support him abroad. He stayed a few years in Charleston, South

Carolina, where gradually, with the support of his influential family, he made portraits of wealthy and famous Americans, including President Monroe. He married Lucretia Pickering Walker of Concord, New Hampshire, in 1821, and they moved to New Haven, Connecticut. The next year he made his first monumental history painting, *The Old House of Representatives*, which included small portraits of 88 members. In 1824, he opened a studio at 96 Broadway in New York, sleeping on the floor for lack of money for a bed. The city of New York commissioned him to paint a heroic-size portrait of General Lafayette, then visiting the White House. On February 10, 1825, Morse happily wrote to his wife in New Haven of his friendship with Lafayette and his progress on the portrait, not knowing that his wife had died two days before, following the birth of their third child. Morse's three children went to live with relatives.

Morse participated in the foundation of the *National Academy of Design* in 1826. He was elected the first president of the academy, a function that he kept until 1845. By 1829, he started a second journey to Europe, this time financed by a number of art patrons, who advanced him money for pictures to be painted in Europe. During this second trip to Europe, he painted the *Gallery of the Louvre*, noted above.

On his return journey from his second tour through Europe, after his discussion with Charles T. Jackson, Morse started making notes on a possible solution for an electrical writing telegraph. The essence of his idea was to use the passage of an electric current through an electromagnet to deflect a pen or pencil in such a way that it could mark a strip of paper passing underneath. He also noted down a telegraph code consisting of numbers corresponding to letters and words. Thus Morse conceived the permanent recording of a message onto paper. These notes, made onboard the *Sully*, still exist among the Morse Papers in the *Library of Congress* in Washington, DC.

Back in the United States, other priorities prevented Morse from immediately realizing his electrical telegraph.[7] He was appointed professor of the literature of arts and design at the *University of the City of New York* (now New York University) on October 2, 1832; the university had been founded just one year earlier. He could not begin work, however, until the fall of 1835, when the main building of the university neared completion on the east side of the Washington Parade Ground, now Washington Square. Morse had a large classroom on the third floor, where he ate, slept, taught, painted, and experimented. He received no salary and depended on fees from his students and the occasional sale of a portrait. Finally, in 1835, in that classroom, he constructed his first electrical writing telegraph. In line with his original profession, his telegraph was an old painting frame on which he had mounted a triangular electromagnetic writing device with a pencil tilting over a moving paper tape. Paper transport was driven by a clock mechanism. The pencil made a succession of V's as it passed across the paper. Figure 6.4 shows a replica of his ingenious construction. For signal transmitting he moved a gliding contact over a sawtoothed bar (shown in the lower part of the figure) with the sawteeth arranged to represent the coding of the numerals 0 and 1 to 9. He used a voltaic pile as the electricity source.

Morse demonstrated his device to some of his friends and acquaintances, among them Leonhard Gale. The scientific knowledge of Gale, professor of chemistry and

---

[7] He completed his *Gallery of the Louvre* and had it exhibited, but as little interest in it was shown, he sold it for $1200. At an auction in 1982, it sold for $3.25 million.

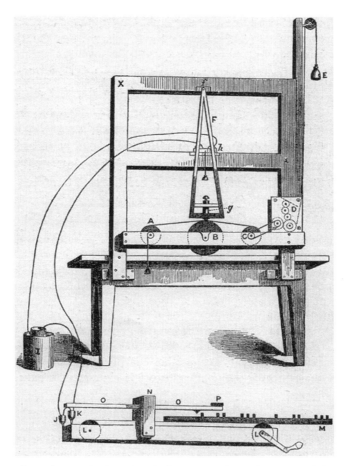

Figure 6.4   Replica of Morse's first electrical writing telegraph. (Scanned from Ernst Feyerabend, *Der Telegraph von Gauss und Weber*, Reichspostministerium, Berlin, 1933, p. 85.)

geology, was of great value to Morse. The first improvement suggested by Gale was to use a battery of voltaic piles. In line with experience gained by Gale's friend, the physicist Joseph Henry, Gale also proposed that the windings on the coil on each arm of the magnet be increased to many hundred turns each.

With those and other improvements to his telegraph, Morse, with the assistance of Gale, demonstrated in his classroom on September 4, 1837, the sending of messages via a wire 550 m long. The message he used, was: "Successful experiment with telegraph September 4th 1837." That demonstration of the not very reliable device resulted in a joint contract of Morse with Gale and Alfred Lewis Vail. Alfred L. Vail (1807–1859), a member of the *Mechanics' Institute*, brought with him two advantages: technical aptitude and a rich father, Judge Stephen Vail, the owner of the Speedwell Iron Works in Morristown, New Jersey.[8] Judge Vail agreed to finance

---

[8] Many of the first American locomotives were built here, as well as the engines for the *Savannah*, the first American steamship to cross the Atlantic Ocean.

further development of the electrical telegraph. On September 23, 1837, a contract was signed committing Alfred Vail to built the instruments and pay the cost of securing patents. Alfred received a 25% interest in the invention.

While Alfred Vail improved the reliability of Morse's instruments, on October 3, 1837, Morse went to Washington to file a *caveat*[9] on the invention. With Alfred Vail constructing the instruments at Morristown, Gale conducting experiments with stronger batteries and greater lengths of wire, and Morse in New York writing a five-dash V-code dictionary of numbers, the work progressed. Alfred Vail also considered extending Morse's code dictionary of numbers by a code of the 26 letters of the alphabet. In November and December 1837 he visited the print shop of Louis Vogt at Morristown to investigate which letters of the alphabet were used most frequently. He then worked out a four-dash V-code for the letters in such a way that the simplest code was for the letters used most.[10]

As the development was taking its time, Judge Vail threatened to withdraw his support when he saw no results. The worried inventors locked themselves in a room in Morristown, avoiding the Judge for about six weeks. Finally, on January 6, 1838, the new apparatus could be demonstrated to Judge Vail. The signal transmitter and the receiving apparatus were placed at opposite ends of a long bench and connected by about 5 km of wire. Morse believed that to ensure an uninterrupted flow of electricity, the diameter of all wiring of a circuit should be the same. Normal line wire was therefore also used for the coil of the electromagnet, which had a total weight of almost 100 kg. To test the new telegraph, Judge Vail handed his son a message to be sent. Morse received the message at the other end of the bench, and to the relief of them all, correctly deciphered the famous historical sentence: "A patient waiter is no loser."

Official demonstrations using about 16 km of wire between transmitter and receiver were then given on January 23 and 24 in Gale's geological cabinet room at New York University and before the *Science and Arts Committee* of the Franklin Institute in Philadelphia on February 8. The next important demonstration was given in Washington for the *House Committee on Commerce*. President Martin Van Buren and his cabinet, including the postmaster general, Amos Kendall (1789–1869), attended the successful demonstration. At the time, a proposal was put before the Congress for the construction of a line between New York and New Orleans using a Chappe-style optical telegraph system. The secretary of the treasury was asked to prepare a report "upon the propriety of establishing a system of telegraphs for the United States." A circular was issued to government officials and other interested parties asking for comments. Morse eagerly replied, explaining the advantages of an electrical system and pointing out that he had sent messages successfully through 10 miles of cable. Moreover, he applied to the government for financing of a 100-mile telegraph line. The chairman of the House committee, Francis Ormond Jonathan Smith (1808–1876), a lawyer from Portland, Maine, arranged a contract with Morse, obtaining a one-fourth interest in Morse's future undertakings. On April 1, 1838, Morse submitted a 6732-word patent application for "a new application and effect of

[9] A declaration of an invention for which a one-year protective period was requested to gain time for further investigation and final patent application.

[10] For example, one V for the letter "e," one V and one dash for the letter "a," and a more elaborate code for the less commonly used letters such, as one dash and three V's for the letter "b."

electromagnetism in producing sounds and signs, or either." Summarizing his patent, he stated: "I specially claim as my invention the use of the motive power of magnetism as a means of operating machinery which may be used to imprint signals upon paper or other suitable material, or to produce sounds in any desired manner for the purpose of telegraphic communication at any distance."

The elaborate application did not result in a patent until June 20, 1840. The patent file sought protection for the following features:

1. The combination of type, rule, lever, etc.
2. The recording cylinder, etc.
3. The types, signs, etc.
4. The making and breaking of the circuit by mechanism, etc.
5. The combination of successive circuits
6. The application of electromagnets to several levers, etc.
7. The mode and process of recording by the use of electromagnetism
8. The combination and arrangement of electromagnets in one or more circuits, with armatures for transmitting signs
9. The combination of the mechanism described, with a dictionary of numbered words

With the U.S. patent still pending and Congress reluctant to support Morse's application for the financing of an experimental line, Morse and Smith went to Europe in June 1838. They went first to England, where they had come too late. The *London Mechanics Magazine* of February 1838 had published an article taken from the American *Journal of Science* describing the Morse invention. Based on that article, a copy of Morse's telegraph had already been made, thus invalidating the invention for a British patent. Tsar Nicholas I of Russia saw in the Morse telegraph potential help for a plot against his regime and thus rejected Morse's proposal. In France, Morse obtained a patent but not permission to operate his system. In Germany in 1845, in line with the prevailing antipatent policy, the invention was considered not to be essential.

While Morse was in Europe, Alfred Vail continued to improve the electrical telegraph. To replace the V's-producing signal sender, he developed a much simpler signal transmitter: a lever-transmitter making and breaking the electrical circuit when it moved up and down, soon generally known as the *Morse key*. With this key the telegraph receiver produced discrete dots instead of the V's and dashes of different lengths. Vail then conceived the dots-and-dashes coded alphabet, which replaced Morse's code of numbers.

6.2.2 Washington–Baltimore Electrical Telegraph Line

Morse returned to Washington in December 1842 to lobby again for the construction of an experimental electrical telegraph line. The Congress was still not much interested, and representatives from Tennessee and Alabama blocked a positive decision. However, encouraging news came from Great Britain, where on July 9, 1839, the world's first electrical telegraph line had begun operation on a 21-km track between the railway stations of London–Paddington and West Drayton, with needle tele-

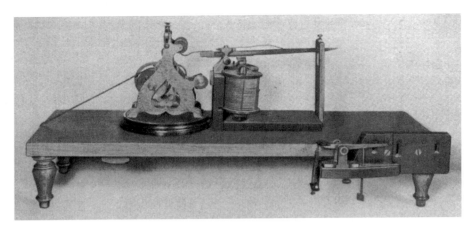

Figure 6.5 Morse telegraph as used for the Washington–Baltimore line. (Scanned from Ernst Feyerabend, *Der Telegraph von Gauss und Weber*, Reichspostministerium, Berlin, 1933, p. 85.)

graphs constructed by Cooke and Wheatstone. In December 1842, Morse demonstrated his telegraph between two committee rooms in the Capitol. On March 3, 1843, by a vote of 89 to 83, Congress passed a bill.[11] On March 14, 1843, Morse received a letter from the secretary of the treasury authorizing him to construct an experimental line between Washington and Baltimore. Morse was appointed superintendent of telegraphs, with Gale, Alfred Vail, and J. C. Fisher as assistants. Vail set to work building the apparatus, and Fisher oversaw preparation of the wire and enclosure of two wires in lead pipes. Figure 6.5 shows the telegraph used for this line.

Morse decided to lay the line underground. Ezra Cornell (1807–1874)[12] obtained the job of making the cable trench and laying the line in it with a special plow that he had developed. When appointed by Morse, he was a plow salesman with the sales rights to a patent for a cable-laying plow. By 1849, he claimed to have built a third of the telegraph lines in the United States. As an early stockholder of major telegraph companies, he became a multimillionaire.[13]

By August 1843, Morse had some 250 km of wire manufactured by the Stephens & Thomas plant in Belleville, New Jersey. The Ohio Railroad gave Morse permission to use the railroad's right-of-way. Work started at Baltimore on October 21. After the laying of about 15 km, work was stopped, however, because the line failed to operate. Reading that Cooke and Wheatstone in England had just shifted from underground installation to pole mounting of line, Morse decided to install the line on poles. As no experience was available, Vail proposed to insulate each wire and bunch the wires together at the poles. Cornell proposed separating the two wires and

---

[11] Seventy Congress members did not to vote at all, "to avoid the responsibility of spending public money for a machine they could not understand."

[12] Ezra Cornell was born on January 11, 1807 in Brooklyn, New York, the eleventh child of a Quaker pottery maker, Elijah Cornell. In 1818, the Cornell family moved to a farm in Ruyter, 70 km northeast of Ithaca, New York.

[13] Cornell, who had had little formal education, wanted to give other young people in similar conditions a better chance. In 1865, he founded, in Ithaca, Cornell University and a public library.

supporting each wire between two glass plates on the pole. Upon advice from Joseph Henry, Morse followed Cornell's proposal.

On April 1, 1844, work started in Washington. Some 500 chestnut poles each 7 m high were erected 60 m apart. Number 16 copper wire was used, insulated with cotton thread treated with shellac and a mixture of beeswax, resin, linseed oil, and asphalt. Two wires were installed in parallel, with all system components in series in a circuit as shown in Figure 6.6a. Replacing one of the wires by copper-plate Earth connections at both stations as shown in part (b) did not lessen the performance. In June 1844, the second wire was used as shown in part (c) for simultaneous telegraphy in both directions. The battery at Baltimore, consisting of acid cells, provided an 80-V electricity source.

The line began operation officially on May 24, 1844. Annie G. Ellsworth[14] sent the first telegram. Knowing how religious Morse was, she selected a quotation from the Bible, Numbers 23:23: "What hath God wrought!"[15] On May 27, 1844 the *New York Tribune* reported under the title "The Magnetic Telegraph—Its Success" that "the miracle of annihilation of space is at length performed." To appreciate the sensational impact of the news, it helps to realize that at that time over 90% of messages were transported by horseback or coach; railway transport had only begun in 1830. Messages from New York needed one day to reach Washington, two weeks to reach New Orleans, and three weeks to reach Chicago.

6.2.3 Pioneering Telegraph Companies

Operation of the Washington–Baltimore electric telegraph line was the responsibility of the postmaster general, Cave Johnson. Morse was appointed as superintendent, with Alfred Vail and Henry Rogers as operators. The first public telegraph office was opened on April 1, 1845, at Seventh Street in Washington, DC. After one year of operation, Johnson reported to Congress: "The importance [of the telegraph line] to the public does not consist in any probable income that can ever be derived from it." Thus, to the great good fortune of the United States, the telegraph was returned to private enterprise, which within 10 years had covered the country with some 50,000 km of telegraph lines. This principle of private enterprise in the United States was to hold good through all subsequent telecommunications technologies, except for a few decades of government-owned satellite communication.

On May 15, 1845, Morse founded the *Magnetic Telegraph Company* together with Postmaster Amos Kendall, Francis O. J. Smith, Ezra Cornell, and Cornell's brother-in-law, Orrin S. Wood. The first objective of the company was to extend the Washington–Baltimore telegraph line to New York. Cornell and Wood went to Boston and New York to raise money. They demonstrated the telegraph to a moderately interested public and slept on chairs for lack of money. Still, work on the line was begun in the same year. Telegraph service between New York, at 120 Wall

---

[14] Annie G. Ellsworth (later Mrs. Roswell Smith, often referred to as "Miss Telegraph") was the daughter of Morse's classmate, the commissioner of patents. She informed Morse on March 4, 1843 of the decision by Congress to build the line.

[15] The same words were used by President Kennedy at the end of the first telephone conversation over a *Syncom* satellite on August 23, 1963, then speaking to the Nigerian prime minister, their voices traveling over a distance of about 72,000 km instead of the 64 km of the first transmission.

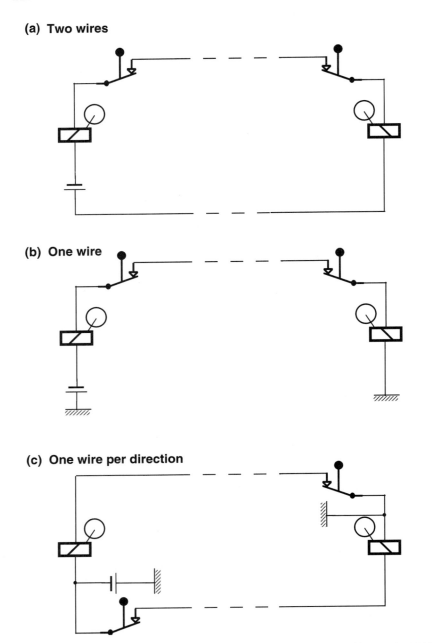

(a) Two wires

(b) One wire

(c) One wire per direction

Figure 6.6 Principles of the Morse line between Baltimore and Washington in 1844.

Street, and Philadelphia began at the end of January 1846. To cross the mile-wide Hudson River, the *Scientific American* (November 25, 1845 issue) proposed supporting the wires with balloons. Cornell, however, laid two lead pipes of his own design across the river at Fort Lee, where the George Washington Bridge now stands. Each lead pipe contained two wires covered with india rubber and cotton saturated with pitch. These lead pipes, which can be considered to be predecessors of submarine cables, had only a short life. Ice swept them away the first winter. Next, two wires were stretched across the river between tall iron towers. The wires were lowered into the water whenever a ship with a tall mast approached; however, they were frequently damaged by the wash of the boats. The line was then built up both sides of the Hudson to a narrow point south of the West Point Military Academy. There the wires were connected to a 122-m mast erected at Manhattan and a second mast at Fort Lee 230 m above the water level of the Hudson. Those wires, too, had a short life, and messengers had to cross the river by ferry to deliver telegrams in New York.

On May 30, 1845, Theodore S. Faxton and John J. Butterfield, both of Utica, obtained a license to build a line between Buffalo, Utica, Albany, and Springfield (later extended to New York). On the way back from Washington, Butterfield met Henry O'Reilly (1806–1886), who immediately got enthusiastic about the prospects of electrical telegraphy.

Henry O'Reilly was born in Carrickmacross, County Monaghan, Ulster, Ireland, on February 6, 1806. He came to the United States in 1816 with his mother and sister while his father was in a debtor's prison. He started work at the *New York Columbian* newspaper. At the age of 17 he became assistant editor of the *New York Patriot*, an organ of the People's Party. Three years later, he became the first editor of the Rochester, New York, *Daily Advertiser.* In 1838 he was appointed postmaster of Rochester by Amos Kendall, who was then postmaster general of the United States. On June 13, 1845, O'Reilly obtained a contract from Amos Kendall giving him the right to "raise capital for the construction of a line with the Morse telegraph from Philadelphia to Harrisburg, Pittsburgh, Wheeling, Cincinnati, and such other towns and cities as the said O'Reilly and his associates may elect, to St. Louis, and also the principal towns on the Lakes." The Morse patent owners were to receive one-fourth of the capital stock and not "connect any Western cities or towns with each other which may have been already connected by said O'Reilly." With this contract, Kendall had intended to give O'Reilly the right to build some lines west from Philadelphia. O'Reilly, however, regarded this contract as authority to organize, build, and manage lines for numerous companies and to establish his own telegraph empire.

Under the contract with O'Reilly, the *Atlantic, Lake, and Mississippi Telegraph Company* was organized on September 14, 1845. The company constructed the second telegraph line in the United States, between Philadelphia and Harrisburg. The line went into operation on January 8, 1846 and was extended to Pittsburgh, where operation began on December 26 of that year.

On July 16, 1845, the *Springfield, Albany, and Buffalo Telegraph Company* was founded with Theodore S. Faxton as president. Cornell built part of the line and became superintendent of the 800-km line between New York and Buffalo. On March 20, 1847, the *Washington and New Orleans Telegraph Company* was founded with David Griffin as president. Charles S. Bulkley built the line. Operation began on July 13, 1848.

Twenty telegraph companies existed in 1850, about half of them in the state of Ohio. A race started among these companies to operate the most profitable telegraph lines. Figure 6.7 shows the major lines that were built within the first five years. The order of completion, the date of opening of telegraph service, the constructor, and the operating company are indicated wherever these data could be traced.

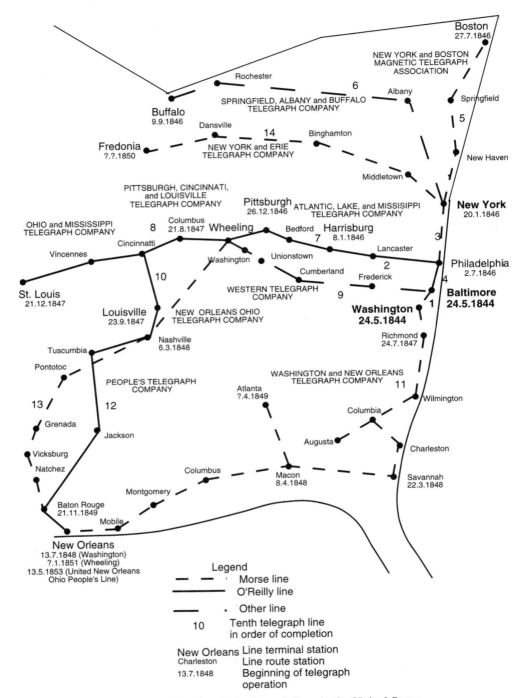

Figure 6.7   First electrical telegraph lines in the United States.

Despite many legal disputes, O'Reilly became the greatest of all pioneer line builders. He constructed over 15,000 km of lines and was the first person to promote the use of social and seasonal greeting telegrams. Unfortunately, O'Reilly increased his debt with each line that he built. He ended his career as a New York Custom House storekeeper and died in poverty at the age of 80 on August 7, 1886.

6.2.4 House Direct Printing Telegraph Systems

In addition to the Morse telegraph system, at least 62 other electrical telegraph systems were developed (according to Judge Woodbury, as mentioned later). Only three systems enjoyed a use in the United States worth mentioning: the chemical telegraph invented by the Scottish inventor Alexander Bain, described in Section 6.4; the direct letter printing telegraph system developed by the self-taught American, Royal E. House; and the direct letter printing telegraph system developed by an American physicist of British origin, David Edward Hughes, which was developed in 1855 and is described in Section 8.2.5.

Royal E. House, one of the founders of *Western Union Telegraph Company*, presented his letter printing telegraph at a Fair at the Mechanics Institute in the City Hall of New York in the fall of 1844, a few months after the Washington–Baltimore line had been placed into operation. House used a sending machine with a pianolike keyboard with 28 keys. The black keys corresponded to the letters A to N and the white keys to the letters O to Z, the period, and the hyphen. Under the keyboard was a revolving cylinder. When the operator pressed a key, it would catch a corresponding tooth in the cylinder and hold it while other parts revolved in alphabetical order until the letter desired was reached. Magnets in the receiving machine moved an equal number of times, and when the letter desired arrived on the type wheel, a blackened silk ribbon and a paper tape were pressed against it, printing the letter. The speed of the House telegraph was around 2600 words an hour. The device used an air compressor, and it required two operators to transmit and two more to receive.

The House telegraph was first used in 1850 by the *New York and Boston Telegraph Company*, later known as the *Commercial Telegraph Company*, also on a line between New York and Boston, to serve newspapers. Furthermore, the House system was used in the state of New York. A major line was built between Buffalo and St. Louis in 1852.

Further major lines to the West were built with House telegraphs. Ezra Cornell and John J. Speed founded the *Erie and Michigan Telegraph Company* to operate a line from Buffalo via Erie, Cleveland, Toledo, Detroit, Jackson, Marshall, Kalamazoo, Michigan City, Chicago, Racine, and Milwaukee. Farmers were so pleased about the line that they provided free poles. The line was ready for operation on April 6, 1848. In 1848, O'Reilly built a line from St. Louis to Chicago. After the two competing lines arriving from the north in New Orleans had been consolidated in 1853, a line was built from this town to Texas via Houston, reaching Galveston by the end of 1854.

House obtained a patent for an electrophonetic telegraph in 1868, but his "talking telegraph" found no application. During the patent battles with Alexander Graham Bell 20 years later, it was noticed that telephony would have been possible with this device. In fact, this retrospective claim for the discovery of telephony was used as an argument in favor of Bell.

6.3   ELECTRICAL TELEGRAPHY IN CANADA

Two years after Morse built the first line in the United States in 1844, the *Toronto, Hamilton, and Niagara Electro Magnetic Telegraph Company*, founded by T. D. Harris and associates, started a telegraph service in Canada between Toronto and Hamilton. Three years later the *Montreal Telegraph Company* was founded with Orrin S. Wood as superintendent. This company bought the Toronto company and extended their line, reaching Montreal on August 3 and Quebec in October 1847. The company started electrical telegraph service with the United States on January 14, 1847 with a line from Ontario to the *New York, Albany and Buffalo* line in Lewiston, New York. A suspended wire crossed the Niagara River. In the same year, other branches served points in Maine, New Hampshire, Vermont, and Michigan.

6.4   ELECTRICAL TELEGRAPHY IN GREAT BRITAIN

The story of the electrical telegraph in Great Britain is basically the story of the needle and pointer telegraphs of Cooke and Wheatstone and is closely connected with the development of railways. Although it is uncontested that Morse is the father of the electrical *writing* telegraph, there are many fathers of the electrical telegraph, and after Morse, Cooke and Wheatstone were the most successful. Further worthwhile contributions in electrical telegraphy came from Alexander Bain, Frederick C. Bakewell, and Frederick George Creed.

6.4.1   Electrical Telegraphs of Cooke and Wheatstone

William Fothergill Cooke (1806–1879), born in Ealing (now part of London), Middlesex, the son of a professor of anatomy, studied at the University of Edinburgh. At the age of 20 he joined the British East Indian Army. After five years of military service in India, he returned to Europe with the intention of entering the profession of his father. He studied anatomy first in Paris and then in Heidelberg. At Heidelberg University while attending lectures in anatomy given by George Wilhelm Muncke, on March 8, 1836, Cooke (Section 6.1) saw a demonstration using a copy of the needle telegraph developed by Schilling von Cannstadt. Cooke had another copy made and took it to England on April 22, 1836. He contacted railway companies and obtained a trial order for an electrical telegraph line from the *Liverpool–Manchester Railway Company*. He met with many difficulties and asked Wheatstone for advice on February 27, 1837.

Charles Wheatstone (1802–1875) was born in Gloucester, where his father had a shop for musical instruments. The Wheatstone family moved to London around 1820, where Charles started making musical instruments. At the age of 21, he published an article in a scientific journal about his experiments with harmonic oscillations of musical tones. Faraday was much impressed by the experiments of Wheatstone and reported frequently about those experiments at the Royal Society. Wheatstone was appointed a professor of experimental physics at King's College, London, in 1834, where in the same year he demonstrated his famous measurements determining the propagation speed of electricity. In 1843, Wheatstone presented various measurements concerning the relation of electrical potential, current, and resistance at the Royal Society. This included a presentation of a differential bridge circuit for measurement of electrical resistances, since then generally known as the *Wheatstone bridge*, although Wheatstone explained correctly that Samuel Hunter Christie had already proposed this method in 1833. Impressed by the extremely high propagation speed of electricity in metallic conductors, Wheatstone was also considering the possibility of signal transmission via metallic lines by the time that Cooke contacted him. He experimented with Cooke's telegraph at King's College when the American professor Joseph Henry of Princeton College visited King's College. Henry could give useful advice on better construction of the electromagnetic coils. Cooke and Wheatstone together then constructed a more reliable five-needle telegraph, including an electromagnetic alarm device, that required six connecting wires. This new device worked by deflecting any two of five astatic compass needle pairs, so that the two intersections of needles pointed to one of 10 possible letters above, and 10 below, their axis. The numbers 1 to 9 and 0 were located on the lower half of the front display, so that only one needle deflection was required for each of these numbers. To operate the transmitter, a small keyboard was used with six pairs of push-buttons. The limitation to 20 possible combinations required that the letters c, j, q, u, x, and z be omitted.

Cooke and Wheatstone applied for a patent on June 12, 1837. A technical specification was submitted on December 12, specifying an astatic needle-pair telegraph with five or four needle pairs, an electromagnetic distant sound-giving alarm, and a fault detection device. The description also included details about the construction of telegraph lines. The patent claimed to cover "improvements in giving signals and sounding alarums in distant places by means of electric current transmitted through metallic circuits." The patent was granted in May 1841 with a priority date of June 12, 1837 as number 7390. This was the first British patent for an electrical telegraph. The text was so general that it virtually secured Cooke and Wheatstone a monopoly position for electrical telegraphy in Great Britain.

Cooke and Wheatstone demonstrated their five-needle telegraph to the directors of the new *London–Birmingham Railway* on a 2.4-km railway track between Euston and Camden Town on July 25, 1837. The six wires were insulated and placed in a steel tube with a diameter of 6 cm. The tube was connected to small wooden poles placed by the side of the rails. The directors were not convinced. Fortunately, the directors of the *Great Western Railways* were more progressive. They commissioned Cooke and Wheatstone to install a telegraph between Paddington Station, the London terminus of their line, and West Drayton, a distance of 21 km. This, the world's first electrical telegraph line, began operation on July 9, 1839.[16] Again, the six wires were placed in a steel tube. The insulation deteriorated very quickly; six noninsulated wires were then installed as an open-wire line mounted on wooden poles. Six porcelain insulators were connected to the poles with metal clamps. Each wire went through one of the insulators. This elaborate, expensive, and unreliable construction forced Cooke and Wheatstone to reduce the number of wires, and thus the number of needles, at the expense of slower operation and the need to apply a code. They constructed a one-needle telegraph requiring two wires and developed a four-unit code with the number and direction of needle deviations used to code the alphabet.

---

[16] One year before the world's first postage stamps, the one-penny black and the two-penny blue, were introduced in England.

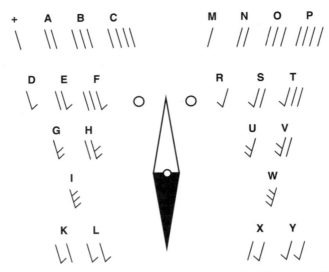

Figure 6.8 Code presentation for the one-needle telegraph of Cooke and Wheatstone.

To assist the operators, they designed a symbolic presentation of the code, shown in Figure 6.8. The code was to be read with respect to the idle position of the needle. Thus, to transmit an "H," the needle had to make the deflections left–left–right–right, or for a "T," left–right–right–right. To enable the operators to "read" the messages by listening, the two anvils on the left and right sides of the needle produced different tones.

Figure 6.9 Three versions of needle telegraphs from Cooke and Wheatstone. (Scanned from Ernst Feyerabend, *Der Telegraph von Gauss und Weber*, Reichspostministerium, Berlin, 1933, p. 75.)

The one-needle telegraph was used successfully when the line from Paddington to West Drayton was extended to Slough in 1843. British patent 10,655 was issued on May 6, 1845 for the one-needle telegraph. The reliable and easily operatable one-needle telegraph became widely used by British railway companies. Some 15,000 of those telegraphs were still in operation toward the end of the nineteenth century. Some even remained in operation until the 1930s. Figure 6.9 shows three versions of the needle telegraph.

In parallel with improvement in the needle telegraph, Wheatstone and Cooke worked on a "step-by-step letter-showing" pointer telegraph, which was easier to handle, albeit at the expense of slower operation, about 15 words per minute. The transmitter used up to 30 keys arranged in a circle. Each key was allocated to a letter and/or a number. Pressing a key caused a pointer to go around until it reached the key pressed. In the receiver, one or two electromagnets would synchronously move a pointer step by step in front of a disk with up to 30 letters and numbers on its edge and stop at the letter or number desired. Figure 6.10 is a contemporary drawing of one of the first pointer telegraph receivers, with two electromagnets and three connecting wires.

At the time, only weak sources of electricity and feeble electromagnets were available. Cooke and Wheatstone developed three successively improved versions of the pointer telegraph in the period 1838–1840. For this first version the electromagnet needed its little energy only to release the needle, whereas the energy required for the movement of the needle was taken from a weight. Three wires were required for

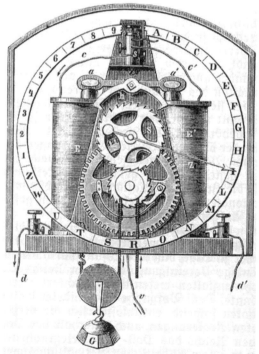

Figure 6.10   Pointer telegraph receiver of Wheatstone and Cooke. (Scanned from *Archiv für deutsche Postgeschichte*, Vol. 2, 1976, p. 6.)

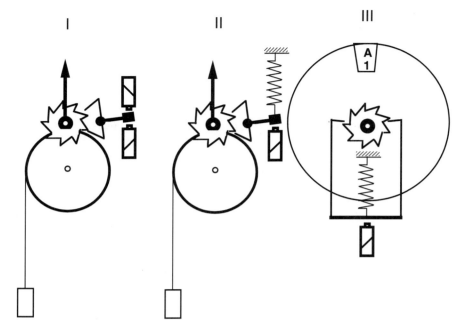

Figure 6.11   Three versions of Cooke and Wheatstone's pointer telegraph.

operation of the two magnets. The second version replaced one of the magnets by a mechanical spring. With this version only two wires were required, but the magnet needed to be stronger to counterbalance the spring. The weight drive of those two versions needed to be reset. In a third version, therefore, with a stronger electro-magnet, the weight drive could be eliminated. Moreover, instead of the needle, a rotating metal disk was connected to the axis of the toothed wheel. The letters and figures transmitted, printed on the edge of the disk, were shown in a little window. Figure 6.11 demonstrates the principle of the three versions.

Wheatstone developed another pointer telegraph in the 1860s, the ABC pointer telegraph (Figure 6.12), which included a transmitter, a receiver, and a crank generator for the line current in one unit. Some 10,000 units of this very reliable telegraph were produced. In 1920, about 1500 ABC pointer telegraphs were still in operation in London alone.[17] Similar versions were very soon produced by Bain, Mapple & Brown, and Nott and Barlow in Great Britain; Breguet and Garnier in France; and Fardely, Leonhardt, Drescher, Kramer, Lewert, Stöhrer, and Siemens in Germany.

Other railway companies soon followed the example of the Great Western Company and installed Cooke and Wheatstone telegraph equipment along their railway tracks. The railway network had a total length of about 12,000 km, of which 3600 km was accompanied by telegraph lines in 1850. The range of potential customers of the telegraph, however, was much larger than that of the railways. A first

---

[17] Whereas Cooke and Wheatstone had patented the needle telegraph, it was Wheatstone and Cooke (in that order, thus indicating who made the major contribution) who applied for a patent for the pointer telegraph at the end of 1839. British patent 8345 was granted on January 12, 1840.

Figure 6.12   ABC pointer telegraph. (Courtesy of Museum für Kommunikation, Frankfurt, Germany.) See insert for a color representation of this figure.

public awareness of electrical telegraphy was caused by the birth of Queen Victoria's second son, Alfred Ernst, at Windsor on August 6, 1844. The news was carried from Windsor to London on the Great Western Railway's telegraph line. The *Times* was on the streets of London with the news within 40 minutes of the official announcement, declaring itself "indebted to the extraordinary power of the Electro-Magnetic Telegraph" for providing the information so quickly.

This public awareness of electrical telegraphy was enhanced dramatically on January 1, 1845, when the telegraph operator at Paddington Station received a telegram from Slough. It informed him that a person by the name of John Tawell, in the garb of a Quaker, wearing a brown greatcoat, was sitting in the last compartment of the second first-class carriage. He had boarded the 7:42 P.M. train to Paddington after having killed his mistress, Sarah Hart. The operator informed the police, who then waited for the murderer at Paddington Station. When John Tawell was hanged, the telegraph had indeed become the talk of London.

Cooke and Wheatstone took advantage of the increasing popularity of the electrical telegraph and sold their patents for £140,000 to the *Electric Telegraph Company*.[18] In the same year, that company began construction of a public telegraph

---

[18] Founded in 1846 by Cooke together with John Lewis Ricardo, a member of Parliament, and prominent financier.

network. This was also the end of the cooperation between the businessman Cooke and the scientist Wheatstone, who had such different characters and intentions that they had started a public dispute about the scope of their contributions in their patents. In 1843, an arbitration committee decided that both men deserved equal rights from the patents: Cooke's merit having been that he brought the electrical telegraph to England and Wheatstone's that he made application practical possible through his scientific knowledge.[19]

6.4.2 Electrochemical Telegraph of Bain

Alexander Bain (1810–1877), born in Watten, Caithness, Scotland, went to London as a watchmaker in 1837. There he constructed an electrical clock in 1840 and an electrochemical telegraph in 1843. The name *electrochemical telegraph* referred to the receiver, where an iron pen rested on a moving paper strip treated with potassium ferric cyanide. When an electric pulse passed through the pen and the treated paper, the potassium ferric cyanide decomposed and left a corresponding mark on the paper. Historically, of more interest is Bain's telegraph transmitter, the first to use a punched tape. The tape was perforated with a dash-and-dot coded message and then placed into the transmitter, where electrical contact made through the perforation sent pulses over the line. Bain obtained a patent for his punching device on December 12, 1846. The perforation was time consuming; however, with later, improved versions, the automatic transmission achieved a speed of 1000 words per minute.

A few railway companies used Bain's electrochemical telegraph. It found only limited application in Great Britain because the Electric Telegraph Company bought the patent rights to prevent competition for the telegraph instruments they made under license from Cooke and Wheatstone. In Austria, Bain's telegraph was used by the *Kaiser-Ferdinand-Nordbahn* in 1846, but within four years it was replaced by Morse telegraphs. Bain was more successful in the United States, although at the cost of expensive lawsuits against Morse. Bain lost most of his income in patent battles and died a poor man despite financial support from the Royal Society.

6.5 ELECTRICAL TELEGRAPHY IN FRANCE

In France, the *Administration du Télégraphe* operated the world's largest optical telegraph system almost exclusively for administrative and military use. The prevailing highly unstable political situation resulting in the revolution of 1848 made it obvious to the French government that electrical telegraphy should also come under their control. It was under those circumstances that Morse came to Paris. He demonstrated his telegraph at the *Institute of France* to Baron von Humboldt, François Arago, and other scientists and obtained a patent on August 18, 1838, the first patent worldwide for his electrical writing telegraph. However, the French telegraph line

[19] This judgment was taken in committee by such famous men as Sir Isambard Kingdom Brunel (1806–1859), who built the first tunnel under the Thames in 1825, and the physicist John Frederic Daniell (1790–1845), who in 1836 developed the constant-voltage copper–zinc galvanic cell called the *Daniell cell.*

 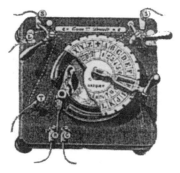

Figure 6.13 Breguet step-by-step telegraph. (Scanned from *Archiv für deutsche Post-geschichte*, Vol. 2, 1976, p. 13.)

administrator, Alphonse Foy, in charge of over 1000 optical telegraph station operators, most of whom were illiterate, doubted that his people could learn the Morse alphabet. In 1839, without further justification, Foy informed Morse that his system would not be accepted in France. Foy then sent a mission to England to study the needle telegraph of Cooke and Wheatstone. He also requested Louis-François Breguet (1804–1883), grandson of Abraham-Louis Breguet and regular supplier of the optical telegraph, to develop and produce an electrical telegraph with needles that reproduced the movements of the semaphore, to facilitate the transition from optical to electrical telegraphy in France. The resulting Breguet–Foy telegraph used two needles, which could show eight different semaphorelike positions. It was first tried out between Paris, Saint Cloud, and Versailles in 1842. In the same year, extensive tests were made with optical telegraph equipment for use at night. The experiments clearly showed that the performance of the electrical telegraph was far superior to that of the optical telegraph. A comparative test of electrical telegraphs was then made along the railway route between Paris (Gare Saint Germain) and Rouen in May 1845. Three different types of equipment were tested: the Cooke and Wheatstone two-needle telegraph, the Breguet two-needle telegraph, and a writing telegraph developed in that year by M. Dujardin of Lille, France. This writing telegraph used a paper-covered cylinder driven by a clock mechanism. A pen connected to the armature of a magnet wrote points and short lines on the paper. As a result of tests that took place on the May 11 and 18, the Breguet equipment was installed in the same year on the Paris–Rouen railway route between Paris and Lille.

Breguet very soon replaced the two-needle and two-wire telegraph with a more advanced one-wire step-by-step telegraph (similar to the Wheatstone and Cooke pointer telegraph), shown in Figure 6.13.[20] The *Breguet telegraph*, also called the *French telegraph*, was standard equipment on French railways for many years. In 1869 it was exported to Japan, where public telegraph service was inaugurated using Breguet's telegraph modified for the use of Japanese characters.

---

[20] For production of the electrical telegraph equipment, Louis Breguet had a fourth floor added to his building at 39 Quai de l'Horloge, Paris, in 1851.

6.6   ELECTRICAL TELEGRAPHY IN GERMANY

In the middle of the nineteenth century, Germany was a conglomerate of states, autonomous territories, and independent towns with different postal systems and relatively little communication between them. Even optical telegraphy was used only by the state of Prussia, because its scattered territories were separated by other states. A different situation arose with the appearance of numerous railway companies in Germany. One year after the Prussian optical telegraph line was put into operation, the first railway transport began in Germany between Nuremberg and Furth. Within 15 years a railway network was constructed with a length of about 6000 km. The optical telegraph was too slow to match the "high speed" of this new means of transport, so that in Germany, too, the electrical telegraph arrived just in time to meet the requirements of the railways. Electrical telegraphy therefore began in Germany in 1843 as a private enterprise for railway and other nongovernment applications. Public electrical telegraphy was begun six years later by a German (the Prussian) government. Neither the Morse telegraph nor the telegraphs of Cooke and Wheatstone were patented in Germany, and only a few units were imported. A dozen German workshops and companies developed their own electrical telegraphs, some of them based on foreign examples.

6.6.1   Railway Telegraph Lines in Germany

The first electrical telegraph line in continental Europe was installed in Germany along a steep railway track between Aachen and Ronheide in 1843. This track was part of the railroad between Aachen and Antwerp constructed in that year. The locomotives available were too weak to climb this section, which had a slope of 1 in 38 m. Therefore, a stationary steam engine was located at Ronheide to pull the train up by rope from Aachen Herbestal over a distance of 2.74 km. To communicate between the stations, two telegraphs were ordered from Cooke and Wheatstone, who supplied a pointer telegraph in a simplified version for six signals only. Figure 6.14 shows the sending (left) and receiving (right) units of this telegraph.[21]

Signaling was made by a combination of two of the five characters, thus providing 25 different signals, followed by a cross. If a signal did not end with a cross, it was a wrong signal and had to be disregarded. This is the first known example of automatic fault indication for electrical telegraphy. By 1855, more powerful locomotives were available, and both the stationary steam engine and the telegraph were taken out of operation.

The second electrical telegraph line in Germany was installed between Wiesbaden and Kastell–Biebrich along the 8.8-km railway track of the Taunusbahn in 1844 with the *typotelegraphen* produced by William Fardely (1810–1869). The line was extended later on a 41.4-km track from Kastell via the stations Hochheim, Flörsheim, Hattersheim, and Höchst to Frankfurt. Fardely was born in Ripon, Yorkshire, England, the son of British–German parents. He moved with his parents to Mannheim in 1820. He acquainted himself with electrical telegraphy in England

---

[21] The six signals probably stand for M, machine; S, seil (German for *rope*); C, convoi (expression used for group a of wagons); T, telegraph; and B, bremse (German for *brake*) with the cross reserved to indicate a quiescent condition and to provide an automatic fault indication.

Figure 6.14 First pointer telegraph used in Germany in 1843. (Scanned from *Archiv für deutsche Postgeschichte*, Vol. 1, 1979, p. 74.)

from 1840 to 1842. Back in Germany he constructed his own electrical needle telegraph, which he baptized the *typotelegraph* and wrote a detailed summary of electrical telegraphy, ranging from Steinheil via Cooke and Wheatstone to his typotelegraph.

The typotelegraph (Figure 6.15) was based on the pointer telegraph of Cooke and Wheatstone but used a weight-driven clock for the pointer movement.[22] The transmission speed was 1000 characters per hour. A single wire with a diameter of 1.5 mm

Figure 6.15 Typotelegraph of Fardely in 1843. (Scanned from *Archiv für deutsche Postgeschichte*, Vol. 1, 1992, p. 58.)

[22] It looked like a pendulum clock and, in fact, was manufactured in a Black Forest clock factory at Furtwangen.

connected the stations. Typotelegraphs were also installed in the German state of Saxony until they were replaced by Morse telegraphs in 1859. In the period 1846–1847, railway companies in the German states of Silesia and Bavaria installed Fardely's typotelegraphs, which remained in operation until the beginning of the 1870s. The first public telegraph line using a typotelegraph was installed in 1851 between Neunkirchen and Ludwigshafen.

The third electrical telegraph line in Germany was installed between Bremen and Bremerhaven in 1846 parallel with the optical telegraph line mentioned in Table 5.1. A two-needle telegraph was used which was constructed by Johann Wilhelm Wendt[23] (1802–1847).

The fourth electrical telegraph line, installed in 1849 between Hamburg and Cuxhaven, used the original Morse telegraph supplied by the American mechanic William Robinson. The line was installed in parallel with the existing optical telegraph line and soon replaced it. Following the American example, an iron wire was used, supported on wooden poles with glass insulators. Friedrich Clemens Gerke (1801–1888), who had gained substantial practical experience as a telegraph inspector on the optical telegraph line, improved the Morse code, which had a lasting influence on its worldwide use (Section 8.10).

By 1850, various railway companies existed in the German states which operated 5856 km of railway tracks. A total of 607 electrical telegraphs were used on those tracks: 237 were made by Kramer, 144 by Siemens & Halske, 135 by Stöhrer, 40 by Bain, 30 by Fardeley, and 15 by Leonhardt. In 1871, the German railway companies agreed to use the Morse telegraph exclusively. This robust system fully met the rigid requirements of railway operation, so that it was used up to 1952 by railways in the German Federal Republic and up to 1964 in the German Democratic Republic.

6.6.2   German Electrical Telegraph Equipment for Public Use

Franz August O'Etzel, the director of telegraphy of the Prussian government and superintendent of the Prussian optical telegraph line between Berlin and Coblenz, encouraged by the famous physicist Alexander von Humboldt, made experiments with a simple electrical telegraph of his own design. King Friedrich Wilhelm IV of Prussia became interested and had O'Etzel's telegraph demonstrated at his castle, Sanssouci, in Potsdam on October 8, 1840. O'Etzel was then asked to construct an experimental line between Berlin and Potsdam. O'Etzel interrupted construction of the line when he learned that Wheatstone had experienced problems with his open-wire lines and that tests had shown that those problems could be solved by placing insulated wires buried in underground iron tubes. Burying the wires, however, would have doubled the cost of the project.

The next development came in 1844, when Carl Gottfried Ferdinand Leonhardt, a clock manufacturer in Berlin, claimed to have developed a more economical solution. He constructed a pointer telegraph similar to the system of Cooke and Wheatstone but using a clock in the transmitter to move a dial to the desired letter of the

---

[23] Wendt had sailed four times around the globe as he changed his profession from sea captain to marine insurance agent at the age of 30. In 1843, he saw Cooke and Wheatstone's one-needle telegraph in operation on the railway track to Slough.

Figure 6.16 Typograph made by the clockmaker Leonhardt, 1844. (Scanned from *Post- und Telekommunikationsgeschichte*, Vol. 2, 2000, p. 61, with permission of the Deutsche Technik-museum, Berlin.) See insert for a color representation of this figure.

alphabet. At each step of the clock an electromagnet was activated and send a pulse over the line. Leonhardt received a patent for his pointer telegraph in 1846. No doubt, Leonhardt's telegraph, which he named *typograph*, was the most elegant electrical telegraph ever made, as can be seen in Figure 6.16. An original typograph is exhibited in the Museum für Verkehr und Technik in Berlin. By royal order, Leonhardt started work on the Berlin–Potsdam line, but he could not solve the problems arising in the production of insulated wires for underground laying. The insulation deteriorated rapidly in the wet soil. Wires were insulated at the time by pressing together two semicircular strips of caoutchouc around the copper conductor. When bending the wire, however, the strips frequently opened and caused short circuits.

At this stage another famous protagonist entered the telegraphic scene, with solutions for improving the speed and reliability of electrical telegraphy and insulation of the wires: Werner Siemens (1816–1892). He was born in Lenthe, near Hannover, the eldest son of 14 children. His parents died at an early age and he took responsibility for his younger brothers and sisters. Since the family lacked the financial means to pay for a university education, in 1835 he joined the Prussian army, where he underwent three years of specialist training in mathematics, physics, chemistry, and ballistics at the *Artillery and Engineering College* in Berlin. After the training, he was raised to the rank of lieutenant. Fortunately, he could devote his attention to scientific tasks and technical inventions, and even attend lectures at the University in Berlin. Together with other young scientists, he founded the *Physical Society*. In 1840 he obtained his first patent. During a period of penal detention for serving as a second in a duel, he used his time to develop an electrolytic method of gold plating. A younger brother, Wilhelm,[24] succeeded in exploiting the commercial potential of that invention in England, thus providing a financial basis for Werner's technical experiments.

In Berlin, Werner Siemens attended a demonstration given with Cooke and Wheatstone telegraphs by Councilor Soltmann in the summer of 1846. He observed that the use of a manual generator to provide the signal current was one of the weaknesses of that telegraph. He first sought to cooperate with Leonhardt, who was not interested, however, and Siemens decided to start telegraph equipment production himself. He rigged up his electric telegraph with "cigar boxes, tinplate, pieces of iron and some insulated copper wire" and demonstrated it to Johann Georg Halske (1814–1890), who he knew from the Physical Society. Halske was born in Hamburg and came with his parents to Berlin in 1827. He became a mechanic and opened a workshop, Böttcher und Halske, for scientific instruments in Berlin in 1844. Halske improved the mechanical design of the new pointer telegraph and produced a few of units in his workshop. Impressed by the prospects of electrical telegraphy, Halske decided to leave his workshop to Böttcher and start cooperating with Siemens. Together they founded the *Telegraphen-Bauanstalt von Siemens & Halske* in Berlin on October 1, 1847. The company started in a 150-m² workshop with 10 employees, three lathes, and a few drilling machines, all driven by human muscle force.

The first product was a pointer telegraph which combined sending and receiving parts in the same unit (Figure 6.17) and required only one wire to connect two telegraph stations, using Earth as a second conductor. Siemens understood that a serious weakness of telegraphy in general was the distortion of pulses on long and faulty lines. To make telegraphy less vulnerable to line problems, he introduced the already

[24]Charles William (born Karl Wilhelm) Siemens (1823–1883) went to Great Britain in 1843. He married Anne Gordon, the sister of Lewis D. B. Gordon, professor at the University of Glasgow and co-owner of the cable factory R. S. Newall & Co. and became a British citizen in 1859. Together with the French engineer P. E. Martin (1824–1915), he developed in 1864 the Siemens–Martin steel production process. He became a member of the Royal Society in 1862, was chairman of the Society of Telegraph Engineers and Electricians, and was knighted by Queen Victoria in the year of his death. He left a large fortune but no children.

Figure 6.17   Siemens's pointer telegraph. (Courtesy of Museum für Kommunikation, Frankfurt, Germany.)

known principle of automatic circuit interruption[25] for moving the pointer to the desired sign (Figure 6.18). The telegraph was tested successfully on the Berlin–Potsdam line in the summer of 1847, together with a similar pointer telegraph constructed by August Kramer (1817–1885) of Nordhausen in 1847. Siemens also found a solution for the insulation of underground cables, described in Section 8.1.2.

King Friedrich Wilhelm issued a royal command on July 24, 1848 to construct two electrical telegraph lines, one from Berlin via Cologne (to replace the optical telegraph) up to the border with Belgium and another from Berlin to Frankfurt. The line to Frankfurt especially was of great importance and urgency because of the election of a German emperor planned to take place in Frankfurt on March 28, 1849. An open competition for the line construction began on November 20, 1847. Nine companies participated in the competition.[26] Kramer was selected to construct the Berlin–Cologne line, and Siemens & Halske obtained the order for the Berlin–Frankfurt line. Robinson also received an order for his Morse telegraph, imported from the United States for test on the lines.[27] A single insulated copper wire was

[25] The principle of automatic circuit interruption had been presented by Johan Philipp Wagner on February 25, 1837 at the Physical Society in Frankfurt. Called *hammer interrupter* by Wagner, it was used primarily for automatic alarm bells. Siemens was first to apply this principle to telegraphy.

[26] The names of the companies were: Brettchen, Drescher, Fardely, Kramer, Leonhardt, Maneri, Moltrecht, Robinson, and Siemens & Halske.

[27] Right-of-way conditions had to be negotiated with the governments of eight German states and nine independent railway companies.

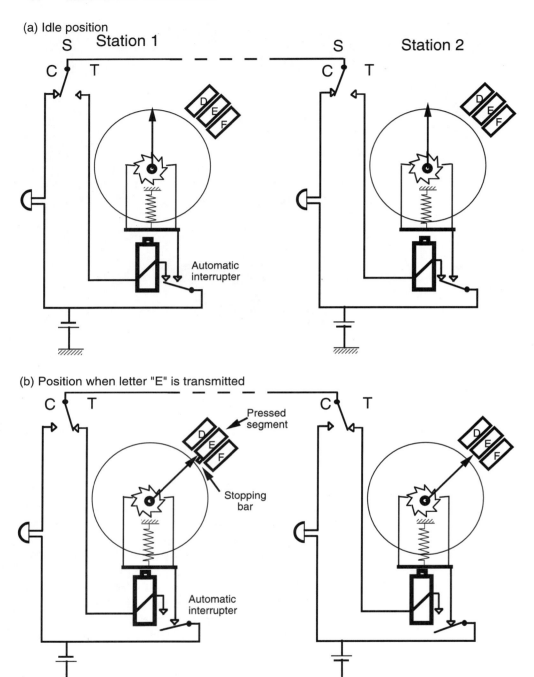

Figure 6.18   Automatic telegraph circuit interruption.

buried along railway tracks between Berlin and Eisenach. Between Eisenach and Frankfurt, no railway existed and an uninsulated copper wire was constructed as an overhead line using porcelain insulators manufactured by the now famous *Meissner Porcelain Factory*. The line had eight repeater stations, where the messages had to be repeated manually. With a length of 500 km, it was at the time the longest European electrical telegraph line.

On March 28, 1849, the first German national assembly convened in the Church of St. Paul in Frankfurt and elected King Friedrich Wilhelm as the first German emperor. Within an hour the election result reached Berlin via the new electrical telegraph line.

The line to Cologne started operation in June 1849. Kramer's telegraph instruments did not meet the operational requirements, however, and were replaced within two years by the more reliable and quicker telegraph of Siemens & Halske. The Siemens pointer telegraph was also installed on routes placed into operation from Berlin to Hamburg in May; to Leipzig, Stettin, and Düsseldorf in June; and between Breslau and Oderberg in October 1849.

A comparative test with Morse and Siemens telegraphs was made on the Berlin–Frankfurt line. The complete speech of King Friedrich Wilhelm from the throne was transmitted by the Morse telegraph within 75 minutes, whereas the Siemens pointer telegraph required seven hours. Taking advantage of the situation that Morse's telegraph was not patented in Germany, Siemens was invited to construct and produce an improved version of the Morse telegraph.[28] In recognition of his valuable invention, Morse received a golden tobacco box from King Friedrich Wilhelm in December 1850.

Halske made a number of mechanical improvements to the Morse telegraph, including relief writing with a steel needle pushing the dots and dashes in the moving paper tape and replacing the weight drive for the pointer with a spring drive. The improved version became the standard for Prussia and later also for the member countries of the Austrian–German Telegraph Union.

Manufacturing rights for the new standard Morse telegraph were also given to the oldest German telecommunications company, the royal purveyor, Lewert, founded in 1800 in Berlin by David Friedrich Lewert (1779–1863). This company then started production of Morse telegraphs in 1851 and manufactured over 5000 units in the nineteenth century. Also, Wilhelm Gurlt (1825–1897) opened an electrical telegraph equipment factory in Berlin in 1853 and supplied Morse telegraph instruments to the Telegraph Administration.[29]

Most of the other German states followed Prussia in 1849 with the construction of telegraph lines. Bavaria constructed a line between Munich and the Austrian town of Salzburg using telegraph instruments developed by Emil Stöhrer in Leipzig in 1846. Stöhrer applied two parallel pins for writing the signal received on a paper tape.

---

[28] Following the prevailing German antipatent policy, the Prussian patent commission had decided in 1845 that Morse's invention was not important enough to be patented.

[29] C. Lorenz bought the company Lewert in 1893 and the company Gurtl in 1915. ITT bought C. Lorenz in 1930 and merged it with Mix & Genest into Standard Elektrik Lorenz (SEL) in 1958. ITT sold SEL to Alcatel in 1987. The company is now called Alcatel SEL.

Figure 6.19   Telegraph lines in Germany before 1850. (© Siemens-Archiv, Munich.)

In Munich, insulated copper wire was placed in clay conduits. In Salzburg, the Austrian and Bavarian telegraph operators worked in the same room so that telegrams from Munich to Vienna could be handed over by the German operator to the Austrian operator for retelegraphing to Vienna. The line, with a length of 142 km, was constructed within two months. Public telegraph service begun on January 15, 1850.

The state of Saxony connected the cities of Leipzig and Dresden with the Berlin–Frankfurt line in the town of Halle in 1849 and opened a public telegraph service on October 1, 1850. The line was extended from Leipzig to Hof, Bavaria, in the same year. By that time, 2000 km of wire was installed in Germany, mainly underground. Figure 6.19 shows the various optical and electrical telegraph lines that were in operation in the German states or planned to be ready in 1850.

The state of Wuerttemberg opened its public telegraph service on April 16, 1851, initially using lines constructed for the railways. Public operation started on the line between Heilbron, Stuttgart, Ulm, and Friedrichshafen. This line was connected with the Bavarian network in Ulm. A second line between Stuttgart and Bruchsal, installed in 1852, was connected in Bruchsal with the network of the state of Baden. The Morse telegraph was used and the lines were constructed using galvanized iron overhead wires. Public telegraph service started in the state of Baden in 1851, also initially using the lines constructed by the railway companies.

6.7   ELECTRICAL TELEGRAPHY IN AUSTRIA

The first electrical telegraph line in Austria (at the time, Austria–Hungary) started operation in March 1847 between Vienna and Brünn (now Brno, Czech Republic). The 154-km-long line run along the Emperor Ferdinand Northern Railway opened eight years earlier. The line was constructed as a trial line and operated so satisfactorily that it was extended by another 200 km, to Prague, in the same year. Almost simultaneously, telegraph lines were constructed between Vienna and Pressburg (now Bratislava, Slovak Republic) and between Vienna and Trieste (now an Italian town). By act of June 22, 1849, the Emperor Franz Joseph I ordered the construction of an electrical telegraph network to cover the entire monarchy. In the same year, on October 3, an agreement for the passage of electrical messages was made between Austria–Hungary and Prussia. Most telegraph lines were constructed along railway tracks and terminated in Vienna. The telegraph offices were located at railway stations or post offices. Public telegraph services began after a few years of exclusive government use in February 1850.

REFERENCES

Books

Aschoff, Volker, *Geschichte der Nachrichtentechnik*, Vol. 2, Springer-Verlag, Berlin, 1995.

Bertho, Catherine, *Histoire des télécommunications en France*, Édition Érès, Toulouse, France, 1984.

Bertho, Catherine, *Telegraphes & telephones de Valmy au microprocesseur*, Hachette, Paris, 1981.

Feyerabend, Ernst, *Der Telegraph von Gauß und Weber im Werden der elektrischen Telegraphie*, Reichspostministerium, Berlin, 1933.

Michaelis, Anthony R., *From Semaphore to Satellite*, International Telecommunication Union, Geneva, 1965.

Oslin, George P., *The Story of Telecommunications*, Mercer University Press, Macon, GA, 1992.

Siemens, Georg, *Der Weg der Elektrotechnik Geschichte des Hauses Siemens*, Vols. 1 and 2, Verlag Karl Alber, Freiburg, Germany, 1961.

Wilson, Geoffrey, *The Old Telegraphs*, Phillimore & Co., Chichester, West Sussex, England, 1976.

150 Years of Siemens: The Company from 1847 to 1997, Siemens, Munich, 1997.

Articles

Andrews, Frederick T., L'héritage du télégraphe, *Revue Française des Télécommunications* (France Telecom), Vol. 77, May 1991, pp. 60–71.

Bernhardt, Manfred, Entwicklungsgeschichte der elektrischen Telegrafie bis zur Einführung des öffentlichen Fernschreibdienstes in Deutschland, *Archiv für deutsche Postgeschichte*, Vol. 1, 1992, pp. 53–60.

Carre, Patrice A., Aux origines des télécommunications d'affaires, *Revue Française des Télécommunications* (France Telecom), Vol. 78, June 1991, pp. 58–67.

Kainz, Christine, Österreichische Postgeschichte-ein Überblick, *Archiv für deutsche Postgeschichte*, Vol. 1, 1979, pp. 111–134.

Korella, Gottfried, Elektrische Telegrafie, *Archiv für deutsche Postgeschichte*, Vol. 2, 1974, pp. 102–105.

Wichert, Hans Walter, Die elektrische Telegrafie in Deutschland, *Post- und Telekommunikationsgeschichte*, Vol. 2, 2000, pp. 59–65.

Internet

www.CCLab.com, American History of Telecommunications, Telecom Global Communications, Inc.; updated March 24, 1998.

PART III

PERIOD FROM 1850 TO 1900

7

EVOLUTION OF TELECOMMUNICATIONS FROM 1850 TO 1900

In the period 1850–1900, optical telegraphy lost its raison d'être and was replaced by the electrical telegraph in most countries within the first 10 years of this period. Various competing electrical telegraph systems appeared on the market. The Cooke and Wheatstone telegraph was used exclusively in England, mainly by British Railways, right into the twentieth century; in Spain; and in improved versions initially also in most other European countries. The far superior Morse system, which could easily be decoded and written in plain language as a *telegram*,[1] finally found worldwide use. National electrical telegraph networks were erected in the industrializing countries and in some of their colonies.

The first international submarine cable was laid in 1851 between England and France, followed in 1852 by cables connecting Ireland with Wales and Scotland and in 1853 with cables connecting England with Denmark and Belgium.

Great events, worthwhile remembering, happened in those early days of electrical telegraphy, such as the laying of the first intercontinental submarine cable between the United States and Europe,[2] the construction of the 2900-km Australian overland telegraph line in two epic years from 1870 to 1872, and an 11,000-km telegraph line between London and Calcutta, completed in 1870. These and other remarkable events are described in Chapter 8.

Electrical telegraphy made news a valuable commodity. Julius Reuter (1816–1899)[3] was probably the first person to recognize and exploit the commercial value of

---

[1] This plain language-telegraphic message, a long description for a small piece of paper, acquired the short name *telegram*, proposed by a Rochester lawyer, E. Peshine Smith, in the *Albany Evening Journal* of April 6, 1852 and in the first issue of *Telegraph Magazine* in October 1852.

[2] Described brilliantly by Stefan Zweig as one of the propitious hours of humankind.

[3] Reuter was born in Kassel, Germany, as Israel Beer Josaphat, in the Jewish faith. In 1845 he was baptized at St. George's German-Lutheran Chapel, Whitechapel, London, and renamed Paul Julius Reuter.

The Worldwide History of Telecommunications, By Anton A. Huurdeman
ISBN 0-471-20505-2   Copyright © 2003 John Wiley & Sons, Inc.

last-minute information. In 1848 he fled from Germany to Paris, where he started as a translator for *Havas* (founded in 1832). In 1850 he opened a news agency in Aachen, Germany, with telegraphic connections to Berlin, Brussels, and Paris. The 150-km telegraph gap between Aachen and Brussels was bridged with a pigeon post service. In 1851 he stopped this service, moved from Germany to London—in those days the center of world information—and began telegraph services on November 13 between London, Paris, and Berlin via the first cross-channel submarine cable. He made a fortune with sales of political and economic news by Reuter's telegrams.[4]

In the United States, six New York newspapers formed the Associated Press (AP) in May 1848 and sold news to various newspapers and trading companies on a subscription basis.[5] In Paris a translation office founded in 1835 changed its name in 1850 to *Agence Havas*, which is now AFP (Agence France Press). The news agencies Reuters, Havas, and Wolfs Telegraphenbureau (founded in 1848 in Berlin) established a worldwide news exchange on January 17, 1870. The word *telegraph*, already used for mail coach services, now became a synonym for *latest information*, and was frequently used as a newspaper name (e.g., the London-based *Daily Telegraph*, founded in 1855), stressing that their information was the latest available.

The major achievement of electrical telegraphy, apart from making news a valuable commodity and substantially improving the security and reliability of railway transportation, has been the creation of an international telecommunications infrastructure; a prerequisite for the development of worldwide telecommunications. At a very early stage it became obvious that construction of an international telecommunications infrastructure required detailed international coordination. The first international agreement concerning border-crossing telegraphy, made between Austria–Hungary and Prussia, resulted in foundation of the Austrian–German Telegraph Union on October 3, 1849. Other bilateral agreements soon followed and finally resulted in foundation of the International Telegraph Union (ITU) in 1865, described in Chapter 13.

Although optical telegraphy was very soon replaced by electrical telegraphy, once the electrical telegraph had spread almost worldwide, the need for rapid transmission of messages was satisfied and there was hardly a desire for a more efficient telecommunications system, such as the telephone, which was developed in this period. Both the French engineer Charles Bourseul, who described the concept of telephony in an article published in 1854, and the German teacher Philipp Reis, who constructed an instrument that reproduced sound electrically, were met with little interest. It was thanks to the perseverance of Alexander Graham Bell that in 1876 the telephone could finally be introduced. The complex events that led to the recognition after a 10-year legal battle that Bell was the inventor of the telephone are described in Section 10.2. Bell was undoubtedly the most successful telecommunications inventor: His telephone has given such an enormous impetus to telecommunications that currently about 1 billion persons can communicate via a worldwide telephone network. By the time that Bell patented his telephone in 1876, a global telegraph

---

[4] On September 7, 1871, the Duke of Saxe-Coburg-Gotha conferred a barony on Julius Reuter, known henceforth as Baron der Reuter. At the end of the twentieth century, Reuter operated a network connecting over 60,000 customer sites worldwide.

[5] With 3500 journalists, AP currently serves 15,000 media enterprises in over 100 countries.

TABLE 7.1   Thirty-Year Period of Major Technical Innovations

Year	Inventor	Country	Invention/Discovery	Domain[a]
1876	Alexander G. Bell	United States	Telephone	te
	Nikolaus Otto	Germany	Principles of internal combustion engine	tr
	Carl Linde	Germany	Refrigeration	va
1877	Elihu Thomson	United States	Electric welding	el
	Thomas Alva Edison	United States	Phonograph	e
	Charles Cros	France	Paléophone	va
	William Bown	Great Britain	Ball bearings	tr
1878	David Hughes	United States	Microphone	te
1879	Thomas Alva Edison	United States	Incandescent electric lamp	el
	William Crookes	Great Britain	Cathode-ray tube	el
	Werner Siemens	Germany	Electric railway	tr
1883	Marcel Deprez	France	Electric power transmission	tr
	Lucien Goulard and John Gibbs	France, Great Britain	Ac transformer	el
1884	Charles Parsons	Great Britain	Steam turbine	tr
1885	Gottlieb Daimler	Germany	Gasoline engine, auto	tr
	William Burroughs	United States	Commercial adding machine	va
1887	Heinrich Hertz	Germany	Electromagnetic waves	te
1888	Emile Berliner	Germany	Flat-disk gramophone	te
	John B. Dunlop	Great Britain	Pneumatic bicycle tire	tr
	Nikola Tesla	Croatia/United States	Polyphase induction motor	tr
1889	Almon B. Strowger	United States	Automatic telephone exchange	te
	Hermann Hollerith	United States	Punch-card tabulating	va
1890	Clément Ader	France	Steam aircraft *Eole I*	tr
1893	Wilhelm Maybach	Germany	Float-feed carburator	tr
1895	Guglielmo Marconi	Italy	Wireless telegraphy	te
	Louis and Auguste Lumière	France	Cinematograph	va
	Wilhelm K. Röntgen	Germany	X-rays	el
1896	Joseph J. Thomson	Great Britain	Discovery of the electron	el
1897	Karl F. Braun	Germany	Cathode-ray oscillograph	el
1898	Rudolf Diesel	Germany	Diesel engine	tr
	Marie and Pierre Curie	France	Radium	va
1899	Michael O. Pupin	Croatia/United States	Pupin coil	te
1900	Ferdinand Graf von Zeppelin	Germany	Zeppelin airship *LZI*	tr
	Max K. E. L. Planck	Germany	Quantum theory	va
1903	Orville and Wilbur Wright	United States	Petrol-driven aircraft *Flyer 1*	tr
1904	John A. Fleming	Great Britain	Electronic tube (diode)	el

(Continued)

TABLE 7.1 *(Continued)*

Year	Inventor	Country	Invention/Discovery	Domain[a]
1905	Albert Einstein	Germany	Relativity theory	va
1906	Lee de Forest and Robert von Lieben	United States, Austria	Triode	el

Source: Adapted from Robert J. Chapuis, *100 Years of Telephone Switching*, Parts 1 and 2, North-Holland, New York, 1982 and 1990.

[a] Transport (tr), telecommunications (te), electricity (el), and various others (va).

network existed—using primarily the Morse telegraph—of more than 1,000,000 km of wire and 50,000 km of submarine cable, connecting more than 20,000 towns all over the world.

Bell's invention of the telephone marks the beginning of a unique 30-year period with an incomparable wealth of technological innovations that all became major components of our contemporary modern civilization. In the short interval between 1876 and 1906, the basic inventions were made that created industrial electrical engineering and determined the future direction of transport, telecommunications, and generally of electricity application.[6] Table 7.1 summarizes those achievements. The most important event after the development of the telephone occurred in 1887, when Heinrich Hertz discovered the existence of electromagnetic waves. Within 10 years of this discovery, Marconi invented the radio (Section 12.4).

Within two years after Bell developed his telephone, switching of telephone lines started with a manual switch at New Haven, Connecticut, in 1878. The accumulation of operational and especially human errors connected with manual switching became so frustrating for funeral director Almon Brown Strowger of Kansas City that he changed his profession and developed the world's first "girl-less, cuss-less, out-of-order-less, and wait-less telephone exchange," which began operation in La Porte, Indiana, on November 3, 1892, starting the era of automatic telephone switching described in Chapter 11.

The nineteenth century closed with an outstanding invention for line transmission. In 1899, Michael Idvorsky Pupin patented the results of his studies on transmission of telephony on long lines.[7] With specially designed coils called *Pupin coils*, placed at 10- to 15-km intervals, the length of nonamplified telephony on underground cable could be extended from 20 km to almost 100 km.

---

[6] The accumulation of inventions was so overwhelming that the head of the patent office in Washington at the end of the nineteenth century recommended closure of the 100-year-old office because "all there was to be invented had been invented."

[7] In fact, in the period between 1886 and 1893, Oliver Heaviside (1850–1925) had published his *telegraph equation*, in which he defined the relation between inductance, capacity, and attenuation of lines. John Stone and George Campbell, working for AT&T, had arrived at conclusions similar to those of Pupin, but much to AT&T's disadvantage, failed to apply for a patent.

8

ELECTRICAL TELEGRAPH SYSTEMS WORLDWIDE

8.1 TELEGRAPH TRANSMISSION TECHNOLOGY

The air was the transmission medium for optical telegraphy; air is abundantly available and did not need to be invented. To enable electrical telegraphy from one point to another, a new technology had to be created: the technology of transmission of electrical signals. This new technology was developed successively for open-wire lines, for underground cable, and for submarine cable. Initially, a single wire was used between the transmitting and receiving sides of an electrical telegraph line, while Earth was used as the return conductor. For simultaneous telegraphing in both directions, two wires in parallel were used, plus Earth as a common return conductor. For long lines in areas with very dry soil conditions, a second wire was introduced for each direction instead of Earth. At the end of the nineteenth century, air also became the transport medium for electrical telegraphy by radio.

8.1.1 Open-Wire Lines

Telegraph signals initially were sent from one station to another via overhead copper lines. Before electrical telegraph signals could be transmitted over longer distances, two major prerequisites had to be met: a suitable way of constructing the lines, and a practical solution for the compensation of signal loss along the line.

The construction of overhead wire lines (usually referred to as *o/w lines*) presented various problems. The wires should combine low electrical resistance with a high mechanical strength. The line support should prevent short circuits between wires even under extreme weather conditions. Copper wires adequately spaced and carried by porcelain or glass insulators fixed to wooden poles provided a practical and eco-

The Worldwide History of Telecommunications, By Anton A. Huurdeman
ISBN 0-471-20505-2   Copyright © 2003 John Wiley & Sons, Inc.

nomical solution applied widely on all telegraph lines in the early years. Copper was chosen for the wires because of its low electrical resistance, albeit at the cost of limited mechanical strength. The valuable copper wire, however, was frequently stolen or damaged by storm and ice loads. The copper was very soon replaced by galvanized iron, to reduce theft and to increase mechanical strength. A further improvement came in 1877, when hard-drawn copper wire with good mechanical strength was invented in the United States. It was first used between Boston and New York in 1884.

To support the wires, impregnated wooden poles were generally used, which lasted 20 years under normal conditions. Cast-iron poles, which easily lasted 50 years, were installed in difficult tropical environments or in areas where suitable wood was not available, for example in Australia. Occasionally, for example in the Philippines, the wires were connected to specially planted living cotton trees with their branches removed. The trees were placed at 30- to 60-m spacing, depending on soil and climatic conditions.

A solution had also to be found for connecting the wires to the poles. In the first installations, the wires were simply led through a hole (covered inside with an insulation material) at the top of the wooden poles. Figure 8.1 shows that this solution was used in 1847 on one of the first telegraph lines in Germany, the Bremen–Bremerhaven line. A better solution, which reduced leakage substantially, involved fixing special glass or porcelain insulators to the poles. Over 100 different versions

Figure 8.1   Early example of open-wire support. (Scanned from *Archiv für deutsche Postgeschichte*, Vol. 1, 1979, p. 75.)

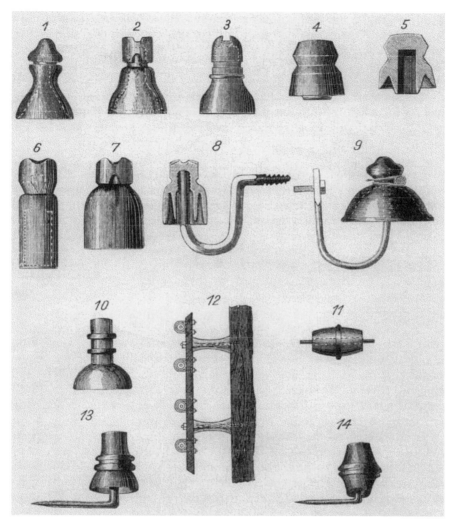

Figure 8.2   Major insulator versions.

were used worldwide, the 14 most common being shown in Figure 8.2. One of the best solutions for low-leakage conductance was the *double-shed porcelain insulator* (numbers 7 and 8 in Figure 8.2), also called the *double-petticoat insulator*, which in a slightly modified form became the worldwide standard.

Both Wheatstone in Great Britain and Henry in the United States presented solutions for the compensation of signal loss over long lines in the early 1830s. During their first experiments on telegraphy, they realized that the electric impulses arriving attenuated at the end of a long line were too weak to operate a telegraph receiver with the prevailing technology. Henry, by experimenting and without knowledge of Ohm's law, was the first to determine the relation between the length and resistance of line wires and the proper proportioning of the battery and electro-magnet of a telegraph receiver. He thus defined in 1830 the basic parameters of electrical telegraphy. This solved the problem for relatively short distances. To cover longer distances, Wheatstone had the idea of adding an additional device. Between the end of the line and the receiver he placed a sensitive electromagnetic device with an armature, which upon receipt of an impulse on the line closed a separate dc circuit powered by a local battery strong enough to operate the telegraph receiver. Henry generally used such a device in 1835 to produce a mechanical effect at a distance, such as ringing a bell. Wheatstone used his device for telegraphy experiments, calling it a *relay*, a word that he took from the practice of British hunters to use fresh horses and dogs, when chasing an animal, to replace those already tired out. The term *relay station* was used later for mail-coach rest areas and in the twentieth century for radio-relay transmission. Relays could be used not only at the ends of lines but also at in-between long-line locations, to receive weak signals and retransmit them at a stronger level through repeater stations. The electromagnetic relay became a general-purpose device to repeat, regenerate, or amplify telegraph signals in route over increasingly long lines. The relay is thus the first piece of line transmission equipment. By the end of the nineteenth century, it had been upgraded to a highly sensitive precision device by Siemens and others and became widely used in telephone switching.

8.1.2   Underground Cable

Open-wire lines are very vulnerable to damage by natural calamities, vandalism, and theft, and of course, cannot be used for large water crossings. As a first solution to the burial of wires—for example, for river crossings—the British engineers William Young and Archibald McNair constructed a lead-sheathed cable. In 1845 they obtained a patent for a cable construction in which a seamless lead sheath was pressed around copper wires that were surrounded by wax.

Valuable material for the insulation of wires came from the Malaysian peninsula around 1840. A British physician, Dr. Montgomery, noticed that the milky fluid tapped from the *Isonandra gutta* tree,[1] when dried, provided a flexible material called *gutta-percha* (from the Malay *gutah* = gum and *percha* = tree). He sent some samples of gutta-percha to London, where they were exhibited at the *Royal Society of Arts* in 1843. The next year, importation started with a quantity of 100 kg. Four years later, Singapore exported 665 tons of gutta-percha. The company SW Silver & Co. of Stratford, East London, was first to use the material to insulate telegraph wires. Samuel T. Armstrong had 5 tons shipped to New York, where he established a factory in Brooklyn to insulate cables. He laid the first cable insulated with gutta-percha under the Hudson River at New York City in 1847. In the same year, William Siemens send some gutta-percha from London to his brother Werner Siemens in Berlin, who recognized the high insulation and protective value of gutta-percha. Together with Johann G. Halske, he constructed a press (Figure 8.3) that produced a seamless gutta-percha coating around copper wires: the prerequisite for reliable underground and submarine cable.

Gutta-percha proved to be an excellent insulation material for underwater cables, but it lost its flexibility if exposed to air and sunshine. Caoutchouc, produced simi-

---

[1] The *I. gutta* tree is a tropical tree approximately 20 m high and 2 m in diameter growing on the Malay Peninsula and in Indonesia.

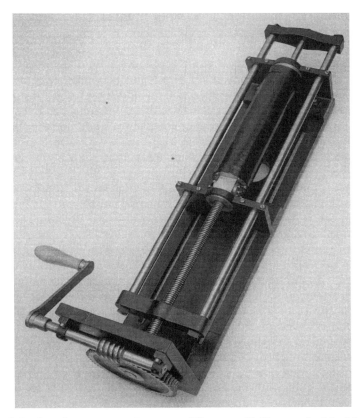

Figure 8.3 Seamless gutta-percha press of Siemens. (© Siemens-Archiv, Munich.)

larly to gutta-percha but from the South American tree *Siphonia cahucu* (also called *Hevea brasiliensis*), proved to have better electrical insulation and especially to be much less sensitive to high temperatures. Seedlings of *S. cahucu*, more commonly called the *rubber tree*, allegedly were smuggled from Brazil to Malaysia, where rubber trees soon covered the ugly mounds of dirt gouged out of the earth to mine valuable tin deposits. The white latex was tapped in hollow coconut cups collected before dawn, dried in the tropical sun, and after conservation with smoke, made transportable as *smoked sheets*. This provided the basic material for bicycle and later for automobile tires, and perfect insulation for telegraph wires and cables.

8.1.3 Submarine Cable

Using cable for electrical telegraphy crossing oceans presented new challenges: salty seawater, an unknown ocean floor, high attenuation of electrical signals over very long distances, transport of a tremendous load of copper and steel, and laying cable smoothly in a deepwater seabed.

Salty seawater, with its good electrical conductivity, can be used as a return conductor, but perfect insulation is then required for the other conductor in the cable. Fortunately, the rubber and tar used initially for the insulation of copper-wire conductors could soon be replaced by the superior gutta-percha, which kept its plasticity even under extreme pressure and at the low temperatures prevailing at the bottoms of oceans.

The assumption that an ocean had a smooth and calm floor proved to be wrong; strong surface currents and stony peaks broke the first cables. Moreover, cable at a depth of 6000 to 8000 m has to withstand tremendous pressures. Strong mechanical armoring was applied, consisting of galvanized steel wires covered by tarred hemp. A brass tape was wound around the copper conductor as additional protection and to prevent signal interruption in case of a wire break. To protect submarine cables against the peaks and surface currents on an ocean floor, they had to be laid not in a straight line but with substantial slack.

Relay amplification of the weak telegraph signal, as applied to terrestrial telegraph lines with local power sources, of course could not be used for submarine cable midway in an ocean. In 1858, the famous British physicist William Thomson (1824–1907) developed a highly sensitive telegraph receiver for submarine telegraphy. Thomson, born in Belfast, Northern Ireland, the son of a mathematician whose lectures he attended from the age of 8, became a professor at the University of Glasgow at the age of 22. In the same year he defined a temperature scale based on the absolute zero-point temperature, later called the *Kelvin scale*. He also discovered the thermoelectric effect (the *Thomson effect*) in 1856, and was knighted Lord Kelvin of Largs by Queen Victoria in 1892. Thomson used a mirror galvanometer with a coil of fine wire wound in a circle, suspended between the poles of a permanent magnet, with a tiny mirror attached to the suspension thread. Instead of sending dots and dashes by closing a Morse key, for a dot the cable operator had to strike a key to the left to connect the cable with the positive pole of the battery, and for a dash he had to strike the key to the right to connect the negative pole. The line transmitting voltage of 50 V resulted in receiving signals of a few millivolts only. A positive or negative signal received caused the mirror to swing slightly to the left or right. A beam of light from an oil lamp was focused directly through the open end of the coil so that the light moved right or left of a line in the center of a small stand. An operator had to call out the dots and dashes, and a second operator wrote and decoded them. Since that did not produce an automatically written record, Thomson solved the problem in 1870 with a siphon recorder. He attached to the galvanometer coil a tiny hollow glass tube not much thicker than a human hair. One end of the tube rested lightly on a moving paper tape and the other end in a small tub of ink. A fine ink mark was thus made down the center of the moving tape. When positive or negative impulses arrived over the cable, the coil swung the siphon above or below a "zero" line, making small mountains and valleys. This siphon recorder was introduced on the transatlantic cable in 1873. After the advent of typewriters, operators "read" the siphon tape as it moved across the tops of their typewriters and typed the messages.

Thomson, applying Fourier's theory on heat conductivity, understood that signaling speed depended on conductor resistance as well as on capacitance. An increase in the diameter of the copper wire substantially reduced the resistance and increased the capacitance only slightly, thus improving the signaling speed.

The initial signaling speed of 10 words per minute could be increased to 15 words per minute with the first siphon recorders, and reached a maximum of 50 words per minute at the end of the nineteenth century with improved cables and automatic (punched-tape) transmission. Duplex telegraphy, conceived by Samuel Morse and J. F. Fischer of Philadelphia in 1842 and introduced on submarine cable by J. B. Sterns in 1878, enabled simultaneous transmission of messages in both directions on a single cable at the speeds noted above.

The fourth challenge, how to lay the heavy cable smoothly on an ocean bed, led to the construction of special cable ships. The first ship used for cable laying was the *Goliath*, chartered to lay a telegraph cable across the English Channel in 1850. It was not built as a cable ship but was converted to carry a cable tank and primitive cable-handling gear. The first submarine cables had a typical weight of 1 kg/m when designed for the ocean floor and 10 kg/m when laid in shallow seafloors with heavy tides and intensive fishing activity. The first transatlantic cable to be laid successfully, in 1866, was 4000 km long and weighed 9000 tons. Only one ship in the world was large enough to carry that load, the 19,000-ton British cargo ship *Leviathan*, built in 1858 by Sir Isambard Kingdom Brunel (1806–1859). Later, its name was changed to *Great Eastern*, when it was used as a passenger ship to transport emigrants to Australia. In 1864 it sat unused at its dock and was obtained by the British cable manufacturers Glass Elliot & Company for laying cable under the Atlantic. The ship had to be converted to accommodate three large tanks in which the cable was coiled and submerged in water to prevent the gutta-percha from drying out. Special paying-out machinery had to be developed and installed on the ship to enable smooth cable laying, even in rough water and varying ocean depths. The *Faraday* (Figure 8.4) was the first ship constructed as a cable ship. The 5000-ton ship was built at the shipyard of Mitchell & Co. in Newcastle-upon-Tyne in 1873–1874 to the design of William Siemens. It served until 1923 and laid 60,000 km of submarine cable.

Figure 8.4 *Faraday*, the first ship designed for cable laying. (© Siemens-Archiv, Munich.)

8.2   ELECTRICAL TELEGRAPH LINES IN THE UNITED STATES

In 1851, the eastern half of the United States was covered with more than 50 telegraph lines. Eleven lines had terminal offices in New York City alone. Thirteen companies operated in the five states north of the Ohio River. All suffered from poor management, differing rules and practices, bad service, and frequent line interruptions. This was the environment in which a great company was born: Western Union.

8.2.1   Western Union

Two men decided to remedy the situation: Judge Samuel L. Selden and Hiram Sibley. They invited a group of Rochester's wealthiest citizens to meet at the Elon Huntington mansion at 100 St. Paul Street to organize a new company that would buy the various nearly bankrupt companies, assume their debts, and unite their lines in a single network. Upon further meetings, "Sibley's crazy scheme" was accepted and the *New York and Mississippi Valley Printing Telegraph Company* was founded on April Fool's Day 1851. After a four-year period of difficult negotiating, a consolidation agreement was finally signed on November 1, 1855. The telegraph lines of Cornell, House, Morse, O'Reilly, Speed, and Wade were consolidated into a new company called the *Western Union Telegraph Company*, usually referred to as *Western Union* (or WU). The formal consolidation came after an enabling act was passed by the New York legislature on April 4, 1856. At the first annual meeting on July 30, 1856, Sibley's brilliant work was recognized by his election as president of Western Union. The new company extended an electrical telegraph network all over the country in cooperation with the railway companies. This cooperation highly simplified right-of-way, operation, and maintenance issues. Within 25 years, Western Union's telegraph networks covered more than 75% of the railway network, and close to 80% of the about 12,000 offices of Western Union were local railway stations that performed double duty as train station and commercial telegraph office.

8.2.2   The Pony Express

In 1848, Mexico ceded the areas that are now California, New Mexico, Nevada, Utah, most of Arizona, part of Colorado, and Wyoming to the United States for $15,000,000. Gold was discovered in California on January 24, 1848, and a period of great expansion started in the West. Within a few years, the population in the West jumped from a few hundred to 300,000. Mail from the eastern United States, transported on muleback or by stagecoach, could easily take a month to reach the West. The need for telegraph lines became obvious. The first electrical telegraph line began operation in San Francisco between Lobos Hill and Telegraph Hill on September 22, 1853, replacing the optical semaphore installed there only four years earlier. The *California State Telegraph Company* established the first line, connecting San Francisco with San Jose, Stockton, Sacramento, and Marysville. The line was built within six weeks by James Gamble with five workers and began operation on October 26, 1853. Gamble, a newspaperman and Mississippi River "steamboater," had gained his telegraph experience with the *Illinois and Mississippi Telegraph Company*.

From Sacramento, a line was put into operation to the gold-mining country at Mormon Island, Diamond Springs, and Nevada City in January 1854. Two years later, the *North California Telegraph Company* extended a line from Marysville to the gold mining center Yreka (now Eureka) on the Pacific coast. The California State Telegraph Company extended their San Francisco–Sacramento line to Los Angeles, where telegraph service was begun on October 8, 1860.

To fill the gap between the telegraph lines in the East and West, a pony express service was opened by William H. Russell between St. Joseph, Missouri (north of Kansas City) via Salt Lake City to Sacramento on April 3, 1860. Russell established the Central Overland California and Pike's Peak Express Company. The company employed 80 young riders[2] with 500 horses. The 3150-km route had 190 stations, each 15 to 40 km apart. A rider took a fresh horse at each station and galloped to three or more stations at an average speed of 16 km/h. At those stations he was to pause only to carry his heavy leather saddle, with the two pockets in which the messages were locked, to another horse. It took 40 riders nine to 10 days and nights to cover the route. A steamer carried the messages on the Sacramento River between Sacramento and San Francisco.

The pony route gradually became shorter while construction of the transcontinental telegraph line progressed, and the pony express stopped when that line was completed on October 24, 1861. The pony express employed a total of 200 riders during those 18 months. Many of the riders could not bear the extreme conditions and stayed only a short time. One of the riders, "Pony Bob" Haslam, rode 600 km in 36 hours when Indians destroyed some stations and killed riders. Another rider once rode 515 km when Indians killed his relief rider. That was the 15-year-old William Cody, who became world famous under his nickname, Buffalo Bill.[3]

Pony express services were also used in other parts of the United States. In 1833, the New York newspapers operated a pony express to get news from Washington. When the Mexican War began in the spring of 1846, a group of newspapers operated a pony express with 60 horses running between New Orleans and New York in six days. That was how the president and his cabinet received the news on April 10, 1847 of the capitulation of Vera Cruz.

8.2.3 First Transcontinental Telegraph Line

Western Union was still in the middle of streamlining its network in 1857 when Sibley proposed constructing a transcontinental telegraph line. To achieve this pretentious goal and fill the missing link, a 3040-km line had to be constructed between Omaha, Nebraska, and Carson City, Nevada, crossing the Rocky Mountains, the Sierra Nevada Mountains, the Great Salt Lake Desert, and vast areas populated by Indians. After three years of lobbying, government funds and the right-of-way on public land were granted by an act of June 16, 1860. The *Pacific Telegraph Company*

---

[2] An advertisement in the San Francisco newspapers read: "Wanted—young skinny wiry fellows not over eighteen. Must be expert riders willing to risk death daily. Orphans preferred. Wages: $25 a week."

[3] William Frederick Cody, born in Iowa in 1846, earned his nickname, Buffalo Bill, a few years later when he was hired to kill buffaloes to feed the crews of the Kansas Pacific Railroad. Within 17 months he had killed 4280 buffaloes.

was incorporated on January 11, 1861 with the objective of constructing the line from Omaha to Salt Lake City, Utah. The California state legislation agreed on March 19, 1861 to absorb the existing western lines and to build a line from Carson City, Nevada, to Salt Lake City.

Construction of the line began in June 1861. Supplies, such as large coils of number 9 (3.3-mm) galvanized wire, insulators, batteries, Morse telegraph instruments, and tools for the eastern part of the line came by steamship up the Missouri River. Supplies for the western part had to come by ship around Cape Horn. The insulators were of the Wade type, consisting of a glass nucleus and a wooden shell, developed by Jeptha Homer Wade, who later became president of Western Union and a multimillionaire. The wooden poles, 15 per kilometer, were made primarily from cedar trees found in canyons, ravines, and along streams. Members of the construction teams were instructed to treat the Indians well. Gifts were given to the Indians, and chiefs of the Sioux and the Snake tribes were given the opportunity to exchange telegrams. The Indians understood that this time their land was respected, so they rarely attacked the telegraph lines. This changed three years later during the Civil War, when the Indians suspected that the "talking wires" were also used to send military information. They started to burn poles and telegraph stations, and hundreds of Indians were killed in counterattacks.

The line was inaugurated on October 24, 1861, eight years before completion of the transcontinental railway. By that time 2250 telegraph offices were in operation nationwide. The first transcontinental message came from Chief Justice Stephen J. Field of California to President Abraham Lincoln. It stressed the important role of telegraphy in the prevailing Civil War situation. Mayor Henry Teschermacher of San Francisco telegraphed Mayor Fernando Wood of New York City: "The Pacific to the Atlantic sends greeting, and may both oceans be dry before a foot of all the land that lies between them shall belong to any other then our united country." The time for sending a message from New York to San Francisco, which had taken six weeks a few years earlier, had now been reduced to minutes.

Its mission having been accomplished, the Pacific Telegraph Company merged with Western Union on March 17, 1864. The transcontinental telegraph line was a major contribution to the unity of the American nation, especially in the difficult days of the Civil War. The social and economic consequences of the transcontinental telegraph line were tremendous and the line became a proud part of the American heritage.

8.2.4 Collins Overland Telegraph Line and the Purchase of Alaska

After completion of the transcontinental line and when an attempt failed to lay a transatlantic cable in 1858 (see Section 8.7.2), Perry McDonough Collins (1833–1900) suggested to Western Union that a telegraph line be established with Europe via Alaska and Siberia.[4] Encouraged by Hiram Sibley, Collins applied to the Congress for support. The House Committee on Commerce advised Congress that it might take many years before an Atlantic cable would operate reliably and proposed to grant $50,000 to support an expedition to find a route across the Bering Strait.

---

[4] Collins was appointed U.S. commercial agent for the Amur River on March 24, 1856. In that capacity he made a long trip across Siberia in 1857.

The *Senate Military Affairs Committee* also reported favorably on February 17, 1862, but the Civil War delayed congressional approval.

Anticipating approval by the Congress, in May 1863, Collins, together with the U.S. minister, Cassius M. Clay, went to St. Petersburg to propose cooperation on a telegraph line from Moscow to San Francisco. The Russian government was very cooperative and agreed to extend a line from St. Petersburg through Siberia to Khabarovsk in East Siberia and from there to Nikolayevsk at the mouth of the Amur River. It also granted a concession for 33 years of a 3200-km route from Nikolayevsk to the Bering Strait. Collins also negotiated in London a right-of-way through the British Columbian part of the proposed route. On July 1, 1864, Congress passed a bill signed by President Lincoln granting Western Union the right-of-way across public land and the support of the U.S. Navy for the planned international telegraph line.

Western Union then immediately started the biggest and most ambitious overland telegraph line project of the time. Collins joined Western Union as a director and became managing director of the *Collins Overland Line Telegraph Company*. The route of the line is indicated in Figure 8.5. Several hundred adventurous and daring men, just returned from the Civil War, were recruited in San Francisco in an army-like way. Both the Russian and the American navies sent a few steamboats for support and protection. An order was sent to Henley & Co. in Great Britain for 8000 km of No. 9 galvanized wire and 80 km of annealed tie wire (to bind the ends of wires) to be shipped to Victoria, British Columbia. For crossing the Bering Strait and the Gulf of Anadyr, 800 km of submarine cable was also ordered in Great Britain.

Sibley and Collins went to Russia to meet Tzar Alexander II on November 1, 1864 with the intention of getting the 33-year concession replaced by a perpetual lease. According to substantial investigations made by George P. Oslin, described in his fascinating book *The Story of Telecommunications*, the Russian officials surprisingly offered to let Western Union buy Alaska instead of paying for a perpetual lease. Western Union was not interested in buying Alaska but immediately got the U.S. government involved. For the Russians, Alaska was a remote territory that they could not easily defend in a war, and they considered their back door better protected if the United States were to own Alaska. An agreement on the transfer of Alaska from Russia to the United States was reached at the beginning of 1867. The price was settled at $7,200,000 for an area as large as all the states east of the Mississippi River. Satisfaction on the proposed sale at the time was greater on the Russian than on the U.S. side. The *New York Tribune* found the proposal to buy "our splinter of the North Pole" valueless. Other papers called it "Walrussia" and "Iceburgia." The U.S. Senate ratified the treaty on April 9, 1867 by a vote of 39 to 12. Official transfer took place on Castle Hill in Sitka on October 18, 1867.

A telegraph line connected Portland with San Francisco on March 1, 1864, and a team started at Portland in July 1865 to extend that line to Seattle. A 60-m fir tree and a 55-m mast were used to suspend the line over the Willamette River. The 1.6-km-wide Columbia River was crossed by cable. The telegraph service with Seattle began on October 26, 1865. Another team started at New Westminster, British Columbia, and reached the Indian village of Hazelton about 1000 km north of New Westminster within one year.

Another team sailed to Petropavlovsk in the Kamchatka peninsula to explore the

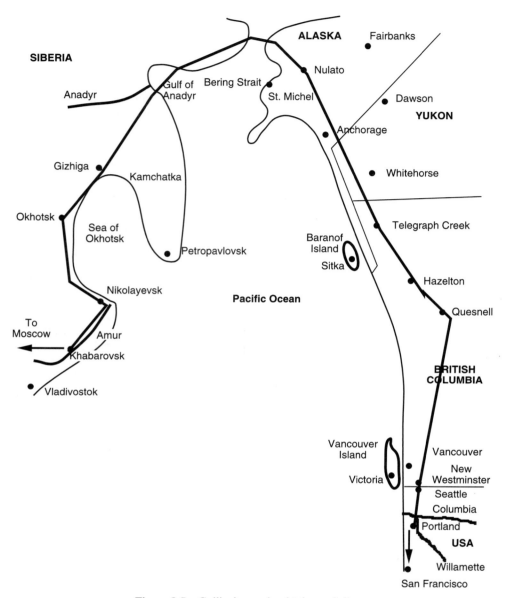

Figure 8.5   Collins's overland telegraph line.

Siberian part of the line. Seven months later, the clipper *Nightingale*[5] arrived with 65 workers, wire, poles, and supplies. The first pole was erected in Siberia on August 24, 1866.

[5] The famous clipper *Nightingale* (named after the opera soprano Jenny Lind, the "Swedish Nightingale," which carried a carved portrait bust of the great and beautiful singer), built at Portsmouth, England in 1851, first used in the China tea trade, was then heavily engaged in the African slave trade and was confiscated and used as a warship by the U.S. Navy; finally, the ship was bought by Western Union.

A further team sailed on July 12, 1865 to St. Michael on the northern coast of Alaska. Here progress was very slow because of deep snow and temperatures down to −45 to −68°C. The ground was frozen to a depth of 1.5 m, so only a few holes could be dug in a day.

Early in 1866, the overland line seemed to have won its race against the transatlantic telegraph cable. On July 27, 1866, however, a second cable was laid successfully in the Atlantic Ocean. Within a few months, this cable proved its reliability, and Western Union decided to stop the work on the overland line on February 27, 1867.

The line from New Westminster to Quesnell remained in operation until 1871, when it was leased to British Columbia. When British Columbia became a Canadian province in July 1871, the Canadian government took over the lease and bought the line from Western Union in September 1880. The *Yukon Telegraph Company* extended the line from Hazelton via Telegraph Creek and Whitehorse to Dawson in 1901. Russia completed the line across Siberia and extended it to provide the first telegraph connection with the Orient. Some names in Alaska still date from the expedition, such as Telegraph Mountain Range, Telegraph Creek, Bulkley River, Bulkley Lake, and Kennicott Lake, Kennicott Glacier, and Kennicott Mountain.[6]

8.2.5   The Hughes Direct Letter Printing Telegraph

David E. Hughes (1831–1900), born in London, emigrated with his parents to the United States at the age of 7. At the age of 19 he became a professor of music at St. Joseph's College in Bardstown, Kentucky. One year later he was appointed professor of physics at the same college. Motivated by the success of Morse, he also investigated electrical telegraphy, and in 1855 he invented the first telegraph system that could print the plain text of a message at both the sending and receiving ends. Hughes introduced a keyboard with 28 keys (later increased to 32), representing the major letters of the alphabet and some other signs for the transmitter. The operator could type the messages in plain language without needing to know the code used for the transmission of the individual characters. The transmitter and receiver both used a wheel with 28 characters on its rim. The wheel was rotating continuously at the high speed of 2 revolutions per second. When the letter or sign desired was over a moving strip of paper, a clutch mechanism activated by an electromagnet pressed the paper to the wheel. Similarly, just as the artist Morse used an easel for his first electrical writing telegraph, the music professor Hughes used a piano keyboard for sending messages with his telegraph (Figure 8.6).

Only the American Telegraph Company and a few other small companies used the Hughes telegraph in the United States. Hughes then left for Europe. He went first to his home country, where, however, he did not find any interest in his invention. In 1858, in France, he received a patent for his telegraph, which was then bought by the French government. Hughes moved to Paris, where he improved his design together with the mechanic Gustave Froment (1815–1864) and started production of his telegraph for the French administration in 1860. Within a few years, this improved direct letter printing telegraph was being used in the telegraph networks of most

---

[6] Charles S. Bulkley, who had constructed the Washington–New Orleans line in 1847, was the engineer-in-chief and the Russian–American naturalist Robert Kennicott, who had been in Alaska for the Smithsonian Institution, was the chief of explorations of the Collins overland telegraph line.

Figure 8.6   Hughes's direct letter printing telegraph in 1865. (Scanned from *Archiv für deutsche Postgeschichte*, Vol. 2, 1978, p. 96.)

European countries: Italy in 1862, England in 1863, Russia and Germany in 1865, Austria in 1867, the Netherlands in 1868, Switzerland and Belgium in 1870, and Spain in 1875. He also sold his patent to the *United Kingdom Telegraph Company* in 1863 and to Siemens in 1866. The ITU, at its second plenipotentiary conference in Vienna in 1868, recommended the Hughes telegraph for major international lines. The German physicist August Raps, working for Siemens & Halske in Berlin, made a further substantial improvement in 1895. He developed a brake arrangement for synchronously operating the transmitter and receiver of the telegraph system (Figure 8.7). With this arrangement the receiver speed is slightly faster than the transmitter speed, thus substantially reducing the danger of losing signals. Upgraded with this brake arrangement, the Hughes telegraph equipment survived until being replaced by teleprinters in the 1930s.

Hughes returned to England in 1877, where he made an important contribution to the improvement of telephony and became a member of the Royal Society in 1880 and later its president. He died in London on January 22, 1900.

8.3   ELECTRICAL TELEGRAPH LINES IN CANADA

The *British North American Electric Telegraph Association*, organized by Frederic N. Gisborne, began building a line from Quebec to the Atlantic in 1847. The line went via River du Loup, 180 km north of Quebec, to St. John, New Brunswick. In St. John the line was connected with a line from Calais, Maine, installed in 1849. From St. John a line was built across Nova Scotia to Halifax, where transatlantic ships dropped in news from Europe. The association was bought by the *Montreal Telegraph Company*, which leased the line between Halifax and St. John to the *American Telegraph Company* in 1856 for their transatlantic traffic.

Figure 8.7 Hughes's direct letter printing telegraph produced by Siemens in 1895. (© Siemens-Archiv, Munich.)

In 1880, upon the encouragement of Western Union in the United States, the *Great Northwestern Telegraph Company*, headed by Erastus Wiman, was founded by an act of the Canadian Parliament. A 99-year contract, signed on July 1, 1881, provided for exclusive interchange of business with Western Union, which owned 51% of its stock. Within one year it leased the lines of the Montreal and Dominion companies, thus providing the first large telegraph network in Canada. The "monopoly" of the Great Northwestern lasted a few years only. In 1885, the *Canadian Pacific Railway* started to build a transcontinental railroad with its own telegraph system along the rail route. Public telegraph service was offered on this system in competition with the Great Northwestern. By the end of 1889, the Canadian Pacific could offer public telegraph service via their network, with a length of almost 10,000 km, from Halifax, Nova Scotia to Vancouver and to Victoria in British Columbia. From Vancouver, the Canadian Pacific also operated a line to Seattle, Washington, and to San Bernardino in southern California.

In 1901, the Dominion Government of Canada began to buy up the existing telegraph companies, with a total network of 56,000 km, and to combine them into the *All-Canadian Government Telegraph Company*, to be operated in connection with the Canadian Post Office Department. In the same year the All-Canadian Government Telegraph Company extended its network by another 3500 km from Vancouver to Dawson in Yukon, in large part using poles constructed for the Collins overland telegraph line in 1865.

8.4 ELECTRICAL TELEGRAPH LINES IN GREAT BRITAIN

By 1852, the Electric Telegraph Company operated some 6500 km of telegraph lines connecting London with over 200 other British towns. At the height of its operations over 1000 young telegraph operators, boys and girls about 15 years old, were employed in London alone. Visitors were allowed to watch their work in the head office at Lothbury in London. The Admiralty operated its own electric telegraph line between London and Portsmouth with instruments from Cooke.

Very soon, two major competitors appeared: the *British Electric Telegraph Company*, operating mainly in the northern part of England and in Scotland, and the *English and Irish Magnetic Company*, with John Brett as its director, operating mainly in Ireland and connecting Ireland with England. John Brett also founded the *Oceanic Telegraph Company* on June 16, 1845, with the intention to participate in a transatlantic cable as soon as this would be technically and financially feasible. The *London District Telegraph Company* was formed in 1859, and the *United Kingdom Telegraph Company*, with lines along public highways, went into business in 1860. All those companies sent telegrams electrically between their own offices only. Messenger boys were still required for the interconnection between the different companies and for the submission of telegrams at the premises of the client. This improved in 1861 when the *Universal Private Telegraph Company* was formed, which provided direct telegraph lines to its clients, one of the first being the royal family.

One of the first major private users of public telegraphy was Lloyd's of London, the insurers. They sent the first telegram in 1845. The telegraph equipment was installed in their London office in 1851. Beginning in 1857, Lloyd's agents at the European ports sent regular telegraphic reports to the head office in London.

Women telegraph operators were employed in Great Britain beginning in 1855. They were considered reliable, loyal, and above all, were paid less than their male counterparts for the same job. Moreover, they "raised the tone of the male staff to a decency of conversation and demeanor."

In Great Britain, as in the United States, electrical telegraphy was begun through private enterprise, which concentrated its activities on the attractive urban regions, strongly neglecting the more rural regions. In 1863, therefore, the government intervened with a *Telegraph Act*, attempting to ensure the correct financial behavior of private companies. Around 1868, the public, and especially the press, became increasingly discontented with the limited accessibility, frequent delays, poor transmission quality, and high rates of the telegraph networks. The government therefore issued the *Telegraph Act of 1869*, in which it decided to nationalize the telegraph companies, except submarine cable operators, and integrate the telegraph service into the national postal organization. On January 28, 1870, the *postmaster general* took control of the national telegraph network.[7] At once the public had access to telegraph service at 1000 postal offices and 1800 railway stations at a national flat rate of 1 shilling for 20 words.

By 1875, the central telegraph office in London was the world's largest telegraph center, connected with 20,000 towns and villages worldwide via 1 million kilometers

---

[7] In this country of fair play, the nationalized companies obtained a generous compensation that heavily indebted the postmaster's budget and enabled those companies to invest in submarine telegraph undertakings.

of landline wires and 50,000 km of submarine cable. Telegraph cables ran directly from Britain to outposts of the British Empire on an *intra-Imperial telegraph network* (retrospectively also called the *Victorian Internet*), with interconnections at key points to the global telegraph network. Messages could be telegraphed from London to Bombay and back in as little as 4 minutes. The office operated over 450 telegraph instruments on three floors. The transport of telegrams between operators on the three floors was made via 68 steam-powered pneumatic tubes.[8]

The Cooke and Wheatstone system survived for a remarkably long time, especially on *British Railways* and in isolated instances right into the twentieth century. None of the other countries, apart from Spain for some time, adopted the Cooke and Wheatstone telegraph. After Cooke and Wheatstone terminated their cooperation in 1846, Wheatstone continued to improve their telegraph. In 1858 he developed his "automatic fast-speed printing instrument," the first telegraph that used a perforated tape for quick, constant-speed transmission of messages. For this purpose he used a perforator with three plungers for "left," "right," and "space."

Cooke left the Electric Telegraph Company when the company was taken over by the government. Wheatstone became a member of the Royal Society in 1855. The Society of Arts awarded both Cooke and Wheatstone the Golden Albert Medal in 1867. Queen Victoria knighted Charles Wheatstone in 1868 and William Cooke in 1869 for their achievements in electrical telegraphy; Sir Charles and Sir William will always be remembered, both inside and outside Great Britain, as two great pioneers of telegraphy.

8.5 SUMMARY OF NATIONAL ELECTRICAL TELEGRAPH ACHIEVEMENTS

The penetration of electrical telegraphy throughout the world cannot be covered within the scope of this book in the same detail as has been described for the United States, Canada, and Great Britain, instead, a country-by-country chronological summary of major achievements is given.

France

1848   Louis Napoléon Bonaparte (the future Emperor Napoleon III) ordered the construction of a national electrical telegraph network.
1851   The first private telegrams were sent in March.
1855   Paris was connected to all prefectural capital towns.
1862   The Hughes telegraph replaced Breguet's pointer telegraph.
1870   Pigeon service with Paris took over the wartime-interrupted telegraphy (Figure 8.8).
1874   Jean-Maurice-Émile Baudot (1845–1903) introduced a five-unit binary code combined with a time-division multiplex system, thus allowing up to 12 messages to be transmitted simultaneously over the same circuit. The Baudot code was eventually standardized by the ITU as International

---

[8] This pneumatic tube system, also used between major buildings in towns, was developed in 1854 by Josiah Latimer Clark, an engineer of the Electric Telegraph Company.

Figure 8.8   Foil with messages transported by a pigeon in 1870. (Scanned with permission of Museum für Kommunikation, Frankfurt, Germany, from *Archiv für deutsche Postgeschichte*, Vol. 1, 1992, p. 16.)

Alphabet number 1 (ITA1). Baudot's name, in the shortened form *Baud*, symbols per second, is used as a unit of the speed of data transmission.

Germany

1876   The telegraph and postal services were united as the *Imperial Post and Telegraph Administration* and Heinrich von Stephan (1831–1897) was appointed as postmaster general.[9] The telegraph network had a length of about 40,000 km, with a circuit length of about 140,000 km, consisting primarily of overhead lines.

1878   A new underground cable network, 2500 km long, connected Berlin with the major German towns (Figure 8.9).

---

[9] Von Stephan had gained an international reputation as one of the main initiators of the World Post Union, which was founded on October 9, 1874. Von Stephan shared this reputation with Montgomery Blair (1813–1883), postmaster general of the United States and initiator of the first International Postal Conference at Paris on May 11, 1863. All European countries, the United States, Egypt, and Turkey were signature members of this union.

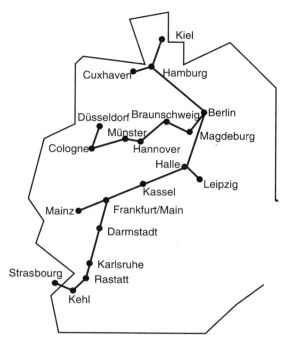

Figure 8.9   Underground cable network in Germany in 1878.

1888   The central telegraph office in Berlin (Figure 8.10), serving 61 lines, became the second-largest European telegraph office.

1899   A total of 13,000 Morse writing telegraphs, 1400 Morse sounder telegraphs, and 700 Hughes plain-language writing telegraphs were used. The total length of the underground cable network was about 50,000 km.

Russia

1851   The St. Petersburg–Moscow electrical telegraph line was established.

1853   The St. Petersburg–Kronstadt electrical telegraph line was established.

1854   The St. Petersburg–Warsaw electrical telegraph line was established.

1855   The telegraph line from Moscow–Kiev to Odessa and Sevastopol on the Black Sea, and between St. Petersburg and Helsingfors (now Helsinki, Finland, then under Russian control) was established. Figure 8.11 shows the Russian state telegraph network with a total length of 11,000 km constructed by Siemens in the period 1853–1855.

1863   The Moscow–Tbilisi (Georgia) telegraph line, 2000 km, crossing the Caucasus was established. In the Caucasus a telegraph network (Figure 8.12) was constructed in the 1860s upon the initiative of the eminent Georgian poet and commander of the Georgian army, Grigola Orbeliani (1804–1883). The figures written by the lines indicate the year of completion. The 256-km submarine telegraph cable laid across the Caspian Sea between Baku and Krasnovodsk contained a single conductor made up of seven twisted copper wires.

Figure 8.10   Hughes's telegraph room in the Central Telegraph Office in Berlin in 1896. (Scanned from *Archiv für deutsche Postgeschichte*, Vol. 2, 1997, p. 100.)

Scandinavia

1853   In Sweden the *Royal Electric Telegraph Administration* was founded. The first electrical telegraph line, connecting Stockholm with Uppsala, was opened to the public. The line was extended in the south up to Helsingbørg, facing Denmark.

1854   In Denmark an electric telegraph line was opened from Helsingør (opposite Helsingbørg) via Copenhagen, Fredericia, Flensburg, and Altona (then Danish territory) near Hamburg, from where a messenger took the telegrams over the border to the German telegraph station in Hamburg.

1855   The first international sea cable was laid between Helsingbørg and Helsingbør and connected the telegraph networks of Sweden, Denmark, and Germany. In Norway the first electric telegraph line was laid between Drammen and Kristiania (now Oslo) and via Göteborg extended to Sweden.

1860   Telegraph communication began from Sweden with Finland and Russia.

1873   The Danish–Norwegian company *Det store nordiske Telegraf-Selskab* offered electrical telegraph service via Fredericia with London, Paris, Berlin, Moscow, Tokyo, and Shanghai.

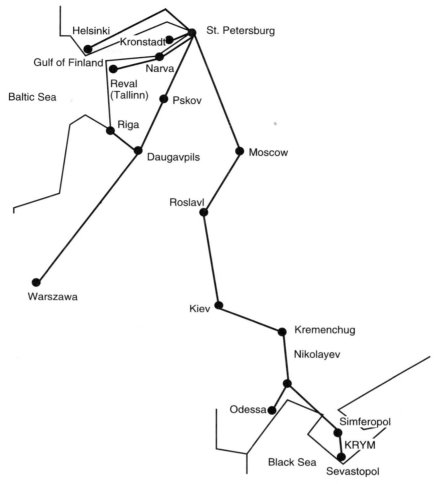

Figure 8.11   Russian state telegraph network in 1855.

Spain

1854   The electrical telegraph line Madrid–Zaragoza–Navarra–Irún, 603 km, was completed and connected at Irún with Biaritz, France. First electrical telegram was sent from Madrid to Paris on November 8.

1863   Seven radial lines connected Madrid with Irún and Santander in the north, Santiago de Compostella in the northwest, Catalonia and Valencia in the east, Andalucia in the south, and in the Extremadura with the Portuguese border.

1874   The Barcelona–Marseille submarine telegraph cable was laid. Initially, the two-needle telegraph of Cooke and Wheatstone and the pointer telegraph of Breguet were used. In 1857, the Morse telegraph was adopted for international communication and the Cooke and Wheatstone telegraph for domestic services.

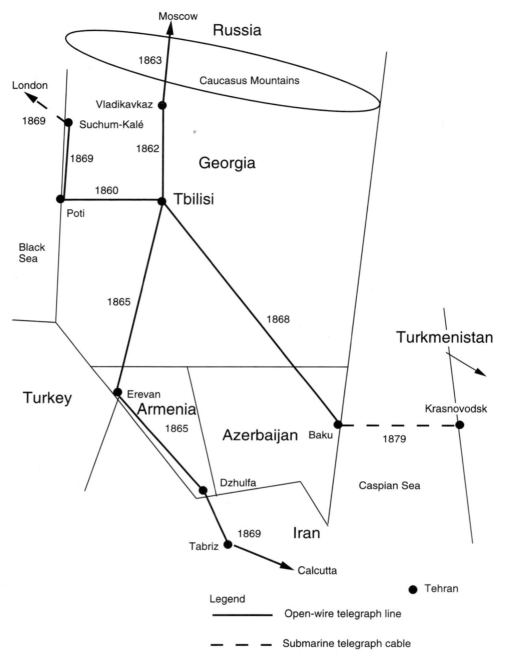

Figure 8.12   Early electrical telegraph network in Georgia.

India

1839   The British physician Sir William Brooke O'Shaughnessy installed an experimental electrical telegraph line near Calcutta with his own telegraph, without being aware of Morse's telegraph. For a part of the line he used the conductivity of the Hugli River in place of a wire. He sent messages by imposing a series of small electric shocks on the operator!

1850   An electrical telegraph line was established in Calcutta between Diamond Harbor and the center of Calcutta.

1854   The Indian Telegraph Act provided the government of India with an exclusive authority for introducing electrical telegraphy service.

1855   O'Shaughnessy was appointed director of telegraphs for India.[10] Within four years a star-shaped network with a total length of 7000 km connected the capital Delhi with Peshawar in the far north (now in Pakistan), Calcutta in the east, Madras in the south, and Bombay (now Mumbay) in the west (Figure 8.13).

1858   The first submarine cable was laid in India between the mainland and the island of Ceylon (now Sri Lanka).

1863   An overland telegraph line with probably the world's longest span was constructed in 1863 near Indore. The line came down from the heights of the Hulner Ghat and crossed the Amar River with a single span 1 km in length.

1895   Another remarkable telegraph line was installed down the Himalayan Mountains in 1895, connecting the town of Gilgit (now in Pakistan) with the Indian telegraph network. The line was carried over the 3500-m Rajdiangan Pass and the 4000-m Burzil Pass. Iron masts about 10 m high supported the strong steel wires over snow heights up to 6 m.

Japan

1869   The first telegraph line was built in Yokohama between the Light Tower Department and the Saibansho government building. This 800-m line was exclusively for government use. Beginning on December 25, public telegraph service was allowed on a one-wire line between the government building in Yokohama and the customshouse in the new Japanese capital of Tokyo.

1870   A telegraph line between Kobe and Osaka was established.

1871   International electrical telegraphy started in Japan when a single-conductor submarine cable arrived in Nagasaki. It was laid in that year by the *Great Northern Telegraph China and Japan Extension* as an extension of the London–Vladivostok Great Northern telegraph cable.

1872   A city telegraph network was installed in Tokyo with a central office in Tsukiji and six branch offices.

1873   Electrical telegraph service began between Nagasaki and Tokyo via a single-wire overland line.

---

[10] By 1861, O'Shaughnessy was surgeon-major and professor of chemistry at the Medical College of Calcutta. He introduced cannabis to Western medicine. He was knighted by Queen Victoria.

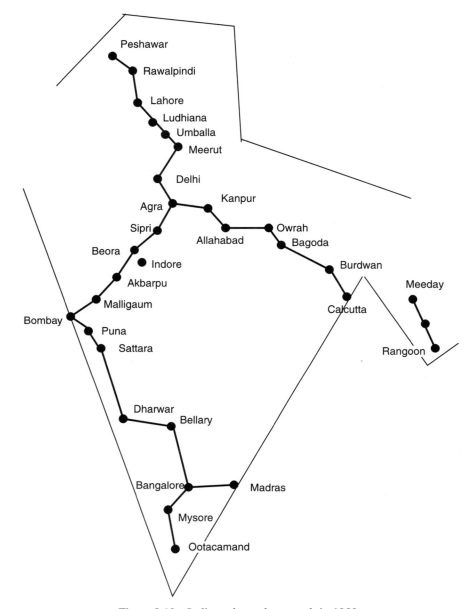

Figure 8.13   Indian telegraph network in 1855.

1874   A submarine telegraph cable connected the isle of Hokkaido with the
       mainland.
1877   Submarine cables were laid between the mainland and the islands of
       Shikoku and Kyushu in southern Japan.
1879   Japan became a member of the ITU. By that time the Japanese telegraph
       network had about 60 offices connected by lines 6000 km long. At the end

Figure 8.14   Breguet telegraph as used initially in Japan. (Scanned with permission of the ITU from Anthony R. Michaelis, *From Semaphore to Satellite*, International Telecommunication Union, Geneva, 1965, p. 64.)

of the nineteenth century, almost all towns and major villages were connected to the national telegraph network, which was then about 24,000 km long.

1897   As a result of the Chinese–Japanese war, Formosa (now Taiwan) became Japanese territory in 1895. A submarine cable was laid to connect Formosa with the Japanese mainland in 1897. By that time, Formosa had 29 telegraph offices, which were connected by an overland network 1000 km long. In Japan, Morse telegraphs and sounders were used mainly on the long-distance lines; the Breguet pointer telegraph was used on the short-distance lines. The Breguet telegraph apparatus was provided with two rows of Japanese hiragana characters (Figure 8.14) in addition to numbers and the letters of the alphabet.[11] Initially, the Morse apparatus and galvanized iron wires were imported from England and Breguet's telegraph from France. Local production of the telegraph apparatus started around 1875.

China

1871   Electrical telegraphy came to China when submarine cable laid by the Great Northern Telegraph China and Japan Extension arrived at landing points in Amoy (now Xiamen, Fujian Province), Hong Kong, and Shanghai.

---

[11] Japanese language writing evolved in the last 10 centuries into two basically different scripts: *hiragana*, a syllabic script with 48 characters for distinct Japanese sound elements, and *kanji*, with 214 basic (originally Chinese) elements combined in over 2000 characters. From the beginning, hiragana script was used in electrical telegraphy. The more complex kanji script was introduced as an alternative for telegrams on February 1, 1994.

1877   The first electrical telegraph line in Tientsin (now Tianjin) between the castle of the governor and the city arsenal was constructed by students of the local mining school.

1880   The *Imperial Chinese Telegraph Company* (ICT) was founded by the Chinese merchant Li Hongzhang in cooperation with the government.

1894   The Tianjin–Shanghai telegraph line was established. By that time, additional landlines had been erected connecting Tianjin and Shanghai with Beijing, Hong Kong, Wuhan, Nanjing, and other cities in the eastern part of China, as shown in Figure 8.15. Transferring the Chinese language into Morse code presented a serious problem. The Chinese language is not based on some 30 alphabetic letters, each basically conveying a particular sound, as in the Latin languages, but on over 50,000 characters, each conveying a different meaning. A Morse code for 50,000 characters would require code units of 17 dots or dashes instead of the six actually used in Morse for the Latin languages. A different approach was made, therefore, in which a four-digit number was given to a set of about 6000 most commonly used Chinese characters: for example, 1800 for *center*, 1801 for *necessity*, 1802 for *inquietude*, 1803 for *preventing*. The numbers were telegraphed and decoded at the receiving end in the Chinese dialect that prevailed in the receiving region. Soon after the first electrical telegraph line had been introduced, a Chinese telegraph code book was published in 1882. This system is still in use, although the present code, modified from the original, lists over 9000 characters. The 250-page official telegraph code book is used like an ordinary Chinese dictionary.

Australia

1854   The first electrical telegraph line was put into operation between Melbourne, Victoria and its harbor town, Sandridge (now Port Melbourne). The line was constructed by the Canadian engineer Samuel McGowan who had studied under Samuel Morse in the United States.

1857   A telegraph line connected Hobart and Launceston in Tasmania.

1858   The Melbourne–Sydney (New South Wales)–Brisbane (Queensland) telegraph line was established. By the end of the nineteenth century, Australia was covered by some 80,000 km of telegraph lines. The major lines are shown in Figure 8.16. The heroic endeavor of building the Australian overland line between Adelaide and Darwin under harsh conditions in a barely explored region, finally resulting in a reliable system that lasted almost a century, is described in Section 8.6.1.

South Africa

1860   The *Cape of Good Hope Telegraph Company Ltd.* installed a 30-km line between Cape Town and Simon's Town. One year later, the same company built a line about 50 km long between East London and King Williams Town, and a 100-km line between Port Elisabeth and Grahamstown in 1862.

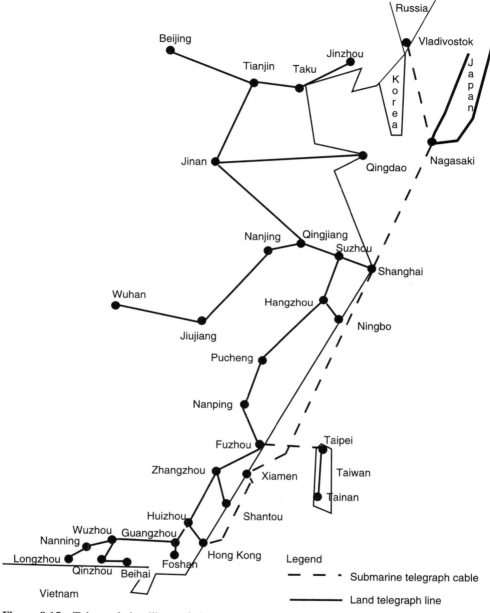

Figure 8.15 Telegraph landlines of the Imperial Chinese Telegraph Co. at the end of the nineteenth century. (Drawing after Claus Seeleman, *Das Post und Fernmeldewesen in China*, Gerlach-Verlag, Munich, 1992.)

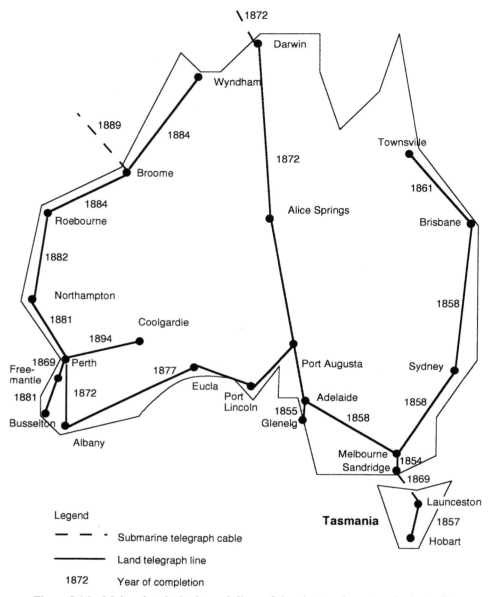

Figure 8.16   Major electrical telegraph lines of the nineteenth century in Australia.

1864   Electrical telegraph operation started between Cape Town and
        Grahamstown, and between Durban and Pietermaritzburg.

1873   The government of the Cape of Good Hope founded the Telegraph
        Department, and British officials took over control of new and existing
        telegraph lines. Extensive line construction continued, primarily along
        railway routes and extended toward the north of South Africa up to
        Rhodesia (now Zimbabwe). The development of electrical telegraph lines

followed the colonial settlement pattern and received its biggest boost with the opening of the Kimberley diamond fields in the 1870s and the Johannesburg gold fields in the 1880s.

8.6 MAJOR TERRESTRIAL TELEGRAPH LINES

8.6.1 Australian Overland Telegraph Line

The story of the Australian overland telegraph line is also the story of Charles Heavitree Todd, who with his pioneering spirit, professionalism, and perseverance motivated some 500 persons to conquer the perils of practically unknown terrain facing the harsh conditions of the arid center and the tropic north of Australia. In the 1860s, terrestrial and submarine telegraph lines were constructed to connect Western Europe with Asian countries. Communication with the Australian continent, however, was still limited to the regular mail service begun in 1852. Mail between Australia and Europe went by boat around the South African Cape of Good Hope in a five-month journey. The opening of the Suez Canal in 1869 slightly reduced the time lag. As described in Section 8.7.3, a submarine cable was constructed that connected Australia with Great Britain in 1870. The landing point for this cable in Australia became the subject of strong competition and controversy between the Australian autonomous colonies.[12] The government of Queensland promoted a landing point at Moreton Bay, north of Brisbane on the eastern coast of Australia, or, alternatively at Normanton on the northern coast. In both cases overland lines to Brisbane would have to pass through swampy jungle. Western Australia proposed to use its capital, Perth, as a landing point. The best prepared proposal, the one finally accepted, came from South Australia. Here the postmaster general Charles Todd had already started making plans for a telegraphic connection with his home country, Great Britain, as early as 1857. Todd proposed using Port Darwin on the northern coast of Australia as a landing point and to construct an overland line all the way down to Adelaide on the southern coast of Australia.

The line proposed by Todd had to pass through a territory that was terra incognita until the Scottish explorer John McDouall Stuart crossed Australia from south to north in 1862. The hardship of the expedition had made him so blind, however, that many of his maps proved to be very unreliable for planning the line. Todd was enthusiastic about Stuart's successful trip, however, and questioned him intensively about his findings. So he learned that the center of Australia was not a large sea, as generally presumed, but a huge desert separated into two parts by a mountain ridge (later called the Macdonnell Ranges). Stuart reported on the existence of a few lakes, rivers, and creeks providing water for people and horses. He also reported that most Aborigines he had met seemed to be friendly. Encouraged by this information, Todd was convinced that the line could be constructed despite the harsh climatic and difficult geographical conditions. In 1862, therefore, he wrote to the governor of South Australia: "The erection of an overland line to the north coast

---

[12] By the Australian Colonies Government Act of 1850, the British government had divided the Australian continent into the autonomous colonies of New South Wales, Victoria, Tasmania, South Australia, Queensland, and Western Australia.

should be regarded as a national work, in the carrying out of which all the colonies could unite."

Charles Heavitree Todd (1826–1910) was the son of a poor wine merchant in Islington (London). At the age of 15, he became a public servant, and although without formal education, he managed to start as a calculator at the Cambridge University Observatory. In 1854, he went to Greenwich, where he was in charge of the galvanic department and got acquainted with the electrical telegraph for transmission of Greenwich Mean Time around Britain. In the same year, the South Australian government asked the colonial office in London to find a qualified person to take charge of astronomical observations and to set up a telegraph line. Todd, the best person available, was offered this job. He got four months to prepare, which he used to acquire telescopes and chronometers, to visit all British telegraph companies, and as a prerequisite for survival in Australia at that time, to marry Alice Bell (1836–1898). Alice had met Charles Todd for the first time when she was 12 and then again a few days after Todd had been offered the job in Australia. She proposed spontaneously to come with him as his wife. She lives on in history, as her name was given to a famous town in the center of Australia: Alice Springs.

After a five-month sailing voyage around the Cape of Good Hope, Charles and Alice Todd arrived in Australia at Port Glenelg on November 4, 1855. In the same month Charles took up his responsibilities as *government astronomer and superintendent of telegraphs for South Australia*. In 1870 he became the postmaster general.

Taking advantage of the competitive situation, the British Australian Telegraph Company agreed to land the submarine cable at Port Darwin provided that the South Australian government would complete the overland telegraph line within 18 months at a penalty of £70 per day of delay. The contract was signed in June 1870, with the completion date fixed as January 11, 1872.

The line would consist of a single strand of No. 8 gauge galvanized iron wire purchased from *Johnson and Nephew* in Manchester. Earth was to be used as the return conductor. Glass insulators came from Germany, and batteries and relays from Britain. Suitable trees for some 40,000 wooden poles were to be found in the local terrain. Insulator pins, made locally from tough eucalyptus ironbark, were to be placed in a hole bored vertically into the top of each pole. A piece of leather was to secure the glass insulator to the insulator pin. A few thousand metal poles, mainly for the humid northern part of the line, were ordered from *Oppenheimer* in London. A hundred Afghan camels with their drivers came from Egypt to transport the roughly 2000 tons of material. A few thousand sheep were sent from Adelaide to the north to provide fresh food for the workers. Todd developed a 5000-word instruction manual for the line workers. Beyond the specific telegraph equipment installation, the manual defined the working time (8.5 hours daily with Sundays off), food rationing (some 3800 calories daily), path clearance (a 4.5-m-wide path to be freed of all undergrowth and overhanging trees), construction of bridges and traversing of creeks, how to find water, how to navigate the way back if lost, as well as special sections on "Health and Morals" and "Men Not To Be Kept Idle."

The overland line (Figure 8.17) had a total length of 3178 km (1975 miles).[13] To retransmit Morse signals on the long line, 11 repeater stations were planned. At each repeater station, two four-room wooden or stone buildings and a fenced plot of land

---

[13] Some sources quote 1800 miles thus 2896 km.

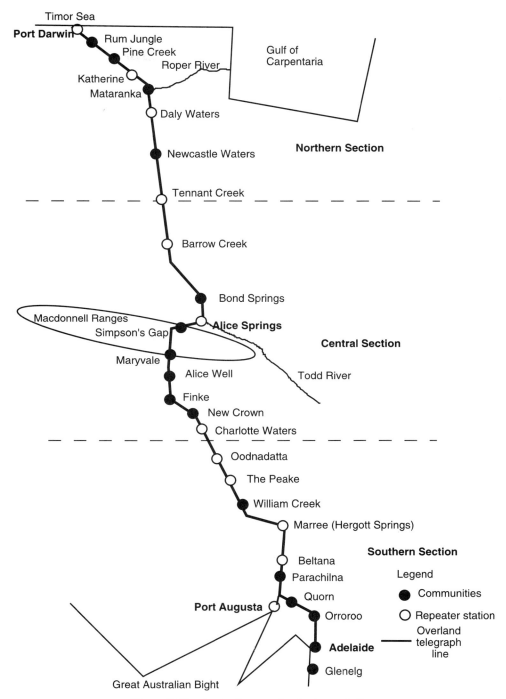

Figure 8.17 Australian overland telegraph line.

to grow vegetables had to accommodate two telegraphers and four line-maintenance workers. A separate powerhouse was to accommodate four large batteries.[14] Most stations in the interior also got a small herd of bullocks to provide fresh meat as well as about 20 horses for transportation. At one time the staff at the Alice Springs repeater station supported themselves with 300 cattle, 200 sheep, a flock of goats, and 70 horses. Later, instead of horses, bicycles were used. Motorcars came into use in 1907.

For construction the line was split into three sections. The northern and southern sections were tendered and supervised by a government overseer. The central section was under the direct command of Todd. The population was so enthusiastic about the project that immediately some 400 volunteers were easily contracted at a minimal payment.

The crew for the northern section left Adelaide by boat to Darwin on August 20, 1870. Three weeks later, 80 men, six officers, the overseer, 78 horses, 10 bullocks, and the telegraph material arrived at Darwin. The first pole was planted at Darwin on September 15, 1870. The prettiest daughter of the Resident officially stated: "In the name of Her Most Gracious Majesty Queen Victoria, I declare this pole well and truly fixed." Moreover, the Resident declared a public holiday for the barely 50 inhabitants living in the town, which had been established exactly 31 years earlier.

The southern section started at Port Augusta, where the first pole was planted on October 1, 1870. This section, about 800 km long, was the shortest and ended at The Peake. The central section started at The Peake, a point that no colonist had passed apart from Stuart and his little team. A small team was sent out in advance to explore the route and find a way through the Macdonnell Ranges. Todd left Adelaide in October 1870 with the last of his teams.

Work on the southern and central sections progressed fairly much according to plan. The major problem on the central section was to find a passage through the Macdonnell Ranges. After weeks of exploration, on February 18, 1871, the overseer, Gilbert McMinn, discovered a narrow gap through the mountains (later called the Simpson Gap), but alas, there was no water for miles around. The second overseer, William Mills, on March 11, found a dry riverbed leading to an area with waterholes and springs north of the mountains in a large valley. Mills baptized his important discoveries Todd River and Alice Springs, respectively, in honor of Charles Todd and his wife.

Work on the northern section became a disaster. Less than 130 km of line had been erected when in November the grossly underestimated rainy season started. Most of the area became a vast lake, some of the workers escaping to little islands where they were deprived of their provisions. Others mutinied and endeavored to return to Adelaide. The supply teams got stuck at the flooded East Finniss River, where at least they must have had a good time, because the place is still named Rum Jungle! A second team was sent from Adelaide. Eighty-seven men were selected from 525 new volunteers. Unfortunately, the new overseer created so much confusion that not a single additional pole had been erected when in November 1871, the cable-laying ship *Hibernia* arrived at Port Darwin with 300 men, who brought the subma-

---

[14] One battery for operation, one for standby, and one for recharging, each consisting of 80 cells enclosed in glass containers over 25 cm high and 10 cm in diameter, were to provide the 120-V line voltage. A fourth, smaller battery bank was used for the sounders and other equipment at the station.

rine cable ashore. On November 19, telegraph communication began between London and Port Darwin, but the overland line still had a gap of 634 km. To safeguard the future of the line and to keep the delay to a minimum, Todd personally took over command in the north section. Three days after the contractual completion date, he left Adelaide on the ocean steamer *Omeo* with 50 men, 80 horses, supplies, and provisions. Upon his arrival at the north and with the rainy season coming to an end, work on the line could start again. The workforce was increased to 300 men, to complete the line before the next rainy season. By mid-June, Todd decided to introduce a pony express for the few remaining gaps and informed the British Australian Telegraph Company that the line was running. Indeed, the "Morse and horse" telegraph service from London started on June 24, 1872. Several international telegrams were already waiting for the pony men, including one saying that Samuel Morse had died in New York on April 2. Todd was devastated that the "father of telegraphy" had not lived to see completion of the line. The pony express lasted only a few days; it could be terminated because the submarine cable between Darwin and Banjoewanji was broken! The last pole was planted on August 9, and the overland line between Adelaide and Port Darwin was officially opened with a closing of the gap at Frews Pond on August 22, 1872; alas the submarine cable was still defective. Locating and repairing the cable defect took almost four months. Finally, on October 22, 1872, the cable was repaired and the mayor of Adelaide received congratulations that the Lord Mayor of London had sent him across a distance of 22,908 km only seven hours earlier. Australia had its first telecommunications link with Britain and the rest of the world. News from Europe arrived at Australia within hours instead of taking months by ocean steamer. Over 500 men had worked on the construction of the line. Despite the difficult working conditions in a harsh and isolated environment, only six workers died. The line had cost about £500,000, four times the original estimate of £120,000. By comparison, some £600,000 had been spent on the Singapore–Port Darwin submarine cable.

The government of South Australia made Todd a Companion of Honor. In 1886, Cambridge University conferred upon Todd the honorary degree of master of arts. In 1889 he became a Fellow of the Royal Society, and he was knighted in 1893. By that time a telegram between London and Melbourne took an average of 160 minutes.

Five years later, Lady Alice Todd died, but Sir Charles continued his work. In 1900, together with his son-in-law, William Bragg,[15] he set up the first Marconi wireless station in South Australia. He retired at the age of 78. By then, in 1905, Australia was covered by some 70,000 km of telegraph lines connecting 300 telegraph stations. He died from gangrene in his left leg on January 29, 1910, when taking sea air at Seawall near Adelaide on *Semaphore Esplanade*.

The overland line had a long life, and some repeater stations became centers of civilization. Alice Springs became the "capital" of the vast central Australian region and is now a major tourist attraction. The Alice Springs telegraph station was restored in the 1960s and developed as a place of historical interest for educational and recreational purposes. The telegraph stations at Barrow Creek and at Tennant Creek became historic sites in the 1990s.

The line operated reliably, at least after many wooden poles had been replaced in

---

[15] William Bragg was winner of the Nobel Prize in Physics in 1915.

the first years by iron poles, as white ants eat the wood and cattle rubbed their backs against the poles. Statistics show that in the five-year period between 1887 and 1892, the line failed a total of only 36 times.[16] Still, these interruptions represented less than 2% of the operational time of the line. In 1898, a copper wire replaced the galvanized iron wire to improve the conductivity, and a second copper wire was added.

Around the end of the nineteenth century, keyboard paper-tape punching machines and automatic Morse transmitters were introduced, increasing operational speed from up to 30 words per minute to up to 400 words per minute. Japanese bombers destroyed the Port Darwin station in 1942, putting the line out of action permanently for the first time in 70 years. After the war the line was fully reconnected, telephones were also operated on the line, and by adding multiplex equipment, up to four telegraph signals could be transmitted simultaneously. Submarine cable landings at other places in Australia (currently, neither Port Darwin nor Adelaide are submarine landing points), cable and radio-relay networks, and satellite Earth stations gradually outdated the line. In 1979, the first major telecommunications route in the world with radio-relay equipment to be powered exclusively by solar energy was opened between Alice Springs and Tennant Creek. The system was later extended along the overland line up to Katherine. An optical fiber cable with a transmission capacity of 2.5 Gbps connecting Adelaide with Port Darwin was installed in 1995, mostly following the route of the overland line.

In 1997–1998, Alice Thomson, the great-great-granddaughter of Lady Alice and Sir Charles Todd and associate editor of the U.K.-based *Daily Telegraph*, together with her husband, traveled through Australia to track the roots of her family and the overland line. She wrote her impressions on past and present Australia in a charming book, *The Singing Line*, published in 1999.[17] Much of the information in this section was taken from that book.

8.6.2   Indo-European Telegraph Line

By the middle of the nineteenth century, India was the most important colony of the British Empire, which owed much of its wealth to this colony. Communications between the two states, however, took a long time. Mail made up at Bombay went by steamer to Suez, continued overland to Cairo, then by canal to Alexandria, and by steamer again to Marseille. There, a summary was made for the immediate attention of the French and British governments. The summary was telegraphed via the French optical telegraph network to Paris. From Paris the summary for London went by mail coach to Boulogne, by steamer to Folkestone, and finally at the end of a four-week journey, arrived by train in London. To improve communication, efforts were made at an early stage of electrical telegraphy to obtain an electrical telegraph connection between India and Great Britain. By 1858, submarine cables existed

---

[16] The interruptions were caused by lightning, storm, bushfire, and short circuits occasionally caused by frogs becoming jammed between wires and iron poles!

[17] The name *Singing Line* refers to the "songlines," or *yiri* in the Walpiri language of the Aborigines, which are tracks across the landscape created by mythical Aboriginal ancestors when they rose out of the dark Earth and traveled, creating mountains, valleys, and waterholes, thus literally singing the land into existence.

between Great Britain and Alexandria in Egypt, and a cable in the Persian Gulf between Fao (or Al Fäw, Iraq) and Karachi (then in British India), with a landing point at the Persian harbor of Bushir. A landline connected Tehran with the submarine cable at Bushir. The *Red Sea and India Telegraph Company* laid another submarine cable from Suez to Aden and Bushir, where it connected with the submarine cable to Karachi. The Red Sea cable, however, was of such poor quality that it became irreparably defective one month after completion in March 1860. A landline was installed, therefore, in Persia from Dzhulfa, at the border with the Caucasus, to Tehran. Thus two combined submarine and landline telegraphic connections existed between Great Britain and India. One line went through Russia from St. Petersburg, via Moscow and Tbilisi to Dzhulfa, and from there to Tehran, Bushir, and Karachi; the other line via southern Europe, Constantinople (now Istanbul), Baghdad, Fao, through the Persian Gulf to Karachi, and via landline to Delhi and Calcutta. Messages on these lines had to be retransmitted 12 to 14 times, partly by operators with a limited command of any written language, let alone English. On average it took one week to send a telegraphic message, and often, if it arrived at all, most of it was seriously garbled.

The construction of a government-owned direct line between Great Britain and India would have required lengthy and complex diplomatic negotiations with the various transit countries. Private enterprise brought a solution. William O'Shaughnessy met William Siemens in 1856 and discussed the feasibility of a direct telegraph line between Great Britain and India. William Siemens convinced his brother Werner of the possibility of such a line, and after some delay, due to an Austrian–Prussian war, Siemens & Halske decided to construct and operate such a consistent line between Great Britain and India. In fact, it became an international family affair involving the Siemens brothers Werner, William, Walter, and Carl: Werner in Berlin, the inventor and head of the company; William in London, director of the cable factory Siemens Brothers Telegraph Works[18]; Carl in St. Petersburg as Siemens representative[19]; and Walter in Tbilisi as German consul and director of a copper mine and metallurgical plant at Kedabeg bought by Siemens in 1864.

Siemens dew up an agreement with Prussia, Russia, and the British–Indian Telegraph Administration in April 1867 in which Siemens obtained the right to set up direct telegraph communications between Great Britain and India and committed itself to start telegraph service within two years of receiving the required concessions. The line was to get the following route (Figure 8.18):

- A submarine cable from Lowestoft, northeast of London, through the North Sea to Emden in northwestern Germany
- A landline via Emden and Berlin in Germany, Torun (then German) and Warsaw in Poland, and Vinnitsa, Odessa, and Kerch in Russia
- A submarine cable through the Black Sea from Kerch to Sukhumi in Georgia

---

[18] Founded as Siemens, Halske & Co. at Woolwich on the Thames in 1858 with the participation of Newall & Co.; in 1865 newly established as a British company called Siemens Brothers Telegraph Works. The factory was confiscated during World War I and sold to the English Electric Company in 1916.

[19] Carl Siemens (1829–1906) was raised to the peerage by Tsar Nicholas II in 1895.

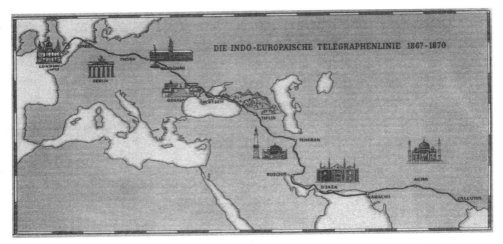

Figure 8.18   Indo-European telegraph line. (© Siemens-Archiv, Munich.)

- A landline via Sukhumi and Tbilisi in Georgia, and via Dzhulfa and Tehran to Bushir in Persia
- A submarine cable between Bushir and Karachi
- A landline between Karachi and Calcutta

An existing submarine cable had to be used between Lowestoft and Emden. The Prussian Telegraph Administration installed the landline between Emden and Torun at the border with Poland. Siemens undertook to build the line from Torun to Tehran. The Teheran–Calcutta link was "upgraded" from already existing but barely functional lines of the British–Indian Telegraph Co. The total link from London to Calcutta had a length of 11,000 km.

Despite thorny negotiations on right-of-way and operating charges with numerous authorities and local potentates, all concessions were obtained within one year. The *Indo-European Telegraph Company* was founded as a shareholder company with its seat at London on April 8, 1868.

The line, with two 6-mm iron wires, was supported by 40,000 dismountable iron poles and 29,000 wooden poles. Wooden poles were used in Germany and Russia; iron poles were installed in Georgia and in Persia.[20] Direct communication over the world's longest telegraph line was not possible with the existing Morse and Hughes telegraph apparatus. Siemens therefore developed a new Morse-type telegraph incorporating a highly sensitive polarized relay that could receive very weak signals and produce the signal received on a punched tape. For retransmitting, Siemens developed a double-T armature which produced positive and negative signal pulses synchronous with the holes in the punched tape (Figure 8.19).

---

[20] One iron pole marked "Siemens Patent London" is still standing after 130 years on the beach at Gagra in Georgia and is now used for local communications. The *ITU Telecommunication Journal*, Vol. 56-XI of 1989, showed a photograph of another pole (with the inscription "Siemens Patent London") still in use in Gagra at that time.

Figure 8.19 Transmitter of the Indo-European telegraph line. (© Siemens-Archiv, Munich.)

The first message was sent on April 12, 1870, from London to Calcutta. The message took 28 minutes, and it was answered within one hour. The service was interrupted for one year when an earthquake on July 1, 1870 destroyed the submarine cable in the Black Sea. As a coastal road had been built in the meantime, an open-wire line was installed to replace the irreparable submarine cable. Apart from an interruption from 1914 to 1921 due to World War I, the telegraph line stayed in operation up to 1931. The Indo-European Telegraph Company was taken over by the British *Imperial and International Communications Ltd.* in 1929 and liquidated in 1932. Sections of the line were kept in service by national operators. Figure 8.20

Figure 8.20   Part of the Indo-European telegraph line in Iran in 1965. (Scanned from *Telcom Report International*, Siemens, No. 6 November–December 1992.)

shows a part of the line that was still in use by the Iranian PTT at the time the photograph was taken in 1965.

8.6.3   Great Northern Telegraph Line

In 1869, three Scandinavian telegraph companies were united to become the *Great Northern Telegraph Company*, with its seat in Copenhagen. In addition to laying submarine cables between Great Britain, Norway, Sweden, and Denmark on the one side and Finland and Russia on the other, the company was entrusted with the challenging task of constructing a trans-Siberian telegraph line from St. Petersburg to Vladivostok. This vast landline was extended with a submarine cable named the *Great Northern Telegraph China and Japan Extension* from Vladiwostok to Nagasaki, Japan and to Shanghai, Amoy (now Xiamen, Fujian Province), and Hong Kong. This intercontinental telegraph link between Europe and East Asia became operational in 1871 and was connected in Great Britain with the transatlantic cables to North America. At Hong Kong the cable was connected with Singapore and Australia via submarine cables operated by the British *Eastern Extension Australasia and China Telegraph Company.*

8.6.4   Central American Telegraph Line

The Central American countries Guatemala, Honduras, El Salvador, Nicaragua, and Costa Rica were connected by an overhead telegraph line in 1879. The line ran from San Juan del Sur in Mexico to San Jose in Costa Rica.

8.7 SUBMARINE TELEGRAPH CABLES

The world's first underwater cable, beneath the Hudson River connecting New York City with Fort Lee, was in operation for a few months in 1845. The cable, with a length of approximately 16 km, consisted of two copper wires covered with cotton and insulated with rubber. It was placed in a watertight lead tube.

8.7.1 European Submarine Cables

Charles Wheatstone proposed operating an underwater cable between Great Britain and France in 1840 at a time when the technology of cable insulation and protection was not yet solved. With the availability of a process for seamless gutta-percha coating around copper wires, in 1849 the brothers John (1805–1863), and Jacob (1808–1898) Brett took up Wheatstone's idea as a starting point for their plan of a transatlantic cable. A first experiment was made in the Strait of Dover with a cable of approximately 3 km between the city of Folkestone and an offshore vessel, *Princess Clementine*, in 1849. Encouraged by the good results, in the next year they selected a route of the shortest distance between Great Britain and France through the Strait of Dover with landing points at Dover and Cap Gris Nez (20 km southwest of Calais in France). They used an unprotected 1.8-mm copper wire insulated with gutta-percha. Lead plates were laid on the wire at 100-m intervals. The cable was laid by the *Goliath*, a freighter converted to carry a tank for the cable and equipped with primitive cable-handling gear. The first and only telegram was sent via this "cable" on August 28, 1850. On the following night, a French fisherman lifted the wire in his net, cut off a piece, and in triumph showed it as a specimen of rare weed with its center filled with gold! It was now understood that a submarine cable needed better protection. In the rope factory of R. S. Newall & Co.[21] in Gateshead on Tyne, 10 iron wires 7.5 mm in diameter were spun around a cable core consisting of four twisted gutta-percha insulated copper wires of diameter 1.65 mm. This cable, with a diameter of 12 cm and a length of 41 km, was laid between St. Margaret's Bay (4 km northeast of Dover) and Sangatte (8 km southwest of Calais, which had a less rocky coast than Cap Gris Nez). Cable laying began on September 25, 1851 and lasted for three days. Official operations started six weeks later on November 13, accompanied by big celebrations in both countries. The cable remained in operation for more than 30 years.

Inspired by this success, very soon further submarine telegraph cables were laid. John Brett established the *Anglo-Irish Magnetic Telegraph Company* with Charles Tilston Bright (1832–1888) as chief engineer. This company first connected Wales at Holyhead across the Irish Sea with Ireland at Howth, northeast of Dublin, in 1852 with cable manufactured by Newall & Co. and laid under the supervision of Bright. One year later a second cable was laid in the English Channel between Ramsgate, England, and Ostende, Belgium. In the same year the Danish peninsula of Jutland was connected with the islands of Fyn and Sjaelland (Copenhagen). In 1854 submarine cables connected Denmark with Sweden, and the Italian mainland was con-

---

[21] Founded by Robert Stirling Newall (1812–1889), born in Dundee. He patented the production of wire rope in 1840 and founded R. S. Newall & Co. together with Lewis B. D. Gordon, professor of engineering at Glasgow University.

nected with Sardinia and Corsica. In Germany in the same year, a submarine telegraph cable 400 m long was laid from Stralsund to the little Baltic Sea island of
Dänholm. One year later that cable was extended by 1200 m to the island of Ruegen.
An international extension to this cable was made in 1865 connecting Ruegen, and
thus Germany, with Sweden at the town of Trälleborg.[22] England and Germany
were connected with a telegraph cable across the North Sea between Cromer, north
of Norwich, and Emden in 1858. Two years, later Denmark was connected to this
cable via the island of Heligoland.

All those cables were laid in relatively swallow water over short distances. The
next challenge was to lay a cable in the Mediterranean Sea with a depth of over
3000 m. John Brett was the first person to accept this challenge. He founded the
Mediterranean Extension Telegraph Company in 1854. With cable from Newall &
Co., he tried to lay a cable in the same year between the Italian island of Sardinia
and Bona on the northern coast of Algeria. He failed, and made a second approach
one year later, but in both cases the cable payed out far too quickly in the deep water
and was at its end long before reaching the Algerian coast. Initially, a brake, which
was operated manually in a trial-and-error way, regulated the speed of paying out
the cable. In 1857, Werner Siemens had developed a cable-laying theory which included a formula defining the required braking force as a relation of sea depth, cable
weight, and desired slack of the cable on the seafloor. Moreover, he constructed a
dynamic paying-out device to keep the braking force at the calculated level. Newall
& Co. then invited Werner Siemens to supervise the third cable laying. The formula
and device proved to be correct, and the third cable was put taken into operation in
1857, and an important experience was gained for the laying of ocean cable.

8.7.2 Transatlantic Telegraph Cables

The story of connecting the Old and New Worlds by submarine telegraph cable is an
epic of courage, enterprise, and perseverance, never before or after experienced in
telecommunications,[23] with Cyrus W. Field (1819–1892) as the major protagonist.
Field, born the son of a paper manufacturer in Stockbridge, Massachusetts, was a
wealthy American businessman who after a decade of successful paper sales, could
afford to retire at the age of 34. After making a six-month trip through South
America, he met Frederick N. Gisborne and soon became obsessed with the idea of
laying a transatlantic cable. Gisborne had founded the *Newfoundland Electric Telegraph Company* in 1852 with the purpose of establishing a telegraph cable between
Halifax and St. Johns in Newfoundland. At that point, nearest the Old World, news
could be collected from ships arriving from Europe and telegraphed to New York.
When asked by Gisborne to support his project, Field is reported to have said: "Why
stop where you do? Why link America only with Newfoundland? Why not lay a
cable on across the Atlantic Ocean and link us with Europe?" Gisborne and Cyrus
Field, with his brother David Dudley Field as legal counsel, went to St. John, and on
April 15, 1854 the Newfoundland legislature passed an act entitled: "Act to incor-

---

[22] This international submarine telegraph cable, 85 km long with three copper wires, remained in operation
until 1927 and experienced only 17 operation interruptions in those 62 years.

[23] Brilliantly described by Stefan Zweig in 1929 and published in 1943 by Bermann-Fischer Verlag A. B.,
Stockholm, under the title *The Propitious Hours of Mankind*.

porate a company under the style and title of the New York, Newfoundland and London Electric Telegraph Company, with the sole and exclusive right to build, make, occupy take or work any line of telegraphs between St. Johns and Cape Ray, or any other points and from any other island, country or place whatsoever during a period of fifty years." Protected by this act, on May 6, 1854, the *New York, Newfoundland & London Electric Telegraph Company* was founded by Cyrus Field with his brother Mathew as construction engineer, Peter Cooper as president, Chandler White as vice-president, Moses Taylor as treasurer, and with Marshall O. Roberts, the founder of the Erie Railroad Company. Samuel Morse, who had encouraged the company with his expert advice that a telegraph signal after a 3240-km undersea trip would still be recognizable, became honorary electrician. Field then went to England to buy a single-conductor submarine cable from the British company Glass Elliot & Company. The company connected New York with Newfoundland in 1856, with 150 km of that cable passing through the Gulf of St. Lawrence. After completion of this cable section, Field again went to England, to convince John Brett and Charles T. Bright, with their technical expertise, to join the enterprise. They agreed and founded the *Atlantic Telegraph Company* (ATC), with Bright as chief engineer. The company issued 350 shares at £1000 each ($5000) of which 250 were sold in Great Britain within two weeks but only 21 in the United States where more confidence was given to Western Union's efforts to construct an overland line through Alaska and Siberia (described in Section 8.2.4).

The famous American professor of meteorology at Lexington University and founder of the field of oceanography, Matthew Fontaine Maury (1806–1873), was the first to make maps of the Atlantic Ocean floor. On an exploration voyage with the U.S. brig *Dolphin* in 1853, he used the Brooke sounder, developed by Lieutenant J. M. Brooke of the U.S. Navy, to measure ocean depth. In 1856, he published his *Physical Geography of the Seas*, in which he indicated a plateau in the middle of the Atlantic Ocean slightly north of the Azores with a depth of around 3000 m only, compared with 4000 to 6000 m at both sides of this plateau.[24] Maury recognized the significance of this flat zone as the preferred passage for Atlantic cables and gave it the appropriate name *Telegraphic Plateau.*

Glass Elliot & Co. produced the cable. It consisted of seven pure copper wires, twisted spirally to form a single conductor, covered with three layers of gutta-percha, a layer of hemp and yarn, and a spiral sheathing of 18 strands of thin iron wire. Half of the cable was stowed aboard a wooden British battleship *Agamemnon*, which was upgraded for the peaceful but difficult task of laying cable on the bottom of the ocean. The other half was stowed on the U.S. frigate *Niagara*. As engineer-in-chief, Charles Bright assembled a staff of engineers and technicians to perform the challenging job, including William Thomson as chief electrician. After two unsuccessful trials, a third attempt was begun on July 17, 1858. The ships met at midocean, where the two cable ends were spliced. The *Agamemnon* returned to the cable landing point of Valentia on the Irish coast, and on August 5, the *Niagara* arrived at Trinity Bay, Newfoundland. After 11 days of testing, on August 16, 1858, the first transatlantic message was sent by the company directors in the United States to those in England, proclaiming in high spirits: "England and America are united. Glory to God on

---

[24] This plateau is part of the *Mid-Atlantic Ridge*, on which Alfred Wegener (1880–1930) based his theory of "drifting continents" published in 1912.

highest and peace on earth, good will toward men!" Two days later, Queen Victoria opened the line officially with a telegram sent to President Buchanan expressing Her Majesty's expectation that the new cable "will prove an additional link between the nations, whose friendship is founded upon their common interest and reciprocal esteem." Big celebrations took place on both sides of the ocean, with banquets and with parades up Broadway. Queen Victoria knighted engineer-in-chief Charles Bright as Sir Charles Tilston Bright. The most important message sent through the cable concerned the signing of a peace treaty between China, England, and France.

It took 16 hours to transmit Her Majesty's 96-word telegram. The telegraph signals faded away due mainly to the still unknown charging and discharging phenomenon caused by the telegraph impulses in the highly capacitive cable. Edward O. W. Whitehouse, a former surgeon who served as chief electrician at Valentia, assumed incorrectly that a higher voltage would improve the signaling. He installed a 1.5-m induction coil, which together with a series of powerful electric cells produced pulses with an estimated potential of 2000 V instead of the previous 600 V, with the disastrous effect that on September 3, 1858 the insulation failed and the cable went silent forever.

Despite the big disappointment about the damaged cable, Cyrus Field soon started preparing a project for a new cable. With 64 ocean crossings, he organized financial and logistic support on both sides of the Atlantic Ocean. In the United States the Civil War delayed his activity, but in Great Britain he obtained the support of another famous submarine cable protagonist, John Pender (details on John Pender appear in Section 8.7.3). On March 17, 1864, John Pender merged Glass Elliot & Co. and the Gutta Percha Company into the *Telegraph Construction and Maintenance Company* (called Telcon), becoming chairman of the company. He joined the Atlantic Telegraph Company with substantial financial involvement together with his partners in Telcon: R. A. Glass (of Glass Elliot & Co.), who became managing director of the ATC, and John Catterton (co-owner of the Gutta Percha Company). A scientific committee with Charles Wheatstone and William Thomson worked out a recommendation for an improved cable. Telcon manufactured this cable with four layers of gutta-percha and a spiral armor of 10 iron wires, making it three times the size of the 1858 cable.

The *Great Eastern*, sited idly at dock in 1864, was converted specially to transport the 4000-km cable, with a weight of 9000 tons, and left from Valentia on July 23, 1865 with one end of the cable connected to a land terminal. James Anderson of the Cunard Line was the captain and William Thomson stayed on board as a consultant to observe the electrical behavior of the cable. He kept constant contact with Great Britain through the cable as it was laid. About midway to Newfoundland, an electrical fault was detected in the cable. In an attempt to cut out the bad portion, the cable snapped and the part that had been laid disappeared. Valiant attempts were made again and again to find and lift the cable, but after nine days, defeat was admitted reluctantly and the *Great Eastern* returned to Ireland. Figure 8.21 shows the *Great Eastern* and the cables used in 1858 and 1866.

Telcon manufactured a new cable with nearly three times the tensile strength of the lost cable and further improved the cable-laying and cable-pickup machinery. The *Great Eastern* left Valentia a second time on July 13, 1866 and arrived at Hart's Content, Trinity Bay, Newfoundland, on July 26, 1866. The cable could take up transatlantic telegraph service, operating at a speed of 3 to 8 words per minute: The

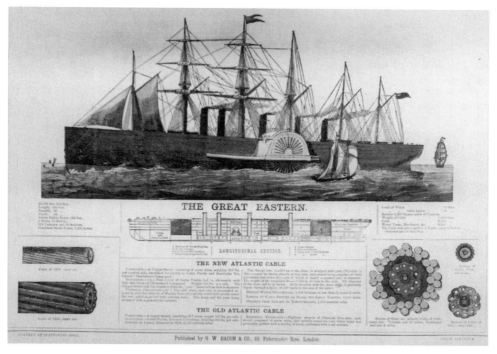

Figure 8.21   *Great Eastern.* (Courtesy of Museum für Kommunikation, Frankfurt, Germany.)

Old and New Worlds were united electrically. This cable remained in operation for six years. On its way back to Europe, the crew on the *Great Eastern* would recover the cable lost on their first voyage, a new cable length was spliced to it, and thus two cables were operated in parallel across the ocean.

The *Great Eastern* laid a third cable in 1868, this time for the *French Atlantic Cable Company.* Baron Emile d'Erlanger and Julius Reuter founded this company in an attempt to achieve independence from the monopoly established by ATC. Telcon and W. T. Henley manufactured the cable, which had a total length of 4785 km. It was laid between Brest on the Atlantic coast of France and St. Pierre's Island south of Newfoundland, at the time a French colony. The cable ship *Chiltern* extended the transatlantic cable from St. Pierre to Duxbury, Massachusetts. Since ATC had exclusive U.S. landing rights, the *Ocean Telegraph Company* was founded, which laid a few miles of cable from shore to a point in the ocean outside U.S. territory, where it connected with the French cable. ATC absorbed the French Atlantic Cable Co. in 1872. Unfortunately, this cable had only a short life. The armoring was not protected sufficiently, so that it rusted and the aggressive teredo, or eunicid, worms could pass through the armoring and eat the gutta-percha insulation.[25] Lifting the cable for repair was not possible with the rusted armoring, and operation had to be terminated in 1876. This was the last transatlantic cable laid by the *Great Eastern*; it laid a cable

---

[25] The appetite of the teredo or eunicid worm for gutta-percha was a major problem in those days, which was solved by adding a brass layer around the insulation.

Figure 8.22   Transatlantic cable at the factory. (Courtesy of Siemens Press Photo.)

between Great Britain and India in 1870 and was then docked in Liverpool, where it rusted away and was scrapped in 1888.

Siemens laid the fourth transatlantic cable in 1874–1875. This was a new venture altogether, and actually the first direct U.S.–British cable. The *Great Eastern* was no longer available, so William Siemens had the world's first genuine cable ship built in Great Britain in 1873–1874, named *Faraday* to honor his scientific supporter, Michael Faraday (Figure 8.4). The cable was made in the cable factory of Siemens Brothers Telegraph Works, which had begun the production of submarine cable only in 1871 (Figure 8.22).

Financing for the project was obtained partially through the Deutsche Bank in Berlin, which was founded in 1870 with Georg Siemens (a distant relative of Werner Siemens) as its first director. The *Direct United States Telegraph Company* (DUS) was established with its seat in London. The *Faraday* left from Ballinskelligs Bay, Ireland, with Carl Siemens supervising the cable laying. The act passed by the New-foundland legislature discussed earlier included clauses about possible nationaliza-tion of the transatlantic line and terminating exclusivity after 20 years. Counting on those clauses, the DUS initially planned a landing at Newfoundland. The New-foundland legislature did not exercise those rights, however, and the DUS did not want to amalgamate with ATC, so the landing had to be shifted to Halifax, Nova Scotia, where a connecting submarine cable was laid to Rye Beach, New Hampshire. Operations began in 1875, two years later, when ATC's cables, which had been laid in 1865, failed. ATC took over the Siemens cable.

In France another attempt was made to avoid ATC's monopoly. The *Compagnie Française du Télégraphe de Paris á New York* was founded by M. Pouyer-Quertier.

Siemens Brothers Telegraph Works was given the order to manufacture and lay the cable on the same route as that of the first French transatlantic cable. The cable was laid successfully and put into operation in 1879, but within one year it was also taken over by ATC.

The next attempt to break the ATC monopoly was made by Jay George Gould. He formed the *American Telegraph and Cable* (AT&C) Company and placed an order with Siemens Brothers to lay two cables between Penzance at Cape Cornwall in Great Britain and Canso in Nova Scotia. The cables were laid by the *Faraday*. Upon completion in 1882, Gould leased the cables to Western Union, which, however, soon joined the cable pool with ATC and thus restored the monopoly.

The first real competition on the transatlantic route began in 1884. James Gordon Bennett, editor of the *New York Herald*, was concerned about the 50 cents per word that had to be paid for telegrams on the transatlantic route. He persuaded John William Mackay (1831–1902) to set up a competing company. Mackay came from Dublin as a poor boy and became a multimillionaire during the California Gold Rush. He established the *Commercial Cable Company* and placed an order with Siemens Brothers to lay two cables between Waterville in Ireland and Canso, Nova Scotia. From Canso two feeder cables were laid to Rockport, Massachusetts, and Rockaway Beach, Long Island, where they connected with the lines of the Postal Telegraph Company, which was majority-owned by Mackay. A price war started in 1884, with the Commercial Cable Company lowering the price to 40 cents per word and Western Union to 12 cents.[26] As both companies suffered losses, they agreed in 1888 on a common tariff of 25 cents per word.

The French cable mentioned above was withdrawn from the pool in 1894 and amalgamated into a new French company, La *Compagnie Française des Cables Télégraphique*. This company had another cable laid, connecting Brest via Cape Cod, Massachusetts, with New York, in 1898. This was the last, and with a length of 5700 km, the longest transatlantic telegraph cable laid in the nineteenth century.

In New York an extension of the transatlantic cable was brought into the Stock Exchange. In London the ATC office was at Throgmorton Street, within steps of the London Stock Exchange. Trading between the two exchanges became so efficient that by the end of the nineteenth century, a broker in New York could telegraph an order to London and within 5 minutes receive a confirmation that the business had been completed.

8.7.3 Submarine Telegraph Cables Connecting Europe Worldwide

During the successful laying of the transatlantic cables, a veritable submarine cable fever developed. In the 1870s, submarine telegraph cables were laid connecting Europe with Africa, Asia, Australia, and South America. Almost all those cables were laid by only a few British companies, and most of those companies were founded on the initiative of one man: John Pender (1815–1896). Born in Scotland, Pender left school early to work in a local cotton and textile factory. He became managing director at 21. Four years later, in 1840, he moved to Glasgow as a cotton merchant and married Marion Cairns, who died when giving birth to their son,

---

[26] The price per word on the transatlantic route developed from $5 in 1866, to $2.50 in 1867, $1 in 1869, and $0.50 in 1875.

James. Pender moved again, now to Manchester, where he formed *John Pender & Company* as a leading distributor of products from Lancashire, Scotland, and India. In 1851 he married Emma Denison and invested in the *English and Irish Magnetic Company*, his first step in a lifelong involvement with electrical telegraphy and international telecommunications. He joined the Atlantic Telegraph Company in 1864, and with the goal of laying the first successful Atlantic cable, he merged Glass Elliot & Co. with the Gutta Percha Company into the *Telegraph Construction and Maintenance Company* (called Telcon) and had the *Great Eastern* converted into a cable ship.

The next challenge taken by John Pender was to connect Great Britain with its major colony, India, by submarine cable. The cable was manufactured by Telcon and laid by the *Great Eastern*. To limit the risks, Pender founded different companies for the construction of specific parts of the submarine cable. First, the *Anglo-Mediterranean Telegraph Company* was founded in 1868 to lay a submarine cable from Malta to Alexandria, Egypt. A terrestrial cable connected Alexandria via Cairo with Suez. In the next year the *Falmouth Gibraltar and Malta Telegraph Company* was founded and laid a submarine cable from Malta via Gibraltar, and Carcavellos near Lisbon, Portugal, to Great Britain, where it landed not at Falmouth on the southeastern coast of Cornwall as originally intended, but at the little village of Porthcurno on the southwestern coast, which eventually became the world's largest submarine cable station, with 14 international telegraph cables.[27] The *British Indian Submarine Telegraph Company* and *British Indian Submarine Extension Company* were also founded in 1869. The first company completed the submarine cable from Suez via Aden to Bombay in 1870. The Extension Company continued the link with an overland telegraph line through the Indian subcontinent from Bombay to Madras. From Madras the company laid a submarine cable to the Malaysian peninsula, with landing points at Pinang, Malacca, and Singapore, where the cable arrived in 1871. By that time Singapore was also connected with Hong Kong via a submarine cable laid by the *China Submarine Telegraph Company*, founded in 1869. In 1871, Hong Kong was also connected with the Great Northern Telegraph line, so that Great Britain had an alternative telegraph line to Southeast Asia. Finally, the *British Australian Telegraph Company Ltd.* was founded in 1870 with the target of expanding this telegraph line to Australia via Batavia (now Jakarta) and Banjoewanji (both Java, Indonesia) and with Port Darwin in Australia. This submarine cable was connected with the Australian overland line in 1872, thus including Australia in this British intra-Imperial telegraph network. After completion of the laying of those submarine cables, the three companies that laid the cable to India merged in 1872 and formed the *Eastern Telegraph Company*, with James Anderson, the captain of the *Great Eastern*, as managing director of the company. The company's head office was at 66 Old Broad Street, London, near the office of the ATC. The other three companies formed the *Eastern Extension, Australasia and China Telegraph Company* in 1873. In 1876, this new company connected New Zealand with Australia by a submarine cable.[28]

---

[27] The last submarine cable was closed at Porthcurno in 1970, exactly one century after opening the station. It is now a museum and the historic archive of C&W.

[28] The Eastern Extension Australasia and China Telegraph Company remained in existence until 1974, when it became part of Eastern Telecommunications Philippines Inc.

After completion of the submarine cables to Asia and Australia, John Pender turned his interest to South America. In 1873 he formed the *Western and Brazilian Telegraph Company*, with the objective of connecting the main ports along the Brazilian coast, and the *Brazilian Submarine Telegraph Company*, to link those port stations with Europe. The "Western" laid and operated a coastal cable connecting Santos, Rio de Janeiro, Bahia (now São Salvador), Pernambuco (now Recife), Fortaleza, São Luis, and Pará (now Belém). The "Brazilian" connected Brazil from Pernambuco via St. Vincent (Cape Verde Islands), Madeira (Canary Islands), and Carcavellos (Portugal) with Great Britain at Porthcurno. The telegraph operation via this cable between Brazil and Europe started on June 23, 1874. When a second cable was laid in parallel a few years later, a serious accident happened between St. Vincent and Pernambuco. The cable ship *Gomer* was rammed by another ship and was the first cable ship to be sunk. The two submarine companies merged to form the *Western Telegraph Company* in 1899.[29]

The British Royal Commission recommended in 1884 that there be a direct submarine cable between the West Indies and Halifax, Nova Scotia. Five years later the British government decided to link the British naval base in Nova Scotia with the British naval harbor at Hamilton, Bermuda. The *Halifax and Bermudas Telegraph Company* was established in the same year and constructed and operated the link. The service started on this cable in July 1890 with a message from the Governor of Bermuda to Queen Victoria. The *Direct West India Cable Company*, founded in 1897, extended the link in the same year from Hamilton via Turks Island to Jamaica.

Africa and Europe were already connected with a telegraph cable in 1854, as described in Section 8.7.1, when a submarine cable was laid between Algeria and Italy. After completion of the Britain–India cable, the *Marseilles, Algiers, and Malta Telegraph Company* was founded in 1870. The cities that gave the name to this company were connected with a submarine telegraph cable in the same year. It then took until 1880 for the next submarine cable to be laid in Africa. In 1879, together with the governments of Portugal, Natal, and the Cape of Good Hope, John Pender founded the *Eastern & South African Telegraph Company*. This company laid a cable in the Indian Ocean which connected Delgao, at the northeastern corner of the Cape of Good Hope, with Mozambique and Zanzibar on the eastern African coast and Aden on the Arabian Peninsula, where it connected with the submarine cable between India and Great Britain.

The *African Direct Telegraph Company* was formed in 1885 by John Pender to lay a cable to the west coast of Africa. A coastal cable was laid from Lagos, Nigeria, via Accra, Ghana, Freetown, and Liberia, to Bathurst (now Banjul), Gambia. From Bathurst a submarine cable was laid to St. Vincent in the Cape Verde Islands, where it connected with the Britain–South America cables. The coastal cable was extended to the south a few years later, connecting Luanda, Benguela, Moçamedes (now Namibe), and Cape Town. To the north the cable was extended via Dakar, St. Louis, and Tenerife to Cadiz in Spain.

Eventually, all the telegraph companies mentioned above, apart from the two connecting Great Britain with the West Indies, merged into the Eastern Telegraph

---

[29] The Western Telegraph Company remained in existence until it lost its concession in 1973 and the national *Emprêsa Brasileira de Telecomunicações* (Embratel) took over its operations. A telegram in Brazil used to be called a "Western."

Company, which then adopted the name *Eastern and Associated Telegraph Companies*. This company became the largest multinational corporation of the nineteenth century, owning about half of the total submarine cable length installed.

8.7.4   Inter-American Submarine Telegraph Cables

Electrical telegraphy in Latin America had an international character right from the beginning, providing communication lines between neighboring countries and with North America and Europe. The first telegraph line established in Latin America probably was the line between Buenos Aires and Montevideo. It included a 40-km cable across the mouth of the River Plate. The line was installed and operated by the *River Plate Telegraph Company*, which was formed for this purpose in 1865 by the Scottish engineer John Proudfoot. The service opened on December 1, 1866 and was the first telegraph service between two countries of South America.

Electrical telegraphy between Latin American countries and the United States developed primarily due to the initiatives of the "father of the cable business between the Americas," James A. Scrymser. In 1866, Scrymser founded the *International Ocean Telegraph Company* with Major General William F. Smith as its president. The company got the exclusive rights for 14 years to establish cables between Florida and the West Indies. The first cable, laid from Punta Rassa, Florida, via Key West to Havana, Cuba, 375 km long, was put into operation on September 10, 1867.[30]

With the support of the International Ocean Telegraph Company, the *West India and Panama Telegraph Company* was formed in 1869 and laid a telegraph cable between the Cuban harbor towns Cienfuegos and Santiago de Cuba and from there across the Caribbean Sea to Colón in Panama. A planned continuation of this cable around the eastern coast of South America to Brazil could not be realized until 1920, when the exclusive contract of the British *Western Telegraph Company* expired.

The International Ocean Telegraph Company came under the control of George Jay Gould, who leased the company for 99 years to Western Union in 1878. Scrymser formed *All America Cables Inc.* in 1884. The West India and Panama Telegraph Company became part of All America Cables in 1907.

Scrymser went to Mexico in 1879, where he founded the *Mexican Cable Company*, later renamed the *Mexican Telegraph Company*. Western Union agreed to route all its telegraph traffic via this new company for 50 years. A 1200-km-long submarine cable was laid from Galveston, Texas, to the Mexican harbor towns of Tampico and Vera Cruz in 1881. Landlines were constructed from Galveston to Brownsville in Texas, from Tampico to Mexico City, and from Vera Cruz to Coatzacoalcos.[31]

From Mexico, Scrymser extended his interest to Central and South America. On May 29, 1879 he formed the *Central and South America Telegraph Company*

---

[30] The Galveston tidal wave, which in 1900 caused 6000 deaths in the south Texas town of Galveston, also damaged the cable between Key West and Havana and interrupted service for 26 days. An alternative cable route was built from New York to Havana and Colón in 1907. After the opening of the Panama Canal in August 1914, a second cable was laid on this route.

[31] Western Union acquired control of Mexican Telegraph in December 1926, and laid another cable from Galveston to Vera Cruz in 1929.

(CSAT). The company laid a terrestrial cable across Mexico between Coatzacoalcos and Salina Cruz (Tehuantepec) linking the Gulf of Mexico with the Pacific Ocean. From Salina Cruz a submarine cable was laid to Libertad (Salvador), San Juan del Sur (Nicaragua), Panama, Buenaventura (Colombia), St. Helena (Ecuador), and Payta and Callao, the port of Lima, in Peru. The line, with a length of about 7500 km between Coatzacoalcos and Lima, began service on October 2, 1882.

From Lima a submarine cable was being operated under an exclusive contract with the British *West Coast of America Telegraph Company*,[32] with landing points at Mollendo in Peru and at Arica, Iquique, Antofagasta, Caldera, La Serena, and Valparaiso in Chile. This contract expired in 1890 and then CSAT extended its line from Lima with a submarine cable via Mollendo to Valparaiso. In Chile, CSAT purchased the *Transandine Telegraph Company*, which operated an almost 2000-km landline over the Andes connecting Chile and Argentina. Buenos Aires was already connected via a coastal submarine cable network with Montevideo, São Polo, and Rio de Janeiro. Beginning in the early 1890s, therefore, direct telegraph communication became possible between North America and the South American countries of Argentina, Uruguay, and Brazil without having to go via Britain.

Scrymser died on April 21, 1918. Two years later, CSAT merged with All America Cables, which together with the Mexican Telegraph Company, became an IT&T subsidiary on April 1, 1926 and was renamed American Cable & Radio in 1939.

In Brazil the first landline telegraph company, the *Companhia Telegrafica Platino–Brasiliera*, was founded in 1872. The construction and operation of landlines in the tropical climate of Brazil proved to be very difficult and unreliable. As late as at the beginning of the twentieth century, a typical out-of-operation time of 210 days per year was reported for the 1300-km-long landline through the less tropical territory between Uberaba (Minas Gerais) and Cuyaba in the Mato Grosso, opened in 1892. The telegraph offices of this and many other lines were mainly in shanty straw huts. The poorly maintained landlines allowed only operation of Morse telegraphs. Around 1910, Baudot and other quick telegraph apparatus and duplex operation could be used on only a few lines around Rio de Janeiro. Right from the beginning, therefore, the major telegraph service was provided by submarine cable laid along the eastern coast of Brazil.

8.8 WORLDWIDE ELECTRICAL TELEGRAPH NETWORK

At the end of the nineteenth century a total of some 90,000 pieces of telegraphy apparatus were in operation, of which 80,000 pieces comprised Morse apparatus; 2300, Hughes apparatus; and the remaining 7700, apparatus for the telegraph systems of Cooke & Wheatstone, Siemens, Breguet, Baudot, and others. The total length of the various overland and submarine telegraph lines worldwide amounted to 2,066,496 km at the beginning of 1895, with a total length of the wires of 5,423,099 km. Table 8.1 summarizes the worldwide development of the electrical telegraphy in the last quarter of the nineteenth century. The length of electrical telegraph networks

---

[32] The West Coast of America Telegraph Company and the River Plate Telegraph Company also became part of the Eastern and Associated Telegraph Companies, controlled by John Pender.

TABLE 8.1   **Worldwide Development of Electrical Telegraphy at the End of the Nineteenth Century**

Continent and Country	Length of National Telegraph Networks (km)			
	1877		1893	
	Lines	Wires	Lines	Wires
The Americas				
United States	141,000	215,000	335,000	1,400,000
Canada	16,121	26,142	51,242	111,221
Mexico	—	—	~20,000	61,000
Brazil	—	—	14,781	31,077
Asia				
Japan	—	—	13,982	39,495
India	29,214	68,783	69,156	216,457
Indonesia	5,655	6,934	8,329	12,028
Oceania				
Australia	25,020	52,278	76,731	145,034
Africa				
Egypt	—	—	3,109	13,375
South Africa	—	—	9,169	22,670
Europe				
Austria	34,087	87,585	44,777	124,373
Belgium	5,174	22,569	7,560	37,990
Bulgaria	—	—	4,819	9,516
Denmark	3,324	8,937	6,205	18,242
France	57,110	150,506	96,125	302,130 (1892)
Germany	54,366	197,784	156,025	550,481
Great Britain	42,008	184,877	54,337	334,244
Greece	—	—	7,651	9,063
Hungary	14,909	49,944	23,601	90,216
Italy	24,088	80,596	42,675	181,181
Luxembourg	—	—	522	1,593
The Netherlands	3,519	12,883	5,539	19,878 (1892)
Norway	8,478	15,108	9,448	18,351 (1892)
Portugal	3,711	8,042	6,830	14,663 (1891)
Rumania	4,142	7,208	5,836	12,879
Russia	94,339	187,526	124,733	244,894 (1892)
Spain	15,489	39,070	39,362	95,811 (1892)
Sweden	10,740	27,809	12,751	37,846
Switzerland	6,507	15,927	8,524	28,701
Turkey	—	—	33,064	51,824

Source: Data from *Statistique générale de la télégraphie. Année 1893*, Berne, 1895.

on the five continents at the end of 1892 are given in Table 8.2. A total of about 100,000 telegraph offices worldwide handled over 350 million telegrams in 1894. Figure 8.23 is a copy of a map of the world drawn in that year by the German Imperial Post, indicating the major electrical telegraph lines that were in operation worldwide at that time.

TABLE 8.2   Electrical Telegraph Networks, Year-End 1892

Continent	Length of Telegraph Networks (km)	
	Lines	Wires
Africa	42,400	60,480
America	549,240	1,825,600
Asia	128,560	309,080
Australia	70,280	138,090
Europe	716,290	2,326,600
Subtotal	1,506,770	4,659,850
Private international submarine lines	258,996	262,272
Total worldwide	1,765,766	4,922,122

Source: Data from Michael Geistbeck, *Weltverkehr*, Herder Verlag, Freiburg, Germany, 1895.

8.9   MORSE, THE FATHER OF ELECTRICAL TELEGRAPHY

Electrical telegraphy began in the United States with the Morse telegraph on May 24, 1844, but who really had invented electrical telegraphy? Soon a multiplicity of patent battles were launched against Morse. In one of those legal cases, Judge Woodbury stated: "Among the sixty-two competitors to the discovery of the electric telegraph by 1838, Morse alone, in 1837, seems to have reached the most perfect result desirable for public and practical use." In 1854, the U.S. Supreme Court decided that Morse's invention "preceded the three European inventions."

It will be correct to state that Joseph Henry provided the theoretical basis, Morse had the vision of the system, Gale made valuable contributions on the electrical design, and Alfred Vail constructed reliable apparatus and developed the dash–dot Morse code and the Morse key. Whereas Morse obtained recognition only slowly and late in the United States, Europe was more generous. In 1858, on the initiative of Napoleon III, Morse was awarded 400,000 French gold francs (equivalent to about $80,000 at the time) from the members of the Austrian–German Telegraph Union as recognition of benefits that his invention had brought to them.[33]

At Morse's eightieth birthday, telegraph people from all parts of the United States gathered in Central Park in New York to attend the unveiling of a bronze statue of Morse with his telegraph. Cyrus W. Field read messages received from all parts of the world. A reception that evening at the Academy of Music was linked by telegraph with cities throughout the United States, Canada, Europe, and Asia. A farewell message was sent "to the telegraph fraternity throughout the world," and Morse took the telegraph key to sign that message with his name in dots and dashes. William Orton, the president of Western Union, said: "Thus the Father of the telegraph bids farewell to his children." Less than one year later, on April 2, 1872, Morse died

---

[33] The Union included Belgium, the Netherlands, Piedmont, Russia, Sweden, Tuscany, Turkey, and the Vatican. Each contributed a share according to the number of Morse instruments in use in each country or region.

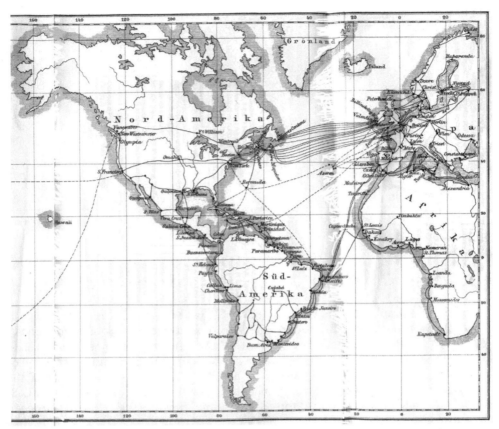

Figure 8.23   Major electrical telegraph lines in operation at the end of the nineteenth century. (Scanned from Michael Geistbeek, *Weltverkehr*, Herder Verlag, Freiburg, Germany, 1895.)

at his home at Locust Grove, Poughkeepsie, New York. By that time Morse signals were sent over more than 1,000,000 km of terrestrial telegraph line and 55,000 km of submarine cable, connecting 20,000 towns and villages in "a global instantaneous highway of thought," the Internet of the nineteenth century.

Coincidently, soon after Morse died, Morse telegraph penetration began to decline. The invention of the telephone, the advance of plain-language writing telegraphs, and the teletypewriter, none of which required specialist skills to operate, gradually replaced the Morse telegraph. In the United Kingdom, the General Post Office (GPO) officially abandoned Morse telegraphy for terrestrial and submarine application in 1932. In the United States that same happened in the early 1960s. In the meantime, however, a new medium, the radio, had provided a renaissance for Morse telegraphy. Wireless transmission of dots and dashes starting at the beginning of the twentieth century, especially for marine applications, survived almost one century, until on January 31, 1999, the Morse SOS distress signal was replaced globally by the satellite-supported Global Maritime Distress and Safety System (GMDSS), described in Section 27.4.4.

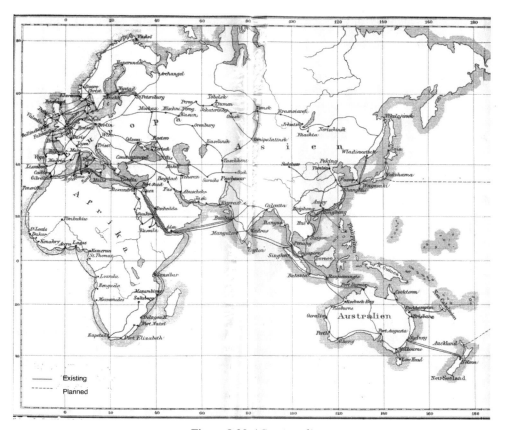

Figure 8.23 *(Continued)*

8.10   MORSE CODES

The dash–dot code, developed for Samuel Morse by Alfred Vail and generally called
the Morse code, applied dashes of different lengths. Friedrich Clemens Gerke, the
telegraph inspector of the Hamburg–Cuxhaven telegraph line (Section 6.6.1),
observed that easily avoidable operational errors were caused by the way the Morse
code was originally conceived. He modified the Morse code by eliminating spaced
dots and applying dashes of constant length equal to three times the length of the
dot.[34] This new Morse code with two distinct signs only, soon called the *Continental
Morse code*, was adopted by the Austrian–German Telegraph Union on October 14,
1851. The ITU at its first meeting in Paris in 1865 (Chapter 13) also recommended
the Continental Morse code, with a few minor modifications, for all telegraph communication between the member countries. In 1938, the Continental Morse code,
with some further minor changes, became the *International Morse code* and was

---

[34] The only exception was the code for zero, which Gerke left at a length of about six dots.

adopted worldwide for submarine, radio, and all international communication. Morse's original code remained in use only on landlines in the United States and Canada. Figure 8.24 shows the three versions of the Morse code for letters and numbers. In addition, there are a dozen other codes for punctuation and operational symbols.

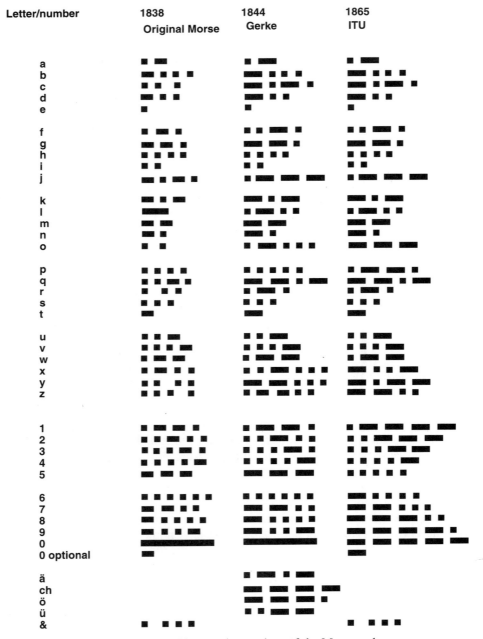

Figure 8.24   Three major versions of the Morse code.

8.11 MORSE TELEGRAPHERS

The Morse telegraph created an entirely new industry. Telegrams became a way of life for both business and individuals, and many large organizations had their own telegraph offices. Electrical telegraphy caused a revolution in world communications, serving virtually every aspect of human activity: government, diplomacy, business, industry, railways, newspapers, military, and the people who had to exchange messages and greetings in everyday life. Also, the mobility of the population migrating within a fast new national territory of the United States spurred intensive use of the new medium. The major actor of this new business was the telegraph operator, usually called a *telegrapher*.

A telegrapher had an esteemed profession; many of them entered the ranks of the emerging middle classes. Telegraphers had their own hierarchy, based on the speed at which a Morse-coded message could be sent and received. First-class operators reached speeds of 45 words per minute. Morse telegraphy also offered women an opportunity to make a reputable career outside their homes. By 1870, a third of telegraphers in the Western Union office in New York, then the largest telegraph office in the United States, were female. Some telegraphers became millionaires: Andrew Carnegie started as a telegraph messenger boy and served as a telegrapher for 12 years; Thomas Alva Edison started as a telegrapher at the age of 17; and Hollywood star Gene Autry, the "singing cowboy," who died in 1998 at the age of 91, was a railroad telegrapher in his youth.

REFERENCES

Books

Bertho, Catherine, *Histoire des télécommunications en France*, Èdition Érès, Toulouse, France, 1984.

Fischer, P. D., *Post und Telegraphie im Weltverkehr*, Ferd Dümmlers Verlagsbuchhandlung Harrwitz & Gossmann, Berlin, 1879.

Geistbeck, Michael, *Der Weltverkehr, die Entwicklung von Schiffahrt, Eisenbahn, Post und Telegraphie bis zum Ende des 19 Jahrhunderts*, reprographic reprint from Herder Verlag, Freiburg, Germany, 1895, by Gerstenberg Verlag, Hildesheim, Germany, 1986.

Godwin, Mary, *Global from the Start: A Short History of Cable & Wireless*, Cable & Wireless, London, 1994. [Leaflet]

Gööck, Roland, *Die großen Erfindungen Nachrichtentechnik Elektronik*, Sigloch Edition, Künzelsau, Germany, 1988.

Hugill, Peter J., *Global Communications since 1884: Geopolitics and Technology*, Johns Hopkins University Press, Baltimore, 1999.

Michaelis, Anthony R., *From Semaphore to Satellite*, International Telecommunication Union, Geneva, 1965.

O'Byrne, Denis, *Telegraph Stations of Central Australia Historical Photographs*, NT Print Management, Alice Springs, Northern Territory, Australia, 1999.

Oslin, George P., *The Story of Telecommunications*, Mercer University Press, Macon, GA, 1992.

Seeleman, Claus, *Das Post und Fernmeldewesen in China*, Gerlach-Verlag, Munich, 1992.

Siemens, Georg, *Der Weg der Elektrotechnik Geschichte des Hauses Siemens*, Vols. 1 and 2, Verlag Karl Alber, Freiburg, Germany, 1961.

Thomson, Alice, *The Singing Line*, Chatto & Windus, Random House Group, London, 1999.

Wilson, Geoffrey, *The Old Telegraphs*, Phillimore & Co., Chichester, West Sussex, England, 1976.

Das deutsche Telegraphen-, Fernsprech- und Funkwesen, 1899–1924, Reichsdruckerei, Berlin, 1925.

150 Years of Siemens: The Company from 1847 to 1997, Siemens, Munich, 1997.

Articles

Andrews, Frederick T., L'héritage du télégraphe, *Revue Française des Télécommunications* (France Telecom), Vol. 77, May 1991, pp. 60–71.

Anon., Das Telegraphenwesen in Japan, *Archiv für Post und Telegraphie*, 1878, pp. 118–121.

Anon., Geschichtliche Entwicklung der Telegraphie in Japan, *Archiv für Post und Telegraphie*, 1900, pp. 762–763.

Bernhardt, Manfred, Entwicklungsgeschichte der elektrischen Telegrafie bis zur Einführung des öffentlichen Fernschreibdienstes in Deutschland, *Archiv für deutsche Postgeschichte*, Vol. 1, 1992, pp. 53–60.

Carre, Patrice A., Aux origines des télécommunications d'affaires, *Revue Française des Télécommunications* (France Telecom), Vol. 78, June 1991, pp. 58–67.

Karbelahsvily, A., From London to the Caspian Sea: the Baku–Krasnovodsk submarine cable, *Telecommunication Journal*, Vol. 57, No. 9, 1990, pp. 630–633.

Nielsen, Arne, Morell, Aus der Post- und Fernmeldegeschichte Dänemarks, *Archiv für deutsche Postgeschichte*, Vol. 2, 1981, pp. 54–80.

Urbitsch, Hans, Tauben als fliegende Postboten, *Archiv für deutsche Postgeschichte*, Vol. 1, 1992, pp. 1–19.

Wichert, Hans Walter, Die elektrische Telegrafie in Deutschland, *Post- und Telekommunikationsgeschichte*, Vol. 2, 2000, pp. 59–65.

Internet

www.CCLab.com, American History of Telecommunications, Telecom Global Communications, Inc.; updated March 24, 1998.

www.cwhistory.com, History of C&W, by Mary Godwin, C&W Curator, Porthcurno, Penzance, U.K.

9

IMAGE TELEGRAPHY

Modern *telefax*, also called *facsimile* (from Latin *facere* = to make and *simile* = similar) or briefly *fax*, is almost as old as electrical telegraphy. In addition to transmission of text, it can transmit all forms of graphics, photographs, logos, and signatures. Electrical telegraphy found worldwide use soon after its beginnings in Great Britain and the United States in the 1840s but disappeared at the end of the twentieth century, due to the superior performance of telefax and e-mail. The history of telefax also started in the 1840s, but although many improvements were made, it found only limited application until modern technology gave it a renaissance, with tremendous growth in the 1980s. Table 9.1 gives a chronology of the various achievements. In the interest of a chronological presentation, the history of telefax is described in three separate chapters: Chapter 9, image telegraphy; Chapter 18, phototelegraphy; and Chapter 31, telefax.

9.1 FACSIMILE DEVICE OF BAIN

The Scotsman Alexander Bain (Section 6.4.2) was a clockmaker who developed a chemical telegraph in 1843. In the same year, he invented a facsimile device. In line with his experience as a clockmaker, he used a clock to synchronize the movement of two pendulums for line-by-line "scanning" of a message. For transmitting messages, Bain applied metallic pins arranged in a binary code on a cylinder made from an electrically nonconductive material. An electric probe that transmitted on–off pulses to the line scanned the metallic pins. The message was reproduced at the receiving station on electrochemically sensitive paper impregnated with potassium ferrite cyanide, similar to that used for his electrochemical telegraph. In his patent description of May 27, 1843 for his "improvements in producing and regulating electric currents and improvements in timepieces, and in electric printing, and signal telegraphs,"

The Worldwide History of Telecommunications, By Anton A. Huurdeman
ISBN 0-471-20505-2   Copyright © 2003 John Wiley & Sons, Inc.

TABLE 9.1   Chronology of Telefax Development

Year	Country	Pioneer	Achievement
Image Telegraphy			
1843	Great Britain	A. Bain	First patent for a telefax device
1847	Great Britain	F. C. Bakewell	First telefax demonstration
1851	Switzerland	M. Hipp	Electromagnetic copy telegraph
1855	Italy	G. Caselli	Pantelegraph
1859	Italy	G. Bonelli	Scanning of relief messages
1861	France	B. Meyer	Helical text reproduction
1866	France	J. J. É. Lenoir	Electrograph patent in Great Britain
1868	Austria	K. Opl	Insulating ink on metal plate
1869	France	G. d'Arlingcourt	Patent for improved electrochemical copy telegraph
1873	Switzerland	G. Hasler	Writing device operating by four electromagnets
1888	United States	E. Gray	Telautograph patent
1894	Italy	L. Cerebotani	Improved pantelegraph
Photo Telegraphy			
1901	Germany	G. Grzanna	Photographic paper
1902	Germany	A. Korn	Use of selenium cells
1906	France	E. Belin	Telegraphoscope/belinograph
1916	Germany	A. Schriever	Photoelectric cells, light rays
1920	United States	AT&T team	Film and light rays
1924	United States	Western Union	Telepix
1925	United States	AT&T team	Telephoto
1926	Germany	A. Karolus	Direct scanning of original
1928	Great Britain	O. Fulton	Fultograph
1928	Japan	Y. Niwa	Phototelegraph
1930	Germany	R. Hell	Patent for phototelegraphy
1934	United States	R. J. Wise	Teledeltos
Telefax			
1948	United States	G. H. Ridings	Desk-fax

he claimed that "a copy of any other surface composed of conducting and non-conducting materials can by taken by these means." The transmitter and receiver were connected by five wires. The experiments conducted with Bain's first device were not very convincing. In 1850, he applied for a patent for an improved version; however, this time he was too late, as his countryman, Bakewell, had obtained a patent for a superior device two years earlier.

9.2   IMAGE TELEGRAPH OF BAKEWELL

The world's first successful telefax transmission was made with an *image telegraph* between Seymour Street in London and Slough in September 1847. This image telegraph was developed by the British physicist Frederick Collier Bakewell. He wrote a message or drew an image with insulating ink (shellac) on a metal foil. In the transmitter the metal foil was wound around a rotating drum driven by a clock mechanism. A threaded rod moved a stylet that followed a tight spiral around the cylinder. The cylinder was connected to a battery and the stylet was connected with the line. An electric current was emitted when the stylet touched the conducting surface, but the current was interrupted whenever the stylet passed over the insulating ink. The receiver used a rotating cylinder synchronized with that of the transmitter. The cylinder was covered by paper impregnated with a solution of potassium ferrite cyanide and was also scanned by a stylet. The white paper turned blue whenever electricity went through the stylet, producing a white copy of the transmitted message or image on a blue background. Transmitter and receiver were connected by only one wire plus Earth. Bakewell received the British patent 12,352 on June 2, 1849. Public demonstrations were given at the Universal Exhibition at London in 1851. Bakewell's device found little use, however; the synchronization was not reliable and the transmission time was too long, but the idea of the recording cylinder led eventually to Edison's phonograph and is still used today in both photo telegraphy and in copying machines.

In 1859, Gaetano Bonelli tried to improve the transmitting device. He endeavored to create messages in relief using electrical contact–producing type with five and later seven metal laminations. His device was not successful either, but contributed to the invention of the phonograph.

9.3   PANTELEGRAPH OF CASELLI

The first telefax machine to be used in practical operation was invented by an Italian priest and professor of physics, Giovanni Caselli (1815–1891). Caselli was born in Sienna, received his ordination in 1836, and became a professor of physics at the University of Florence in 1849. In 1851, he founded the journal *La Recreatione*, in which he explained physics to the general public. In an effort to improve existing electrical telegraph devices, he conceived of the *pantelegraph*, a telegraph that would write everything.

Caselli understood that deficient synchronization was the major drawback of the telefax devices of Bain and Bakewell. He solved this problem by introducing a regulating clock which at each oscillation of its pendulum, made or broke the current for magnetizing the pendulum regulators, thus ensuring that the transmitter scanning and receiver writing stylets moved strictly in step. To provide the time base, the pantelegraph used a huge pendulum with a weight of 8 kg mounted on an imposing frame with a height of 2 m. Instead of one message on a rotating drum, two messages were written directly with insulating ink on two separate fixed metal plates. One plate was scanned during movement to the right of the pendulum and the other plate during the movement to the left, so that two messages could be transmitted in one cycle. The reproduction appeared in blue on a white background. Figure 9.1 shows an early model produced by Caselli.

In 1856, the Duc of Tuscany financed some of the experiments. For lack of further interest in Italy, however, Caselli went to Paris in 1857, where he received the enthusiastic support of the physicist Léon Foucault (famous for his pendulum

Figure 9.1   Pantelegraph of Caselli. (Scanned from *Archiv für deutsche Postgeschichte*, Vol. 1, 1995, p. 59.)

experiments in the Panthéon in 1851) and from Gustave Froment. The latter improved Caselli's model and the physicist Alexandre Edmond Becquerel (1820–1891) demonstrated the improved pantelegraph at the *Academie of Science* in May 1858. Louis Napoléon Bonaparte, by then Emperor Napoleon III, attended a demonstration on January 10, 1860 and ordered the use of the pantelegraph in the French national telegraph network. The pantelegraph was tested successfully on the Paris–Amiens telegraph line, 140 km long, in 1861 when the composer Gioacchino Rossini sent his signature from Paris to Amiens. A further experiment was made on a line 800 km long between Paris and Marseille. This test was also successful, so that the pantelegraph became accepted for use on the French telegraph network by law on April 24, 1864. Official operation started on the Paris–Lyon line on February 16, 1865 and was extended to Marseille in 1867. Typically, 40 twenty-word telegrams could be transmitted per hour.

The pantelegraph was also introduced in England on a line between London and Liverpool in 1863. Caselli received a U.S. patent in the same year. One year later, the Russian Tsar Nicholas I used the pantelegraph between his palaces at St. Petersburg and Moscow. Unfortunately, an economic crisis stopped operation of the pantelegraph in England in 1864, and the Franco–Prussian war of 1870–1871 interrupted pantelegraph service in France forever. Caselli returned to Italy, where he died in Florence in 1891. An original pantelegraph is exhibited at the Musée des Techniques in Paris.

Figure 9.2 Autographic telegraph of Bernhard Meyer. (Scanned from *Archiv für deutsche Postgeschichte*, Vol. 1, 1995, p. 59.)

9.4 AUTOGRAPHIC TELEGRAPH OF BERNHARD MEYER

After the Franco–Prussian war, the pantelegraph was replaced in France by the *autographic telegraph*, developed by the French telegraph operator Bernhard Meyer (1830–1884) in the 1860s. Meyer used a drum with a helical edge which enabled a transmission speed about twice as fast as that of the pantelegraph. In 1865 he obtained a patent for his autographic telegraph (Figure 9.2), which was used on several French lines beginning in 1871.

Bernhard Meyer was also the first to produce a multiplexer capable of sending four telegraph signals simultaneously over the same line. His multiplexer was first used on the Paris–Lyon line in 1872 and was also used in Austria, Germany, and Switzerland. Just before he died, Meyer was the first to use perforated tape for the retransmission of Morse signals, in 1884.

9.5 TELAUTOGRAPH OF ELISHA GRAY

The American physicist Elisha Gray (1835–1901), born in Barnesville, Ohio, made various improvements in electrical telegraphy beginning in the early 1870s. He first

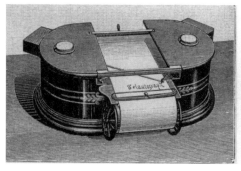 

Figure 9.3 Telautograph of Elisha Gray. (Scanned from *Archiv für deutsche Postgeschichte*, Vol. 1, 1995, p. 59.)

developed a universal private line automatic printer, used mainly by private companies on leased lines. On July 31, 1888, he received a patent for a machine he called the *telautograph*. This was the first facsimile machine in which a stylet was controlled by two bars: one moving horizontally and the other vertically, a predecessor of the X/Y coordinate plotter. The movement of the transmitting stylet produced coded pulses on the line, which controlled X/Y movement of the stylet in the receiver. Figure 9.3 shows both the transmitter and the receiver unit. Obviously, Gray had already mastered advanced industrial design. Gray founded the Telautograph Company, which at the end of the twentieth century still manufactured telefax apparatus.

The telautograph was modified by Foster Ritchie at the end of the nineteenth century and called the *telewriter*. It could be operated on a telephone line, enabling simultaneous copying and speaking. When called in absence of a subscriber, the telewriter switched on automatically and reproduced a handwritten message. The telewriter was used in England and in Germany, where the company *Mix & Genest* (now Alcatel SEL) obtained a license for local production.

REFERENCES

Books

Gööck, Roland, *Die großen Erfindungen Nachrichtentechnik Elektronik*, Sigloch Edition, Künzelsau, Germany, 1988.
Oslin, George P., *The Story of Telecommunications*, Mercer University Press, Macon, GA, 1992.

Articles

Andrews, Frederick T., L'héritage du télégraphe, *Revue Française des Télécommunications* (France Telecom), Vol. 77, May 1991, pp. 60–71.
Barnekow, Rolf, and Manfred Bernhardt, Die Vorläufer der Telefaxgeräte, *Post- und Telekommunikationsgeschichte*, Vol. 1, 1995, pp. 57–62.
Wenger, P.-A., The future also has a past: the telefax, a young 150-year old service, *Telecommunication Journal*, Vol. 56, No. 12, 1989, pp. 777–782.

10

TELEPHONY

10.1 EVOLUTION LEADING TO TELEPHONY

Whereas various scientists had investigated electrical telegraphy in the seventeenth and eighteenth centuries, very few people made investigations into the electrical transmission of sound. It was Robert Hooke, the great English scientist, who made the first suggestions as to how speech might be transmitted over long distances. A German "doctor of world-wisdom and teacher of mathematics and physics," Gottfried Huth suggested acoustical telephony in his little book, *A Treatise Concerning Some Acoustic Instruments and the Use of the Speaking Tube in Telegraphy*, published in Berlin in 1796. Huth proposed that during clear nights, mouth trumpets or speaking tubes should be used to pass messages from tower to tower. Although his proposal was impractical, his fame is assured by the sentence in his book: "To give a different name to telegraphic communication by means of speaking tube, what could be better than the word derived from the Greek: *Telephone?*"

In 1844, in London *telephones* were used consisting of a foghorn working on compressed air conveying signals 4 or 5 miles by four alternate notes. These tones could be given out separately, played like those of a cornet, and prolonged while a finger remained at a note. Charles Wheatstone, coinventor of the electrical telegraph, also used the name *telephone* for his *enchanted lyre*, used for transmitting sound from one room to another.

Hardly noticed by the scientific world, first a French and then a German person made practical proposals for telephony. In France, Charles Bourseul (1829–1912), born in Brussels, after having served in the army in Algeria, became a telegraph official in Paris in 1849. He was soon promoted to subinspector of the telegraph lines and rose to be the director of post and telegraphs. As a subinspector he published an

The Worldwide History of Telecommunications, By Anton A. Huurdeman
ISBN 0-471-20505-2 Copyright © 2003 John Wiley & Sons, Inc.

article in the August 26, 1854 issue of *L'Illustration* in which he explained his ideas about the possibility of transmitting speech by means of electricity ("transmission électrique de la parole"). The article was mentioned in other magazines, including the German weekly *Didaskalia* issued in Frankfurt on September 28, 1856. Bourseul tried to convince his superiors of the prospects of speech transmission, but they rejected his ideas as "une conception fantastique."[1]

Philipp Reis (1834–1874), a teacher in Friedrichsdorf near Frankfurt, Germany, was first to construct a device that transmitted sound by means of electricity, in 1860. Reis was born in Gelnhausen, near Frankfurt, the son of a baker. He lost his mother before he was a year old and his father when he was 10. He was then sent to a boys' institute called Garnier, in Friedrichsdorf, where he learned the English and French languages. At 14 he went to Frankfurt, where he undertook a commercial apprenticeship and studied physics, mathematics, chemistry, and the Latin and Italian languages. After a one-year military service and further apprenticeships, he married in 1858 and returned to the Institute of Garnier as a teacher of physics, chemistry, mathematics, and the French language. Knowing about electrical telegraphy, Reis had been looking for a way to transmit sound electrically since his youth. He might have read the article of Bourseul in *Didaskalia*. At Garnier, he had the time and facilities to find a solution with a device that he called a *telephon*. As a transmitter, he used an animal membrane stretched over a wooden cone which had the shape of an ear. An electrical contact, consisting of brass foil connected to the membrane and a platinum wire, formed part of an electrical battery circuit with a receiver. Sound caused the membrane to vibrate and to make and break the electrical contact, thus producing current pulses. The receiver consisted of a coil with six layers wound around a knitting needle 15 cm long. The current pulses caused rapid magnetization and demagnetization of the knitting needle, which reproduced the sound. In his first experiments, this sound was amplified by placing the receiver on a violin.

Reis, a member of the Physical Society of Frankfurt[2] since 1851, gave a first demonstration of his telephon to that society in the lecture hall of the Senckenberg Foundation at Frankfurt on October 26, 1861. A second demonstration was given there on November 16. Professor Boettger, president of the society, confirmed in the *Polytechnical Notes* of the society in 1863 that music and songs could be recognized and that conversation should be possible with an improved version.

Reis made 10 improved versions of the transmitter and four versions of the receiver, in which the receiver was placed on top of a wooden case needle instead of on a violin. For signaling between two instruments, he added a Morse key to each instrument and an electromagnet to the transmitter. Figure 10.1 shows Reis with the seventh version of the telephon, which he demonstrated on May 11, 1862 at the Freie Deutsche Hochstift (a society of science and arts founded in 1859) in Frankfurt. Figure 10.2 shows the final versions, produced in the workshop of J. Wilh. Albert & Sohn[3] in Frankfurt, in 1863. Albert produced a small quantity, which went to

---

[1] Nothing was heard of Bourseul for a long time, until in 1907 he requested that the French Post Administration grant him a pension increase for being the inventor of the telephone. The matter was investigated and, indeed, Bourseul got an annual pension increase of 3000 francs.

[2] The Physical Society was founded in 1817 and was part of the Johann Wolfgang Goethe University.

[3] The son of Valentin Albert, who had assisted Cooke with his first electrical telegraph. This workshop later became part of the company Hartmann & Braun.

Figure 10.1   Philipp Reis with his telephon transmitter, 1861. (Scanned from A. P. Koppenhofer, *Als Philipp Reis das Telefon Erfand*, Geiger Verlag, Horb am Neckar, Germany, 1998, p. 103.)

laboratories in Germany and abroad for further experimentation. In the enclosed operating instructions, Reis recommended using the instruments for repeating his interesting experiments for the reproduction of tones between distant stations. Copies of the telephon were made in a few other workshops, such as those of Rudolph Koenig in Paris, William Ladd in London, Mitchel Yeates in Ireland, and Hauck in Vienna. While in Russia to promote his plain-language writing telegraph, David Edward Hughes received a telephon from Reis, which he demonstrated to Tsar Alexander II in Petersburg in 1865. An American physicist, Henri van der Weyde, read about the telephon in a textbook on physics in 1866 and made his own version, which he presented at the Poly-technical Club of Philadelphia in 1868. Koenig sent a telephon to the Smithsonian Institution in Washington, DC, in 1874, which was shown to A. Graham Bell one year later. Reis demonstrated the tenth version of his telephon to the emperor of Austria–Hungary, Franz-Joseph, and the king of Bavaria, Maximilian II, at Frankfurt on September 6, 1863. Despite this honor, Reis, said to "just be an autodidact," unfortunately was not accepted by his contemporary German scientists, who did not consider his invention a serious matter. In particular, Johann Christian Poggendorf (1796–1877), publisher of the *Annals of Physics and Chemistry*, refused to publish an article about the telephon in the annals. Disillusioned and ill, Reis died of tuberculosis in 1873, convinced that he had

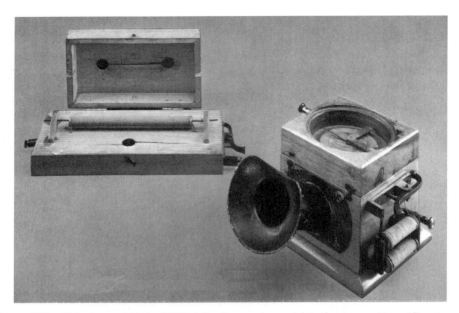

Figure 10.2   Telephon of Reis, 1863: left, the receiver; right, the transmitter. (Courtesy of Museum für Kommunikation, Frankfurt, Germany.) See insert for a color representation of this figure.

given humankind a great invention. In the meantime, at least in his home country, he is considered to be the inventor of the telephone. Moreover, the British scientist Silvanus Thomson, in his 1883 publication *Phillip Reis: Inventor of the Telephone* (Figure 10.3), confirmed: "The honor, to have transmitted the human voice by means of electricity first, owes to Reis."

10.2   THE TELEPHONE OF ALEXANDER GRAHAM BELL

10.2.1   Alexander Graham Bell, the Father of Telephony

The most vital step from telegraphy to telephony made by a single person is documented in *patent 174,465* issued by the U.S. Patent Office on March 7, 1876 in response to an application submitted on February 14, 1876 by Alexander Graham Bell, Salem, Massachusetts, on the subject "Improvement in Telegraphy." There are at least two remarkable peculiarities concerning this most valuable single patent ever issued, which is now generally accepted as the basis for the beginning of the telephone era:

1. The patent application was filed at 14:00, exactly two hours before a caveat was submitted by Elisha Gray (1835–1901)—cofounder of Western Electric Manufacturing Co.—regarding a device similar to that constructed by Reis and capable of transmitting the human voice.
2. The six-page patent application from Bell explicitly refers to telegraphy only and mentions neither the word *telephone* nor *speech*.

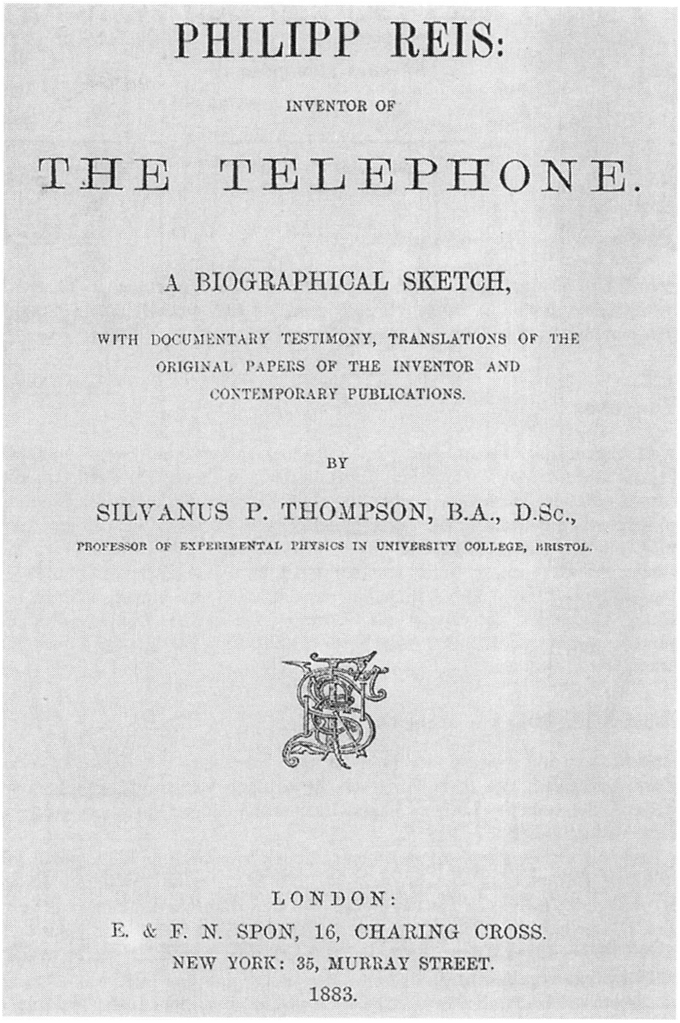

PHILIPP REIS:

INVENTOR OF

THE TELEPHONE.

A BIOGRAPHICAL SKETCH,

WITH DOCUMENTARY TESTIMONY, TRANSLATIONS OF THE
ORIGINAL PAPERS OF THE INVENTOR AND
CONTEMPORARY PUBLICATIONS.

BY

SILVANUS P. THOMPSON, B.A., D.Sc.,

PROFESSOR OF EXPERIMENTAL PHYSICS IN UNIVERSITY COLLEGE, BRISTOL.

LONDON:
E. & F. N. SPON, 16, CHARING CROSS.
NEW YORK: 35, MURRAY STREET.
1883.

Figure 10.3 Thompson's book in defense of Reis. (Scanned from A. P. Koppenhofer, *Als Philipp Reis das Telefon Erfand*, Geiger Verlag, Horb am Neckar, Germany, 1998, p. 97.)

Bell describes in his application a method of simultaneous operation on a single line of a number of telegraph instruments, each tuned to a different resonance frequency. Figure 10.4 is a reproduction of the sixth page of Bell's patent application, illustrating the parallel operation. After a long, detailed description of the proposed *improvement in telegraphy*, it is the following claim, made at the very end of the text, which secured for Bell, despite 600 lawsuits, the world's first legally confirmed telephone patent: "I claim, and desire to secure by Letters Patent ... 5. The method of, and apparatus for, transmitting vocal or other sounds telegraphically, as herein described, by causing electrical undulations, similar in form to the vibrations of the air accompanying the said vocal or other sounds, substantially as set forth."

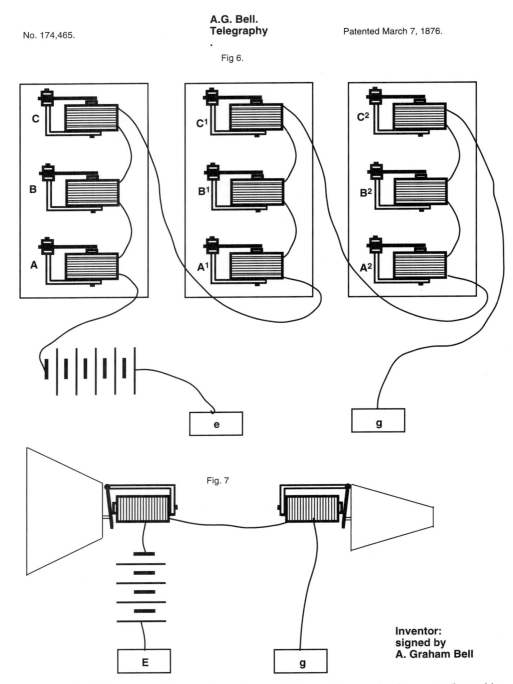

Figure 10.4 Bell's proposal for *vocal or other sound transmission* as simultaneous telegraphic signals on one line. (From A. A. Huurdeman, *Guide to Telecommunications Transmission Systems*, Artech House, Norwood, MA, 1997, Figure 1.6; with permission of Artech House Books.)

10.2.2   Early Days of Bell in Great Britain

Alexander Graham Bell (1847–1922) was born the second son of Alexander Melville Bell and Eliza Grace Symonds Bell on March 3, 1847, in Edinburgh. His father, his two brothers, his uncle, and his grandfather had taught the laws of speech in the universities of Edinburgh, Dublin, and London. For three generations the Bells had been professors of the science of elocution. Grandfather Alexander Bell invented a system for the correction of stammering and similar defects of speech. Father Alexander Melville Bell was the dean of British elocutionist, the author of *Standard Elocutionist* and another dozen textbooks on elocution, and the inventor of an ingenious sign language which he called "Visible Speech," in which letters of the alphabet were represented by certain actions of the lips and tongue. As a boy, A. Graham constructed an artificial skull from gutta-percha and india rubber, which, when enlived by a blast of air, would pronounce several words in an almost human manner. At 14 he graduated from Edinburgh's Royal High School and then attended lectures at Edinburgh University and at the University College of London. At 17 he became a resident master of music and elocution in the Weston House Academy at Elgin, County Moray, Scotland. In 1865 the Bell family moved to London, where in 1868, A. Graham became his father's assistant and assumed full charge, while his father lectured in Boston, Massachusetts.

In London, A. Graham Bell met two famous persons who gave direction to his future scientific experiments on telegraphy and telephony: Alexander J. Ellis and Sir Charles Wheatstone. Ellis was the president of the *London Philological Society*. In 1870 he translated the *Lehre von den Tonempfindungen* (The Sensation of Tone), written by Hermann von Helmholtz (1821–1894) in Berlin in 1862. Bell read the book by Helmholtz in the original German. Ellis showed Bell how Helmholtz had kept tuning forks in vibration by the power of electromagnets and blended the tones of several tuning forks together to produce the complex sound of the human voice. At the home of Sir Charles Wheatstone, Bell saw an ingenious talking machine that had been made by Baron de Kempelin. From those meetings, it appeared to Bell that it should be possible to develop a musical telegraph, for example, with a piano keyboard, so that many messages could be sent at once over a single wire.

10.2.3   Bell's Telephone Experiments in the United States

The sudden death of A. Graham's older brother Melville, three years after the death of his younger brother Edward, both of whom died from tuberculosis, an illness which then also affected A. Graham, prompted the family's move to Canada in July 1870. A. Graham settled with his parents at Brantford, Ontario, where his health improved rapidly. In Brantford he taught "Visible Speech" to a tribe of Mohawk Indians.

On April 1, 1871, Bell left Canada and arrived in Boston to become a teacher at the Clarke School for the Deaf, a job offered to him by the Boston Board of Education. In 1872 he opened a School of Vocal Physiology for teachers of the deaf. One year later he became a professor for vocal physiology at Boston University.

This was at the time that Edison invented his quadruplex telegraphy system. Bell expected that a musical or harmonic telegraph system would provide more parallel telegraph channels for a lower investment. His professional success in Boston would not have left any time for experiments on telegraphy or telephony, but fortunately, Thomas Sanders, the father of one of his pupils, 5-year-old Georgie Sanders, offered to let Bell stay with the Sanders family at Salem, 25 km from Boston, and use the cellar of the house as his workshop. Here he made several experiments on vibrations of sound and its electrical transmission. With the assistance of a Boston friend, Clarence J. Blake, an aural surgeon, he constructed a "phonautograph" with an ear of a deceased man. Placing a straw between the eardrum and a piece of moving smoked glass, he produced tiny curvatures on the glass when speaking in the ear.

A second pupil who became a decisive factor in Bell's career was Mabel Hubbard, born in 1857, who had lost her hearing through an attack of scarlet fever when a baby. Her father, Gardiner Green Hubbard, was the founder and president of the Clarke School of the Deaf and an influential lawyer who had secured gas for Cambridge in 1853 and pure water and street railways for Boston. Green Hubbard went to Washington in October 1874 to conduct a patent search. He noted that no invention had been registered similar to Bell's proposed musical or harmonic telegraph. He and Sanders then agreed to provide Bell with financial support. On February 27, 1875, Bell made an agreement, later called the *Bell Patent Association*, with Sanders and Hubbard in which the latter two each committed himself to finance 50% of Bell's further experiments on telegraphy in return for equal shares for all three from any patents that Bell would develop. However, both Sanders and Hubbard disliked Bell's experiments on telephony, especially with the ear, and forced him to concentrate on telegraphy experiments; otherwise, they treatened to stop their financial support. Even more, considering that Bell and Mabel Hubbard had fallen in love with each other, Green Hubbard made it clear to him that "if you wish my daughter, you must abandon your foolish telephone."

After settling this contract, Bell moved his workshop from the cellar in Salem to 109 Court Street, Boston, where he rented a room from Charles Williams, a manufacturer of electrical supplies. He obtained the help of Thomas A. Watson, a young repair mechanic and modelmaker; both lived nearby in two cheap rooms. Figure 10.5 depicts Bell and Watson experimenting in the workshop at Boston.

A prototype was made of a harmonic telegraph that could send a few telegraphic signals simultaneously. Electromagnets kept steel-spring armatures in vibration, each on a different frequency, thus producing intermittent currents with different pulse rates on the transmission wire. With this prototype, Bell went to Washington on March 1, 1875 to consult a patent lawyer as well as Joseph Henry.

Henry, then 78, famous and secretary at the Smithsonian Institution, inspected the harmonic telegraph, just as he had helped Samuel Morse almost 40 years earlier. Henry showed to Bell the telephon of Reis, which was exhibited at the Smithsonian Institution. Henry encouraged Bell to concentrate on the electrical transmission of speech instead of harmonic telegraphy, stating: "You are in possession of the germ of a great invention, and I would advise you to work until you have made it complete." Upon Bell's remark that he did not have the necessary electrical knowledge, Henry replied, "Get it!" Encouraged by Henry, Bell continued both his telephony experiments and the improvement of the harmonic telegraph. In a letter to Hubbard, he expressed his goal: "If I can get a mechanism which will make a current of electricity vary in its intensity as the air varies in density when a sound is passing through it, I can telegraph any sound, even the sound of speech." On February 25, 1875, he submitted his first patent application for an apparatus "transmitting two or more tele-

Figure 10.5   Bell and Watson experimenting in the Boston workshop. (Scanned from *Archiv für deutsche Postgeschichte*, Vol. 1, 1981, p. 75.)

graphic signals simultaneously along a single wire." The resulting patent, 161,739, was granted on April 6, 1875.

While Bell and Watson continued experimenting with their harmonic telegraph, a breakthrough in telephony was made on June 2, 1875. In one of the telegraphs, a steel-spring armature, malfunctioned. When Watson tried to repair it, he adjusted a screw too tightly, so that the spring pulled back and produced a twang. Bell, working in another room, heard from the telegraph in front of him the sound of that twang. For the first time, sound produced in one telegraph was transmitted electrically to and reproduced in a second telegraph. Bell understood that instead of an intermittent pulsating current, as desired for the harmonic telegraphy, the tight screw had caused a continuous pulsating current which transmitted the sound of the twang. Encouraged by this result, but diverting from his contract with Sanders and Hubbard, the next day Bell had Watson build his first telephone, popularly called the *gallows tele-*

Figure 10.6 Bell's first telephone, 1875. (Courtesy of Museum für Kommunikation, Frankfurt, Germany.)

phone, for its distinctive frame. Figure 10.6 shows a model manufactured from the original.

In this telephone, basically a single diaphragm made of gold beater's skin replaced the steel-spring armatures. The sensitivity of this diaphragm was not satisfactory, and 10 successive months of experimenting produced only inarticulate noises. Under heavy time pressure in view of the alleged progress being made by Elisha Gray and Gardiner Hubbard's denying him Mabel unless he was successful, Bell decided to go ahead with a new patent application. To satisfy his financers, that application was still titled "Improvement in Telegraphy," but almost hidden, included "transmitting vocal or other sounds." He submitted his second patent application on February 14, 1876 and obtained his patent 174,465 on March 7, his twenty-ninth birthday.

Having obtained this patent, but still not satisfied with his insensitive telephone, Bell made another experiment using a liquid transmitter to try out the principle of varying resistance. He connected a membrane to a wire floating in an acid-filled metal cup. A wire attached to the membrane was connected with a battery; a second wire connected to the cup led to a distant receiver. Speaking in front of the membrane, the wire would move in the acid and vary the electrical resistance as a function of the speech. This varying current was sent to a receiver, where an electromagnet activated a membrane. On March 10, while Bell experimented with this liquid transmitter connected by wire with a receiver—probably a gallows telephone—which Watson observed in another room in the basement at a distance of about 6 m, an accident happened. Bell spilt some acid from a battery on his cloth and spontaneously asked for help, calling: *"Mr. Watson, come here, I want you,"* a phrase that was heard by Watson not via the corridor, floors, or walls separating the two rooms but at his telephone receiver.

Bell realized that a liquid transmitter was an impractical solution. He went back once more to his electromagnetic experiments along the lines described in his patent. With a more sensitive diaphragm, a stronger permanent magnet, and a better coil,

Figure 10.7   Bell's butterstamp telephone, 1877. (Courtesy of Museum für Kommunikation, Frankfurt, Germany.) See insert for a color representation of this figure.

Bell and Watson managed to produce a transmitter that transferred the human voice into a pulsating electrical current that reproduced the speech in a receiver of construction similar to that of the transmitter without requiring an external source of electricity. This telephone was popularly called the *butterstamp telephone* (Figure 10.7) because it resembled the wooden stamps then commonly used for impressing designs on pats of butter. This telephone could be used alternately as transmitter when speaking and as receiver when listening. For calling, the diaphragm was initially tapped with a pencil; later, Watson mechanized the calling by incorporating a tiny hammer, which struck the diaphragm when a button was pressed. On August 1, 1878, Thomas Watson filed for a telephone calling patent. Similar to Henry's doorbell, a hammer operated by an electromagnet struck two bells.

10.2.4   Bell's Telephone: "It DOES Speak"

Bell's new telephone was first demonstrated in May 1876 at the *American Academy of Arts and Sciences* and the *Massachusetts Institute of Technology* (MIT). During the following six weeks, the butterstamp telephone was shown at the Philadelphia Centennial Exhibition (memorializing the Declaration of Independence of the United States of America on July 4, 1776 in Philadelphia). Hubbard was one of the centennial commissioners. By his influence Bell's telephone was exhibited on the stand of the Department of Education, albeit on a small table, in a narrow space between a stairway and a wall. Bell had no money to go to the exhibition, but Mabel Hubbard was going, and on the railway station in a dramatic farewell, she insisted that Graham join her, so he finally jumped on the moving train and arrived in Philadelphia as a "blind" passenger. It became his most successful trip.

After much trouble, Green Hubbard had obtained a promise that the judges of the exhibition would include a few minutes for the telephone in their special inspection tour on June 25, 1876. After spending hours inspecting the wonders of the exhibition, such as gas lighting, running water supply, sewage, the musical telegraph of Elisha Gray, and various printing telegraphs shown by Western Union, it was around seven o'clock when the 50-person delegation of judges, scientists, reporters, and other officials arrived at the Department of Education. Tired and hungry, they hardly looked at the telephone, made some jokes at Bell's expense, and wanted to leave the exhibition quickly when suddenly, Dom Pedro II, Emperor of Brazil from 1840 to 1889, with his wife Empress Theresa and a bevy of courtiers, entered the room. Dom Pedro recognized Bell and exclaimed "Professor Bell, I am delighted to see you again!" The judges at once forgot their tiredness and wondered who this young inventor was who was a friend of an emperor. Dom Pedro had once visited Bell's class of deaf-mutes at Boston University and initiated the first Brazilian school for deaf-mutes in Rio de Janeiro.

Bell demonstrated his telephone to the emperor. Dom Pedro held the receiver to its ear as Bell spoke at the distant end. Highly surprised, the emperor exclaimed *"My God! It talks!"* Now, after imperial appreciation, Bell got the attention he deserved. Judge William Thomson (since 1892, Lord Kelvin of Largs) from England, at the time the world's foremost electrical scientist, then also tried the telephone, and declared: *"It DOES speak,"* it is the most wonderful thing I have seen in America. With somewhat more advanced plans and more powerful apparatus, we may confidently expect that Mr. Bell will give us the means of making voice and word audible through the electric wire to an ear at hundreds of miles distant." Suddenly, the "toy" of the exposition became the sensation of the exposition. A mayor of a U.S. city was so impressed by Bell's telephone that he predicted, "I can see the time when every city will have one!" Quite an underestimation, and yet for many African villages, still a dream. The judges stayed the next three hours with Bell. Bell's telephone became the star of the centennial. Bell was given a certificate of award for his invention, which in the words of William Thomson was "the greatest marvel hitherto achieved by the electrical telegraph."

Despite, or because of, this scientific praise, the press and the business world considered the telephone a scientific toy. The *Times of London* even alluded to it as the "latest American humbug" and explained why speech could not be sent over a wire. The *New York Herald* wrote: "The effect is weird and almost supernatural."

Green Hubbard understood that the telephone needed publicity. He borrowed a telegraph wire between New York and Boston for half an hour, and Bell sent a tune from Boston over the 400-km line in the presence of William Thomson. The operator in New York confirmed good reception of *Yankee Doodle*. Shortly afterward, while visiting his parents in Canada, Bell sent songs and quotations from Shakespeare by telephone over a 13-km telegraph line. One of the first positive press reports appeared in the *Boston Adviser* on October 19, 1876, about a three-hour telephone conversation between Bell and Watson over the Boston–East Cambridge telegraph line on October 9.

Bell then started a series of 10 lectures, at $100 a lecture, the first money he received for his invention. The first lecture was given in Salem on April 3, 1877, for an audience of 500 people. A wire was installed between Boston and the lecture hall in Salem. From Boston, Watson sent messages to various members of the audience. Requests to repeat his lecture came from Cyrus W. Field, veteran of the first transatlantic telegraph cable, and many others. At a lecture for 2000 people in Providence, a band playing in Boston was heard. An audience in Boston heard *The Marriage of*

Figaro sung in Providence. In New Haven, 16 Yale professors stood in line, hand in hand, similar to the 200 Carthusian monks in 1746, while a telephone conversation was held through their bodies.

10.2.5   Bell Telephone Company

The first permanent telephone line was erected in April 1877 between the workshop of Charles Williams in Boston and his home in Somerville, a distance of about 5 km. One month later, a friend of Williams, E. T. Holmes, who operated a burglar-alarm system in Boston, installed telephones in five banks. The network was operated as a trial telephone system free of charge by day and as a burglar alarm by night. The five telephones were connected to a switch in Holmes's office; thus was born the first private telephone exchange. Soon afterward, Holmes took his telephones out of the banks and started a real telephone business among the express companies of Boston. Further business telephone networks were opened in New Haven, Bridgeport, New York, Philadelphia, and Detroit.

On July 11, 1877, Bell, now famous, married Mabel Hubbard. On July 9, two days before the couple married and left on their honeymoon trip to Europe, the three members of the patent agreement formed the *Bell Telephone Company of Massachusetts* (BTC). At first, the company had only one full-time employee, Thomas Watson, as superintendent, but a few days later, R. W. Devonshire was hired to keep the books. The company's 5000 shares of stock were distributed as follows: Alexander Graham Bell, 10 shares; Mabel Bell, 1497 shares; Gardiner Green Hubbard, 1387 shares; Gertrude Hubbard, 100 shares; Thomas Sanders, 1497 shares; Thomas Watson, 499 shares; and C. E. Hubbard (Gardiner's brother), 10 shares.

A few months later, on October 6, the weekly journal the *Scientific American* published a front-page article on Bell's telephone (Figure 10.8). When the Bell Telephone Company was formed, only 778 telephones were in use, all manufactured in Charles Williams's little shop, and the firm desperately needed additional capital, so shortly after Mabel and Graham Bell left for Europe, Gardiner Hubbard offered all rights in the telephone for $100,000 to Western Union. William Orton, the president of Western Union, refused the offer with the rhetorical question: "What use could this company make of an electrical toy?"[4] Orton requested Chauncey M. Depew to appoint a committee to investigate Hubbard's offer. That committee reported: "We found that the voice is very weak and indistinct. We do not see that this device will ever be capable of sending recognizable speech over a distance of several miles. Mr. Hubbard's fanciful predictions, while they sound rosy, are based on wild-eyed imagination and lack of understanding of the technical and economic facts of the situation, and a posture of ignoring the obvious limitations of his device, which is hardly more than a toy. We do not recommend its purchase." Within one year, despite this expert's advice and ignoring Bell's patent, Western Union started operating its own telephone system.

In the meantime, BTC developed a new strategy of encouraging agents and other companies to develop the telephone business under BTC license. A first successful approach was made when Thomas Sanders convinced a group of people to invest in

---

[4] Orton disliked Hubbard because he had attacked Western Union in 1868 and lobbied for a government-financed telegraph postal system under his control.

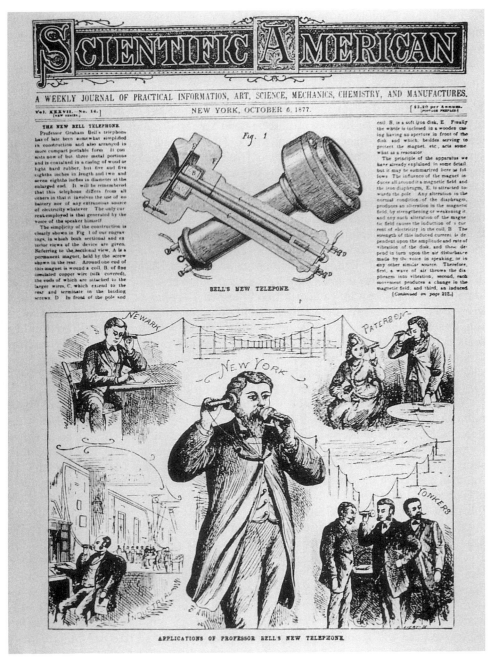

Figure 10.8 One of the first scientific publications on Bell's telephone. (Courtesy of Museum für Kommunikation, Frankfurt, Germany.)

the development of telephony in the urban northeast. They formed the *New England Telephone Company* on February 12, 1878. This was the first of numerous companies which obtained for royalty payments, exclusive rights to Bell patents in specific geographical areas.

To bring in additional investors, BTC was reorganized and with increased capital newly incorporated under the same name in Massachusetts on June 30, 1978. One month later, as a further improvement, most fortunate for BTC, Theodor Newton Vail (1845–1920) was persuaded to give up his job as head of the government mail service, with authority over 3500 postal employees, and join BTC as managing director.

Theodor N. Vail's great-uncle was Judge Stephen Vail, father of Alfred Vail, the partner of Morse. At the age of 19, Theodor Vail worked at several Western Union telegraph offices in New York. A few years later he became a mail clerk on western trains. He systematized the routing of mail so well that he was called to Washington in 1873 to improve the railway mail. In 1876 he was appointed general superintendent of the government mail service. In that function, Vail met Green Hubbard, who President Hayes had appointed as the head of a commission on mail transportation. Hubbard, impressed with Vail's capabilities and connections, offered him management of the BTC.

Along with O. E. Madden, who was recruited from the *Domestic Sewing Machine Company*, Vail brought professional management to BTC. At the end of 1878, BTC managed to get William H. Forbes, a Boston financier with considerable business experience, onto the board of directors. Forbes worked out another reorganization plan embracing all Bell interests in a single company, the *National Bell Telephone Company* (NBTC), incorporated on March 13, 1879 with Forbes as head of the board of directors. By that time, over 10,000 Bell phones were in service, all leased[5] from NBTC. The first telephone exchanges were opened at 82 Nassau Street in March 1879, and a few months later at 97 Spring Street in New York.

10.2.6 Bell's Honeymoon Trip to Europe

Mabel and Alexander Bell made their honeymoon trip to Europe with a few telephone sets in their luggage. They first visited Great Britain, where William Henry Preece (1834–1913, Sir William from 1899), then assistant engineer in chief of the General Post Office (GPO), made the first public demonstration of Bell's telephone to a British Association meeting in Plymouth in September 1877.[6] The *Bell Telephone Company Ltd.* was established in London and opened its first public telephone exchange at 36 Coleman Street in the City of London on April 2, 1879. Manufacturing rights for Bell's telephone were given to the *Consolidated Telephone Construction and Maintenance Company* and the *India Rubber, Gutta Percha and Telegraph Works Company*.

---

[5] Leasing was introduced by Hubbard to protect patent rights. Hubbard learned about leasing at the Gordon–McKay Shoe Machinery Company, where he had been an attorney. The leasing was used until June 1968, when the FCC in the *Carterfone* decision allowed non-Bell equipment to be legally attached to Bell System lines.

[6] Preece made the presentation on behalf of Bell, who was suffering from a short illness. Two years later, Preece reported officially to a House of Commons committee that "the telephone might be good for the Americans, but not for the British, who have plenty of messenger boys."

Thomas Alva Edison also came to Great Britain, where he founded the *Edison Telephone Company of London Ltd.* In 1880, the Bell and Edison companies merged into the *United Telephone Company*, which then used the best of the common patents.

David E. Hughes, who had returned to his native London in 1877, made an important contribution to telephony. In that year, Edison and Blake had developed telephone transmitters using carbon as a variable resistance. Hughes proved that the variation in the resistance in response to speech was not caused by compression of the carbon as generally understood, but was due to a "microphonic effect" at the junctions between the carbon parts of the transmitter. He presented his findings at the *Royal Academy of London* on May 8, 1878. In the interest of a wide application of his discovery, he did not seek a patent but exceptionally, made the results of his experiments freely available.[7] Within a short time, almost every telephone manufacturer brought out a different mechanical solution and patented its own version of Hughes's carbon transmitter.

Louis John Crossley was the first to apply Hughes's ideas. Crossley's transmitter consisted of four carbon pencils loosely placed between carbon blocks in a diamond formation. For many years the company Blakey and Emmott in Halifax produced telephones for the Post Office using Crossley's device. William Johnson was the next person to design a carbon transmitter. He used only two carbon pencils, a solution that found only limited application. The next developer of a carbon transmitter in Britain was a clergyman, the Reverend Henry Hunnings of Boltby, Yorkshire. Allegedly not knowing about Hughes's publications, he developed and patented in 1878 the first carbon granule transmitter.[8] Frederic A. Gower made another approach. Whereas others endeavored to develop small and light receivers to be held to the ear, Gower, in 1879, made his receiver large and heavy, kept it inside a cabinet, and conveyed the sound to the user's ears via two flexible tubes. His *gooseneck telephone* was adopted by the Post Office as the successor to the telephone of Blakey and Emmott. It was widely used in France, Portugal, and Japan, and also by the railways in the Great Britain. Gower initially used electromagnetic transmitters, but from 1880 onward he used various carbon transmitters.

The penetration of the telephone in Britain was greatly hampered by the Telegraph Act of 1869, by which the Post Office had acquired a monopoly on all telegraphic communications. The high court confirmed in 1889 that the telephone was a telegraph within the meaning of the act, which limited the activities of the private telephone companies.

Bell went from Britain to France, where among others he met Antoine Breguet and his father, Louis F. C. Breguet. They became good friends and the Breguets obtained four licenses from Bell for the production of telephone sets in France. The first telephones in France were then installed between the laboratory and the workshop on two different floors of the house of the Breguets at 39 Quai de l'Horloge, Paris. In 1792, production of semaphores had started in this building; in 1842,

---

[7] In fact, a carbon transmitter needed careful adjustment, good protection against vibration, and adequate means to prevent packing of the granules.

[8] Hunnings used a platinum diaphragm with a brass plate slightly behind it and the intervening space filled with powdered coke. The powder was soon replaced by slightly larger particles, generally referred to as granules.

Abraham-Louis Breguet's son Louis-François (1804–1883) changed production to electrical telegraph equipment; and in 1878, production of telephones started. The building (Figure 3.4) is thus the world's oldest place of telecommunications equipment production. Antoine Breguet presented his telephone to the *Académie Française des Sciences* in 1878. The Breguet company produced telephones for a few years only. Antoine Breguet died in 1882, his partner Alfred Niaudet in 1883, and Louis-François Breguet, nearly 80 and unable to bear these misfortunes, died a fortnight after Niaudet. It appears that Breguet's telephone production facility was taken over by Clement-Agnes Ader (1841–1925), better known for his 50-m flight with his *Eole 1*, a steam engine–driven airplane made in 1890. Ader developed his own version of the Hughes transmitter using an arrangement of 10 carbon pencils. He installed the first telephone network in Paris in 1879. This telephone network and the Breguet–Ader telephone production became part of a company named *Sociète Générale des Telephones* (SGT), founded on December 10, 1880 (with Clement Ader as cofounder), which operated the first telephone networks in France. At that time, France had 3039 telephone subscribers in Paris and 1812 subscribers outside Paris. The French government decided in 1889 that the "law on telegraph lines No. 6801" dating from 1837 should also apply to telephony and established a state monopoly for operation of the domestic telephone network under the responsibility of *L'Administration des Postes and Télégraphes*. The commercially more risky external network, including that connecting the colonies with the mother country by means of submarine cables, was left in private hands, mainly by SGT. Subsequently, in 1893, SGT had to transfer its telephone network with 11.000 subscribers to the government and continued under its new name, *Sociète Industrielle des Téléphones* (SIT),[9] only production of telephone equipment.

Germany was not on Bell's honeymoon itinerary, but accidentally, two telephone sets that he had brought to Great Britain arrived a few weeks later on the desk of the general postmaster, Heinrich von Stephan (1831–1897) of the *German Imperial Telegraph Administration* in Berlin. After reading about Bell's telephone in the *Scientific American* of October 6, 1877 (Figure 10.8), Von Stephan wrote on October 18 to George B. Prescott of the Western Union to inquire whether Western Union had made trials with this new device. Before an answer could have arrived, Henry C. Fischer, chief of the London main telegraph office, visited von Stephan on October 24, and as a curiosity, brought the two telephone sets that he had obtained from Bell.[10] Von Stephan, who had a financial problem in extending the German telegraph network, immediately recognized the value of this device as an effective, low-cost substitute for the Hughes and Morse telegraphs, which required skilled operators. With his famous dynamism he staged an immediate trial, evaluation,

---

[9] SIT was bought by CGE in 1932 and renamed *l'Industrie des Téléphones* (L'IT), and in 1946 *Compagnie Industrielle des Téléphones* (CIT). In 1970, CIT merged with the Compagnie *Alsacienne de Construction Atomique et de Télécommunication* (Alcatel, founded in 1920) and still exists under the name Alcatel CIT. Consequently, Alcatel is probably the company with the world's longest tradition of industrial production of telecommunication equipment, starting in the watch workshop of Abraham-Louis Breguet at 39 Quai de l'Horloge, Paris, in 1792.

[10] Von Stephan returned the two telephones after the successful trials to Fisher, who then presented them in 1889 to the German Imperial Post Museum in Berlin, where they are still exhibited. This museum, established on August 24, 1872, was the world's first telecommunications museum. Other postal museums were opened in 1876 at St. Petersburg, in 1878 at Budapest, and in 1889 at Vienna.

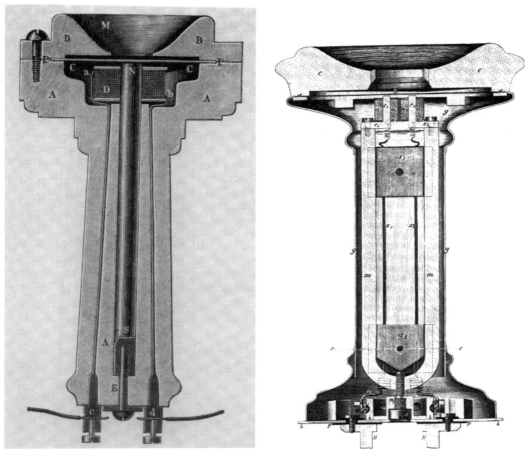

Figure 10.9   Cross sections of the Bell (left) and Siemens (right) telephones in 1878. (Scanned from *Archiv für deutsche Postgeschichte*, Vol. 1, 1986, pp. 37, 39.)

authorization, and implementation program. Bell had not patented his telephone in Germany,[11] so von Stephan requested German companies to produce German versions of Bell's telephone. Siemens started telephone production in November 1877[12] at a rate of 200 telephones per day, Mix & Genest (now in Alcatel SEL) followed two years later. On November 26, Bell, from Great Britain, wrote to Siemens: "Gentlemen, it is rumoured here that you are manufacturing and selling telephones in Germany. As the inventor of the articulating telephone I write to ascertain the facts of the matter. Requesting the favour of an early reply. Yours truly, Alexander

[11] Germany had no patent tradition. Although patent letters had been issued in England and Italy since the fifteenth century and patent protection was introduced in the United States in 1787, in Germany the first patent law came into force as late as July 1, 1877. Bell married 10 days later and might not have been informed of the possibility of securing a patent in Germany.

[12] Siemens obtained a first patent on December 14, 1877 and a second for his electrodynamic telephone (with horseshoe magnet) in March 1878.

Figure 10.10   Telephone of Siemens with rattle (left) and whistle (right) calling, 1878. (Courtesy of Museum für Kommunikation, Frankfurt, Germany.)

Graham Bell." On November 29, Werner Siemens confirmed "the facts of the matter" and clarified: "As you have failed to patent your lovely invention in Germany, we will continue the production, but please inform us in which countries you have a patent so that we can refuse orders from those countries; we have already declined orders from England, Austria, and Belgium."

Siemens improved Bell's telephone in 1878 by replacing the rod magnet with a horseshoe magnet[13] with two large pole shoes, which gave good performance up to 75 km. Figure 10.9 shows the inside of the original butterstamp telephone of Bell and the telephone of Siemens. Instead of separate signaling, Siemens added a rattle or a whistle for calling (Figure 10.10). By the end of the nineteenth century 9789 telegraph offices were equipped with a telephone. After the successful introduction of the telephone for telegraph services, von Stephan decided in 1880 to introduce the telephone for public telephony, too. Public telephone service started in Berlin on January 12, 1881, very modestly with eight subscribers. In fact, the opening of the net-

---

[13] Bell had also used horseshoe magnets in 1876, and Watson designed a version using laminated steel. In 1880, improved single-pole receivers with laminated magnets were produced, but it was not until the 1890s that Bell receivers with *double-pole* (horseshoe) *magnets* came on the market.

Figure 10.11 Wall-mounted telephone, as used around 1885. (Courtesy of Museum für Kommunikation, Frankfurt, Germany.)

work had been planned for April 1, with 48 subscribers, including nine telephones for the stock exchange, but was suddenly advanced in order not to lose the ranking of being the first town in Germany with telephony to Mulhouse. That industrial town in the Alsace (from 1871 to 1918, German territory) had obtained a telephone concession upon the initiative of a local industrialist and member of (the German) parliament, Auguste Lalance (1830–1920), and opened its telephone network on January 24, with 71 subscribers. Figure 10.11 shows a typical telephone used at that time with a butterstamp receiver resting on the line-hook and a butterstamp transmitter built into a wall-mounted box. Local networks were also installed in Hamburg, Frankfurt, Breslau, Cologne, and Mannheim in 1881 and in another 27 towns until the end of

Zur Benutzung der
Fernsprechstelle ist der
Einwurf eines 10-Pf.-
Stückes erforderlich.
Wenn die Leitung
nicht frei ist, fällt
durch einen Druck auf
den weissen Knopf die
Münze wieder heraus.

Gesprächsdauer:
3 Minuten.

Act.Ges.Mix & Genest, Berlin.
D.R.-Patent a.

Figure 10.12   First coin-box telephone in Germany, 1891. (Courtesy of Alcatel SEL.)

1883. Most of those networks also included public telephone boxes (Figure 10.12). Interconnections between the local networks were installed from 1883 onward. This was first limited to distances below 100 km. By mid-1880, however, the butterstamp telephone, used so far for transmitting and receiving, was replaced by a carbon transmitter and an electromagnetic receiver that enabled telephone service over much longer distances. Long-distance service then started between Berlin and Hamburg (280 km) in 1887 using an overhead line with two bronze wires with a diameter of 3 mm. Bronze wires with a diameter of 5 mm were used for longer lines: for instance, for the 1192-km Berlin–Paris line taken into operation on August 6, 1900.

In 1889, all telegraph lines were opened for public telephony and telephones were installed at about 5000 post offices, so that suddenly, countrywide telephone access was available. An act of April 6, 1892 confirmed to the state exclusive rights of telegraphy and telephony. Figure 10.13 shows a typical desk-type telephone with two receivers and a crank for the built-in signaling generator as used at the end of the nineteenth century, when about 230,000 telephone lines were installed in a network

Figure 10.13 Desk telephone with two receivers, around 1900. (Courtesy of Alcatel SEL.)

with a total length of 830,000 km, almost 70% of the telephone lines were still overhead (Figure 10.14).

10.2.7 Telephone Developments in Sweden

The Swedish engineer Hopstock had the honor of demonstrating Bell's telephone to King Oscar II in the summer of 1877. A few months later, IBTC founded an affiliated company in Sweden and installed the first line between the telegraph office and the Grand Hotel in Stockholm. Because of lack of progress, three telegraph inspectors (Bratt, Lybeck, and Recin) established the *Stockholm Telephone Company* in 1880. Within a few months, the company became SBTC (Stockholm Bell Telephone Company) and an affiliate of ITBC. At the same time, a number of private associations formed cooperatives to set up telephone networks in smaller towns. To remain independent from SBTC, those cooperatives bought their telephones and manual exchanges from a new company founded by Lars Magnus Ericsson (1846–1926). Ericsson, born on a little farm in Vegerbol in southwestern Sweden, left home at 14, to work as a smith's apprentice across the border in Norway. At 20 he became

Figure 10.14 Overhead line construction in Berlin in 1882. (Scanned from *Archiv für deutsche Postgeschichte*, Vol. 1, 1981, p. 130.)

a skilled instrumentmaker in the telegraph workshop of A. H. Öller in Stockholm. At night he studied English, German, mathematics, and metallurgy. Upon Öller's recommendation he obtained a government grant allowing him to work and study electro technology in Germany (with apprenticeships at Siemens in Berlin) and in Switzerland from 1872 to 1875. Back in Sweden he founded an electrical engineering workshop in April 1876 under the name L. M. Ericsson & Co. His companion was Carl Johan Anderson, a colleague from his time at Öller's workshop. The initial business of the company was the manufacture and repair of telegraph instruments. Repair work on the telephones imported by SBTC from the United States encouraged Ericsson to develop and manufacture his own telephones of superior quality and more elegant design. He produced his first telephone in 1878. In the same year he married Hilda Simonsson, his future partner in life and business.

Ericsson also designed his own version of a Hughes transmitter, a granule transmitter of compact construction. To permit adjustment, he used a screw, or helic—hence its name, *helical transmitter*. He combined the transmitter and receiver in a handset in 1884 and made it the company's trademark for many years.

The Swedish Telegraph Administration *Telegrafverket* (later named Televerket) also established a telephone network in Stockholm in 1881, initially for exclusive government use but opened to the public in 1889. In the meantime, another Swedish telephone protagonist appeared on the scene: Henrik Tore Cedergren (1853–1909). After unsuccessful attempts to obtain reduced rates from SBTC, Cedergren founded a rival company, the *Stockholm Allmänna Telefonaktiebolag* (SAT, Stockholm Public Telephone Company) in April 1883. A period of strong competition began. Televerket acquired most of the small operating companies throughout Sweden and installed competing networks in towns that already had a private operator. Moreover, both Televerket and SAT established their own manufacturing facilities. The real winners of this competition were the Swedish telephone subscribers, who at the end of the nineteenth century paid the lowest telephone charges in Europe, on the order of 10 to 30% of those in Great Britain, France, Italy, and Spain, and enjoyed the highest teledensity, 1.45.

10.2.8 Biggest Patent Battle on Telecommunications

Back in the United States, Bell faced big problems. For 17 months no one disputed Bell's claim to be the original inventor of the telephone. But after the success at the centennial, and after Bell had explained his invention before more than 20,000 people, after several hundred articles appeared in newspapers and scientific magazines, after everybody could read the text of Bell's patent, and after the BTC gradually extended its operations, and especially now that even Western Union had given weight to the telephone, many persons suddenly claimed to have developed a telephone before Bell. A persistent patent war started, comprising 600 lawsuits and lasting for 11 years. Anyone who possessed a telegraphic patent in which expressions such as "talking wire," "voice," or "sound" were used now saw the chance to claim rights to telephony. The most serious claims came from Elisha Gray and Amos Emerson Dolbear when Bell fought against Western Union's illicit use of his patent.

Gray had developed a device for "transmitting musical tones by electricity," for which he obtained a patent in the United States in February and in Great Britain in October 1874. In the same year, Western Union bought the device. Gray continued to investigate harmonic telegraphy and telephony. On February 14, 1876, he filed a caveat at the U.S. Patent Office two hours after Bell had filed his famous patent. The record book for that day shows as the fifth entry: "A. G. Bell, $15" and as the thirty–ninth entry: "E. Gray, $10." In his caveat, Gray states: "It is the object of my invention to transmit the tones of the human voice through a telegraphic circuit, and reproduce them at the receiving end of the line, so that actual conversations can be carried on by persons at long distances apart." Although this caveat contained a clear description of the phenomenon of telephony, Gray did not built a model of this telephone, and said in a letter to Bell written on March 5, 1877: "I do not claim even the credit of inventing it [the telephone]." Within one year he changed his mind and strongly supported Western Union in the most serious lawsuit against Bell (Section 10.2.8).

Amos Emerson Dolbear (1835–1901), born in Norwich, Connecticut, and from 1874 professor of physics and astronomy at Tufts College, in 1877 wrote a booklet entitled *The Telephone: An Account of the Phenomena of Electricity, Magnetism, and Sound as Involved in Its Action with Directions for Making a Speaking Telephone.* In it he reported on several years of experimenting, which resulted in the construction of a speaking telephone with a U-shaped permanent magnet. In the same year he wrote to Bell: "I congratulate you, Sir, upon your great invention, and I hope to see it supplant all forms of existing telegraphs." One year later, however, he too claimed that Bell's patent of March 7, 1876 was awarded improperly since the device described would not work and that Phillip Reis's unpatented device of 1860 would work just as well. A Reis telephon was tried in open court and was reported as an utter failure: of the hundred words spoken into the telephon, fewer than eight could be guessed at correctly. Judge Lowell pronounced: "A century of Reis would never have produced a speaking telephone by mere improvement of construction. It was left for Bell to discover a new art: that of transmitting speech by electricity. To follow Reis is to fail; but to follow Bell is to succeed." In fact, whether or not he was aware of it, Lowell was right: because as speech produces a fluctuating wave of continuous character, the Reis telephon, using a diaphragm to make and break an electrical circuit, transmitted a pitch of sound (thus, an uncoded digital signal was produced), which could not produce intelligible speech with the technical means prevailing. The Bell telephone varied the strength of a current as a function of speech in a continuous circuit, and thus transmitted the entire sound and could produce intelligible speech.

The longest lawsuit, lasting nearly four years, was initiated by Daniel Drawbaugh, an ingenious but apparently not very inventive mechanic who lived in a country village near Harrisburg, Pennsylvania. As a subscriber to the *Scientific American*, he had imitated more than 40 inventions and exhibited them as his own. In 1884 he claimed to have invented a telephone complete with switchboard before 1876, but for lack of money he had not filed a patent. Over 500 witnesses were required finally to prove his swindle. He tried again in 1903, claiming that he, rather than Marconi, had discovered radio transmission.

Bell had to defend his patent in 600 costly lawsuits of various natures, including five at the U.S. Supreme Court. Fortunately, Bell was defended by two master lawyers: the conservative and dignified Chauncy Smith, who had great experience, and the quiet James J. Storrow, who had an encyclopedic memory. When Storrow became a lawyer of Bell's, he first spent an entire summer in his country home in Petersham studying the laws of physics and electricity. The BTC established its own patent department in 1879 under the control of the systematic and convincing Thomas D. Lockwood. Those three lawyers worked as a perfect team against over 50 eminent lawyers, and with the exception of two trivial contract suits, they never lost a case. Smith died one year later in the courtroom while accusing a lawbreaker. In the next 30 years, Lockwood applied for some 80,000 patents for BTC. In 1884, the U.S. Patent Office began an 18-month investigation of all telephone patents and confirmed: "It is to Bell that the world owes the possession of the speaking telephone."

10.2.9   Battle of David Against Goliath

At the time, Western Union was the most powerful electrical company in the world. It had 400,000 km of telegraph wire over 160,000 km of route and was supplying its customers with various kinds of printing telegraphs and dial telegraphs, some of which could transmit 60 words a minute. These accurate instruments, it believed, could never be displaced by such a scientific oddity as the telephone. It continued to believe this until one of its subsidiary companies, the *Gold and Stock Telegraphy Company*, reported that several of its machines had been superseded by telephones from BTC. William Orton, in an effort to get rid of BTC, requested its chief electrical expert, Frank L. Pope, to investigate the validity of Bell's patents. Pope made a six-month examination, he bought every book in the United States and Europe that was likely to have any reference to the transmission of speech, employed a professor who knew eight languages to translate the foreign-language books, interviewed various experts, and visited libraries and patent offices. In his final report he concluded that the Bell patents were valid and stated that there was no way to make a telephone except Bell's way. He advised purchase of the Bell patents.

Pope's qualified report was disregarded, and Western Union decided to start a telephone operation themselves. BTC then had 3000 telephones installed. On December 6, 1877, Western Union created the *American Speaking Telephone Company* (ASTC), with George Walker as president, Norvin Green as vice-president, and Orton as a director. ASTC decided to use telephone receivers based on developments of Elisha Gray, Amos E. Dolbear, and George M. Phelps, and a carbon telephone transmitter developed by Thomas Alva Edison. Tests with those devices were made in 1878 between New York, Philadelphia, and Washington. The first central telephone exchange was opened in Western Union's office at 198 Broadway in New York on August 1, 1878.

Now began the most difficult time for BTC: how to compete with Western Union, which had a superior telephone transmitter, a host of agents, a network of wires, and above all, thousands of customers. As a first improvement, a more sensitive, variable-pressure contact transmitter developed by the German immigrant Emile Berliner in 1876 replaced Bell's weak telephone transmitter. Berliner applied the principle of variable resistance, basically using an iron diaphragm touching a steel ball. He filed a caveat on April 14, 1877 and joined BTC in the following September. In Bell's Boston laboratory, Francis Blake, Jr. of the U.S. Geodetic Survey improved the transmitter of Berliner by replacing the steel ball by carbon in a way differing from Edison's approach.

In the meantime, Western Union had started another attack; Gray was introduced as the original inventor of the telephone. BTC, now NBTC and with Vail in command, responded courageously with a patent infringement suit in the U.S. Circuit Court in Boston on September 12, 1878. Court hearings during the following year convinced George Gifford, Western Union's most experienced patent counsel, that further litigation would be a waste of time and money. Western Union, which by then had found that operating telephones on its telegraph lines was hardly possible, suddenly offered to compromise out of court. A committee of three from each side was appointed. After further months of disputation, an historical agreement was signed on November 10, 1879 in which:

- Bell was confirmed as the original inventor of the telephone.
- Western Union sold its complete telephone network to NBTC, consisting of telephone networks in 55 cities with 56,000 subscribers. NBTC agreed to pay a royalty of 20% on the fees of those subscribers over the period of the 17 years that the Bell patents were still valid.
- NBTC agreed not to operate a telegraph network in the United States.
- Western Union gave NBTC use of its telephone inventions and of its rights-of-way for pole lines, but retained the right to use telephones in its own business.

David won the battle against Goliath and within a few years itself became a Goliath. New telephone companies operating under NBTC license were established all over the United States. To prevent competitors from acquiring those companies, NBTC endeavored to keep the majority of shares of those companies. This required substantial capital; therefore, on April 17, 1880, NBTC was divided into the *American Bell Telephone Company* (ABTC), with additional capital and Vail as general manager, with the *International Bell Telephone Company* (IBTC) as a holding company for foreign activities. IBTC's first foreign company was the *Bell Telephone Manufacturing Company* (BTM), founded at Antwerp, Belgium, in 1882 (now Alcatel BTM).

10.2.10   Pioneers Leave the Telephone Business

Having the telephone firmly established after four years of struggle and ABTC under the professional control of Vail and others, the pioneers left the Bell companies. Hubbard became the president of the *National Geographic Society* (founded in 1888); he died in 1897 and was succeeded the next year by Bell in this function. Bell established an educational pamphlet for this society which his future son-in-law and future president, Gilbert M. Grosvenor, transformed into the society's unique journal, the *National Geographic.*

Sanders, who had spent his entire capital of about $110,000 in Bell telephone development and initial operations, sold his shares for almost $1 million but lost most of it in a Colorado gold mine. Williams sold his factory to the Bell Company in 1881 and became a millionaire. Watson left the Bell Company also as a millionaire in 1881. He established a shipyard near Boston in which were built half a dozen warships for the U.S. Navy.

Bell, since 1879 a resident of Washington, DC, continued to experiment on communications. Among other achievements, together with Charles Summer Tainter, he developed a photophone for transmission of sound on a beam of light. In 1880 the French government honored him with the Volta Prize. He used the prize of about $10,000 to establish the Volta Laboratory, where in association with Tainter and his cousin, Chichester A. Bell, he invented the graphophone, which recorded sound on wax cylinders and disks. A significant part of his royalties went into the *American Association to Promote the Teaching of Speech to the Deaf*, since 1956 called the *Alexander Graham Bell Association for the Deaf.* In 1882 he obtained U.S. citizenship, but three years later he went back to Canada. There he acquired land at Baddeck on Cape Breton Island in Nova Scotia. In surroundings reminiscent of his early years in Scotland, he established his home at Beinn Bhreagh (Gaelic for "beautiful

Figure 10.15 Inauguration of the New York–Chicago telephone line by A. Graham Bell on October 18, 1892. (Scanned with permission of the ITU from Catherine Bertho Lavenir, *Great Discoveries: Telecommunications*, International Telecommunication Union, Geneva, 1990, p. 39.) See insert for a color representation of this figure.

mountain"), complete with research laboratories. There he developed sonar detection, solar distillation, and hydrofoil crafts. His hydrofoil HD-4 attained a world speed record of 113 km/h (70.86 mph) on September 9, 1919. Bell had 18 patents granted in his name and another 12 shared with collaborators. These include 14 for the telephone and telegraph, four for the photophone, one for the graphophone, five for aerial vehicles, four for hydro airplanes, and two for a selenium cell. Figure 10.15 shows Bell, on October 18, 1892, inaugurating the New York–Chicago telephone line. With a length of 1500 km, this was the World's longest telephone line.

At his Nova Scotia home, on the last day before he died (August 2, 1922) Bell consulted by telephone his physician, who lived 60 miles away in Sydney. During his funeral service on August 4, at 6:25 P.M. eastern time, at the moment Bell was laid to rest, the 15 million telephone lines served by the Bell System in the United States and in Canada were silent for 1 minute. His burial site was blasted from rock on a spot he had selected on the crest of Beinn Bhreagh Mountain, near Baddeck. Bell had been in failing health for several months and eventually succumbed to progressive anemia. Half a year later, on January 3, 1923, Mabel Bell died.

10.3 COMPANIES WITH COMMON BELL ROOTS

Soon after the pioneers left the telephone business, telephony became big business. Giant telecommunications companies with worldwide operations evolved from the modest Bell companies of the 1880s. In the United States these were the Bell System companies: the telephone manufacturer Western Electric; the world's largest telephone operator AT&T; and the telephone technology standard-setting Bell Telephone Laboratories. Outside the United States the companies Northern Electric in Canada and Nippon Electric Company in Japan were founded by Bell System companies and became their major competitors. From Cuba came a most remarkable small company in 1920, which, thanks to the Bell System, became the world's largest multinational telecommunications company, IT&T (International Telephone & Telegraph Company, named ITT since 1958), which, however, on January 1, 1987 collapsed like a supernova and disappeared, giving birth to a new multinational company named Alcatel.

Those companies heavily determined the course of telecommunications and as such reflect the history of more than one century of telecommunications. To cover that fascinating part of industrial history, however, is beyond the scope of this book.

10.4   WORLDWIDE INTRODUCTION OF TELEPHONY

When Bell had developed his telephone, he found much skepticism, and the initial acceptance of his telephone was rather low in almost all countries. Once installed, the merits of the telephone became obvious and the teledensity soon increased exponentially. It took 20 years until 1 million telephones were installed; then this amount doubled within four years. At the end of the nineteenth century, access to a telephone was available in most urban centers of industrialized countries. Telephone connections between urban centers existed in most of those countries. Official statistics about the development of telephony did not exist, but the Bell and Edison companies carefully registered the sales of telephones while the relevant patents were in effect to collect royalties. An approximate summary of the worldwide distribution of telephones at the end of the nineteenth century is given in Table 10.1, which shows that 60% of telephones were used in the United States, 35.5 percent in Europe, and 4.5% in the rest of the world. The year of telephone introduction of all countries that used the telephone at the end of the nineteenth century is given in Table 10.2.

10.5   INTERNATIONAL TELEPHONY

International telephone operation began in North America on border-crossing lines between Detroit, Michigan, and Windsor, Ontario, Canada on January 20, 1881, and between Brownsville in Texas and Nuevo Laredo in Mexico on February 26, 1883. In South America international telephone operation began in October 1889

TABLE 10.1   Worldwide Distribution of Telephony at the End of the Nineteenth Century

Year	U.S.	Europe	Rest of World	Total
1880	47,900	1,900	—	49,800
1885	147,700	58,000	11,800	217,500
1890	227,000	177,000	31,500	435,500
1900	1,355,000	800,000	100,000	2,255,000

Source: Data primarily from *Bell System Technical Journal*, March 1958.

TABLE 10.2   Worldwide Introduction of Telephony during the Nineteenth Century

Year	The Americas	Europe	Rest of World
1876	United States, Brazil	—	—
1877	—	Belgium, France, Germany, Great Britain, Switzerland, Sweden	—
1878	Canada, Jamaica	Italy	Australia, New Zealand
1879	Chile	The Netherlands	Senegal, Singapore
1880	Mexico	Norway	South Africa
1881	Argentina, Guatemala	Austria, Denmark, Hungary, Russia	China, Egypt, India, Thailand
1882	—	Malta	Mauritius
1883	—	Czech Republic, Poland, Portugal, Ukraine	—
1884	Barbados, Nicaragua	—	Myanmar
1885	Uruguay	Luxembourg, Spain	New Caledonia
1886	Bermuda, Trinidad & Tobago	—	—
1888	Peru	—	—
1889	—	—	Japan
1890	—	Iceland	Ghana, the Philippines, Vietnam
1891	—	—	Malaysia, Tunisia
1892	Colombia	—	Tanzania
1894	—	Finland	Benin, Togo
1895	—	—	Lesotho
1897	—	—	Ethiopia
1898	—	—	Fiji
1899	—	—	Macau, Republic of Congo (Brazzaville)

Source: Data from Gerhard Basse, Die Verbreitung des Fernsprechers in Europe, Nordamerika, Lateinamerika, Afrika, Asien und Ozeanien, *Archiv für Post und Telegraphie*, Vol. 1, 1977, pp. 58–103 and Vol. 1, 1978, pp. 24–93.

between Buenos Aires, Argentina, and Montevideo, Uruguay via a line that included a 90-km telephone cable through the Rio de la Plata.

In Europe the first international telephone communication originated in Switzerland. An experimental telephone connection between Switzerland and Italy was made on January 6, 1878, when Michele Patocchi,[14] telegraph inspector at Bellinzona (then capital of Ticino), operated the telephone over a telegraph line from his office to the telegraph office in Milan. The first commercial international telephone communication began on August 1, 1886 between the Swiss town of Basel and the town of St. Louis in the Alsace via a single iron wire with a diameter of 3 mm. In the same year the line was extended from Basel to nearby Liestal in Switzerland and from St. Louis to Mulhouse (20 km northwest of Basel) in the Alsace. The territory

[14] Patocchi made various experiments with the telephone and published his results. He spoke, for example, from Bellizona with a colleague in Luzern (165 km) on Christmas day, 1878, and he presented Donizetti's opera *Don Pasquale* at the Teatro Sociale to an audience in another hall in Bellinzona.

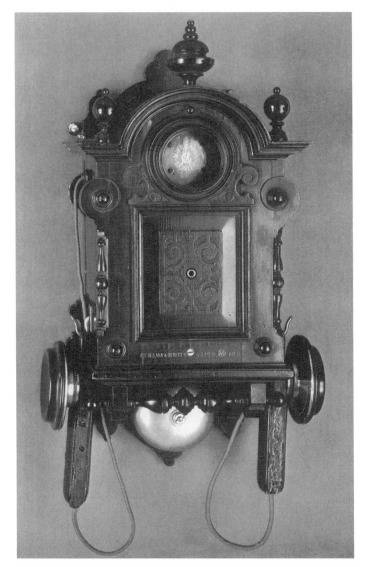

Figure 10.16 Wall telephone, 1888. (Courtesy of Museum für Kommunikation, Frankfurt, Germany.) See insert for a color representation of this figure.

of Alsace, which was moved repeatedly between France and Germany, at that time was German territory and became a part of France again after World War I. Thus both Germany and France may see this as their first telephone communication with Switzerland. Unfortunately, the line was interrupted from September 30, 1887 to October 5, 1892, due to political tensions between Switzerland and Germany. With improved relations in 1892, further border-crossing international telephone connections were established between Kreuzlingen, Switzerland and Konstanz, Germany on August 25 of that year and a little later between Basel, Switzerland and nearby

Figure 10.17   Luxury table telephone used by the Vatican, 1900. (Courtesy of Museum für Kommunikation, Frankfurt, Germany.) See insert for a color representation of this figure.

Loerrach, Germany. The first international long-distance line between Switzerland and Germany was opened in 1898 between Basel and Frankfurt (320 km), followed in 1900 by Basel–Stuttgart on January 20, and Basel–Berlin (900 km) on April 20. Further international telephone lines were installed from Switzerland to France and Austria at the end of the nineteenth century.

Commercial telephone service between Belgium and France began on February 24, 1887 with a line between Brussels and Paris (300 km). Experimental telephone communication between those two towns was made via the existing telegraph line on May 16, 1882. Regular telephone service between London and Paris, a distance of 500 km, began on April 1, 1891 via the first submarine cable for telephony.

International telephone connections between Austria and Germany were established in 1892 between Brengenz in Austria and Lindau in Germany and between Berlin and Vienna on December 1, 1894. Telephone communication between Germany and Denmark began with the opening of a line between Hamburg and Copenhagen on October 8, 1895. Telephone communication between Germany and Belgium started in the same year between Cologne and Brussels and in the following year with the Netherlands. The longest international telephone line installed in the nineteenth century was between Berlin and Budapest via Vienna. Telephone operation on this line with a length of 1300 km started on September 1, 1897.

Figure 10.18 Luxury telephone handset, late 1890s. (Courtesy of Museum für Kommunikation, Frankfurt, Germany.) See insert for a color representation of this figure.

The only international telephone line installed in the nineteenth century outside the Americas and Europe was probably the line between the West African countries Togo and Benin, where in 1894 a telegraph line in Togo between the capital Lomé and Aného (45 km east of Lomé) was extended from Aného to Grand Popo in Benin (then called Dahomey) and used for telephony.

10.6   THE ART OF TELEPHONE SETS

The telephone was a new product for which no predecessor existed. The shape of the product had to meet its function. The shape of Bell's butterstamp telephone clearly evolved from the requirement to accommodate a long magnet rod and a cylindrical diaphragm positioned perpendicularly. Basically, all subsequent telephones designed in the United States followed this pattern of functionality. In Europe more consideration was given to the fact that the telephone had a prestige value and should fit into its milieu elegantly. This led to the production of remarkable pieces of art. Teams of the finest craftsmen made special telephone sets, in gold, ivory, engraved steel, and carved wood, occasionally even with art nouveau decorations, which found their place in palaces, the offices of presidents, and in the villas of the upper class. The book *Vintage Telephones of the World*, written by P. J. Povey and R. A. J. Earl, once curators of the British Museum, and published by the IEE in 1988, shows far over 100 of such rare and unusual telephones of outstanding design.[15] From the selection of those telephones, it appears that the Swedish company L.M. Ericsson and the German company Mix & Genest in particular employed very capable craftsmen who drew their inspiration from many different sources, so that the resulting products had

---

[15] The book *Telefone 1863 bis heute*, published by the German Museum Foundation of Post and Telecommunications in 2001, shows and describes 350 telephones that are in the possession of the foundation, including the three shown in Figures 10.16 to 10.18.

such nicknames as *ashtray, biscuit barrel, coffee mill, cotton reel, Eiffel Tower, spider, Singer sewing machine*, and *steam engine*. The production of these artistic luxury telephones stopped with World War I, when in addition to all other disastrous effects, hardly any customer was left who could still afford luxury telephones.[16] An impression of this past glory is given in Figure 10.16, which shows a carved wooden wall telephone inspired by Black Forest cuckoo clocks, produced by Mix & Genest in 1888; in Figure 10.17, which shows a gold-plated table telephone produced for the Vatican by Siemens & Halske in 1900; and in Figure 10.18, which shows a luxury telephone handset for an extension station produced by Mix & Genest in the late 1890s.

REFERENCES

Books

Feyerabend, E., *50 Jahre Fernsprecher in Deutschland, 1877–1927*, Reichspostministerium, Berlin, 1927.

Gold, Helmut, et al., *Telefone 1863 bis Heute*, Museumsstiftung für Post und Kommunikation, Edition Braus, Frankfurt, Germany, 2001.

Gööck, Roland, *Die großen Erfindungen Nachrichtentechnik, Elektronik*, Sigloch Edition, Künzelsau, Germany, 1988.

Horstmann, Erwin, *75 Jahre Fernsprecher in Deutschland, 1877–1952*, Bundesministerium für das Post- und Fernmeldewesen, Bonn, Germany, 1952.

Koppenhofer, A. P., *Als Philipp Reis das Telefon Erfand*, Geiger Verlag, Horb am Neckar, Germany, 1998.

Oslin, George P., *The Story of Telecommunications*, Mercer University Press, Macon, GA, 1992.

Povey, P. J., and R. A. J. Earl, *Vintage Telephones of the World*, IEE History of Technology Series 8, Peter Peregrinus, London, 1988.

Siemens, Georg, *Der Weg der Elektrotechnik Geschichte des Hauses Siemens*, Vols. 1 and 2, Verlag Karl Alber, Freiburg, Germany, 1961.

Terrefe Ras-Work, *Tam Tam to Internet: Telecoms in Africa*, Mafube Publishing and Marketing, Johannesburg, South Africa, 1998.

Winston, Brian, *Media Technology and Society—A History: From the Telegraph to the Internet*, Routledge, London, 1998.

Young, Peter, *Person to Person: The International Impact of the Telephone*, Granta Editions, Cambridge, 1991.

Das deutsche Telegraphen-, Fernsprech- und Funkwesen, 1899–1924, Reichsdruckerei, Berlin, 1925.

Articles

Basse, Gerhard, Die Verbreitung des Fernsprechers in Europe und Nordamerika, *Archiv für Post und Telegraphie*, Vol. 1, 1977, pp. 58–103.

---

[16] In the last decades of the twentieth century, L. M. Ericsson again made luxury gold-plated telephones, primarily for wealthy customers in the Middle East.

Basse, Gerhard, Die Verbreitung des Fernsprechers in Lateinamerika, Afrika, Asien und Ozeanien, *Archiv für Post und Telegraphie*, Vol. 1, 1978, pp. 24–93.

Green, E. I., Telephone, *Bell System Technical Journal*, Vol. 37, March 1958, pp. 289–324.

Klein, Wolfgang, Pioniere des Fernsprechwesens, *Archiv für Post und Telegraphie*, Vol. 1, 1977, pp. 4–15.

Müller-Fischer, Erwin, and Otfried, Brauns-Packenius, Zeittafel zur Geschichte des Fernsprechers, 1852 bis 1945, *Archiv für Post und Telegraphie*, Vol. 1, 1977, pp. 16–34.

Internet

www.CCLab.com, American History of Telecommunications, Telecom Global Communications, Inc.; updated March 24, 1998.

www.privateline.com/TelephonyHistory, Telephone history, by Tom Farley.

11

TELEPHONE SWITCHING

11.1 MANUAL SWITCHING

Switching started in the 1870s as telegraph switching, with a selection device to connect the various telegraph instruments of a telegraph office with the few lines available for interconnection with corresponding national and international telegraph offices. Switching at these offices was done primarily to enable telegraph companies to provide service conveniently on a time-shared basis between offices and to facilitate the connection to a given destination of telegraph instruments on lines in working order. Telegraph messages were served manually on a store-and-forward basis and required little switching.

A separate annunciator system with a calling device at the user's premises was used for initiating messages between users. Messengers responded to these signals to pick up written messages that were delivered to and transmitted from the telegraph office. Less frequent users of the service had no calling device, but simply brought their messages to the telegraph office, or, later, used the telephone.

In early telephone networks, switching from a calling telephone subscriber to the line of the subscriber called was also done manually. The subscriber calling used a ringer (a little generator) with a crank attached to the telephone set to attract the attention of the central office operator, to whom the caller gave the name of the subscriber being called. The operator, in front of a switchboard, would first ring the subscriber being called and then through-connect the two lines on the switchboard by means of a patch cord with plugs (also called *jacks* after the French-Canadian inventor, Mr. Jack) at both ends.

The first manual telephone exchange was installed at New Haven, Connecticut, on January 28, 1878, with a switchboard serving 21 subscribers (Figure 11.1). The

The Worldwide History of Telecommunications, By Anton A. Huurdeman
ISBN 0-471-20505-2   Copyright © 2003 John Wiley & Sons, Inc.

Figure 11.1   World's first telephone exchange, New Haven, Connecticut, 1878. (Scanned from *Archiv für deutsche Postgeschichte*, Vol. 1, 1977, p. 97.)

operator had to rotate four small brass arms to interconnect the wires on the switchboard. Two arms connected the two subscriber wires to the switchboard, the third arm connected the operator to the two subscribers, and the fourth arm was used to ring the subscriber being called. After this first manual switchboard, many others followed. These switchboards were made almost exclusively in small workshops operating on an artisan basis. Figure 11.2 shows such a manual exchange used by the German Telegraph Administration beginning in 1881. A more sophisticated exchange used around 1900 is shown in Figure 11.3.

Most "inventors" of manual exchanges remain anonymous. However, an engineer employed by the Western Electric Manufacturing Company, C. E. Scribner, is said to have taken out over 500 patents. He helped to develop a multiple switchboard, which became a necessity when the number of subscribers began to exceed what one operator could handle. To avoid double seizing of a subscriber line, an operator had

Figure 11.2   First manual exchange used in Germany, 1881. (Courtesy of Museum für Kommunikation, Frankfurt, Germany.)

Figure 11.3   Manual exchange used in Germany around 1900. (Courtesy of Alcatel SEL.) See insert for a color representation of this figure.

to inform the other operators about busy lines. For that purpose, messenger boys ran from one position to another distributing "busy bulletins" or just shouting the names of the busy subscribers. In 1883, Scribner simplified the procedure by inventing a system for testing the accessibility of a subscriber line on the multiple positions in an exchange by touching the sleeve of the jack with the plug tip of the operator's answering cord. On the multiple positions of a manual exchange, two separate banks of jacks were then provided:

1. Those that were to answer a subscriber's call, which were connected to a limited number of operator positions only
2. Those used to set up a call to a subscriber, which were connected to all operator positions

In 1880 the Western Electric Manufacturing Company brought out its famous *Standard* model. The Standard consisted of a panel with keys and pairs of cords that were plugged into jacks and had counterweights for easy handling and restoration to the idle position after use. Each pair of cords was attached to a ringing key and to an annunciator shutter indicating the end of the call. It was possible to join several boards, each capable of connecting 200 to 300 subscribers.

Figure 11.4   First manual exchange used in Amsterdam, 1881. (Scanned from *Archiv für deutsche Postgeschichte*, Vol. 1, 1977, p. 95.)

The Standard was exported from the United States to so many countries that the brand name became an international synonym for manual exchange. They were also manufactured and exported by the Western Electric subsidiary at Antwerp in Belgium. Figure 11.4 shows such an exchange installed by the Dutch Bell Telephone Company.

The introduction of a pair of wires instead of a single metal wire, and Earth return, to reduce crosstalk led to substantial alterations in manual exchanges. Relays were transferred to racks at some distance from the operator's panels, with wiring between the racks and panels. An early switchboard of that type, termed *metallic*, without an Earth return but with two wires per subscriber, was installed in the New York City Cortland Street office in 1887 with a capacity of 10,000 lines.

Manual switching created a new occupation for women. The first operators were boys, but their lack of discipline, impoliteness, and clumsiness added substantially to initial operating problems. Female telephone operators, with their natural attentiveness, polite manners, and high level of concentration, improved the service significantly at lower cost. The first woman operator, Miss Emma M. Nutt, began work on September 1, 1878 at the Boston exchange.[1] In 1902, the New York Telephone

---

[1] The *Kansas City Star* of December 31, 1899, wrote: "To become a 'hello' girl the applicant must not be more than 30 years old or less than 5 feet 6 inches tall. Her sight must be good, her hearing excellent, her voice soft, her perception quick and her temper angelic. Tall, slim girls with long arms are preferred for work on the switch boards to reach over all of the six feet of space allotted to each operator. It is said that girls of Irish parentage make the best operators. They are said to be quicker with their fingers and their wits and control their tempers admirably."

Company started a school, the first of its kind in the world, for the education of female telephone operators. Out of about 17,000 applicants annually, fewer than 2000 were accepted and given one year of training, usually followed by employment. Marriage was a valid reason for dismissal.

A subscriber's call was indicated at the exchange by the drop of an annunciator shutter. The shutters took up substantial space and were noisy. The incandescent electric lamp, invented by Edison in 1879, used in a small size beginning in 1894 (first used in Chicago), replaced the shutters. Use of these small lamps coincided with the introduction of a common battery (for all subscribers) at the exchange. The first common battery (CB) exchange was brought into service in Lexington, Massachusetts, in December 1893. Until the introduction of the CB system, all subscribers had to have a battery and a hand-operated magnetogenerator in their homes, with all their inconvenience.

A disadvantage of the initial CB system was the problem of crosstalk between subscriber lines. Two patents solved this problem. Scribner, as chief engineer at Western Electric, introduced the Stone–Scribner bridged impedance system. Hammond V. Hayes, chief engineer of AT&T, added an induction coil in the dc power supply.

In Europe the CB system was introduced at Bristol, England, in 1900 and in the same year in Adlershof near Berlin, followed by Brussels in 1902 and the Hague in the Netherlands in 1903. It served 15,000 subscribers. By 1910, the largest CB exchange in the world served 60,000 subscribers in Moscow. Figure 11.5 shows such a manual telephone exchange as used in Berlin beginning in 1891.

11.2   EVOLUTION LEADING TO AUTOMATIC SWITCHING

During the first 10 to 15 years of telephony, a number of inventors conceived the idea of replacing operators and their plug-and-cord switchboards with automatic

Figure 11.5   Manual telephone exchange in Berlin, 1891. (Scanned from *Archiv für deutsche Postgeschichte*, Vol. 1, 1977, p. 25.)

installations. The brothers Daniel and Thomas Connolly, together with Thomas J. McTighe,[2] invented the first automatic telephone exchange in Great Britain. A model was shown at the Exposition International d'Electricité at Paris in 1881 and currently occupies a place of honor in the National Museum of History and Technology, in Washington, DC. However, a practical application was not made with this exchange. At the same exhibition in Paris, two French inventors, Leduc and Bartelous, showed automatic switching machines, which were also never used.

The first system to be used came from a British engineer, Dave Sinclair. In 1883 he developed "a mechanism by which a subscriber to a branch exchange could be connected to any other on the system by an operator at a control exchange and without the intervention of an operator at the branch exchange." British patents 3380 and 5964 were granted in 1883, and patent 8541 in 1884. Sinclair's system was used in Coatbridge, Scotland, and can be considered as a precursor to the semiautomatic systems introduced around 1910. Models of the system are exhibited in London in the Museum of the Institution of Electrical Engineers and in the Science Museum in South Kensington.

George Westinghouse in the United States also applied for a patent for a semiautomatic system in 1879.

In Sweden, Henrik Tore Cedergren developed automatic switching equipment for five subscribers in 1885.

In Italy, in the autonomous Vatican State, then under the reign of Pope Leo XIII, G. B. Marzi developed and installed an automatic switching exchange in 1886. Ten telephone stations were operated for several years via this exchange in the offices of the Holy See.

In Russia, automatic telephone switching equipment was developed successively in 1887 by K. I. Mostsisky, in 1893 by S. M. Apostolov (U.S. patent in 1893), and in 1894 by M. F. Freidenberg.

In Hungary, Ferenc Puskás developed automatic telephone switching equipment that began operation on May 1, 1881.

None of the above-mentioned inventions resulted in a follow-up. For automatic telephone switching to happen, the world of telecommunications had to wait for an undertaker in the United States. The accumulation of operational and, especially, human errors connected with manual switching became a nightmare for Almon Brown Strowger (1839–1902), owner of a funeral home in Kansas City, Missouri. Strowger, borne in Penfield, New York, initially was a teacher in his hometown. He volunteered as a trumpeter for northern troops in the American Civil War from 1861 to 1865. After the war he intended to become a teacher again, but after some years he settled down as an undertaker in Kansas City. To enhance his prestige and enhance his business, he became the first person in town to have a telephone. Unfortunately, so the story goes, his major competitor had considerable political influence in Kansas City, and not only did he obtain a telephone, but he had a girlfriend working as a telephone operator at the local exchange and was a personal friend of the director of the local telephone company. As a result, Strowger got few orders, and even when a good friend of his died, the "telephone mafia" succeeded in placing the funeral arrangements in the hands of Strowger's competitor. Strowger

---

[2] They obtained U.S. patent 222,458 on September 10, 1879, only one year after the first manual telephone exchange was installed at New Haven.

looked for ways to circumvent the "hello girls." He analyzed the way the telephone operators worked. He noticed that to connect a desired telephone line (e.g., number 43) the operator would take the jack of the patch cord up to row four and move along that row to position three, where she would place the jack in the socket. From those observations, Strowger conceived of the principle of an automatic telephone exchange in which a telephone caller, by sending electrical impulses from a telephone, could direct a selector at the exchange, step by step, to the desired subscriber. Thus, when calling line 43, a selector would first move to row four and then to position three in that row.

Strowger's innovation, initiated a development in automatic switching as follows:

- *In the nineteenth century:* the Strowger electromechanical direct-control step-by-step systems (covered in Section 11.3)
- *In the first half of the twentieth century:* electromechanical indirect-control systems, also called common-control systems, and electromechanical crossbar switching systems (covered in Sections 16.3, 16.4, and 29.7)
- *In the second half of the twentieth century:* electronic switching systems and digital switching systems (covered in Sections 29.3 and 29.4)

11.3 STROWGER SYSTEM

Strowger received U.S. patent 447,918 on May 10, 1889 for his automatic telephone exchange. Patents were also applied for in Great Britain on May 6, 1891, and in Germany on June 27, 1892. In this first patent, Strowger described his automatic selector for telephone-line banks of 100 contacts arranged semicylindrically in two versions:

1. 100 contacts in 10 groups of 10 contacts arranged one group after the other in a horizontal plane. The selector arm first moves in big steps to the desired decade and then in small steps to the desired unit.
2. 100 contacts in 10 rows arranged one above the other, each row with 10 contacts. An arm moves along the contacts, first vertically to the desired row and then horizontally for the 10 units in that row.

11.3.1 Strowger's First Operating Exchange

Strowger had great difficulty both in making an operating model and in finding a manufacturer. Neither the Bell Telephone Company nor Western Electric was interested in his revolutionary device, which would dominate automatic telephone switching for more than half a century. Therefore, in 1891, with the help of an enterprising nephew, Walter S. Strowger, and with the financial support of businessman Joseph Harris from Chicago, Strowger finally decided to launch his own company, the *Strowger Automatic Telephone Exchange Company.* The first model of Strowger's selector (version 1 of his patent) was presented at the International Exhibition in Chicago in 1893, where it found great interest, especially from representatives of foreign telephone operators. Strowger's invention was publicized enthusiastically as the "girl-less, cuss-less, out-of-order-less, and wait-less telephone."

The world's first automatic telephone exchange with Strowger equipment went into operation in La Porte, Indiana, on November 3, 1892, operated by the *Cushman Telephone Company*. This independent company had replaced an affiliate of the Bell Telephone Company which earlier had installed a manual telephone exchange in La Porte. In 1890, there was a lawsuit between the two companies, and the judge ruled that the telephone equipment of the Cushman Telephone Company infringed the then-still-valid Bell patents and had to be removed. In July 1892, the municipal authorities of La Porte, which had been deprived of telephone services by this judicial ruling, allowed the Cushman Telephone Company to install another exchange, this time to be provided by the Strowger Automatic Telephone Exchange Company. Thus, La Porte got the first automatic exchange in public service in the world, or as reported by the *Chicago Herald*, "the first telephone exchange without a single petticoat."

The first exchange was manufactured with 80 selectors, for 80 telephone subscribers, and each selector had 100 positions arranged in a horizontal plane as described above as version 1 of the patent. The 100 positions of all 80 selectors needed to be through-connected by wiring similar to that on manual switchboards to enable equal access by each subscriber to any other subscriber.

At the subscriber's premises two local batteries were required: one for calling and another for speaking. Five wires, in addition to the Earth return, were needed between each subscriber and the exchange: three wires for calling, one for speaking, and one for call release. To call a subscriber, the calling subscriber had to press two buttons: one button for the decades and another for the units. The buttons were pressed the number of times equal to the value of each digit.

11.3.2   Strowger's Up-and-Around Switch

Encouraged by this first success, in 1893 Strowger extended his company making an excellent choice of collaborators, who went on to make names in the domain of automatic switching development: Frank A. Lundquist and the two brothers John and Charles Erickson. These three engineers of Scandinavian origin had attempted unsuccessfully to set up their own business and now brought their technical knowledge into the Strowger company. They first developed a 90-line experimental exchange with the line contacts arranged on a number of banks above each other, each bank positioned horizontally with 10 parallel wires. An equal number of shafts, also arranged in a horizontal plane, constituted the selecting arms. This design, known as both the *piano wire board* and the *zither*, was soon abandoned because it required a large number of relays and caused severe crosstalk between parallel wires.

In 1895, the Erickson brothers, together with a new engineer, Alexander E. Keith, began work on the first version cited in Strowger's patent. They constructed the two-motion, vertical and horizontal movement selector, that became famous as the Strowger system and is still in use in some remote exchanges after more than 100 years. Basically, this selector consists of:

- A semicircular row or bank of 10 contacts selected by the rotation of an arm or *wiper*
- A vertical arrangement of 10 of these rows, one above the other

---

TECHNOLOGY BOX 11.1

The Strowger Up-and-Around Selector

The Strowger selector consists basically of selector arms moving in front of contact banks. Figure 11.6 shows the movement control of a selector arm in front of one contact bank, with 10 levels each of 10 contacts, thus giving a total of 100 positions. Each subscriber line has three wires, two for speaking and one for signaling. The arrangement shown in the drawing corresponds to one wire only. A complete Strowger selector therefore consists of three contact banks arranged one above the other and each explored by its own wiper.

The Strowger selector is controlled directly by the dialing pulses, which the caller's dial generates for each of the various digits. The wiper with brush c (and two other wipers, not shown) are solidly connected to two ratchets:

1. A vertical ratchet connected to the armature of relay H, for an initial vertical upward movement. During this movement, brush c does not touch the contacts.

2. A half-circle ratchet wheel connected to the armature of relay D, for the subsequent rotary horizontal movement.

Thus, to select line 46, relay H attracts its armature four times and thus the three wipers are moved upward to level four of the contact banks. Then relay D activates its armature six times so that the three wipers move horizontally to contact six on the fourth level.

- A selecting arm choosing first one of the 10 rows of contacts (a *level*) by an upward movement and then one of the 10 contacts in a row by a horizontal rotary movement

A patent for this selector was applied for on December 16, 1895 and finally granted as U.S. patent 638,249 on December 5, 1899. Technology Box 11.1 gives a concise description of the Strowger up-and-around selector, and Figure 11.6 shows the mechanical arrangement of the selector.

The first exchange with this two-motion Strowger selector was brought into service, again at La Porte, in June 1895. Soon after, various other exchanges of this type, with capacities varying between 200 and 400 lines, were installed in Michigan City, Indiana; Albuquerque, New Mexico; Trinidad and Manchester, Iowa; Rochester and Albert Lea, Michigan; Albion, New York; and Milwaukee, Wisconsin.

The Strowger Company made the next improvement in the calling device in 1896, when the pushbuttons for calling were replaced by a dialing disk. This first dial, for which a patent was applied for on August 20, 1896 and granted under U.S. patent 596,062 on January 11, 1898, used 10 projecting vanes instead of fingerholes. The vanes were replaced by fingerholes some five years later. The dial had three wires,

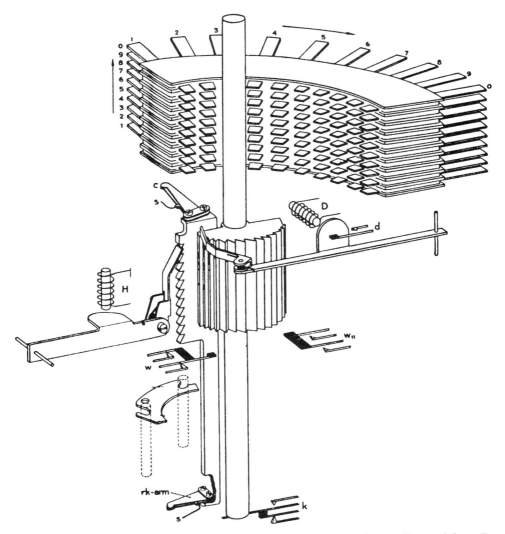

Figure 11.6 Mechanical arrangement of a two-motion Strowger selector. (Scanned from D. van Hemert and J. Kuin, *Automatische Telefonie*, 5th printing, Corps technische Ambtenaren, 1933/1953, p. 13.)

one for the hundreds digit, one for the tens digit, and the third for the units digit. The expression *dial telephone* was commonly used in the United States to signify automatic telephone service.

In 1896, subscribers connected to the private exchange of the Milwaukee Town Hall had the honor of being first in the world to be equipped with a dial telephone. The world's largest automatic exchange in the nineteenth century was put into service in Augusta, Georgia in 1897. In this 900-line exchange, two *selection levels* were introduced:

1. *Group selectors* (one for each subscriber), giving access to a group of 100 subscribers
2. *Line selectors* (one for each group of 100 lines), also called *assignment selectors* or *final selectors*, to select the required subscriber in the group of 100

Thus for each subscriber a simple 10-position one-movement group selector instead of a 100-position two-movement selector was used.

Almon B. Strowger retired from his company for health reasons in 1896. He moved to Florida, where he died February 26, 1902 in Greenwood.

REFERENCES

Books

Chapius, Robert J., *100 Years of Telephone Switching (1878–1978)*, Part 1: *Manual and Electromechanical Switching (1878–1960s)*, North-Holland, New York, 1982.

Oslin, George P., *The Story of Telecommunications*, Mercer University Press, Macon, GA, 1992.

12

RADIO TRANSMISSION

12.1 EVOLUTION LEADING TO RADIO TRANSMISSION

Radio transmission, invented at the end of the nineteenth century, has a straight line of evolution over an 80-year period. It began in 1820, when the Danish physicist Hans Christian Oerstedt discovered the electromagnetic field caused by electric current during the course of a lecture he was giving at the University of Copenhagen. The self-educated British scientist Michael Faraday, having verified Oerstedt's findings, predicted the existence of electromagnetic waves in his paper "Experimental Researches in Electricity" in 1832. In contrast to the prevailing theories about magnetism as instantaneous action at a distance, Faraday envisaged the existence of *lines of force* emanating from, or terminating on, electric charges or magnetic poles.

In 1832 the American scientist Joseph Henry, professor of natural history at Princeton College, Princeton, New Jersey, noticed that the sudden discharge of a Leyden jar caused a short current of oscillatory nature, which induced a current in a secondary circuit placed at some distance. He reported on his research to the American Philosophical Society in 1843, as follows: "The discharge of the Leyden jar is not correctly represented by the single transfer of an imponderable fluid from one side of the jar to the other; the phenomenon requires us to admit the existence of a principal discharge in one direction, and then several reflex actions backward and forward, each more feeble than the preceding until the equilibrium is obtained. A single spark on the end of a circuit in one room produced an induction sufficiently powerful to magnetize a needle in a parallel circuit of wire placed in the cellar beneath." Henry thus established the phenomenon of *induction*, without, however, drawing any immediate practical consequences. The German technician Heinrich Daniel Rühmkorff (1803–1877), in his workshop at Paris in 1851, applied the induc-

The Worldwide History of Telecommunications, By Anton A. Huurdeman
ISBN 0-471-20505-2   Copyright © 2003 John Wiley & Sons, Inc.

tion principle to generate high voltages. He constructed a coil with two windings galvanically insulated from each other: a primary winding with a low number of turns and a secondary winding with a large number of turns. An intermittent low-voltage current sent through the primary winding induced a high voltage in the secondary winding. For this coil, later called the *Rühmkorff coil*, which became the basis for future experiments with spark bridges, he received the Volta Prize from Emperor Napoleon III.

The British physicist William Thomson followed up the oscillation theory of Henry and provided the theoretical basis for the construction of oscillators with the *Thomson oscillation quotation*, which he defined in 1853. The German physicist Berend Wilhelm Feddersen (1831–1918) produced practical confirmation of Thomson's oscillation theory. Feddersen recorded sequences of photographs from sparks that were reflected on a rapidly moving concave mirror. In 1857 he submitted a paper "On the Electric Discharge of the Leyden Jar" in which he described how oscillating currents can be generated when an electrical condenser is discharged into a conductor.

An American dentist, Mahlon Loomis (1826–1886), was on the verge of inventing radio telegraphy as early as 1866 when he demonstrated the transmission of signals between two mountains in the Blue Ridges range, at a distance of 22 km. Loomis assumed that Earth was surrounded by a layer of static electricity, which he called the *static sea*, through which electrical waves could be propagated. He used a kite held by a wire loop. A rectangular copper-wire aerial of dimensions 40 by 40 cm was attached to the kite and connected with the two upper ends of the wire loop with a length of 180 m. At the lower side of the wire loop, one end was connected with Earth and the other to a galvanometer. One such arrangement, which he called an *aerial telegraph*, was placed on top of Cohocton Mountain and a second on top of Beorse Deer Mountain in October 1866. Disconnecting and connecting the wire to the galvanometer at one station caused a clear deflection on the galvanometer at the other station. Loomis repeated the demonstration two years later for scientists and members of Congress in Washington with aerial telegraphs on two ships anchored at a distance of 3 km in the Chesapeake Bay. He received U.S. patent 129,971 on July 30, 1872 for his aerial telegraph. This was the world's first patent issued for wireless telegraphy. Although Loomis obtained some financial backing and founded the *Loomis Telegraph Company* in the same year, financial misfortune, a fire, and public distrust stopped his radio activities. He became a mineralogist at the *Great Magnetic Iron Ore Company* in Mount Athos, near Lynchburg, Virginia, in 1877. Before he died, he wrote to his brother: "The time will come when this discovery will be regarded as of more consequence to mankind than Columbus's discovery of a new world. I have not only discovered a new world, but the means of invading it. My compensation is poverty, contempt, neglect, forgetfulness."

A Scottish physicist, James Clerk Maxwell (1831–1879), subjected Faraday's hypothesis to a rigorous mathematical analysis and summarized the various concepts of electricity known at his time and widened them substantially, resulting in a comprehensive theoretical system of electrodynamics that comprised electric and magnetic phenomena. In his paper "A Dynamical Theory of the Electro-Magnetic Field," presented to the Royal Society on December 8, 1864, he presented mathematical formulas for propagation of the hypothetical electromagnetic waves. In 1873 he published the two-volume *Treatise on Electricity and Magnetism*, in which he formulated a theory that electromagnetic waves are of the same nature as light, differing only in wavelength, and thus would have the same typical characteristics as to propagation speed, polarization, reflection, and refraction. By purely mathematical reasoning, Maxwell claimed that all electrical and magnetic phenomena could be reduced to stresses and motions in a medium which he called the *ether*. An Irish physicist, George Francis FitzGerald (1851–1901), professor at Trinity College, Dublin, was the first to strongly support Maxwell's theory; in contrast, William Thomson did not agree with the theory. FitzGerald presented his supporting ideas to the Royal Society in January 1879 in a paper entitled "On the Electro-Magnetic Theory of Reflection and Refraction of Light," in which he showed concurrence of Maxwell's theory with the latest theories on light. He also proposed a method by which electromagnetic waves might be produced by discharging a condenser. A British physician, Oliver Joseph Lodge, and others began experiments to investigate the validity of that theory. In Germany, the *Academy of Science* in Berlin introduced a competition in 1879 to prove particular aspects of the validity of Maxwell's theory, a contest finally won by Heinrich Hertz in 1889. Hertz made a series of experiments in the period 1886–1889 in which he proved the existence of electromagnetic waves and their commonality with light. Within 10 years after Hertz had proved the existence of electromagnetic radiation, Marconi invented the radio.

12.2   EXPERIMENTS OF HEINRICH HERTZ

Heinrich Rudolf Hertz (1857–1894) was born in Hamburg on February 22, the eldest son of G. F. Hertz, a prominent lawyer in a family of successful merchants. His mother was the daughter of a physician. After two years of civil engineering apprenticeships in Frankfurt and Dresden and military officer's training in Berlin, at the age of 20 he went to the Polytechnic in Munich to pursue a career as a civil engineer. Going through the study syllabus, he obtained the impression that civil engineering would not satisfy his personal abilities, and after consulting his father,[1] he decided to devote his time to the study of natural science. One year later he went to Berlin, where he studied under the already famous physicians Hermann von Helmholtz (1821–1894) and Gustav Kirchhoff (1824–1887). In 1880 he obtained a doctoral degree magna cum laude on the subject of magnetic induction in rotating balls. In the same year he was appointed assistant to von Helmholtz, who made Hertz aware of the necessity to prove Maxwell's theory that electromagnetic forces are propagated through space with the same speed as light, and that, in fact, light itself is an electromagnetic phenomenon. An integral component of Maxwell's theory asserted that an electromagnetic force would give rise to a displacement current in any nonconductor subjected to an electromagnetic field. Confirmation of this phenomenon would be an important step toward a proof of Maxwell's theory. The *Academy of Science* in Berlin therefore offered a prize for research on this phenome-

---

[1] As a striking example of the paternal respect prevailing at that time, here is a quotation from Heinrich's letter to his father seeking his approval to change from civil engineering to natural sciences: "... and so I am asking, dear Papa, not so much for your advice as your decision ... if you tell me to study natural science, I shall take it as a great gift from you and I shall do so with all my heart. I do believe that this will be your decision, ... but if you consider it best for me to continue along the road on which I have set out but in which I do no longer believe, I shall do that, too, and do it without reservation."

non. Helmholtz invited Hertz to undertake such research. Hertz realized, however, that neither the required means of detecting the presence of electromagnetic waves nor the apparatus to generate the required high frequencies was available, but he kept the challenge in his mind. In 1883 he became a lecturer in theoretical physics at the University of Kiel. One year later he published his first paper related to electromagnetism, entitled "On the Relations between Maxwell's Fundamental Electromagnetic Equations and the Fundamental Equations of the Opposing Electromagnetics," in which he claimed the superiority of Maxwell's theory over the "opposing action-at-a-distance" theory, which still prevailed in continental Europe. The paper brought him offers of three professorships. Attracted by the facilities of the laboratories, he accepted an appointment as a professor of experimental physics at the oldest German technical high school at Karlsruhe (founded in October 7, 1825, as Polytechnic School, renamed Technical High School in 1885), where he succeeded Karl Ferdinand Braun. Hertz began lectures on electrotechnology and meteorology in Karlsruhe on April 1, 1885. The next year he married Elisabeth Doll, daughter of a professor of geodesy at the same school. Among the facilities at Karlsruhe he found a pair of *Knockenhauer Spirals* (coils of wire embedded in spiral tracks cut into the surface of circular wooden disks, also called *Reiss spirals*). While experimenting with those coils in the autumn of 1886, he observed a small spark passing between the terminals of one of the coils whenever he discharged a Leyden jar through the other coil. He realized that this phenomenon was due to the occurrence of oscillatory currents at a very high frequency induced in the spiral. This encouraged him to start a series of experiments in response to the academy's competition.

Hertz made his decisive experiment in 1888 and reported his success to the academy (paraphrased here in a much shortened form) as follows:

I constructed a mirror by bending a zinc sheet 2 m long, 2 m broad, and 0.5 mm thick into the desired parabolic shape over a wooden frame of the exact curvature. The height of the mirror was thus 2 m, its aperture 1.2 m and its depth 0.7 m. A primary oscillator was fixed in the middle of the focal line. The wires which conducted the discharge were led through the mirror; the induction coil and the battery cells were accordingly placed behind the mirror so as to be out of the way. I then constructed a second mirror, exactly similar to the first, and attached a rectilinear secondary conductor (as resonator) to it in such a way that two wires of 50 cm length lay in the focal line, and the two wires were connected to a sparkgap behind the mirror so that the observer could adjust and examine the sparkgap without obstructing the course of the waves.

The primary oscillator was supplied with current from three accumulators via the induction coil, and gave sparks 1–2 cm long. The small sparks induced in the secondary conductor were the means used for detecting the electric force in space. In the rooms at my disposal I could perceive sparks behind the second mirror up to a distance of 16 m between the two mirrors.

From the mode in which our ray was produced and the reflection and polarization experiments made, we can have no doubt whatever that it consists of *transverse electric and magnetic oscillations: those in the vertical plane are of an electrical nature, while those in the horizontal plane are of a magnetic nature.*

The experiments described appear to me, at any rate, eminently adapted to remove any doubt as to *the identity of light, radiant heat, and electromagnetic wave motion.*

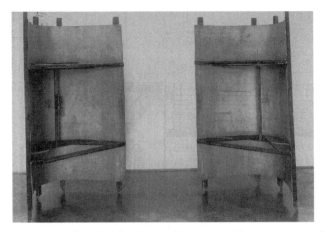

Figure 12.1 Experiments of Heinrich Hertz. (Scanned with permission of the ITU from Anthony R. Michaelis, *From Semaphore to Satellite*, International Telecommunication Union, Geneva, 1965, p. 118.)

Hertz summarized the fact that the object of his experiments was to test the fundamental hypotheses of the Faraday–Maxwell theory, and that the result of the experiments confirmed the hypotheses and terminated the old action-at-a-distance philosophy. The first radio-relay antennas made by Heinrich Hertz (Figure 12.1) still exist and are exhibited in the *Deutsche Museum* in Munich together with devices for proving the propagation, reflection, refraction, and polarization of the electromagnetic waves, and the radiation pattern of the electromagnetic waves as drawn by Hertz.

The sparks produced by the waves received at the second mirror were extremely small, and Hertz reported: "It appears impossible, almost absurd, that these tiny sparks should be visible, but in a completely dark room they are visible to a relaxed eye." How difficult it must have been in those days to detect the electromagnetic waves becomes obvious when reading in the report to the academy: "Acting on friendly advice, I have tried to replace the sparkgap in the secondary conductor by *a frog's leg prepared for detecting currents*: but this arrangement, which is so delicate under other conditions, does not seem to be adapted for these purposes."

The electric sparks from the primary oscillator radiated trains of damped electromagnetic waves with a wavelength of around 60 cm (thus at a frequency of about 500 MHz). Hertz thus discovered that electromagnetic waves with a very short wavelength can—like light—be directed in a narrow beam toward a receiving station that is in a direct line of sight and sufficiently close for the signal to remain strong enough to be detectable.

In addition to his report to the academy, Hertz presented a detailed description of his experiments in a paper entitled "Electromagnetic Waves in Air and Reflection," published in May 1888. In this paper, Hertz gave due credit to the achievements of Lodge and Fitzgerald, stating:

> I may here be permitted to record the good work done by two English colleagues who at the same time as myself were striving towards the same end. In the same year in which I carried out the above research, Professor Oliver Lodge, in Liverpool, investigated the

theory of the lightning conductor, and in connection with this carried out a series of experiments on the discharge of small condensers which led him to the observation of oscillations and waves in wires. Inasmuch as he entirely accepted Maxwell's views and eagerly strove to verify them, there can scarcely be any doubt that if I had not anticipated him he would have succeeded in observing waves in air, and thus also in proving the propagation with time of electric force. Professor Fitzgerald, in Dublin, had some years before endeavored to predict, with the aid of theory, the possibility of such waves, and to discover the conditions for producing them.

The discovery of electromagnetic waves was well received by the scientific world and hailed as the settlement of a great scientific controversy: the confirmation of Maxwell's theory and disproving the action-at-a-distance theory. Hertz gained high national and international recognition. In Germany he was offered an appointment as professor of theoretical physics at the University of Berlin or another for experimental physics at the University of Bonn. Still being interested in experiments, he accepted the appointment at Bonn in 1889. In the same year he was accepted into the exclusive circle of Corresponding Members of the Royal Prussian Academy of Science and received the Prix Lacaze of the Académie des Sciences, Paris. The next year he received the Rumford Medal from the Royal Society in London and the Mateucci Prize of the Societá delle Scienze in Napoli, followed in 1891 by the Bressa Prize from the Academy of Torino, in Italy.

Following the discovery in 1888, considerable technological progress was still required, especially in the generation of very high frequencies, before practical radio transmission systems could be realized. Hertz did not even expect that electromagnetic waves could ever be used for transmission of the human voice, which in fact was not possible with spark-generated damped electromagnetic waves. In Bonn, Hertz left the study of electric waves to others and returned to investigations on the discharge of electricity in rarefied gases and then turned his attention to a treatise on mechanics. But he could not complete that treatise. In the summer of 1892, he suffered from severe sinusitis, which eventually led to chronic blood poisoning, of which he died on January 1, 1894.[2]

12.3   RADIO TRANSMISSION FROM THEORY TO PRACTICE

Following the discovery of electromagnetic waves was another period of basic investigations. Verifying repetitions of Hertz's experiments were made by Edouard Sarasin (1843–1917) and Lucine de la Rive (1834–1924) in Geneva, by Antonio Giorgio Garbasso (1871–1933) and Emil Aschkinass (1873–1909) in Berlin, by Jagadis Chunder Bose (1858–1937) in Calcutta (transmitting 2.5 km over the River Hoogly in 1898), by Ernst Rutherford (1871–1931) first in New Zealand and in 1895 at Cambridge, and by Augusto Righi (1850–1920) in Bologna.

---

[2] Hertz left a young widow and two very young daughters. In 1938, for political reasons, Mrs. Hertz, together with her daughters Johanna, a doctor of medicine, and Mathilde, a doctor of physics, escaped to the U.K., where they settled in Cambridge. They took numerous original manuscripts of Heinrich Hertz with them which are now part of the National Collection in the Science Museum in London. Other major documents concerning Heinrich Hertz are at the Deutsche Museum in Munich. Mrs. Hertz died in London in 1941.

The first experiments with electromagnetic waves reportedly were made by Tesla in 1889 in the United States. Nicola Tesla (1856–1943), born at Smilian, Yugoslavia, after studying at Graz and Prague and working at telegraph companies at Budapest and Paris, came to the United States in 1884, where for some time he worked for Edison. He left Edison because of discrepancies about an award for an invention and in 1889 opened his own laboratory doing high-frequency and high-tension projects. Two years later he produced the *Tesla transformer* for high voltages and began construction on high-frequency radiation stations. The Tesla transformer was a dynamo with 384 poles, which made 1600 rotations per minute and thus generated a frequency of $384 \div 2 \times 1600 \div 60 = 5100$ Hz. He installed two such stations up to 30 km apart with the intention of achieving wireless electrical energy transportation between stations instead of using high-voltage overland lines. Experiments at this still rather low frequency were not successful, and Tesla stopped them without considering starting experiments for signal transmission.

A first practical result came from Edouard Eugène Desiré Branly (1844–1940), born in Amiens, France. He became a doctor of physics and medicine and was successively professor of physics at the Lyceum of Bourges, at the Sorbonne in Paris, and from 1866 at the Institute Catholique in Paris. In 1891 he rediscovered the cohesion effect on small particles under the influence of electricity.[3] Based on this effect he constructed a device, later called a *coherer*, for detecting electromagnetic waves, consisting basically of a tube filled with iron filings which coalesced and thereby substantially reduced their resistance upon being subjected to electromagnetic waves. A galvanometer in series with a coherer and a battery thus showed the presence of an electromagnetic wave. Branley demonstrated the coherer for the *Academy of Science* in Paris and published his results in *La Lumière Èlectrique* in May 1891. Like Hertz, however, Branley was a pure scientist who was not interested in practical applications. But the coherer was of great importance in the first decade of radiotelegraphy, and Branley received the Nobel Prize in Physics in 1921.

The first prediction on the use of electromagnetic waves for telegraphy was made by William Crookes (1832–1919, Sir William after 1897) in an article "Some Possibilities of Electricity" in the *Fortnightly Review* of February 1892, predicting "... the bewildering possibility of telegraphy without wires, posts, cables, or any of our present costly appliances."

The next important event was a lecture given on June 1, 1894 to the Royal Institution in London by Oliver Joseph Lodge (1851–1940). Lodge, born at Penkhull, Staffordshire, was the first of nine children; his grandfather had 25 children, and he had 12 children. At the age of 30 he became professor of experimental physics at the University College, Liverpool. In the aforementioned lecture, Lodge paid homage to Hertz, who had died five months earlier. In his lecture he presented mainly the results of his own experiments for detection of Hertzian waves in particular with a practical form of Branly's coherer.[4] Lodge coined the name *coherer* (from the Latin *cohaere* = stick together) for this device. The cohesion of the iron filings in Branley's coherer, when subjected to electromagnetic waves, allowed the passage of current from an

---

[3] The cohesion effect had been noticed by Guitard in 1850, by Samuel Alfred Varley in 1866, by Lord Rayleigh in 1879, and by Calzecchi Onesti in 1884, but had not yet been applied.

[4] Lodge had already discovered a device for detecting electromagnetic waves in 1889 and used it in various devices constructed in 1891–1892. In 1893, however, he found that Branly's coherer performed better.

auxiliary power supply to operate a relay that reproduced transmitted Morse signals. The coherer had to be tapped, however, after each electromagnetic signal to separate the filings and prepare them to react to the next electromagnetic signal. Lodge improved the coherer by adding a device that shook the fillings between spark receptions and prepared them to react to the next radio-frequency signal. His lecture, including additional material, was published and widely disseminated in a book entitled *The Work of Hertz and Some of His Successors*. The lecture was attended by a telegraph engineer, Alexander Muirhead, who also recognized the practical value of electromagnetic waves for the transmission of telegraphic signals. He provided Lodge with telegraphic equipment, such as Kelvin's highly sensitive siphon galvanometer, and a Morse key for a repetition of his lecture to be given to the British Association for the Advancement of Science at Oxford in August 1894. Three years later, Lodge, with his understanding of the phenomenon of electrical resonance, introduced the principle of selective tuning to a common frequency for a transmitter and corresponding receiver by variation of the inductance of the oscillating circuits, which he called *syntony*. His patent 11,575, for which he applied in May 1897, served as the fundamental basis of all future radio equipment. Lodge could have gone into history as the inventor of radiotelegraphy, but like Hertz and Branley, he did not envisage a commercial application.

In Russia, Alexander Stepanovitch Popoff (1859–1906), born the son of a priest, received education in an ecclesiastical seminary school and planned to become a priest but changed his interest to mathematics and entered the University of St. Petersburg in 1877. He graduated with distinction in 1883 and joined the teaching faculty of the university to lecture in mathematics and physics in preparation for a professorship. His interest changed to electrical engineering, which, however, was not taught at colleges, so he became an instructor at the Imperial Naval Torpedo School at Kronstadt, near St. Petersburg. After reading of Lodge's experiments, he began experiments with radio in 1895. He presented a paper "On the Relation of Metallic Powders to Electric Oscillations" at a meeting of the *Russian Physical-Chemical Society* on April 25, 1895. In July of the same year, assuming that thunderstorm lightning should radiate electromagnetic waves, he connected a coherer attached to an ink recorder with a lightning conductor on the roof of the Institute of Forestry in St. Petersburg and detected the occurrence of thunderstorms at ranges up to 50 km. Figure 12.2 shows the receiving equipment, which he called *an apparatus for the detection and registration of electrical oscillations*. He described the results of his experiments and the means he used for recording them in a long paper addressed to the *Journal of the Russian Physical-Chemical Society* in January 1896. He ends that paper with the words: "In conclusion I may express the hope that my apparatus, when further perfected may be used for the transmission of signals over a distance with the help of rapid electric oscillations, as soon as a source of such oscillations possessing sufficient energy will be discovered." Two months later, on March 12, he gave another demonstration before the same society, but then turned his interest to Roentgen rays, which were discovered in that year. Substantial confusion exists regarding this demonstration. Some sources claim that at that meeting Popoff transmitted the words "Heinrich Hertz" in Morse code with his apparatus over a distance of 250 m. In the "conclusion," quoted above, however, Popoff clearly expresses his hope that his apparatus may be perfected for transmission of signals if a source of oscillations possessing sufficient energy were to be discovered, which makes it

Figure 12.2   Popoff's first radio receiver, 1895. (Scanned with permission of the ITU from Anthony R. Michaelis, *From Semaphore to Satellite*, International Telecommunication Union, Geneva, 1965, p. 121.)

unlikely that he discovered that source and perfected his apparatus in the following two months. To clarify this discrepancy, Charles Susskind of the University of California carried out exhaustive investigations into all the relevant contemporary records. He presented his results in a long paper entitled "Popoff and the Beginnings of Radiotelegraphy" in the *Proceedings of the Institute of Radio Engineers*, Vol. 50, in 1962. In this paper he draws the conclusion that the records show that Popoff did not transmit intelligence at the demonstration on March 12, nor on any other occasion before mid-1896. The reference to mid-1896 was important because by that time Guglielmo Marchese Marconi had transmitted radio signals successfully over a few kilometers.

12.4   THE RADIO INVENTED BY MARCONI

Guglielmo Marchese Marconi (1874–1937) was born the son of wealthy cosmopolitan parents at their Palazzo Marescalchi in Bologna. His Italian father was a landowner and silk merchant, his Scottish-Irish mother, born Annie Jameson, was related on her father's side to the Irish whiskey distillers Jameson and on her mother's side to the Scottish whiskey distillers Haig. Guglielmo was educated first in his native town and then in Florence. He failed the entrance exams to the Italian Naval Academy and to the University of Bologna, so he went to a technical school and got further education at home. Fortunately, a neighbor was Augusto Righi, one of the first to experiment with shorter wavelengths than those used by Hertz. Righi lectured on Hertzian waves at the University of Bologna. To this purpose Righi constructed a spark bridge with three sparkgaps in a row, with four spheres, two large ones in the middle and one half as large at each side at adjustable distances. Righi allowed

Marconi access to his lectures and to some of his instruments. In 1894, Marconi started to experiment at his father's estate, Villa Grifone, near Pontecchio, Bologna. He used a threefold spark bridge from Righi with an induction coil operated by a Morse key at the transmitting side and detected with a homemade coherer. He transmitted dots and dashes all across the room and later in the garden. Marconi's coherer consisted of an evacuated glass tube with a length of 6 cm and a diameter of 4 mm, containing two silver electrodes with a 0.5-mm gap filled with fine filings of 96% nickel and 4% silver. To increase the performance, he attached an elevated metal object to one end of the spark bridge, which thus functioned as an antenna, and to the other end he connected a metal plate buried in the ground. The coherer was connected similarly with an antenna and a ground plate. This grounding was an original idea from Marconi, so far not used in any of the above-mentioned experiments. He placed reflectors around the antenna to concentrate the radiated energy into a beam, and he improved his apparatus, which increased the transmission range of the radio signals to about 2.4 km by 1896. His brother Alfonso assisted Guglielmo and used his gun to confirm the good receipt of signals.

The Italian Ministry of Post did not believe that the untried radio technology invented by Marconi could improve on the established telegraph. In February 1896, therefore, Marconi left for England accompanied by his mother, who had influential relatives in London, in particular her elder sister's son, Colonel Henry Jameson-Davis. After a few months in a private hotel, they settled at 21 Burlington Road, St. Stephen's Square. Within a few days of his arrival, Marconi made a patent application that was filed at the patent office on March 5, 1896 as number 5028, entitled "Improvements in Telegraphy and in Apparatus Therefore." Apparently, something went wrong with this application, and a second application was submitted on June 2 with the assistance of Fletcher Moulton, a distinguished patent lawyer, and paid for by Colonel Jameson-Davis. This application was accepted and patent 12,039/1896 became the first patent for wireless telegraphy ever granted. In the meantime, certainly with the support of Colonel Jameson-Davis, two letters were written:

- On March 30, a letter written by A. A. Cambell-Swinton, a leading consulting electrical engineer, to William Henry Preece, then chief engineer of the Post Office, in which he introduces the young Italian as somebody "who has got considerably beyond others in the field of Hertzian waves"
- On May 20, a letter written by Marconi to the Secretary of State for War Affairs, in which he offers his device as a system for radio control of torpedoes or other unmanned vessels

Marconi met Preece and was given the opportunity to demonstrate his apparatus to officials of the Post Office, the Navy, and the Army in the summer. A field trial was made between two Post Office buildings in the City of London, and on distances up to 2.8 km on Salisbury Plain in September. The decisive field trial took place in May 1897, over a distance of 5.3 km between Lavernock Point, 8 km south of Cardiff, and the island Flat Holme in the Bristol Channel. This location was chosen because a reliable means of communication for coordination of the test existed between those two points in the form of a parallel-wire system developed earlier by Preece. A large drum of sheet zinc was connected on top of a 35-m mast at both sites. These drums, with dimensions of 1.8 m length and 0.9 m diameter, functioned as

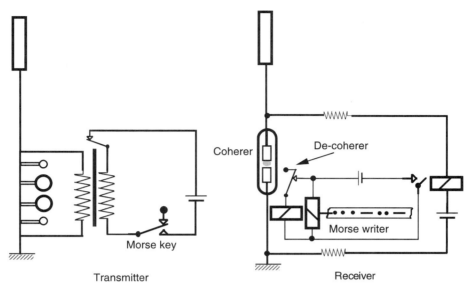

Figure 12.3 Transmitter and receiver of the Bristol Channel trial, May 1897.

antennas and were connected by gutta-percha-covered aluminum wire to the transmitting apparatus at Flat Holme and to the receiving apparatus at Lavernock Point. Figure 12.3 shows a circuit diagram of the trial arrangement, and Figure 12.4 shows the transmitting apparatus at Flat Holme. The system operated at a frequency of about 2.5 MHz, with a transmitting power between 10 and 20 W. The two coils shown in the receiver circuit, then called *inductance rolls*, prevented the high-frequency energy from the antenna flowing off through the battery. The decoherer actually was an electric bell without bells and with a little hammer connected to the automatic interrupting armature.

Figure 12.4 Marconi's transmitter, 1897. (Scanned with permission of the ITU from Anthony R. Michaelis, *From Semaphore to Satellite*, International Telecommunication Union, Geneva, 1965, p. 122.)

The material was partly provided by the Post Office and partly by Marconi. The trials were prepared and carried out by Post Office engineers in cooperation with Marconi. The official trial took place on May 13, and 14, in the presence of William Preece and his principal technical officer and successor, John Gavey; Major C. Penrose of the War Office; Major G. A. Carr and Captain J. N. C. Kennedy of the Royal Engineers; Viriamu Jones of the University College, Cardiff; and Adolf Slaby. Slaby, a professor at the University of Berlin, was reluctantly invited by the Post Office upon personal request of the German Kaiser Wilhelm II. The reception of the signal, the Morse code "V" (three dots and a dash), was good. Encouraged by the good results, another test was made across the Bristol Channel over a distance of 14.5 km from Lavernock Point as transmitting station, to Brean Down on the Somerset coast as receiving station, on May 18. A wire attached to a kite that was connected to an approximately 100-m mast was used as an antenna at Brean Down.

The tests were convincing and on June 4, Preece presented to the royal institution the results of the tests, together with a basic explanation of Marconi's wireless telegraph system, which attracted considerable publicity, both in England and abroad. Impressed by those results, the Italian government extended Marconi an immediate invitation, and in the same month, tests were made with Marconi's radio between a land station at La Spezia and Italian warships at distances up to 19 km.

Back in England, Marconi proposed his patent to the Post Office. Obviously upon instructions of his superiors, Preece surprisingly offered a bargain price of £10,000 only. This poor offer persuaded Marconi to follow his cousin's advice, and on July 20 he founded the *Wireless Telegraph and Signal Company Ltd.*,[5] based in London, with Jameson-Davis as its first managing director. The secretary of the Post Office and his legal advisers disagreed that the Post Office should support a private company in developing an invention, and Preece was formally instructed to refrain from further experiments with Marconi. From that moment, the relation between the British Post Office and Marconi was strained for many years. Looking for an alternative source, the Post Office made another series of tests across the Bristol Channel in the period from October 1899 to April 1900 with a detector made by the Hungarian inventor Bela Schaefer and improved the apparatus for their own use.

Deprived from his contact with the Post Office, and under the Telegraph Act of 1863 forbidden to operate a revenue-earning service in Britain or in British territorial waters, Marconi looked for alternative solutions. Fortunately, the Telegraph Act allowed private companies to send messages for its own use and for the use of others as long as no direct charge was made for messages handled. Under those conditions the *Dublin Daily Express* used his equipment for on-the-spot reporting on the Kingston yacht race in 1898. A reporter on board the *Flying Huntress* in the Irish Sea telegraphed its observations via Marconi's transmitter to the shore at Kingston, whence they were telephoned to the newspaper's Dublin office. The next occasion was a knee injury of the Prince of Wales. The Prince cured his knee on a royal yacht in Cowes Bay and kept daily contact via Marconi radio with his mother, Queen Victoria. Figure 12.5 shows Marconi with his transmitter in 1898 with one spark bridge only. Communication beyond British territorial waters came in 1899, when Marconi's equipment was used successfully for wireless telegraphy across the English Channel over a distance of 51 km between South Foreland in England and Wimer-

[5] The name of the company was changed to Marconi's Wireless Telegraph Company Ltd. in 1900.

Figure 12.5 Marconi with his transmitter, 1898. (Scanned from *Archiv für deutsche Post-geschichte*, Vol. 2, 1988, p. 166.)

eux in France (Figure 12.6). In the same year, British battleships exchanged messages via Marconi equipment over distances up to 121 km. In September of that year, Marconi equipped two U.S. ships off Sandy Hook, New Jersey, to report to newspapers in New York about an international yacht race for the America's Cup. This achievement attracted worldwide attention and encouraged Marconi to found the *Marconi Wireless and Telegraph Company of America* with support of British financiers on November 22, 1899. In the same year the American liner *St. Paul* was the first ship to be equipped with radio. The U.S. Navy installed Marconi radio on a shore station at Babylon, Long Island, on the battleship *Massachusetts*, on the cruiser *New York*, on the torpedo boat *Porter*, and on the SS *Philadelphia* in 1900.

Based on the experience above, Marconi made various improvements to his radio, including frequency tuning and the inductive coupling of the spark circuit to the antenna (which, however, had already been patented in Great Britain by the German physicist Braun, described below) and obtained his famous patent 7777 on April 26, 1900.[6] In that year, according to an ITU source, the world's first permanent radio service was opened connecting the Sandwich Islands to Hawaii. In the same year, on May 15, the first public radiotelegraphy service in Europe was opened with ocean steamers en route to or from the harbor of Bremenhaven, Germany. For this service, Marconi equipment was installed on a light tower on the island of Borkum, and at a distance of 35 km on the lightship *Borkum Riff*, which was located near the steamship route.

The first radio emergency call from a ship was sent on April 29, 1899, when the captain of the British lightship *East Goodwin* telegraphed over a distance of 22 km to

[6]The U.S. Supreme Court overturned U.S. patent 7777, indicating that Lodge, Tesla, and John Stone appeared to have priority in the development of radio-tuning apparatus.

Figure 12.6   Antenna for crossing the British Channel, 1899. (Scanned from *Archiv für deutsche Postgeschichte*, Vol. 2, 1988, p. 166.)

the light tower on South Foreland: "We have just been run into by the steamer *RF Matthews* of London. Steamship is standing by us: our bows very badly damaged."

12.5 RADIOS OF MARCONI'S COMPETITORS

Information about the successful operation of Marconi's radio was an incentive for many scientists and engineers to obtain similar or better results. In Russia, Popoff, one of the first, improved his apparatus and made wireless ship-to-shore telegraphy over a distance of 10 km for the Russian Navy in 1898. The next year, he visited wireless stations in France and Germany and together with the French engineer Ducretet, developed radiotelegraphy equipment which enabled telegraphy over a range up to 50 km. Popoff found little support from the Russian government and returned to St. Petersburg as a professor at the *Electro Technical Institute*, of which he later became the director.[7]

---

[7] In Russia, on May 7, 1945, the fiftieth anniversary of the invention of the radio by A. S. Popoff was celebrated in the Bolshoi Theater and it was announced that in the future, May 7 would be celebrated as "Radio Day."

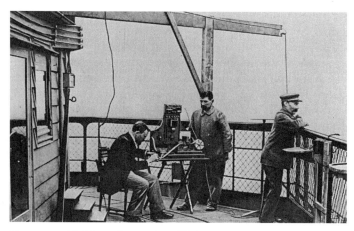

Figure 12.7 Ducretet with his radio on the Eiffel Tower, 1898. (Scanned with permission of the ITU from Anthony R. Michaelis, *From Semaphore to Satellite*, International Telecommunication Union, Geneva, 1965, p. 126.)

In France, Eugène Ducretet constructed radio equipment which he used for experiments in the autumn of 1898. He transmitted radio signals from the Eiffel Tower which were received near the Panthéon on November 5, 1898. Figure 12.7 shows the station on the Eiffel Tower.

In Germany, Adolf K. H. Slaby (1849–1913), encouraged by Kaiser Wilhelm II and with the assistance of Count George Wilhelm Alexander Hans von Arco (1869–

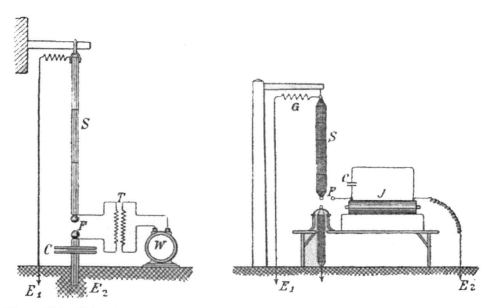

Figure 12.8 Slaby–Arco transmitter. (Scanned from *Archiv für Post und Telegraphie*, No. 8, April 1901, p. 255.)

1940), developed the Slaby–Arco radio system. Slaby, born the son of a bookbinder at Berlin, studied at the Trade School in Berlin, lectured in mathematics and mechanics at the Trade School in Potsdam, and became the first professor in 1883 and later the director of the newly established field of electrotechniques at the Technical University in Berlin. The Slaby–Arco system used a spark oscillator, a coherer, and a Morse telegraph writer. It operated at about 250 kHz. Figure 12.8 shows two versions of the transmitter, at the left with a dynamo as spark generator and at the right with a high-voltage inductor, which included an automatic interrupter if connected to a dc source. Contrary to Marconi, who installed the antenna isolated from Earth, Slaby and Arco connected the upper end of the antenna to Earth via a coil. This coil (G) ensures that during charging with the low-frequency current, all the energy is accumulated in the capacitor (C), whereas for the high-frequency spark current the coil presents a good insulation from Earth so that a maximum of energy is radiated. The receiver was connected to the same antenna arrangement in which the coil reduced the influence of atmospheric disturbances. An official demonstration attended by Kaiser Wilhelm II was given in Berlin over a distance of 1.3 km on August 27, 1897. Production of this radio system started in 1898 in the Radio-Telegraphy Department, established by Slaby and Arco within the Allgemeine Elektricitäts-Gesellschaft (AEG, founded in 1887 from the Deutsche Edison Gesellschaft, which was founded in 1883). Experiments were made with transportable stations on land and on warships. With a 300-m-long wire as antenna connected to a balloon, a distance of 21 km was achieved in 1897 and up to 60 km one year later.

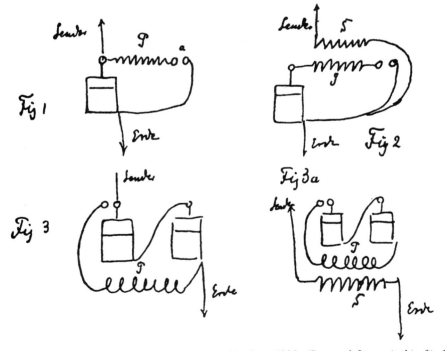

Figure 12.9 Drawing from Braun's patent application, 1898. (Scanned from *Archiv für Post und Telegraphie*, No. 8, April 1901, p. 561.)

Up to 48 km could be covered with an antenna wire length of 30 m. In the nineteenth century, Slaby–Arco concentrated their activities mainly on military applications.

The German physicist Ferdinand Braun (1850–1918), professor of physics at the University of Strasbourg, made an important improvement in 1898. Instead of connecting the antenna directly to a spark oscillator, he developed a spark oscillator circuit which was connected to the antenna inductively instead of galvanically. This substantially delayed the damping of the transmitter. Figure 12.9 shows the original drawing which Braun made for his patent application. Figures 1 and 3 correspond to the direct connection to the antenna as used by Marconi, Fig. 3a shows the inductive connection, and Fig. 2 shows a combination of both solutions.

The coherer circuit of the receiver was also connected to the antenna by means of a transformer that improved the efficiency of the coherer. Braun received German patent 111,578 for "Telegraphy without Directly Connected Wire" on October 14, 1898 and British patent 1862 on January 26, 1899. With this solution Braun increased the range of the transmitter by at least a factor of 3 over that of the open spark oscillator circuit of Marconi. Braun made a further improvement in 1899 when he replaced the coherer by a crystal detector. Successful tests were made in 1899 over a distance of 62 km between Cuxhaven and the island of Heligoland. Braun founded the *Funkentelegraphie GmbH* at Cologne in 1898, and the next year *Prof. Braun's Telegraphie GmbH*, in Hamburg, also called Telebraun. Both companies became part of the *Gesellschaft für drahtlose Telegraphie System Prof. Braun und Siemens & Halske*, founded in Berlin on July 27, 1901.

REFERENCES

Books

Garratt, G. R. M., *The Early History of Radio from Faraday to Marconi*, Institution of Electrical Engineers, London, 1994.

Gööck, Roland, *Die großen Erfindungen Radio: Fernsehen Computer*, Sigloch Edition, Künzelsau, Germany, 1988.

Hugill, Peter J., *Global Communications since 1884: Geopolitics and Technology*, Johns Hopkins University Press, Baltimore, 1999.

Huurdeman, Anton A., *Radio-Relay Systems*, Artech House, Norwood, MA, 1995.

Magie, William Francis, *A Source Book in Physics*, McGraw-Hill, New York, 1935.

Michaelis, Anthony R., *From Semaphore to Satellite*, International Telecommunication Union, Geneva, 1965.

Nesper, Eugen, *Die drahtlose Telegraphie*, Verlag von Julius Springer, Berlin, 1905.

Oslin, George P., *The Story of Telecommunications*, Mercer University Press, Macon, GA, 1992.

Articles

Beyrer, Klaus, Hertz und die Anfänge der Funkentelegrafie, *Archiv für deutsche Postgeschichte*, Vol. 2, 1988, pp. 155–168.

Blanchard, Julian, Hertz, the discoverer of electric waves, *Bell System Technical Journal*, Vol. 17, July 1938, pp. 327–337.

Jentsch, Die Fortschritte der Funkentelegraphie, *Archiv für deutsche Postgeschichte*, No. 8, April 1901, pp. 251–422.

von Ellisen, Hans-Joachim, Drahtlose Telegrafie mit gedämpften Wellen, *Archiv für deutsche Postgeschichte*, Vol. 2, 1993, pp. 25–52.

von Kniestedt, Joachim, 100 Jahre Funkpatent von Ferdinand Braun, *Archiv für deutsche Postgeschichte*, Vol. 1, 1999, pp. 47–49.

13

INTERNATIONAL COOPERATION

Telegraphy was initially a national affair whereby each country used its own equipment and procedures. Early telegraph messages arriving at a country's border needed to be set down on paper and reentered manually in the telegraph network of the neighbor country. A first agreement for electrical border passing of telegraph messages was made between Austria–Hungary and Prussia on October 3, 1849. The agreement settled technical interface conditions, including an improved Morse alphabet (developed by Friedrich C. Gerke; Section 8.10) as a common standard, as well as rates and procedures so thoroughly that it was used as an example for similar agreements between many other European countries.

The bilateral agreement between Austria–Hungary and Prussia was soon followed by similar agreements between Prussia and Saxony in 1849 and between Austria and Bavaria in 1850. These four states went a step further and created the *Austrian–German Telegraph Union* in Dresden in 1850. The telegraph network of the union (Figure 13.1) was 6870 km long and had a total wire length of 7870 km; thus most lines still used Earth as the return conductor. All lines in Austria were overhead lines. Some of the lines in Prussia and Saxony were underground but suffered serious insulation problems and had to be replaced by overhead lines in 1851.

The German state of Wuerttemberg joined the union in 1851, followed by the Netherlands and the German state of Hannover in 1852 and the German states of Baden and Mecklenburg in 1854. At a meeting of the union in Vienna in 1855, it was decided that the international telegraph lines should be physically through-connected, or otherwise, two Morse telegraphs would operate in an automatic back-to-back repeater version, thus getting rid of the manual reentering of messages at the frontiers. Other meetings followed; in Berlin in 1853, in Munich in 1855, and in Stuttgart in 1857. At this last conference a sensible step was taken which laid the

The Worldwide History of Telecommunications, By Anton A. Huurdeman
ISBN 0-471-20505-2   Copyright © 2003 John Wiley & Sons, Inc.

Figure 13.1 Network of the Austrian–German Telegraph Union in 1850. (Scanned from *Archiv für deutsche Postgeschichte*, Vol. 1, 1994, p. 35.)

pattern for all subsequent conferences on communications, still used today: the introduction of conventions and regulations. All international provisions of a rigid nature, such as legal relations and tariff agreements, were defined in *conventions*. Other provisions, more likely to be altered, such as technical matters and the use of the telegraph by the public, were embodied in *regulations* annexed to conventions. The Austrian–German Telegraph Union was terminated in 1872, after formation of the German Empire.

France settled several bilateral telegraph agreements with Belgium in 1851, with Switzerland in 1852, with Sardinia in 1853, and finally, with Spain in 1854. Delegates from these five countries met in Paris in 1855 and founded the *West European Telegraph Union*. Their provisions were identical to those of the Austrian–German Telegraph Union, with the exception of lower telegraph rates and an agreement on languages. It was agreed that French, German, Italian, and Spanish could be used as well as English, although Great Britain was not a member. Between 1859 and 1861 the following independent states also joined the West European Telegraph Union:

FERNIQUE ✴ PHOT.   31 , RUE DE FLEURUS , PAR

Danemark	Bavière	Norvège	Wurtemberg	Belgique	Portugal	Secrétaire	Suisse Bade	Turquie	Prusse	Italie	Grèce	Secrétaire	Espagne
Faber	de Weber	Nielsen	de Klein	Vinchent	Damasio	de Lavernelle	Curchod Roppen	Agathon Effendi	de Chauvin	Minetto	Manos	Dupré	de Hacar

Pays-Bas	Bavière	Hanover	France	Belgique	France	Suède	Espagne	Russie	Autriche
Staring	de Dyck	Gauss	Jagerschmidt	Fassiaux	V.te de Vougy	Brandström	Sanz	G.l Mjr de Guerhard	de Wattenwyl

Figure 13.2   Participants of the foundation conference of the ITU in Paris, 1865. (Courtesy of Museum für Kommunikation, Frankfurt, Germany.)

the two Sicilies, Denmark, Luxembourg, Norway, the Vatican, Portugal, Russia, Sweden, and Turkey. This finally led to the foundation of the International Telegraph Union in Paris in 1865.

Upon the invitation of Emperor Napoleon III, a conference took place in the Salon de l'Horloge at the Palais d'Orsay in Paris with the French Minister of Foreign Affairs, M. Drouyn de Lhuys, as chairman. Delegates to the conference came from the following 20 sovereign states: Austria, Baden, Bavaria, Belgium, Denmark, France, Greece, Hamburg, Hannover, Italy, the Netherlands, Norway, Portugal, Prussia, Saxony, Spain, Sweden, Switzerland, Turkey, and Wuerttemberg. The conference started on March 1, 1865; Figure 13.2 shows the participants. De Lhuys proposed successfully to draft a single convention combining the Austrian–German and the bilateral West European telegraph agreements. At the end of the convention, de Lhuys, expressing the spirit of the discussions, put the question thoughtfully: "We have met here as a veritable Peace Conference. Although it is true that war is frequently caused by a mere misunderstanding, is it not a fact that the destruction of one of the causes makes it easier for nations to exchange ideas and bring within their reach this prodigious means of communication, this electric wire which conveys thoughts through space at lightning speed, providing a speedy and unbroken link for the scattered members of the human race?"

On May 17, 1865 the Convention of the International Telegraph Union was signed. This was the first international agreement concerning most of Europe since the Peace of Westphalia in 1648.[1] The French language was the official language of

---

[1] Ending the Thirty Years' War in Europe. In contrast, the Universal Postal Union was founded in 1874 and the International Union of Railways as late as 1922.

Figure 13.3 First issue of the *Journal Télégraphique*, 1869. (Scanned with permission of the ITU from *ITU News*, Vol. 7, 2000, p. 14.)

the conference. A uniform rate of tariffs was agreed upon, with the exception of Russia and Turkey, which were allowed to charge higher rates for the easternmost parts of their territories. The French gold franc was accepted as the monetary unit. The private telegraph companies in states that were members of the new union were asked to conform to its rules.

The United States could not participate at the convention because private companies operated the telegraph networks.[2] Despite the absence of the United States, the *Telegraph Regulations* drawn up at the conference stated that the Morse telegraph, then widely preferred, was adopted provisionally for use on international lines. Any telegraph administration could become a member of the union. Private telegraph operating companies, however, could send "representatives" only. At the time of the foundation of the union the total length of the telegraph lines of the member states was 500,000 km, and the total number of telegrams sent had been 30 million.

The second Plenipotentiary Conference of the union took place at Vienna in 1868.

[2] Moreover, the Civil War had just come to an end on Palm Sunday, April 9, 1865, and six days later, President Lincoln was murdered.

Apart from admitting Persia and India as members (India was represented by Great Britain, although herself not yet a member), the major achievement of that conference was the setting up of a permanent bureau, charged with the routine administrative work of the union. This bureau was located at Berne, Switzerland, until 1948, when it moved to Geneva. At this conference the Hughes telegraph was recommended for communication on the major international lines.

As another important event with long-lasting effects the conference decided that "the Director [of the ITU] shall be responsible for editing a special journal, using information which he shall collect, in order to keep the administrations abreast of any progress made in the field of telegraphy." This decision resulted in the publication of the *Journal Télégraphique*,[3] first published on November 25, 1869 (Figure 13.3). The magazine was published in French, and beginning in 1948 in three languages (French, English, and Spanish). In fact, the Journal Télégraphiqué had three predecessors: *the National Telegraph Review*, published since 1853 in New York; the *Zeitschrift des Deutsch-Österreichischen Telegraphenvereins*, published since 1854 by the German–Austrian Telegraph Union, and the *Annales Télégraphiques*, published since 1855 by the Imperial French Telegraph Administration.

Great Britain became a member at the third plenipotentiary conference of the ITU in Rome in 1871 after the GPO had acquired the exclusive right to transmit telegrams within Great Britain in 1869. At the same meeting, Japan sent its first observer.

The last plenipotentiary conference of the union in the nineteenth century was held at St. Petersburg in 1875. There the important decision was taken that technical experts of the members' telegraph administrations would be responsible for keeping the telegraph regulations up to date. These experts were to meet periodically at *Administrative Conferences*, so called because they represented their own telegraph administrations. Delegates from the members of the Union with diplomatic powers were to be called upon only if and when it became necessary to revise the convention (for the first time in 1932 in Madrid). A total of 52 countries and 25 private companies were members, of the union at the end of the nineteenth century.

REFERENCES

Books

Michaelis, Anthony R., *From Semaphore to Satellite*, International Telecommunication Union, Geneva, 1965.

Articles

Reindl, Josef, Der Deutsch–Österreichische Telegraphenverein und die Entwicklung des Telegrafenwesens zwischen, 1850 und 1871, *Archiv für deutsche Postgeschichte*, Vol. 1, 1994, pp. 30–46.

---

[3] The name of this highly informative telecommunications magazine was changed to *Telecommunication Journal* in 1934. It changed to *ITU Newsletter* in January 1994. Since 1996 the magazine is published with 10 annual issues as *ITU News*.

PART IV

PERIOD FROM 1900 TO 1950

14

EVOLUTION OF TELECOMMUNICATIONS FROM 1900 TO 1950

Whereas the nineteenth century, in the context of telecommunications, was the century of telegraphy, the twentieth century started with tremendous progress in the domains of telephony, line transmission, and radio transmission, due primarily to discoveries by an American, an Austrian, and a British scientist.

The British scientist, John Ambrose Fleming (1849–1945), born at Lancaster, Lancashire, after studying at the universities of London and Cambridge, joined the Edison Electric Light Company in London in 1882. During a visit to the United States in 1884, Edison demonstrated to Fleming the *Edison effect*: the accumulation of free electrons near the filament of ordinary electrical lamps, discovered by Edison one year earlier. Back in England and (the first) professor of electrical engineering at the University of London, Fleming also advised Marconi and learned about the shortcomings of the coherer as a detector of electromagnetic waves. He started research on the Edison effect[1] and decided to search for a better solution with Edison's incandescent lamps. He placed a platinum cylinder around the filament of the lamp and connected this cylinder with a galvanometer. An electromagnetic wave transmitted through the lamp was rectified and produced a clear deflection on the galvanometer. Fleming understood that his device, soon called the *diode*, in addition to detecting electromagnetic waves in radio equipment, could be used for rectifying alternating currents. He submitted a patent application (British patent specification 24,850) for a "Two-Electrode Valve for the Rectification of High-Frequency Alternating Currents" on November 16, 1904. He thus discovered the diode, which replaced Branly's coherer as a more reliable and more sensitive device for the detection of electromagnetic waves.

---

[1] Sir Joseph John Thompson (1856–1940) discovered that the Edison effect consisted of a stream of electrons from a negative pole (the cathode) to a positive pole (the anode) in 1897.

The Worldwide History of Telecommunications, By Anton A. Huurdeman
ISBN 0-471-20505-2   Copyright © 2003 John Wiley & Sons, Inc.

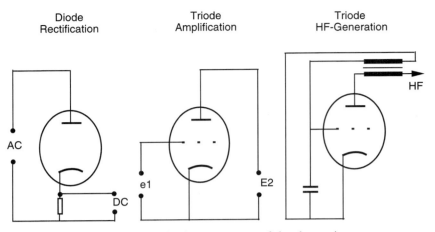

Figure 14.1   Three basic components of the electronics era.

A revolutionary development enabling the electrical amplification of weak signals for the first time, came in 1906. Independently, an Austrian scientist and industrialist, Robert von Lieben (1878–1913), and an American physicist, Lee de Forest (1873–1961), working for Western Electric in Chicago, added a grid between the cathode and the anode of the diode. Von Lieben, searching for a means of amplification of weak telephone signals on long lines, developed his "incandescent valve with amplification," for which he applied for a patent at the Imperial patent office in Vienna on March 3, 1906. De Forest, searching for a means of amplifying weak radio signals, applied for a patent on October 25, 1906 for "a three-electrode valve as a device for amplifying feeble electric currents, the amplification being achieved by using a voltage on the intermediate electrode [the grid] to control the plate current." He called his device the *Audion*. Thus the triode was created and the ***electronic era*** started, as well as a long legal dispute between Fleming and de Forest.[2]

In 1912, independent of each other, an Austrian physicist, Alexander Meissner (1883–1958), working in the Telefunken laboratories in Berlin, and Lee de Forest discovered and patented the electronic high-frequency generator by feeding back part of an amplified signal to the grid of the Audion. De Forest sold his patents to AT&T. In Germany, Lieben, three months before he died, sold his patent to a Lieben Consortium formed by the companies AEG, Siemens, Telefunken, and Felten & Guilleaume. Figure 14.1 shows the principle of the three basic subsequent developments: diode, triode, and high-frequency generator.

Industrial series production of electronic valves (tubes) with a reasonable lifetime became possible in 1913 when the U.S. physicist and chemist Irving Langmuir (1881–1957) developed a method of obtaining a very high degree of evacuation of air and gas from the interior of electronic valves by means of a mercury high-vacuum pump. Langmuir also reduced the power consumption of electronic valves by introducing a cathode made of a thorium–wolfram alloy. It soon appeared that the

---

[2] Initially, the U.S. courts held that the de Forest's addition of the grid was dependent on Fleming's work. In 1943, however, two years before Fleming died, the U.S. Supreme Court decided that the original patent for Fleming had always been invalid.

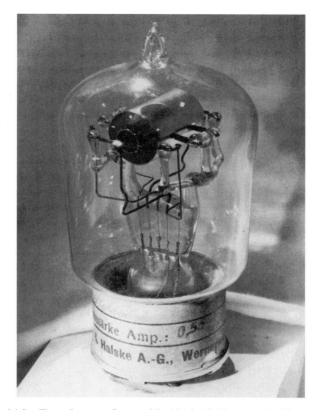

Figure 14.2   Tetrode manufactured in 1916. (© Siemens-Archiv, Munich.)

circuit components connected to the anode reduced the amplification factor of the triode. A German physicist, Walter Schottky (1886–1976), working for Siemens, solved this problem in 1916 by adding a second grid, called a *screen grid*, between the anode and the first grid. Thus a four-electrode valve called the *tetrode* was created (Figure 14.2). Schottky obtained German patent 300,617 for his "grid-amplifier valve based upon a space-charge and screen grid principle" on May 31, 1916. The positive effect of the screen grid, however, was combined with undesired production of electrons from the anode. This problem was solved in 1926 when a Dutch physicist, Bernardus Dominicus Hubertus Tellegen (1900–1990), working for Philips Gloeilampenfabrieken in Eindhoven, added another grid between the screen grid and the anode and thus created the pentode.[3] A hexode followed in 1932, a heptode in 1933, and a few years later, finally, a valve with eight electrodes called the octode. In its various versions the valve became the primary tool in electronic development until it was replaced by a superior device, the transistor (Section 24.1). The development

---

[3] This was Tellegen's first patent in a series of 57 patents. In the period 1946–1966, he was professor extraordinary of circuit theory at the University of Delft. In 1957 he published a network theory known as *Tellegen's theorem.* He was elected a member of the Royal Academy of Sciences of The Netherlands in 1960.

of the transistor took place in this period; the major effect on telecommunications, however, came in the second half of the twentieth century.

Electronics are not a domain of telecommunications; the effect of electronics on telecommunications, however, has been vital for the development of new devices, systems, and technological approaches, such as:

- Long-distance telephony
- Multiplexing of telephone and telegraph channels
- Radiotelephony with continuous-wave transmitters
- Radio-relay transmission

AT&T and other telephone operators immediately recognized the significance of the triode. The Pupin coil had already extended the operational length of a good telephone overhead line to about 2000 km. Now adding an electronic amplifier equipped with one or more triodes every 30 km made it possible to construct long landlines in nationwide telephone networks. Following the example of the United States, worldwide telephone penetration began, as described in Chapter 15. With the increased use of telephony, switching was improved and evolved from manual switching to complex automatic electromechanical switching systems, as described in Chapter 16.

The spark generators developed for radiotelegraphy in the nineteenth century were replaced at the beginning of the twentieth century by arc generators and frequency alternators, which produced more constant waves. The triode enabled the development of high-frequency wave generators, which very soon replaced the arc and frequency alternators and improved the radiotelegraphy substantially and made radiotelephony possible, as described in Chapter 17.

A decisive step toward a better use of physical circuits could also be obtained with the advent of electronic tubes, the realization of filters and resonance circuits, and the consequent possibility of continuous electronic generating of various frequencies. With those building blocks, *carrier frequency* transmission, also called *multiplex*, could be developed as described in Chapter 20. Telephone multiplex systems with three channels and four channels were operated on the long open-wire line systems of AT&T around 1918. A 10-channel voice-frequency telegraphy system was taken into operation between New York and Pittsburgh in 1923. Velocity-modulated electronic valves and very-high-frequency circuit engineering enabled the development of radio-relay systems in the early 1930, as described in Chapter 21.

Other major achievements in the last decade before World War II were the development of phototelegraphy and teleprinters, covered in Chapters 18 and 19; coaxial cable transmission, covered in Chapter 20; and cryptography, the subject of Chapter 22.

15

WORLDWIDE TELEPHONE PENETRATION

15.1 WORLDWIDE TELEPHONE STATISTICS

About 2 million telephones were in use worldwide at the beginning of the twentieth century, of which 1.4 million were in the United States and about 0.5 million in Europe. Telephone penetration had hardly started in Asia and Africa and took another three decades before it reached beyond the capitals of those continents. Basically, the telephone penetration depended greatly on the degree of industrialization, education, linguistic and dialectic diversities, social development, and political circumstances. The figures on the telephone penetration at the beginning of the twentieth century are approximate, because the annual statistics on the worldwide development of telephony issued by the ITU at that time were not complete. Private companies operated most telephone networks, and several of those companies were not eager to reveal their figures to their national telegraph administrations, which, as members of the ITU, collected the data. In 1910, AT&T began its own, more complete worldwide evaluation and publication of telephone penetration under the title *Telephone Statistics of the World*.[1] At that time, 10 million telephones were in use worldwide. The AT&T "statistics" became an annual publication and was authoritative for over half a century. Table 15.1 summarizes the information on telephone penetration at the end of the year 1910. The table shows that by 1910 almost 70% of all telephones were used in the United States. This uneven worldwide distribution of telephones can also be seen from Table 15.2, which gives the teledensity (number of telephones per 100 inhabitants) of the capitals and other major towns for January 1,

---

[1] AT&T published *Telephone Statistics of the World* with the remarkable subtitle: "Compiled by statistician, American Telephone and Telegraph Co., from government reports, telephone company reports, and personal correspondence. While all the statistics may not represent the absolute facts, they are believed to be the best available and substantially correct."

The Worldwide History of Telecommunications, By Anton A. Huurdeman
ISBN 0-471-20505-2   Copyright © 2003 John Wiley & Sons, Inc.

TABLE 15.1   Worldwide Telephone Penetration, 1910

Continent and Country	Telephones (thousands)	Teledensity (no. /100 population)	Percent of Total
The Americas			
Canada	284	3.8	2.5
United States	7,596	8	67.4
Central America	27	—	0.3
South America	86	—	0.8
Total Americas	7,993	—	71.0
Europe			
Austria	113	0.4	1.0
Belgium	47	0.6	0.4
Denmark	95	3.5	0.9
France	233	0.6	2
Germany	1,069	1.6	9.5
Great Britain	649	1.4	5.8
Italy	70	0.1	0.6
The Netherlands	65	1	0.6
Norway	63	2.6	0.6
Russia	151	0.25	1.3
Spain	25	0.08	0.2
Sweden	187	3.4	1.6
Switzerland	79	2.1	0.7
Rest of Europe	121	—	1.1
Total Europe	2,967	—	26.3
Other			
Africa	34	—	0.3
Asia	149	—	1.3
Oceania	118	—	1.1
Total worldwide	11,272	—	100

Source: Data reported by AT&T, 1911.

1926. The worldwide number of telephones reached the 20 million mark in 1922, the year that Alexander Graham Bell died, the 50 million mark in 1939, and 75 million in 1950. This increase was not gradual but followed largely the economical and political situation over that period, as shown in Figure 15.1, which as a typical example, depicts the telephone growth in Germany. A steep increase in the number of telephones at the beginning of the twentieth century reflects the strong industrial and colonial activities of that time. World War I, with its heavy destruction in Western Europe, interrupted telephone development and reduced the number of subscribers. The Great Depression of 1929 in the United States affected the entire world in the following years and caused a second large reduction and implementation delay of telephony up to about 1937. Finally, the disastrous World War II brought telephone penetration, in Germany and many other industrialized countries, back to the level of the 1920s. Table 15.3 shows the teledensity just before World War II for countries with a teledensity above 1.

TABLE 15.2   Teledensity in Capitals and Other Major Towns, January 1, 1926

Teledensity	Cities	Teledensity	Cities
31	San Francisco	10	Berlin
28	Stockholm	8	Munich
27	Washington	7	Auckland, Havana
24	Los Angeles, Toronto	6	Brussels, London, Osaka, Rotterdam, Tokyo
23	New York	5	Amsterdam, Antwerp, Budapest, Vienna
21	Chicago	4	Buenos Aires, Glasgow, Prague, Warsaw
17	Montreal	3	Beijing, Dublin, Milano, Moscow, Rome, St. Petersburg
16	Copenhagen	2	Shanghai
15	Oslo	1	Naples
13	Zurich		

Source: Data from E. Feyerabend, *50 Jahre Fernsprecher in Deutschland, 1877–1927*, Reichspostministerium, Berlin, 1927.

15.2   TELEPHONE PENETRATION IN THE UNITED STATES

At the beginning of the twentieth century, 1,355,000 telephones were installed in the United States, of which 554,000 were operated by over 6000 independent companies. Competition became so intense that in many cities two or more companies operated in parallel without a facility of interconnection between those networks.[2] Around 1905, AT&T accepted interconnection between their local networks and competing local independents. In 1913 the independents also got access to AT&T's long-distance lines under the Kingsbury agreement. In 1914 about 50% of subscribers of the independents had access to the Bell System, and in 1930, over 99%.

The Graham Act of 1921 confirmed and legalized the Kingsbury agreement but also allowed AT&T to buy "duplicating telephone companies." This situation was slightly improved for the independents in 1922 when AT&T issued the Hall Memorandum. In that memorandum, E. K. Hall, vice-president of AT&T, confirmed its commitment to F. B. MacKinnon, president of USITA, "not to purchase or consolidate with connecting of duplicating companies except in special cases in which USITA would be given notice at least 30 days before the parties conclude any formal agreement." In fact, AT&T slowed down the acquisition of independents and it was not until 1945 that AT&T bought the last duplicating company, the Keystone Telephone Company in Philadelphia.

The application of Pupin coils from 1900 enabled telephone communication from New York to Kansas City and New Orleans. The invention of the triode by Lee de Forest in 1906 enabled the installation of voice-frequency amplifiers on long routes so that transcontinental telephony became possible in 1915. At the opening of the New York–San Francisco route, in New York, Bell again spoke the words "Mr.

[2] New York, exceptionally, was operated exclusively by AT&T.

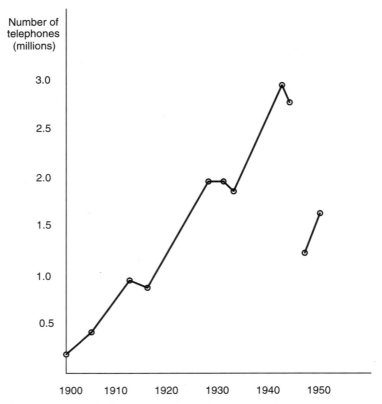

Figure 15.1 Telephone development in Germany from 1900 to 1950.

Watson, come here, I want you!" to his former assistant 4800 km away in San Francisco, using a model of his first telephone.

The telephone subscribers enjoyed automatic telephone switching from most independent companies, whereas AT&T started this service only in October 1922, at the Pennsylvania office in New York. In 1928, at the fiftieth anniversary of the telephone, only 20% of the Bell telephone network had automatic operation. By 1934, 50% of the total telephone service was automatic. It was not until 1940 that in New York the last manual exchange, Murray Hill–2, was replaced by an automatic exchange.[3] Herbert Hoover was the first president of the United States with a permanent telephone on his desk, in 1929. Previous presidents either had a telephone temporarily or had to use a booth outside the executive office.[4]

The telephone was particularly interesting for farmers. Special rural multiparty lines served from six to as many as 15 farmhouses along a maximum 65-km-long line.

---

[3] Automation was 97% in 1960, 99.9% in 1968, and the last manual exchange, in Maine, was retired in 1978.

[4] President Garfield was the first U.S. president to possess a telephone in his house (in 1878, when he was still a member of Congress).

TABLE 15.3   Worldwide Telephone Teledensity, 1940

Teledensity	Countries	Teledensity	Countries
16.6	United States	5.1	Belgium
14.3	Sweden	4.8	Finland
12.8	Canada	3.9	France
11.9	Denmark	3.7	Austria
11.2	Switzerland	3.0	Argentina
10.5	New Zealand	2.0	Japan
8.5	Australia, Norway	1.7	Chile, Hungary
7.0	Great Britain	1.6	Italy
5.3	Germany	1.4	Czechoslovakia
5.2	The Netherlands	1.3	Spain

Source: Data from E. Horstmann, *75 Jahre Fernsprecher in Deutschland, 1877–1952*, Bundespostminis-terium für das Post- und Fernmeldewesen, Bonn, 1952.

By 1910, some 2 million of the total 7.5 million telephones were used by farmers, corresponding to 25% of farmers. Most of those rural systems used magneto signaling; a subscriber could recognize a call for him by a coding of long and short rings. Most systems were operated on a not-for-profit cooperative basis. About 50% were also connected with the Bell System. In 1920, 86% of farmers were connected by telephone in the then mainly agricultural state of Iowa. A popular request of farmers at the time was to get "good roads and telephones." The situation deteriorated during the Great Depression. In 1940, fewer farmers had telephones than in 1920. To improve the situation, the Rural Electrification Act was issued in 1949, the *National Telephone Cooperative Association* (NTCA) formed in 1954, and the Rural Telephone Bank established in 1971.[5] Table 15.4 summarizes the telephone development in the United States over this period.

TABLE 15.4   Telephone Development in the United States

Year	Number of Telephones	Teledensity
1900	1,355,000	1.76
1910	7,635,400	8.20
1920	13,329,400	12.39
1930	20,202,000	16.34
1940	21,928,000	16.52
1950	43,004,000	28.09

Source: Data from *Bell System Technical Journal*, Vol. 37, March 1958.

[5] Currently, NTCA has about 1000 members, mainly cooperative telephone operators, a few regional or statewide telephone associations, and rural equipment and service providers.

15.3 TELEPHONE PENETRATION OUTSIDE THE UNITED STATES

The telephone penetration outside the United States took place primarily in Australia, New Zealand, Japan, and in Europe. In Australia, long-distance telephony started between Melbourne and Sydney in 1907 via a 1000-km line along the railway. The line was then extended by another 2200 km to Townsville. In 1950, the number of telephones had increased in Australia from 37,600 in 1904 to 1,110,000, and in New Zealand from 8700 in 1900 to 370,000.

In Japan, a remarkable achievement was a phone call over 1583 km between Tokyo and Sasebo (in the southwestern corner of Japan, near Nagasaki) in 1905. Four years later, the number of telephones reached 100,000. The entire 83,000-telephone system in Tokyo went out of service at the time of the Great Kanto Earthquake in 1923, when downtown Tokyo burned to the ground and about 74,000 persons died. Within four weeks 14,000 telephones were back in operation. The need for reconstruction provided the opportunity to replace manual switches with automatic switching equipment made in Japan. Despite the earthquake, the number of telephones in Japan increased from 340,000 in 1920, to 556,000 in 1925, to 1 million in 1934, and to about 1.3 million before World War II. The telephone service was still unsatisfactory, however. As an example, it took about a full day before a long-distance call between Tokyo and Osaka actually went through.

The Japanese telecommunications network suffered heavily from World War II destruction; only 540,000 telephone lines were still usable in 1945. Immediate reconstruction began after the war with active support from the United States; engineers from Western Electric and Bell Telephone Laboratories instructed Japanese engineers and executives on network reconstruction and equipment quality manufacturing. The prewar level of telephone infrastructure was reached again in 1950, with 1.7 million telephone lines.

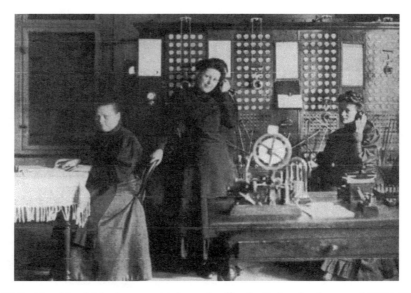

Figure 15.2 Combined telephone and telegraph office in Switzerland, around 1900. (Scanned from *Archiv für deutsche Postgeschichte*, Vol. 1, 1977, p. 122.)

TABLE 15.5   Telephones Outside the United States, 1925

Telephones	Countries
1000–2500	Bolivia, Costa Rica, Guatemala, Honduras, Iran, Israel, Kenya, Mozambique, Myanmar, Nigeria, Senegal, Sudan, Thailand
2500–5000	Ecuador, El Salvador, Greece, Iraq, Portugal, Zimbabwe
5000–10,000	Bulgaria, Canada, Luxembourg, Malaysia, Morocco, Panama, Peru, Singapore, Tunisia
10,000–50,000	Chile, Colombia, Cuba, Egypt, Hong Kong, Indonesia, Ireland, Korea, the Philippines, Romania, Turkey, Uruguay, Venezuela
50,000–100,000	Belgium, Brazil, China, Hungary, India (including Pakistan), Mexico, Poland, South Africa, Spain
100,000–500,000	Argentina, Australia, Austria, Czechoslovakia, Denmark, Finland, Italy, The Netherlands, New Zealand, Norway, Sweden, Switzerland, USSR, Yugoslavia
500,000–700,000	France, Japan
1,500,000–1,700,000	Germany, Great Britain

Source: Data from Gerhard Basse, Die Verbreitung des Fernsprechers in Europe, Nordamerika, Lateinamerika, Afrika, Asien und Ozeanien, *Archiv für Post und Telegraphie*, Vol. 1, 1977, pp. 58–103 and Vol. 1, 1978, pp. 24–93.

In Europe, at the beginning of the twentieth century, the manual telephone and telegraph offices were often combined, as shown in Figure 15.2. The first automatic telephone service began at Hildesheim, Germany, on July 10, 1908. In Great Britain, by virtue of the government's decision of 1899, the public *General Post Office* (GPO) absorbed all private telephone operating companies in the period ending January 1, 1912.[6] Automatic telephone service was introduced in the same year. It took until July 20, 1933, however, for the City Telephone Exchange of London to open automatic service. This was the world's busiest exchange, with 11,400 lines and 100,000 calls per day.

Statistics published on telephone penetration in Europe in January 1, 1912 show that London, with 220,782, had the most telephones, followed by Berlin with 133,876, Paris with 84,500, Stockholm with 73,200, Hamburg with 65,000, Vienna with 52,355, Copenhagen with 46,000, Glasgow with 42,300, St. Petersburg with 39,600, Moscow with 36,900, and Warsaw with 25,830. The teledensity in the European capitals varied from 21.1 for Stockholm, 6.1 for Zurich, 3.0 for London, 1.9 for Rome and 0.4 to Athens.

By the middle of the 1900–1950 period, a total of 25 million telephones were used worldwide, of which 16.7 million were in the United States and thus 8.3 million were in all countries outside the United States. An approximate distribution of those 8.3 million telephones, for all countries with more than 1000 telephones, is given in Table 15.5.

---

[6] The *Telephone Corporation of Hull*, set up in 1902, and the *Telephone Council* of the state of Guernsey were not absorbed by the GPO. Similarly, the telephone network on the island of Jersey was not nationalized, but operated by the National Telephone Company on behalf of the state of Jersey.

REFERENCES

Books

Brooks, John, *Telephone: The First Hundred Years*, Harper & Row, New York, 1975, 1976.

Feyerabend, E., *50 Jahre Fernsprecher in Deutschland, 1877–1927*, Reichspostministerium, Berlin, 1927.

Horstmann, Erwin, *75 Jahre Fernsprecher in Deutschland, 1877–1952*, Bundesministerium für das Post- und Fernmeldewesen, Bonn, Germany, 1952.

Articles

Basse, Gerhard, Die Verbreitung des Fernsprechers in Europe und Nordamerika, *Archiv für Post und Telegraphie*, Vol. 1, 1977, pp. 58–103.

Basse, Gerhard, Die Verbreitung des Fernsprechers in Lateinamerika, Afrika, Asien und Ozeanien, *Archiv für Post und Telegraphie*, Vol. 1, 1978, pp. 24–93.

Green, E. I., Telephone, *Bell System Technical Journal*, Vol. 37, March 1958, pp. 289–325.

16

ELECTROMECHANICAL TELEPHONE SWITCHING

Telephone switching at the beginning of the twentieth century was still primarily manual switching. Even in the United States, only the Independent telephone companies introduced automatic switching; AT&T stayed with manual switching. After World War I, however, worldwide introduction of automatic switching was begun, primarily using the Strowger system. To ease the transition from manual to fully automatic switching, the semiautomatic Lorimer and panel systems were developed in the 1910s. Fully automatic systems appeared around 1920 derived from the Lorimer system, such as the rotary and the LME 500-point system. Crossbar switching was used in Sweden before World War II and found worldwide use after the war.

16.1 WORLDWIDE INTRODUCTION OF THE STROWGER SYSTEM

16.1.1 Strowger System in the United States

After Almon Strowger retired from his company, the company name was changed to the *Automatic Electric Company* (Autelco) in 1901, with Alexander E. Keith as its technical director.[1] Development of the Strowger system (Section 11.3) continued with the introduction of improved group selectors that selected the thousands and hundreds and then gave access to the line selectors. This solution was first applied in New Bedford, Massachusetts, for an exchange with a capacity of 4000 subscribers. A similar exchange for 10,000 subscribers (a capacity exceeding that of any manual contemporary exchange) was installed in Chicago in 1903, including the facility of call metering.

In the following year, the principle of *preselection* of a free common group selector for a number of subscribers was introduced and used by Autelco at the Wilmington,

---

[1] In 1955, Autelco merged into the General Telephone and Electronics Corporation (GTE).

The Worldwide History of Telecommunications, By Anton A. Huurdeman
ISBN 0-471-20505-2   Copyright © 2003 John Wiley & Sons, Inc.

Delaware, exchange. The principle of preselection, substantially reducing the number of group selectors required depending on the simultaneous telephone traffic expected, had been patented in 1894 by Romaine Callender, owner of the Callender Company in Brantford, Ontario, Canada. Preselection operates very quickly and unnoticed by the subscriber from the moment that he or she lifts the handset until beginning to dial. Telephone switching based on the Strowger system is usually called *step-by-step switching*. Operation of a step-by-step switch for 10,000 lines is shown in Figure 16.1 and explained in Technology Box 16.1.

In May 1905, the first automatic exchange operated entirely from a *central battery* was installed in South Bend, Illinois. The other battery initially installed at a subscriber's premises and intended solely for the selecting process had already been eliminated in 1895 at the second exchange in La Porte. This power supply system was called a *common battery* in the United States and a *central battery* in Britain and elsewhere (both names were abbreviated to CB).

In 1908, for an exchange in Pontiac, Illinois, the system of two wires only between subscribers and exchange was introduced for battery power supply, dialing, and transmission of speech current. Dial pulses now corresponded to intermittent interruptions of the subscriber line.

In 1910, about 200,000 subscribers of the Independent companies were connected to some 130 automatic telephone exchanges supplied by Autelco. The already mighty American Bell Telephone Company was not interested in this non-Bell product[2] or in automatic switching and continued to serve its customers by manual and semi-automatic exchanges. This changed in 1916, when AT&T had taken over so many independent telephone operators that operated Strowger exchanges that AT&T had to reconsider their resistance to automatic telephone switching. Subsequently, AT&T concluded a patent agreement with Autelco for the manufacture of Strowger exchanges by Western Electric. Moreover, in 1919, a further agreement was made between AT&T and Autelco concerning the direct supply of Strowger exchanges by Autelco to telephone companies of AT&T. The first exchange supplied by Autelco under this agreement opened its service in Norfolk in 1919. For many years to come, the majority of step-by-step exchanges for AT&T were manufactured by Autelco.

In the beginning of the 1920s, to meet requirements for interexchange routing of calls, Autelco designed the *director system*. Basically, the system consists of a register–translator that receives and stores the dialing pulses from the subscriber and translates them to a new series of pulses that control the selectors of the local exchange as well as the relevant selectors in the corresponding exchange(s) in case of trunk calling. With this system, Autelco could offer a similar flexibility for interexchange routing as was inherent with the indirect-control panel and rotary systems that were developed in the meantime. The first exchange with the Strowger director system was brought into service in Havana (Cuba) in 1924.

16.1.2   Strowger System in Canada

In Canada, the first Strowger exchanges were introduced in 1883 in London, Seaforth, Mitchell, and Arnprior in Ontario, and in Terrebonne in Quebec. Unfortu-

---

[2] Thus, Bell demonstrated the first example in telephone switching of the *N.I.H. factor* ("not invented here," thus, "not implemented here").

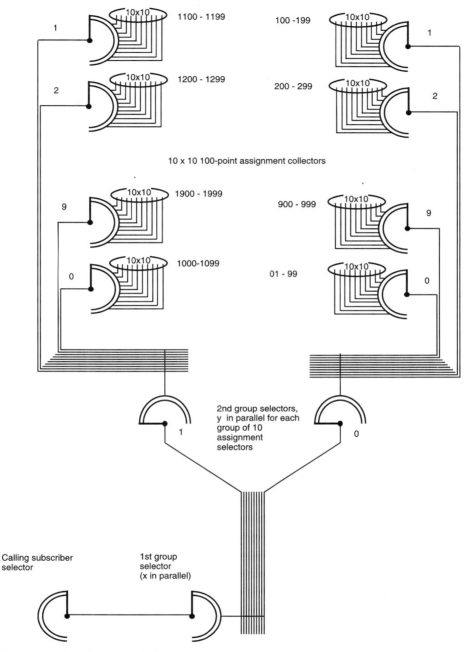

Figure 16.1 Step-by-step switching arrangement for 10,000 subscribers. (From A. A. Huurdeman, *Guide to Telecommunications Transmission Systems*, Artech House, Norwood, MA, 1997, Figure 1.8; with permission of Artech House Books.)

TECHNOLOGY BOX 16.1

Step-by-Step Switching

With the basic 100-position selector, a maximum of 100 subscribers can be served. For exchanges with more than 100 subscribers, 10-point group selectors have to be inserted before the assignment selector. With one group selector the capacity of an exchange increases to 1000 subscribers, with two group selectors to 10,000, with three group selectors to 100,000, and with four group selectors to 1,000,000 subscribers. Figure 16.1 shows how 10,000 subscriber lines can be served with two group selectors. To prevent one single calling subscriber from blocking the exchange for further simultaneous calls, a number of group selectors are always operated in parallel. In Figure 16.1 this is indicated with x and y, whereby $x > y$ and the value of x and y depends on the traffic load of the exchange.

Source: Adapted from A. A. Huurdeman, *Guide to Telecommunications Transmission Systems*, Artech House, Norwood, MA, 1997; with permission of Artech House Books, 1997.

nately, history relates that none of these remained in operation for longer than several weeks. The first Strowger exchange really to operate in Canada was the one in Whitehorse, Yukon, thus also the most northern (61°N), brought into service in 1901.

16.1.3   Strowger System in Japan

Japan's public telephone service began in 1890 with the opening of manual exchanges in April at Tokyo (Figure 16.2) and in June at Yokohama with equipment imported from BTM, Belgium. Service between Tokyo and Yokohama started on December 16. The first common battery switching system was installed at the Kyoto exchange in 1903. From 1916 onward, the administration became interested in automating local exchanges, and comparative studies were performed on the various systems available worldwide. Step-by-step systems were considered best for Japan because of their simple mechanism, easy serviceability, apparent reliability under earthquake conditions, and popularity, being by far the most widespread system at that time.

Following the Great Kanto Earthquake[3] in 1923, it was decided to take advantage of the reconstruction of the telephone network and to introduce automatic service. Two different systems were chosen:

1. The *American Strowger system* (in Japan called the *type A system*), mainly for the Tokyo area, with the first exchange in operation at the Kyobashi branch in Tokyo by January 1926

[3] In which some 140,000 persons died and which completely devastated the Tokyo and Yokohama areas, together with almost all of their telephone network plant.

Figure 16.2 Telephone exchange at Tokyo, 1899. (Scanned from *Archiv für deutsche Post-geschichte*, Vol. 1, 1978, p. 71.)

2. The *German Siemens rotary system* (in Japan called the *type H system*), installed initially in the Yokohama area, with the first exchanges in operation at the Chojamachi branch and at the Yokohama main office by March 1926

Long-distance calling was introduced in 1927. Equipment for the exchange in Kyoto was still imported Strowger equipment, whereas Siemens supplied the equipment for the exchanges in Osaka and Kobe. Local production of both the A and H types started in 1934. Interworking between the A and H systems, however, caused more problems than anticipated. In 1935, therefore, development was begun of a purely Japanese step-by-step system, called the *T system*. The first exchange of the T system made in Japan was installed at Nara near Kyoto in 1940. World War II, however, stopped further development.

In 1938, according to the last international statistics published before World War II, Japan, with a population of 72 million, had 1 million subscriber lines, of which 350,000 were served automatically by 132 exchanges. After the war, Japan turned to crossbar switching, with the first switch installed in 1955.

16.1.4 Strowger System in Germany

With the initial reluctance of AT&T to install automatic switches, Autelco diverted their attention to Europe, where they were successful in Germany beginning in 1900, and in 1912 in Great Britain. The German Imperial Post, still open for good ideas coming from the United States (as previously for Bell's telephone), signed a contract for a trial exchange. First trials with an exchange for 400 lines delivered from Chicago took place on May 21, 1900 in Berlin. Encouraged by the good results, local production under license from Autelco began at the German weapons factory company Ludwig Loewe & Co.[4] for a commercial exchange with a capacity of 1200

---

[4] In 1901, Ludwig Loewe & Co. obtained the patent rights for Germany and all of Europe, with the exception of France and the U.K.

Figure 16.3 Hildesheim automatic telephone exchange. (Courtesy of Museum für Kommunikation, Frankfurt, Germany.)

subscribers for the town of Hildesheim. This, the world first automatic public telephone exchange outside the United States, was put into operation on July 10, 1908, serving 900 subscribers. Figure 16.3 shows the exchange and Figure 16.4 a telephone used for this network.

In 1907, in view of the technical problems at the weapons factory, the German Imperial Post contacted Siemens & Halske. After the Hildesheim exchange was in operation, an agreement was made between the four parties involved (Autelco, Loewe & Co, Imperial Post, and Siemens) to shift production and further development of Strowger equipment in Germany to Siemens & Halske. The technical staff involved at the weapons factory was taken over by Siemens in their *Wernerwerk* factory in Berlin.

In 1909, Siemens & Halske installed the first automatic exchange designed according to the specifications of the Imperial Post at Dallmin near Berlin. The exchange was for 20 subscribers only and still working on local batteries but incorporating preselectors from its own improved design as a 10-point rotary switch instead of Strowger's complicated plunger. This was followed in 1910 by a 1000-line exchange at Altenburg, in Thüringen, incorporating central batteries and preselection. The Bavarian Post installed a 2500-line exchange at Munich–Schwabing in December 1909 (Figure 16.5).

It was also in Bavaria that the world's first automatic long-distance service was introduced. This took place in the district of Weilheim in 1923. Twenty-two exchanges situated in a radius of some 30 km around this town were interconnected, as shown in Figure 16.6. Three major innovations were incorporated for the network:

Figure 16.4   Telephone used at Hildesheim, 1908. (Courtesy of Museum für Kommunikation, Frankfurt, Germany.)

1. A *time-charging system* with *time-zone counting*, in which at the end of the call, the subscriber call meter received a number of pulses, depending on the duration of the call and the area of the subscriber called

2. The introduction of 50-Hz ac instead of dc signaling for the transmission of line signals to avoid interference from electrified railway lines in the vicinity of trunk lines

3. A subscriber numbering system with a unique code number for each subscriber in the entire district

In 1927, Siemens made important improvements in the Strowger switch. Instead of the lifting, turning, returning, and dropping movements of the original Strowger switch, Siemens constructed a lifting and turning switch called *Heb-Dreh-Wähler* (HDW, lift-turning switch). With this switch, at the end of a call, the selector arm—which in the Strowger system used to return the same way it came—continued to the end of the contact bank, whereupon the wiper set dropped and returned horizontally

Figure 16.5 Munich–Schwabing automatic telephone exchange. (© Siemens-Archiv, Munich.)

to its initial rest position. In this way, equal wear and tear of the contacts was obtained and prevented oxidation of seldom-used contacts. Figure 16.7 shows at the bottom the original Strowger switch as used in the Hildesheim exchange, and at the top the improved two-motion switch.

16.1.5 Strowger System in Great Britain

In Great Britain all exchanges were still manual in January 1912 when the General Post Office (GPO) took over the telephone networks of private operators. The GPO decided to introduce automatic telephone switching using various systems available on the market. The first automatic exchange was put into operation in London (Epsom) on July 23, 1912 with a 1500-line exchange supplied by the *Automatic Telephone Manufacturing Company Ltd.* (ATM) in Liverpool.[5]

As an early example of the policy of local production in the interest of national industrial telecommunications development with a minimum of foreign currency,[6] in 1922 the GPO decided to standardize on a single automatic British national system

---

[5] An additional 18 automatic telephone exchanges, supplied by five different companies, were installed in the period between 1912 and 1923. ATM supplied nine exchanges, Siemens Brothers Ltd. supplied five, Standard Telephone and Cables Ltd. supplied two Western Electric rotary exchanges (from their factory in Antwerp), and Lorimer and Relay Automatic Telephone Company (belonging to the Marconi Group) each supplied one exchange.

[6] This policy was soon followed by Sweden and Russia and some 60 years later was used again by Brazil, China, India, Taiwan, South Africa, and other countries.

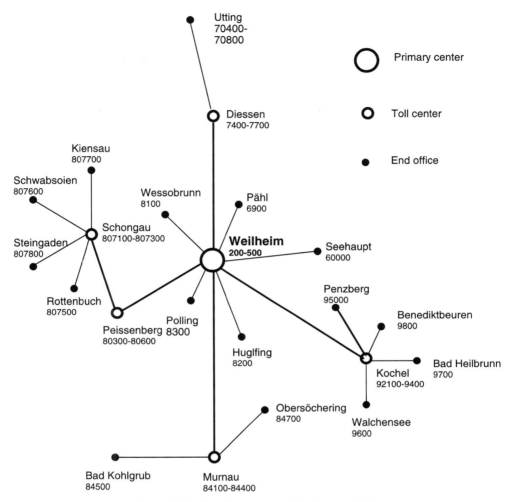

Figure 16.6   Trunk network of Weilheim, 1923.

to be produced in Britain, based on the Strowger system in two versions: with director facilities for the major towns, and without director facilities elsewhere. The foreign patents were obtained and pooled by British manufacturers. The standardization related basically to:

- Uniform subscriber sets with dial and equal pulse trains
- Two-wire loop without an Earth return
- Standardized battery voltage
- Rotary preselectors with 25 positions

The first Strowger–director exchange made in Britain was installed in Holborn in
1927. The British Strowger equipment was also used widely in the Commonwealth

Figure 16.7   Two versions of the Strowger switch. (© Siemens-Archiv, Munich.)

countries and was exported to many Latin American countries. The last Strowger public telephone exchange equipment was installed in the U.K. in 1985.

16.1.6   Strowger System in Austria

In Austria the first automatic telephone exchange was installed at Graz in 1910. The exchange, using the Strowger system, served 2000 private subscribers and 1200 corporate subscribers. Instead of a dial at the subscriber's set, a number-setting device called the *Dietl calling device* was used. The Dietl device enabled subscribers to set up the desired number by means of levers before activating the automatic switching process by turning a call-beginning handle. Although this was a more elaborate process then with dialing, subscribers presumably made fewer calling faults, and if there were any, they would notice that they themselves were to blame and not the apparatus.

16.1.7   Strowger System in Sweden

Around 1930, L.M. Ericsson developed an *XY* step-by-step switch. Instead of the two-motion upward/rotary movement of the Strowger selector, the two-motion movement of the XY selector took place in two directions perpendicular to one another, thus along axes X and Y—hence the name XY *switch*. Like the Strowger switch, the XY switch had a bank of contacts with a capacity of 10×10 lines.

16.2 AUTOMATIC OR SEMIAUTOMATIC SWITCHING?

Strowger's step-by-step two-motion automatic switch, although very reliable in operation, experienced strong competition at the beginning of the twentieth century from the steadily improving manual switching exchanges. Even then, a steady transition from manual switching to automatic switching could not be taken for granted. The question of automatic versus semiautomatic was an issue of heated debate between supporters of the two different switching solutions.

In 1902, in the United States, the independent telephone companies used automatic switching widely for their approximately 1.1 million subscribers, whereas the various Bell System telephone companies slowly changed from manual switching to semiautomatic switching for their approximately 1.3 million telephone subscribers. The *National Telephone Exchange Association of the United States*, in which since 1880 all American telephone-operating companies were represented, organized annual meetings, seminars, and symposia, where the issue of automatic versus semiautomatic switching was discussed. Following the U.S. example, a first *International Congress of Telegraph and Telephone Engineers* took place in Europe in Budapest in September 1908, followed by a second in Paris in 1910. By then, automatic switching in Europe was used only in the German Empire, the Kingdom of Bavaria, and in Austria–Hungary. As a big surprise to the European delegates—used to progressive ideas coming from the United States—John J. Carty (1861–1932), chief engineer of AT&T, strongly advocated semiautomatic service, which did not entail changing the subscriber's telephone set for one with a dialing disk and kept the network investment under better control. Carty stated: "The automatic system is not simple for the subscriber, who must take his telephone from the hook and perform a number of manual operations depending on the character of the call he wishes to make. Then he must press a button, which if all goes well rings the subscriber desired."[7] A third conference was planned to be held in the fall of 1914 but had to be canceled as World War I began in August 1914, interrupting the discussion of automatic versus semiautomatic switching.

Around the end of World War I, AT&T also came to the conclusion that the tremendously increasing urban traffic in large towns, with the subsequent routing of calls in tandem by telephone operators in a number of exchanges, could not continue to be handled manually. The labor shortage in the United States in the immediate postwar years, with its unprecedented increase in labor cost, was an additional incentive toward the decision taken in 1920 to start with the installation of full automatic exchanges. The first automatic exchange in the Bell system was then commissioned in October 1922, with a panel switch at the Pennsylvania office in New York. Beginning in 1923, the panel system was produced at the rate of about 100,000

---

[7] John J. Carty obviously did not anticipate the tremendous increase in telephone service which 40 years later provoked the statement by one of his successors that "if calling trends persisted, AT&T would eventually need to hire half the women in the United States to serve as operators connecting phone calls."

lines per year up to 1926, and then increased rapidly to 400,000 lines a year by 1931. The cities of New York, Chicago, Kansas City, Buffalo, and Philadelphia were all equipped with panel exchanges by 1924.

Following the example of AT&T, a worldwide wave of telephone switching automation began. The four major suppliers of telephone switching equipment competed aggressively for market shares. The switching systems of those suppliers were not compatible, and as standardization of telecommunication systems interface was still in its infancy, the company that obtained the order for the first telephone exchange in a country could almost be sure subsequently to be given orders for a supply of their equipment for all exchanges in the same national network. The four companies that dominated the market were:

- The *International Automatic Electric Corporation* of Chicago, with its major affiliate, the *Automatic Telephone Manufacturing Co. Ltd.* in Liverpool, U.K., supplying the Strowger system
- The *International Western Electric Corporation* of New York, belonging to AT&T before 1925 (after 1925 belonging to IT&T and then named the International Standard Electric Corporation of New York), with major affiliated companies BTM in Antwerp and *Standard Telephone and Cables* (STC) in London, supplying the rotary system
- Siemens & Halske in Berlin, supplying their Strowger rotary system
- L.M. Ericsson in Stockholm, supplying the LME 500-point system

The relative strengths of those companies can be understood from the size of the workforces they employed around 1925. Western Electric claimed to have 13,000 workers (53%); Siemens, 6000 workers (24.2%); L.M. Ericsson, 3800 workers (15.4%), and Automatic Electric, 1800 workers (7.4%). Automatic telephone switching was introduced in three phases:

- *Phase 1:* between exchanges in the same urban area, called *junction* or *interoffice switching*
- *Phase 2:* between exchanges in the same country, called *toll switching* in the United States, *trunk switching* in most other countries; also called *long-distance switching*
- *Phase 3:* between telephone networks of different countries, called *international direct dialing* (IDD)

Phase 1 took place between 1920 and 1930 in the great urban centers all over the world. An understanding of the worldwide penetration of the various switching systems can be obtained from Table 16.1, which lists the introduction of automatic switching in chronological order up to the year 1930.

Phase 2 had its early beginning in October 1923, when the first automatic interexchange operation for a network other than that of a large city was brought into service in the Weilheim region of Bavaria. The panel system used exclusively by the AT&T network, introduced in 1921, covered both phases 1 and 2. Application of long-distance switching in the other countries started around 1930 in Switzerland but due to World War II, became widely used only around the 1970s. Continuation

TABLE 16.1   Chronology of the Worldwide Introduction of Automatic Switching

Year	Country	Town	Switching System
	First-Generation Pioneering Systems		
1892	United States	La Porte, IN	Strowger
1901	Canada	Whitehorse	Strowger
1905	Canada	Toronto, etc.	Lorimer
1908	Germany	Hildesheim	Strowger
1909	Germany	Dallmin/Potsdam	Siemens
		Munich	Siemens
1910	Austria	Graz	Strowger
	Germany	Altenburg	Siemens
	Hungary	Budapest	Strowger
1912	Australia	Geelong	Strowger
	Great Britain	Epsom (London)	Strowger
1913	France	Nice	Strowger
1914	Great Britain	Darlington	Rotary
		Hereford	Lorimer
	Sweden	Stockholm	All-relay
1915	Great Britain	Accrington	Strowger
		Chepstow	Strowger
		Newport	Strowger
	United States	Newark, NJ	Panel
1916	Great Britain	Blackburn, Paisley	Strowger
		Portsmouth	Strowger
		Dudley	Rotary
		London	All-relay
1917	Switzerland	Zurich	Rotary
1918	Great Britain	Leeds	Strowger
		Grimsby	Siemens
1919	Great Britain	Stockport	Siemens
	Second-Generation Full Commercial Systems		
1920	The Netherlands	The Hague	Rotary
1921	Great Britain	Hurley, Ramsey	Siemens
	Norway	Oslo	Rotary
1922	Finland	Helsinki	Siemens
	France	Fontainebleau	All-relay
	Germany	Berlin	Siemens
	Great Britain	Fleetwood	All-relay
	India	?	All-relay
	New Zealand	Oamaru	Rotary
	United States	Omaha, NE, etc.	Panel
		New York, etc.	Panel
1923	Austria	Vienna	Siemens
	Belgium	Brussels	Rotary
	Chile	Valparaiso	Strowger
	Cuba	Havana	Strowger
	Denmark	Copenhagen	Rotary
	Germany	Wertheim (trunk)	Siemens

(Continued)

TABLE 16.1 *(Continued)*

Year	Country	Town	Switching System
	Luxembourg	Luxembourg	Siemens
	The Netherlands	Rotterdam	LME
	Great Britain	Southampton	Siemens
1924	Argentina	Buenos Aires	Strowger
	China	Shanghai	LME
	France	Bordeaux	Strowger
		Lyon, Nice	Strowger
		Dieppe	LME
	Germany	Munich, Dresden	Siemens
		Leipzig, Hamburg	Siemens
	India	Bombay	Strowger
	Iraq	Basra	Strowger
	Italy	Verona	LME
	South Africa	Port Elizabeth	LME
	Sweden	Stockholm	LME
	Switzerland	Geneva	LME
1925	Italy	Roma, Torino	LME
1926	Egypt	Cairo	Strowger
	Japan	Tokyo	Strowger
		Yokohama	Siemens
	Mexico	Mexico City	LME
	Spain	Madrid	Rotary
1927	China	Shanghai	Rotary
	Great Britain	London	Strowger
	Ireland	Dublin	Strowger
	Romania	Bucharest	Rotary
1928	France	Paris etc.	Rotary
	Hungary	Budapest	Rotary
1929	Russia	Rostov-on-Don	LME
1930	Brazil	Rio de Janeiro	Rotary
	Hong Kong	Hong Kong	Strowger
	Turkey	Ankara	Rotary

of the implementation of automatic switching after World War II is covered in Section 29.2.

16.3   ELECTROMECHANICAL INDIRECT-CONTROL SYSTEMS

Indirect-control switching systems were developed originally for semiautomatic operation to eliminate most manual operations at the exchange and—contrary to direct-control step-by-step systems—without shifting the control function to the subscriber. Around 1910, semiautomatic switching, which eliminated most of the manual apparatus in the exchange, replacing it with automatic equipment controlled not by the subscribers but by the operator, was a valid alternative to automatic switching, for the following reasons:

1. It relieved the subscriber of any additional effort and thus reduced erroneous operation.

2. It did not involve any change in the apparatus installed on the subscriber's premises.

3. It required little operator intervention time, and thus—although requiring more telephone operators than were required for automatic operation—the number of operators could be reduced considerably compared with manual operation.

The major development of semiautomatic switching was made in North America, where successively the following systems appeared in service in the period from 1910 to 1915:

- The automanual and all-relay systems in the United States for independent operators
- The Lorimer system in Canada
- The panel system in the United States for AT&T
- The rotary system in Europe and also in the United States for AT&T

Whereas the automanual system remained semiautomatic, the other three systems became fully automatic. In fact, during AT&T's period of preference for semiautomatic switching, both the panel and rotary semiautomatic systems were offshoots of the non-AT&T fully automatic Lorimer system and beginning in 1923 were upgraded to fully automatic operation. Other fully automatic systems were developed in Sweden: the LME 500-point system, and a similar system in Switzerland: the Hasler Hs 31 system.

16.3.1 Automanual and All-Relay Systems

Development of the automanual system started in 1906 with a patent received by Edward E. Clement, an industrial property lawyer in Washington, DC. The North Electric Company of Galion, Ohio, produced and installed the system. Charles H. North founded the company in Cleveland, Ohio in 1884, and it is claimed to be the country's oldest manufacturer of equipment for the independent telephone industry. Early in the twentieth century, Charles North went into partnership with Ernst Faller, a German citizen living in New York City, who in 1901 received U.S. patent 686,892 for his automatic *self-acting telephone system*. In 1907, North and Faller joined forces with Clement to produce the automanual system. With the automanual system, subscribers could use conventional telephone sets without a dialing device. Upon lifting the handset, the calling subscriber activated a double-search operation in the telephone exchange: first to find the calling line and then to find a free operator. An all-relay circuit was used for this line finder. The free operator's set was switched automatically and put into a listening position so that the subscriber could mention the number of the line desired. The operator entered that number on a keyboard (which was much quicker than dialing on a rotating disk), and pressing a *startup key* started a pulsing device sending pulses to the line selectors of the exchange. Like the Strowger selector, the North Electric selector had two movements, but with a rotary movement and a subsequent shift on a horizontal axis only.

Automatic equipment allowed metering and disconnection of a line at the end of a call without operator intervention. Once a call was set up, the operator could handle the next call. Thus the period of intervention by an operator, and accordingly the period of seizure of the circuits between the operators and the automatic equipment was reduced to a minimum. Consequently, compared with manual operation, the number of operators was considerable less. It was claimed that in the automanual service an operator could handle 1500 calls per hour compared with 230 to 250 in a manual exchange. The first automanual exchanges were installed in Ashtabula and Lima, Ohio, in 1914. Western Electric acquired the manufacturing rights in 1916.

The North Electric Company developed another telephone switching system around 1913, which used relays exclusively and hence was called the *all-relay system*. A first and unique application was the use of an all-relay solution for the line finders in the automanual exchange installed at Lima. A fully automated version of the all-relay system was installed as a private exchange at the Galion High School in 1920. The first public exchange of this type was installed at Copley and River Styx, Ohio, in 1929. It was then called the CX (city exchange) and installed in several thousand small public exchanges served by independent companies. The all-relay CX system owed its success to its high dependability and its very low maintenance requirements. In 1951 the North Electric Company was taken over by L.M. Ericsson to adapt and manufacture L.M. Ericsson crossbar equipment.

16.3.2 Lorimer System

Three Canadian brothers, George William, James Hoyt, and Egbert Lorimer, worked in their early youth in a company set up in Brantford, Ontario, Canada, by Romaine Callender. Callender, who was a music teacher and organ builder, after building an automatic organ player, designed an automatic switching system for telephony. He designed three different systems, patented successively in 1892, 1894, and 1896. Callender was not successful with his switching adventures and the Lorimer brothers took over the Callender Company, which in 1899 became the *American Machine Telephone Company*. On April 24, 1900, the Lorimer brothers filed a patent for the *Lorimer system*, which was a further development of Callender's third system.

Over the next 10 years the characteristics of the Lorimer system were destined to influence the design of an entire range of systems known as *indirect-control systems*. The main innovations of the Lorimer system were the following:

- The selectors had one (rotary) movement only.
- A number of selectors were installed horizontally one above the other, all driven by a central motor with a vertical shaft.
- Selector rotation was controlled by electromagnets that connected/disconnected a selector from the rotating vertical shaft.
- A number-setting device was used by subscribers instead of a dial.
- "Revertive" pulses, created during rotation of the selector, were sent back to the number-setting device, which then counted down to zero and initiated disconnection of the selector from the driving shaft. The selector thus stopped on the

Figure 4.2 Volta explaining the voltaic pile to Napoleon, as painted in 1897 by G. Bertini. (Courtesy of Musei Civici Como.)

Figure 5.4 Restored optical telegraph station at Chatley Heath.

Figure 6.12 ABC pointer telegraph. (Courtesy of Museum für Kommunikation, Frankfurt, Germany.)

Figure 6.16 Typograph made by the clockmaker Leonhardt, 1844. (Scanned from *Post- und Telekommunikationsgeschichte*, Vol. 2, 2000, p. 61, with permission of the Deutsche Technikmuseum, Berlin.)

Figure 10.2 Telephon of Reis, 1863: left, the receiver; right, the transmitter. (Courtesy of Museum für Kommunikation, Frankfurt, Germany.)

Figure 10.7 Bell's butterstamp telephone, 1877. (Courtesy of Museum für Kommunikation, Frankfurt, Germany.)

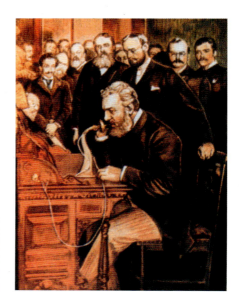

Figure 10.15 Inauguration of the New York–Chicago telephone line by A. Graham Bell on October 18, 1892. (Scanned with permission of the ITU from Catherine Bertho Lavenir, *Great Discoveries: Telecommunications*, International Telecommunication Union, Geneva, 1990, p. 39.)

Figure 10.16 Wall telephone, 1888. (Courtesy of Museum für Kommunikation, Frankfurt, Germany.)

Figure 10.17 Luxury table telephone used by the Vatican, 1900. (Courtesy of Museum für Kommunikation, Frankfurt, Germany.)

Figure 10.18 Luxury telephone handset, late 1890s. (Courtesy of Museum für Kommunikation, Frankfurt, Germany.)

Figure 11.3 Manual exchange used in Germany around 1900. (Courtesy of Alcatel SEL.)

Figure 20.1 German cable ship *Stephan* at Yap, 1905. (Scanned from *Archiv für deutsche Postgeschichte*, Vol. 1, 1982, p. 28.)

Figure 25.1 Typical radio-relay repeater station in Thailand. (Courtesy of Gerd Lupke.)

Figure 25.15 Inside view of a self-supporting radio-relay tower in Malaysia, 1975. (Courtesy of Alcatel SEL.)

Figure 27.7   Satellite HS 601 HP made ready for launch, 1999. (Courtesy of Boeing Satellite Systems.)

Figure 27.10   Ocean satellite launching platform, 1999. (Courtesy of Sea Launch.)

Figure 27.16   Satellite de-orbiting with a Terminator Tether. (Courtesy of Tethers Unlimited.)

Figure 32.7 UMTS prototype handset, 2000. (Courtesy of Siemens Press Photo.)

Figure 33.4 Public telephone in the center of Bogotá, 1970.

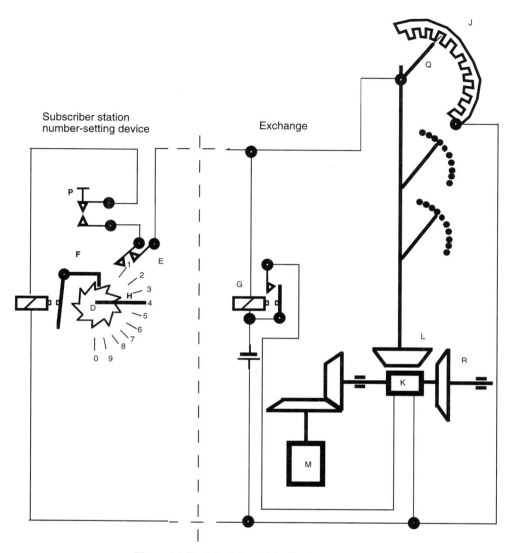

Figure 16.8 Principles of the Lorimer system.

position corresponding to the digit marked by the subscriber on the number-setting device.

The Lorimer system is shown in Figure 16.8 and described concisely in Technology Box 16.2. Although ingenious and trend setting for a range of indirect-control systems that competed successfully over half a century with direct-control step-by-step systems, the Lorimer system found only very limited application. A few Lorimer exchanges were installed in Canada, at Toronto, Brantford, and Peterboro, in 1905. A few test prototypes were installed in Europe. Two in England, one of them in Hereford, was put into service on August 1, 1914, just before World War I broke out. Two Lorimer exchanges were installed in France, in Paris and in Lyon, and

TECHNOLOGY BOX 16.2

Lorimer System

The operation of the Lorimer system is explained briefly, with Figure 16.8 showing:

(a) *Subscriber station*, with the number-setting device consisting of:
  - A series of two to four levers H (one for the units, one for the tens, one for the hundreds, and one for the thousands). The drawing shows one lever only, with which the subscriber composes the desired number (in the example, line 4).
  - A switch P, which the subscriber has to push down after having set the number desired. (This switch thus has a function equivalent to that of lifting the handset from its hook in today's telephones.)
  - An electromagnet with armature F, which moves ratchet D with lever H one step back at the reception of each revertive pulse from the exchange.
  - The interrupter E, which interrupts the line selection as soon as H has returned to its idle position.
(b) *Exchange* with:
  - The motor K, which drives wheel R continuously.
  - The relay G, which energizes the electromagnet K as soon as the subscriber has pushed switch P.
  - The electromagnet K, which upon being energized by relay G attracts wheel R toward wheel L and starts the rotation of the selector shaft.
  - The contact arm Q, which rotates simultaneously with the two selector wipers and generates the revertive pulses by falling into the notches of the toothed contact disk J.

another in Italy, in Rome. Then the Lorimer brothers' American Machine Telephone Company disappeared. In line with the American proverb that "it's the pioneers that get the arrows and the settlers that get the land," the Lorimer patent was purchased by Western Electric in 1903 and became the starting point for a successful AT&T family of indirect-control switching systems.

The initial efforts of Western Electric in the domain of automatic telephone switching were directed toward developing an automatic system for rural areas where the cost of operators could not be covered by the limited traffic volume. As early as 1884, Western Electric took out a number of patents for a *Village System*, which involved the party-line concept but without a central exchange. The various lines in that system were all looped through each subscriber's station, from which calls could be set up by a system of keys manipulated by the subscribers themselves. This meant, for example, that 20 to 40 subscribers could be served with only four telephone lines. In 1900, another rural system was worked out for serving 20 to 100 subscribers, this time from an automatic exchange using single-level selectors. The calling subscriber had to move a rotary arm over a dial with, in compliance with the capacity of the exchange, 20 to 100 positions. About 40 of those exchanges were installed between 1902 and 1904.

Subscriber dissatisfaction and the extension limitations of the one-level selection in 1902 brought AT&T to a change in policy: manual operation for small offices and semiautomatic operation for large offices. Western Electric was requested to develop a project for a 10,000-line exchange. Having obtained the Lorimer patent, Western Electric decided to improve that system for use on large semiautomatic exchanges. In 1906, two development teams from Western Electric were given the task of developing an exchange for use in large cities. Although working independently and almost in competition, both teams regularly compared their ideas and agreed on common features, but they followed different approaches to the mechanical realization, finally resulting in two different systems, the panel and rotary systems, with the following features in common:

- The number-setting device of Lorimer's system is shifted from the subscriber to the exchange.
- The subscribers' stations are equipped with a dial similar to that used for the Strowger systems.
- A *register-translator*, also called a *sender*, registers the decimal dialing pulses of a calling subscriber and translates these pulse sequences into a new series of pulses (not necessary on a decimal basis) controlling the successive operations required at the exchange, as well as in the corresponding exchange(s) in case the subscriber called is connected to another exchange.
- A line finder connects a calling subscriber automatically upon lifting the telephone handset with a free position in the register, or in semiautomatic operation, with a free operator.
- A number of selectors are clutched individually to the permanently rotating shaft of a common motor.
- Each selector sends revertive pulses to the register, which are used to control the rotation of the selector.
- The contact banks of the selectors are no longer limited to groups of 10 lines (as was required in direct-control systems) but are extended to 100 or more lines, thus substantially reducing both the number of selectors and the probability of finding a busy line.

16.3.3 Panel System

In designing the panel system, particular attention was given to finding a solution that enabled economical industrial production and minimized expensive time-consuming wiring work during installation on site. This led to a construction, obviously derived from a manual switchboard, whereby the line jacks are replaced by multiple accessible metal strips placed horizontally one above each other in a flat contact bank, in front of which vertically moving selector rods replace the vertical movement of the operator's arms. In total, 500 telephone lines arranged in five separate panels one above the other were served by 60 selector rods, 30 at each side of the panels, so that up to 60 of the 500 lines can be busy simultaneously. Popularly expressed, the panel system can be compared with a 500-story skyscraper served by

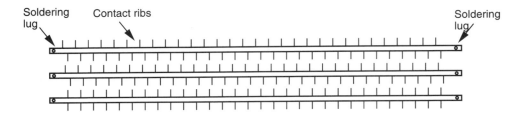

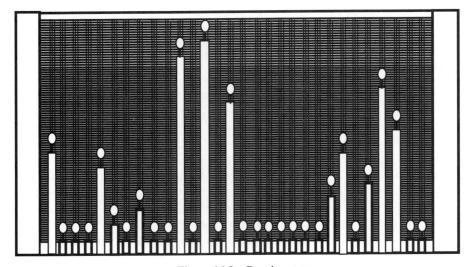

Figure 16.9   Panel system.

60 elevators. Figure 16.9 shows one row of three metal strips (for one subscriber) and the front view of a panel with 30 selector rods.

In 1912, the studies conducted on two fronts—panel and rotary—by AT&T demonstrated the potential superiority of the panel system for serving large bundles of trunk circuits between exchanges, a necessity expected to be met in view of the rapid increase in telephone density in the large cities. It was decided, therefore, to concentrate on perfecting the panel system rather than using the rotary system.

After a test installation in 1912 (a private exchange at the headquarters of Western Electric), two semiautomatic exchanges were put into operation in 1915 in Newark, New Jersey. It was not until November 1918 that AT&T decided to go over to fully automatic panel exchanges to serve large cities. Automatic switching between the telephone lines of all subscribers on the national telephone network required that each telephone subscriber have a specific number, which was to be dialed to move the selectors in the exchanges to the subscriber called. Initially, this number could correspond with the order of application of new subscribers of the local exchange. With the introduction of national automatic switching, a code number for each local network had to be added to the local number.

In 1917, an AT&T engineer named Blauvelt had the simple but brilliant idea of combining digits and letters on the dialing disk. AT&T could then envisage seven-

TABLE 16.2 Combinations of Letters and Digits

Digit	Letters					
	Berlin	Copenhagen	London	New York	Paris	Sydney
1	A	Exchange	—	—	—	A
2	B	ABD	ABC	ABC	ABC	B
3	C	EFG	DEF	DEF	DEF	E
4	D	HIK	GHI	GHI	GHI	J
5	E	LMN	JKL	JKL	JKL	L
6	F	OPR	MN	MNO	MN	M
7	G	STU	PRS	PRS	PRS	U
8	H	WXY	TUV	TUV	TUV	W
9	J	ÆØ	WXY	WXY	WXY	X
0	K	—	O	Z	OQ	Y

digit dialing whereby the first three digits represented the first three letters of the name of the town (or town area) of the relevant exchange (e.g., BOS for Boston). This combination of letters and digits found international application. Table 16.2 gives a summary of the combinations that have been used in six major capitals.

The first fully automatic panel exchange was brought into service in Omaha, Nebraska, in 1921, soon followed by the first panel exchange in New York City, the

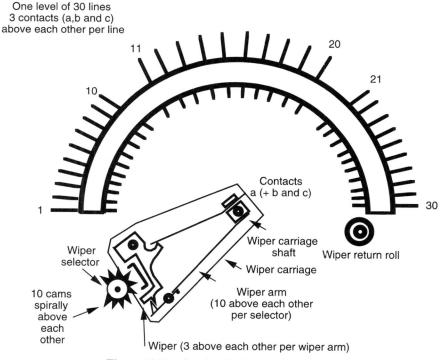

Figure 16.10 One level of a rotary selector.

Pennsylvania exchange, which began operating on October 14, 1922. From then on telephone automation by the panel system took off at high speed in New York and other large cities. At the peak of its deployment, toward 1958, when crossbar systems were advancing, too, some 7 million subscribers in 26 large cities were served by panel exchanges. The panel system remained in exclusive use by the Bell System companies in the United States; no single unit has been exported. From the end of the 1950s, the number of panel exchanges diminished steadily as they were replaced after 30 or 40 years of satisfactory operation by modern, mainly crossbar exchanges. In 1977, only 0.4% of Bell System subscribers were still connected to a panel exchange, and in the early 1980s the era of the panel system came to an end.

16.3.4   Rotary System

With its cylindrical selectors, the rotary system is more clearly descended from the Lorimer system than is the panel system, with its flat selectors. Moreover, the contact banks of the rotary system are explored by a horizontal rotary movement, not as in the panel system by a vertical upward movement. Unlike the Lorimer and Strowger systems, the rotary selection is no longer on a decimal basis but is based on 30 lines per level. Ten levels, each of 30 lines, are arranged in an arc one above the other, resulting in 300 lines per selector (200 lines in an early version). Figure 16.10 shows one level of the switch, and Technology Box 16.3 concisely describes the rotary system.

TECHNOLOGY BOX 16.3

Rotary System

The rotary selector has 10 contact banks, one above the other, each with 30-line contact sets. One set consists of three contacts one above the other: two for the two line wires and one for a test wire per line. Figure 16.10 shows one level of contacts together with its corresponding wiper carriage. Ten wiper carriages are connected with the common wiper carriage shaft one above the other, each at the level of its corresponding line contacts. Here also three wipers are located one above the other for the three contacts of one line. All 10 wiper carriages rotate together. In front of the 10 wiper carriages is a wiper selector with 10 spirally arranged cams. This wiper selector can take 10 different positions, in each one of which one cam will be in line with the wipers of the wiper carriage, corresponding to the level desired and then releases the three wipers of that wiper carriage only. At the end of a rotation, a fixed disk pushes the wipers back to their idle positions. Thus, at any rotation one wiper carriage will explore only 30 line contacts.

The two vertical shafts of the wiper selector and of the wiper carriages are driven by a common motor and controlled by electromagnetic clutches. The revertive pulses, which enable the translator-register to detect the position adopted by the shafts and to stop their rotation at the position desired, are generated by the notches on a wheel at the top of each shaft, over which passes a roller or a set of contact springs.

The rotary selector was originally called the *McBerthy selector*, after the head of the Western Electric research team that researched the system. When in 1911 AT&T decided to adopt the panel system, which with 500 lines had a larger selector bank than the rotary system and thus was considered more economical for large U.S. cities, the rotary system was considered more adequate for the European continent, which still had very modest telephone densities. McBerthy went to Europe and terminated development of the rotary system in the same year at BTM in Antwerp, Belgium.

The first rotary exchange, made in Belgium, was installed in Landskrona, Sweden, in 1912 and the second in Angers, France, in 1913. These first two exchanges were still semiautomatic. The first fully automatic rotary exchange, with a capacity of 2800 lines, was then put in operation in Darlington, England, on October 10, 1914, one day after the German occupation forces requested BTM to close its factory. Fortunately, a great number of drawings, tools, and machines could be shipped to Western Electric's sister company, STC, in Great Britain. Also, many employees of BTM could escape from occupied Belgium and contribute to a continuation of development and production of rotary exchanges during World War I, initially at STC and later also at the Western Electric plant in Hawthorne, Illinois, near Chicago. So despite the war, rotary exchanges were installed in 1916 at Dudley in England, with 400 lines, and in 1917 in Zurich–Hottingen, Switzerland, and in Bergen, Norway.[8] The exchange in Zurich, commissioned on July 29, 1917, with 7000 lines was the largest manufactured so far. It was initially operated semiautomatically but was upgraded in 1922 to a fully automatic exchange with 10,000 lines, the first in Switzerland, and was extended successively to a final capacity of 40,000. After the war, in 1919, a 9000-line semiautomatic exchange was put into operation in Marseille, France. It was fully automated in 1927.

In March 1919, the machine tools were returned to BTM and production of rotary equipment could be resumed in Belgium. The exchanges already ordered before the war for South Africa and New Zealand could finally be delivered in 1922. The exchange for Oamaru in New Zealand was put into service in January 1922 and remained in operation until 1972: 50 years of service, a record of longevity.

The first automatic rotary exchange in the homeland of BTM was put into operation in Brussels in 1922. This was the first of a series of exchanges of the same type, which made the Brussels network fully automatic by 1928.

Rotary exchanges were then introduced in a large number of major European cities, such as in the Hague in 1920; Oslo in 1921; Copenhagen in 1923; Geneva in 1924; Basel in 1926; Bucharest in 1927; Paris and Budapest in 1928; and Barcelona, Liège, and Antwerp in 1931. Rotary exchanges outside Europe were installed in Shanghai in 1927 and in Rio de Janeiro in 1930. Rotary exchanges appeared in a series, 7A to 7EN[9] of steadily improved versions, summarized in Table 16.3.

The first application of rotary switches for long-distance switching was made in Switzerland in 1933 between the cities of Basel and Zurich. Long-distance switching with Strowger exchanges was introduced almost simultaneously between Berne and

---

[8] Which was fully destroyed shortly after commissioning by the great fire of Bergen.

[9] The denomination "7" for the rotary system family bears no relation to a previous series of switching equipment but is simply derived from the administrative fact that No. 7 was the code number for Western Electric equipment intended for use *exclusively outside the United States*.

TABLE 16.3   Rotary System Series

Version	Year	Basic Modification	First in Service
7A	1911	—	Landskrona, 1912
7A1	1922	Gear-drive clutch instead of magnetic friction, 100-point line finder (instead of 60), double contacts	Nantes, France, 1925
7A2	1927	More compact design, 200-point line finder, improved register access	Bucharest, 1930
7B	1927	Level selector replaced by an additional control circuit	Barcelona, 1931
7B1	1952	Various mechanical improvements	Enghien, France, 1954
7D	1929	Forward pulses instead of revertive, auxiliary rotary level marking switch	Zürich, 1930
7E	1953	Cold-cathode tubes for controlling the selectors by means of level marking with different electrical potentials and phases, multifrequency signaling	Scheveningen, The Netherlands
7EN	1963	Tubes replaced by transistors and wired circuits by printed circuits, simultaneous rotation of first and second line and register finder, one-step selector outlet control instead of in three steps	Belgium, 1965

Lausanne, whereas the first application of interworking between a rotary exchange in Geneva and a Strowger exchange in Lausanne was also introduced at that time.

The life span of the rotary system (i.e., the manufacture of equipment under that name) lasted from 1910 to about 1975. The vast majority of rotary exchanges were manufactured in the BTM plant in Antwerp. Forced by national interests for local production of vital telecommunication equipment, however, rotary equipment was also produced in other factories that belonged to IT&T, in France, Spain, Switzerland, Italy, the Netherlands, and Norway. At the end of the production period the rotary system had been supplied to all continents in the quantities shown in Table 16.4.

16.3.5   Uniselector System in France

Thomson Houston in France developed a rotary step-by-step uniselector switching system in the period 1924–1927.[10] The system, called the *R6 system* (R for *rotatif*), was a direct-control system designed for exchanges with a capacity of 2000 to 4000 lines. The first of these exchanges was put into service in Troyes in 1928, followed by some 30 others between 1928 and 1939.

---

[10] Shortly before the company was taken over by ITT and renamed Compagnie Générale de Constructions Téléphonique (CGCT).

TABLE 16.4   Number of Lines Using the Rotary System Worldwide

	Type				
	7A	7B	7D	7E	Total
Subscriber lines (millions)	4.6	0.8	3.1	1.2	9.7
Trunk circuits (thousands)	56	28	94	167	345

16.3.6   LME 500-Point System

The LME 500-point system is indirectly another offspring of the Lorimer system. Around 1909, the Televerket considered the possibility of automating their telephone networks in Stockholm and Göteborg. They sent their engineers Axel Hultman and Herman Ollson on a mission to the United States, then the leading country in telephone deployment. Upon their return to Sweden the two engineers recommended that an entirely automatic system adapted to Swedish conditions from the American panel and rotary systems would be the preferred solution. Televerket followed their recommendation and requested Hultman and L.M. Ericsson jointly to develop a prototype system. This study resulted in 1918 in a prototype installation of the LME 500-point exchange with a two-movement rotary/radial selector. The contact bank of each selector consisted of 25 frames of 20 lines arranged horizontally in the shape of a quarter-circular fan with 25 radii. To select the desired frame, the selector wiper arm made a rotary movement. To select the desired line within a frame, the selector wiper arm made a radial movement along the frame selected up to the line desired. Each selector thus could explore $25 \times 20 = 500$ lines. The system is shown in Figure 16.11 and described concisely in Technology Box 16.4.

The first automatic LME 500-point exchange was not installed in Sweden but in The Netherlands. A 5000-line exchange was put into service on May 10, 1923, in Rotterdam, at the exchange of the *Rotterdam Municipal Telephone Administration*. In the same year, two exchanges were installed in Norway at Hamar and at Christiansund.

The first LME 500-point exchange in Sweden was commissioned in 1924 at the Norra Vasa exchange in Stockholm and then adopted by Televerket as the standard system for all major exchanges in Sweden.

By 1920 many administrations and countries had already chosen another automatic system for their large cities. As a result, the LME 500-point system was introduced mainly in countries that did not yet have automatic exchanges or where telephone operations were divided among several operators, such as in The Netherlands, Norway, Italy (Verona in 1924), France (Dieppe in 1924), and in the USSR in 1927.

Outside Europe, the LME 500-point system was installed in China (Shanghai), in South Africa in 1924, and in Mexico City in 1926. For more than 50 years, successive versions of the LME 500-point system have been installed serving over 5 million subscriber lines all over the world. Reportedly, one of the merits of the LME 500-point system was the limited maintenance required, due to its robust mechanical structure and the easy replacement of faulty units.

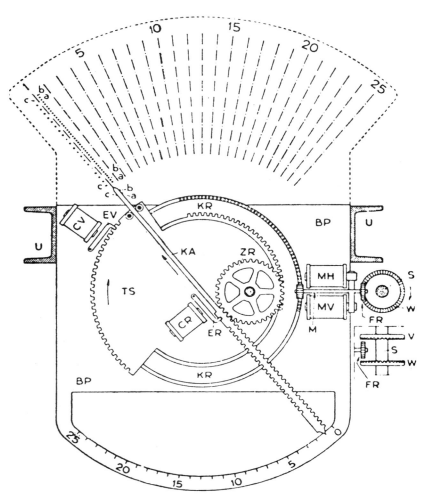

Figure 16.11   Mechanism of the L.M. Ericsson 500-point selector. (Scanned from D. van Hemert and J. Kuin, *Automatische Telefonie*, Corps Technische Ambtenaren, 5th print., 1933/ 1953, p. 13.)

16.3.7   Hasler Hs 31 System

In the early 1920s, when automatic telephone switching was beginning to expand in Switzerland, Hasler S. A. of Bern began to make telephone switching equipment. Hasler obtained manufacturing licenses from L.M. Ericsson, first for an all-relay exchange based on a patent obtained by Gotthilf A. Betulander in Sweden in 1912 and then an exchange for rural application: type Hs 25 with 25-point rotary line finders derived from the LME 500-point system.

In 1931, Hasler developed its own system type Hs 31, a 100-point two-motion (rotary/radial) selector in design and mechanism similar to the LME 500-point system but much slower (30 instead of 200 steps per second). The exchange capacity was initially limited to a maximum of 2000 lines but in 1934 extended to 10,000.

TECHNOLOGY BOX 16.4

L.M. Ericsson 500-Point System

The LME 500-point system (Figure 16.11) consists of flat selectors, of which 40 to 60 are placed horizontally one above the other in a rack. A common continuously rotating shaft (S in the drawing) drives all selectors of a rack. The operation of the selector can be summarized as follows:

1. An electromagnet with coils MH and MV and an armature with a gear wheel at each end controls the two selector movements. Depending on the position of the armature, the first gear wheel FR can be coupled to the upper or lower of the two superimposed driving wheels W. In the rest position, FR is disengaged from both wheels.

2. The second gear wheel FR, mounted solid with the first, is permanently meshed with the toothed outer rim of a horizontal ring KR. The latter has a second set of teeth on its inner rim which engages with a toothed wheel ZR, which is also horizontal and eccentric in relation to the ring: It is this toothed wheel that engages with the rack on the wiper arm KA. Finally, a selector disk TS, mounted on the same shaft as the ring KR but not solid with it, is fitted with two rollers between which the wiper arm KA slides. Locking of magnet CV, which is in a spatial fixed position, stops the selector disk TS when its armature is at rest. A second locking magnet, CR, mounted on the selector disk TS, stops the wiper arm rack in relation to TS when its armature is at rest.

3. When the rotary movement starts, owing to excitation of the clutch magnet MV, the magnet CV is excited, freeing the selector disk TS while CR remains at rest, so that the toothed wheel ZR, the wiper rack, and the selector disk TS are locked with the ring KR and turn with it about the vertical shaft of the selector.

4. When the rotary movement (i.e., the radial selection) must be stopped in the desired direction, that is, when the end of the wiper arm is opposite the desired contact frame, CV drops back, immobilizing the selector disk TS, and at the same time, CR is excited, freeing the rack; since KR continues to rotate, the toothed wheel ZR, the shaft of which has been immobilized by the stoppage of TS, begins to turn and causes a radial movement of the wiper arm which advances into the frame until it reaches the desired line among the 20 lines on the frame, where it stops.

5. If magnet MH is excited instead of magnet MV, the reverse movements are produced.

6. Group selectors and final selectors have a rest position to which they return at the termination of the selection. The line finder has no rest position. At the end of the call, the wiper arm returns to a rest position, leaving the contact frame with which it had established a connection, but remaining opposite to it.

TECHNOLOGY BOX 16.4 *(Continued)*

7. Revertive pulses are sent to the register by means of a toothed ring (not shown on the drawing) mounted on the selector disk TS. The selector is controlled by a register that converts the decimal numbers dialed by the subscriber into a fixed nondecade notation for the setting of the 500-point selector.

The Hs 31 was replaced by a new system, Hs 51, a register system using rotary uniselectors, introduced in 1953 at Wohlen (Argovia). By 1958, over 70% of Swiss exchanges used Hasler equipment, serving some 550,000 lines, 45% of all Swiss subscribers.

16.3.8   Automatic Switching Systems in the USSR

Since 1927, manufacture of the LME 500-point system under L.M. Ericsson license took place in the USSR in the Krasnaya Zara factory in Leningrad. The first exchange produced in this factory was commissioned at Rostov-on-Don in 1929. At the end of World War II, as most of the telecommunications infrastructure had been destroyed, including the Krasnaya Zara factory, it was decided to reconstruct the plant and to change to production of a 10-step type ATS 47 switch based on a pre-war German model. Production of the ATS 47 started in 1947. An improved version ATS 54 was designed in 1954. During a 30-year period some 11 million subscribers lines were installed with the ATS equipment.

16.4   CROSSBAR SWITCHING

Crossbar switching, as the name suggests, operates a lattice of rectangular crossed bars. The switch operates magnetically in such a way that it is free from moving brushes and sliding contacts. Crossbar switching originates in a development made by the Western Electric engineer J. N. Reynolds in 1913.[11] The basic operation principle is shown in Figure 16.12 and explained briefly in Technology Box 16.5. There are two main divisions in a crossbar switch: the *control subsystem*, which establishes the talking *path* within the application of a *marker*, and the *switching network subsystem*, with crossbars. With crossbar switching the dial pulses are stored in a *register* temporarily.

The Swedish engineer Gotthilf Ansgarius Betulander was first to construct a crossbar switch in 1919. Basically, he made two major modifications to Reynolds's switch:

1. The function of the bars was reversed: The horizontal bars became the selecting bars and the vertical bars the holding bars.

[11] U.S. patent 1,131,734 of March 16, 1915.

Vertical selecting bar

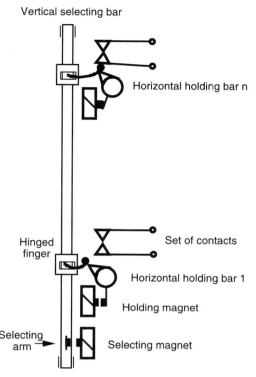

Horizontal holding bar n

Set of contacts

Hinged
finger

Horizontal holding bar 1

Holding magnet

Selecting
arm

Selecting magnet

Figure 16.12 Principles of the crossbar switch of J. N. Reynolds, 1913.

2. Flexible steel wire fingers replaced the cams and rollers.

The basic advantages of Betulander's crossbar switch and thus of crossbar switching in general are:

1. Single point per line used for both outgoing and incoming calls (thus no need for multiplying lines)
2. Easily adaptable for common control systems
3. High transmission quality, due to high contact pressure
4. High operation speed
5. Robust mechanical construction hardly requiring any preventive maintenance

Betulander realized that the development effort still required for making his switch suitable for large-capacity exchanges would surpass his financial resources, so he sold his little company to L.M. Ericsson and rejoined Televerket in 1920.

By then, Televerket, inspired by AT&T's change of policy toward fully automatic switching, decided to launch a large-scale program of automation of its local networks: in particular, those in the large cities. Four competing systems were taken under real service tests:

TECHNOLOGY BOX 16.5

Crossbar Switch of J. N. Reynolds

The crossbar switch, as conceived by J. N. Reynolds in 1913, uses a small number of vertical selection bars, and in front of each vertical selection bar a larger number of horizontal holding bars. Figure 16.12 shows one selection bar with 1 to n horizontal bars. The vertical bars perform the selection function (like the wiper selector shaft of a rotary switch). When the selecting electromagnet attracts the selection arm of a vertical selection bar, that selection bar makes a small rotation whereby rollers at the end of the hinged fingers are brought under all the sets of contacts at the intersection of that vertical bar and its corresponding horizontal bars 1 to n.

The horizontal bars act as holding bars. When one of these is rotated a half turn by its holding electromagnet, cams on the holding bar, located at each intersection point with a vertical selection bar, turn upward and close the set of contacts located at that intersection point where a roller is brought under the set of contacts.

Once the contact has been made, the selecting magnet and the vertical selecting bar can return to their home position; the hinged roller remains trapped between the cam and the set of contacts (as shown for holding bar n) as long as the horizontal holding bar is kept in active position by activation of the holding magnet during the entire duration of the call. Thus, once a call has been established, the vertical selection bar can be activated again for further call selection by means of other hinged fingers that are not yet engaged.

---

1. Step-by-step Strowger-type switch of Siemens
2. Rotary system of BTM
3. LME 500-point system
4. LME crossbar system developed by Betulander

The crossbar switch was given the modest role of serving Sweden's medium- and very small capacity rural exchanges in a modified step-by-step version manufactured initially by a small factory belonging to Televerket and later by L.M. Ericsson. The first crossbar switch, with a capacity of 7500 lines, was installed in Sundsvall in 1926, followed by Limhamn in 1930 and Malmö in 1932. Nine similar exchanges entered into service between 1937 and 1941. The continuation of crossbar switching after World War II is covered in Section 29.1.

16.5   PRIVATE SWITCHING

Private switching refers to switching effected on the premises of a public network subscriber to support the internal communication within the subscriber's own organization and the external communication with the public network in a time-sharing mode among users of the private switch. The first private switch was installed in 1880

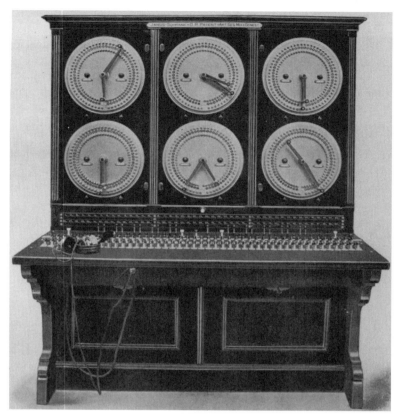

Figure 16.13 Europe's first private exchange. (Scanned from company presentation: Aktien-Gesellschaft Mix & Genest, Telephon- und Telegraphen-Werke, Berlin, 1902, p. 39.)

in Dayton, Ohio, with one main line connectable to seven extensions. In Europe. private switching began in 1900 in Germany with the Janus switch of the company Mix & Genest (now Alcatel SEL). Like the Roman God Janus, the "two-headed" switch could "look" into the public telephone network as well as into the private telephone network. Figure 16.13 shows this private manual exchange.

A manual private switching device is usually referred to as a *private branch exchange* abbreviated as PBX, whereas an automatic private switching device is called a PABX. The world's largest PBX was installed at the Pentagon in Washington in 1942 with 13,000 internal lines and 125 operator positions.

REFERENCES

Books

Chapius, Robert J., *100 Years of Telephone Switching (1878–1978)*, Part 1: *Manual and Electromechanical Switching (1878–1960s)*, North-Holland, New York, 1982.

Feyerabend, E., *50 Jahre Fernsprecher in Deutschland, 1877–1927*, Reichspostministerium, Berlin, 1927.

Gööck, Roland, *Die großen Erfindungen Nachrichtentechnik Elektronik*, Sigloch Edition, Künzelsau, Germany, 1988.

Libois, Louis-Joseph, *Genése et croissance des télécommunications*, Masson, Paris, 1983.

Oslin, George P., *The Story of Telecommunications*, Mercer University Press, Macon, GA, 1992.

Siegmund, Gerd, *Grundlagen der Vermittlungstechnik*, Decker's Verlag, Heidelberg, Germany, 1993.

van Hemert, D., and J. Kuin, *Automatische Telefonie*, Uitgegeven door de Vereeniging van hoger personeel der PTT, Corps Technische Ambtenaren, 5th print., 1933/1953.

17

HIGH-FREQUENCY RADIO TRANSMISSION

17.1 EVOLUTION OF RADIO TECHNOLOGY

17.1.1 Spark Radio Transmitters

The twentieth century began with a spectacular achievement in radiotelegraphy, so far claimed to be impossible by the scientific world. The Morse signal "S" was transmitted across the Atlantic Ocean on January 1, 1901. The experiment was made by Marconi with the assistance of John Ambrose Fleming. A multiple-wire 48-m-high antenna was installed at Poldhu, Cornwall, in England (Figure 17.1), while at Signal Hill, St. John's, Newfoundland, an aerial wire was connected to a kite that stood 130 m high. Fleming operated the transmitter at Poldhu; Marconi, 3500 km away, listened to his receiver at St. John's and confirmed, via Atlantic cable, good receipt of the radio signal. The *New York Times* of December 15, 1901 reported with admiration, "Wireless Spans the Ocean." Contrary to the prevailing theory that electromagnetic waves could be propagated only in a straight line, Marconi demonstrated that radio propagation far beyond the horizon was possible. Two physicists, Oliver Heaviside (1850–1925) in England and the American Arthur Edwin Kennelly (1861–1939), also of British origin, independent of each other in 1902 attributed Marconi's success to the existence of an ionized layer in the upper atmosphere that would reflect radio waves. As described in Chapter 27, their assumption of the existence of an ionized layer proved to be correct: with the addition that there exists not one but at least three such layers. The very low frequency radio waves used by Marconi, however, did not reach those layers, but followed Earth's curvature due to Earth's gravitation.

Telegraphy between Europe and Newfoundland was still the exclusive right of

Figure 17.1   Radio station, Poldhu. (Scanned with permission of the ITU from Anthony R. Michaelis, *From Semaphore to Satellite*, International Telecommunication Union, Geneva, 1965, p. 130.)

the Atlantic Telegraph Company. With the assistance of Alexander Graham Bell, Marconi erected a new station in 1902 at Table Head, Glace Bay, in Nova Scotia 3800 km from Poldhu, followed in 1903 with a station at Cape Cod, Massachusetts, 4800 km from Poldhu. On those stations a funnel-shaped antenna was fixed between four 71-m towers placed in a 70-m square. An ac generator produced 50 kW. The long-wave transmitter operated at a frequency of about 328 kHz (915 m), and the spark voltage varied from 20 to 100 kV.

The first radiotelegraphic transmission of a complete text was achieved on December 17, 1902. Radiotelegraphy between Great Britain and the United States started on January 19, 1903 with an exchange of greetings between King Edward VII and President Theodore Roosevelt. Commercial radiotelegraphy began in 1907 between Glace Bay and Clifden in Ireland, although still limited to press and business messages.

In April 1909, Robert Edwin Peary (1856–1920) sent his radio telegram "I found the North Pole." King George knighted Marconi in 1914. By that time Marconi began installation of 13 long-range radiotelegraphy stations for the British Navy at Ascension Island; the Falkland Islands; Banjul in Gambia; Ceylon, Durban, and Port Nolloth in South Africa; Demerara in British Guiana (now Guyana); the Seychelles; Singapore; St John's (Canada); Aden; Hong Kong; and Mauritius. All stations were in operation by 1916. Marconi made another landmark radiotelegraph transmission in 1918 when a message from his long-wave station at Caernarvon, North Wales, was received in Australia over a distance of 17,700 km.

17.1.2 Squenched Spark Radio Transmitter

In Germany both the *Allgemeine Elektricitäts-Gesellschaft* (AEG) and the *Gesellschaft für drahtlose Telegraphie System Prof. Braun und Siemens & Halske* produced radio equipment with spark transmitters. They were so heavily involved with patent suits that Kaiser Wilhelm II forced them on May 27, 1903, to merge their radio departments into a new company initially called *Gesellschaft für drahtlose Telegraphie m.b.H.* The telegram code of the new company was "Telefunken"; this was soon also used as the short name of the company, and in 1923 the name of the company was changed to *Telefunken. Gesellschaft für drahtlose Telegraphie.*

The spark transmitters used so far produced strongly damped waves, were extremely loud, and produced a high level of ultraviolet radiation that was dangerous for the eyes. The inductive coupling of the spark circuit to an antenna circuit as introduced by Ferdinand Braun in 1898 reduced the damping, but then a substantial part of the energy was lost by a continued oscillation between the two circuits over the duration of each spark. The German physicist Max Wien (1865–1935), born at Königsberg, solved those problems in 1905. Wien developed a special spark bridge consisting of a series of copper disks separated from each other by a gap of 0.5 mm, which produced very short silent sparks of high energy that was extinguished automatically on the first zero level. This caused a pulse excitation in the secondary circuit, which then produced a wave of constant frequency at slowly diminishing amplitude. Figure 17.2 shows the various wave behaviors of the Braun and Wien systems.

Wien's squenched spark transmitter, which increased the efficiency of the transmitter and enabled the generation of higher frequencies, and Braun's crystal detector became the basis for the radio equipment produced by Telefunken. Instead of the

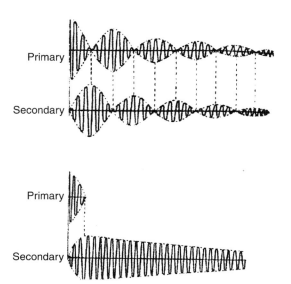

Figure 17.2 Antenna radiation waves according to Braun (above) and Wien (below). (Scanned from *Archiv für Post und Telegraphie*, No. 8 April 1901, p. 562.)

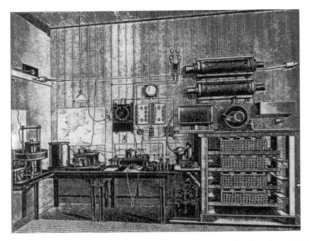

Figure 17.3 Coastal radio station, Scheveningen, The Netherlands, 1905. (Scanned from *Archiv für deutsche Postgeschichte*, Vol. 2, 1993, p. 35.)

loud bangs of the spark transmitter, which could be heard over kilometers,[1] the new Telefunken transmitter produced a decent tone between 500 and 2000 Hz; the equipment was therefore called the *System der tönenden Funken* (the system of the sounding sparks). Coastal radio stations with Telefunken radio equipment operating at wavelengths between 600 and 2000 m were installed at Scheveningen in The Netherlands (Figure 17.3); Montevideo, Uruguay; and Norddeich[2] in Germany in 1905. A large 80-kW radio station was erected in the same year in a swampy area (thus with good grounding) at Nauen near Berlin with a screen antenna carried by a 100-m guyed mast; the mast was increased to 200 m in 1911. On March 30, 1912, the mast collapsed in a storm and was replaced immediately with a 260-m mast. Similar Telefunken stations were built in 1911 at Sayville near New York,[3] Cartagena in Colombia, and in the German colonies at Kamina in Togo, West Africa, at Windhoek in Southwest Africa, and on the isle of Yap in the Pacific Ocean[4] in 1913. At Nauen the antenna power was increased to 375 kW, which made it the most powerful at that time. The stations operated between 6000 and 17,000 m (50 to 17 kHz). Figure 17.4 shows an aerial photograph of the station with the antennas, and Figure 17.5 shows the transmitter room around 1914.[5]

[1] Cynics claimed that the spark transmitter could be heard beyond the distance received.

[2] By request of Kaiser Wilhelm II, the station at Norddeich was installed with equipment produced by Telefunken to replace the station at Borkum (installed in 1900), where a Marconi operator had refused to accept a message from the Kaiser sent by Telefunken equipment on board the German ship *Hamburg* during a trip in the Mediterranean Sea in 1905.

[3] In 1911, Telefunken founded the *Atlantic Communication Company (Telefunken System of Wireless Telegraphy* in New York. The Sayville station was on Long Island.

[4] Smaller stations for radiotelegraphy between the islands in the Pacific Ocean under German protection were also installed in 1913 on the Marshall island Nauru, at Bitapaka near Rabaul on the island Neu-Pommern (now New Britain) in the Bismarck Archipel, and in 1914 at Apia on the island Samoa (now under U.S. protection).

[5] On August 2, 1914, a message was sent repeatedly from Nauen uriously stating "a son is born," but the message was understood by radio operators on ships of the German merchant fleet. One day later, when Germany declared war on France, the ships had arrived at a safe German port.

Figure 17.4   Radio station at Nauen, Germany, around 1914. (Courtesy of Siemens Press Photo.)

Telefunken became the major competitor of Marconi, but after a few years of heavy patent battling, an agreement on patent sharing was made in 1911. They even established a common operating company, the *Deutsche Betriebsgesellschaft für drahtlose Telegraphie* (DEBEG; German operating company for wireless telegraphy) in January 1911, owned 55% by Telefunken and 45% by Marconi. Two years later, Marconi and Telefunken also founded a common company in France called *Société Anonyme de Télégraphe sans Fil* (SA TSF).[6] DEBEG took over the radiotelegraphy of the entire German mercantile marine. In the same year, Marconi acquired control over the Russian company *Wireless Telegraphs and Telephones* and the American company *United Wireless*. Marconi also made an agreement with Western Union on the use of their transcontinental telegraph circuits for a "Wireless Girdle Round the World," which he planned to establish but was stopped by World War I.[7] Telefunken continued its export activities during the war and supplied, for instance, the equipment for a radio network in South America. A central station at Cachendo near Arequipa, Peru, opened radio service on December 8, 1917, with Lima as well as other stations with Telefunken equipment in Argentina, Brazil, Bolivia, Chile, and Uruguay at distances up to 2200 km and passing partly over the 6000-m-high Andes.

[6] TSF merged in 1957 with the Société Française Radio Électrique into the Compagnie Sans Fil (CSF), which is now part of Alcatel.

[7] Marconi and Braun met each other for the first time when both received the Nobel Prize in Physics on December 11, 1909. Upon the beginning of World War I, Marconi started a patent suit in New York against Telefunken, with the aim of closing the Sayville station. Braun was sent to New York to defend his patent. War conditions forced him to remain there, and having suffered from cancer for many years, he died in New York on April 20, 1918.

Figure 17.5   Transmitter room of the radio station at Nauen, Germany, around 1914. (Courtesy of Siemens Press Photo.)

International radiotelegraphy with Germany began again after World War I, in November 1918, when a message received at Nauen from New Brunswick, New Jersey, asked: "Will you accept commercial business messages from USA?" Radiotelegraphy with squenched spark-transmitting equipment over a distance of 20,000 km was achieved in the same year between Nauen and a station in New Zealand.

17.1.3   Poulsen Convertor Arc Radio Transmitter

The damped waves produced by spark transmitters were good enough for the transmission of pulsed Morse signals, but continuous waves are required for the transmission of speech. A Danish physicist, Valdemar Poulsen (1869–1943), born at Copenhagen, was first to develop a continuous-wave transmitter in 1902.[8] Instead of a spark bridge he used an arc lamp to generate electromagnetic waves. The arc lamp, consisting basically of two carbon electrodes subjected to high voltage, was invented in 1821 by the British inventor Humphry Davy and used for lighting. A British physicist, William du Bois Duddell (1872–1917), discovered in 1899 that arc lamps could also be used for the generation of frequencies up to about 1 MHz, albeit for a short duration only, as the carbon electrodes burned out quickly. Poulsen solved that problem by encapsulating the electrodes in a glass tube filled with hydrogen. The positive electrode was made of copper instead of carbon and was water-cooled. A

---

[8] In 1898, Poulsen obtained the world's first patent for a voice recorder, which he called the *telegraphon*. For lack of money and interest, the telegraphon was never put into production.

slowly rotating carbon rod was used as a negative electrode. A pair of electromagnets deionized the gap between the two electrodes to obtain quick and exact extinguishing of the arc.

Transmitters using the Poulsen convertor arc generator were introduced by the Amalgamated Radio Telegraph Company Ltd. in 1906 and were used extensively before and during World War I. Poulsen convertor arc radios were also produced under license by the *Federal Telegraph Company* of California and the German company *C. Lorenz AG*. Radio equipment with the Poulsen convertor arc generator was used for radiotelegraphy until 1907, when Poulsen managed to transmit speech with his radio over a distance of 270 km. Further development of radiotelephony with Poulsen convertor arc transmission was made by C. Lorenz AG. This company developed a special arrangement of 12 parallel microphones that were inserted into the antenna feeder circuit directly to obtain amplitude modulation of the transmitter frequency. This arrangement limited the output power to about 5 kW. The first experiment was made in cooperation with the German Navy on their SMS *Berlin* in 1908. Lorenz–Poulsen marine radios with a transmitting power of 1.5 to 4 kW were installed on Navy ships in the following years and obtained good radiotelephone audibility at short distances. Lorenz achieved radiotelephony over a distance of 370 km between Lyngby in Denmark and Berlin in 1909.

An Italian physicist, Quirino Majorana (1871–1957), found another interesting solution for modulation of the Poulsen transmitter. He developed a hydraulic microphone (Figure 17.6) in which the acid density of water—and thus the electrical resistance—varied as a function of speech. Sound waves of speech in front of the membrane are conducted via the flexible tube (A) to the tube (T), where they push sulfuric acid drops to the glass container (B) and thus modulate the current in the primary antenna circuit.[9] The radiotelephone experiments of Majorana took place mainly between the coastal stations and the torpedo boat *Lanciere* of the Italian Navy. In 1909 he achieved radiotelephony over a distance of 420 km between Rome and Monte San Giuliano near Trapini on the island of Sicily.

In the United States, C. F. Elwell, after studying electrical engineering at Stanford University, left for Europe in 1909. He returned the same year with an option for use of the Poulsen convertor arc in the United States and founded the *Poulsen Wireless Telephone and Telegraph Company*. Elwell made experimental radiotelephone transmission between Stockton and Sacramento in early 1910. *United Wireless* in Sacramento deliberately jammed his experiments, but this gave Elwell the experience that telegraphy was much less affected by interference than was telephony, and one year later he founded a new radiotelegraph operating company, the *Federal Telegraph Company*. This company started regular radiotelegraph service between San Francisco and Honolulu (3850 km) in 1912, albeit at night only, to avoid the high atmospheric interference in daytime, but at a rate substantially below the prevailing submarine cable rates.[10] Night service was also established between Stockton and Chicago, whereas a day-service radiotelegraph network was operated on a chain

---

[9] Majorana had used his powerful microphone successfully in 1905 on a metallic telephone line between Rome and London. At that time, this was not possible with other microphones on the still nonamplified telephone lines.

[10] Due to the radiotelegraph service, the press rate fell from 16 cents to 2 cents and the minimum number of words per day (night) could be increased from 120 to 1500.

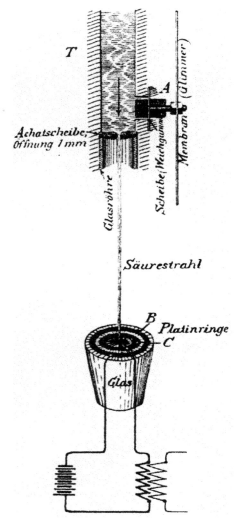

Figure 17.6 Hydraulic microphone of Majorana, 1909. (Scanned from *Archiv für Post und Telegraphie*, No. 24, 1901, p. 297.)

connecting Seattle, Portland, Medford, San Francisco, Los Angeles, Phoenix, El Paso, Fort Worth, Kansas City, and Chicago. The U.S. Navy had installed a 100-kW frequency alternator transmitter developed by Fessenden (Section 17.1.4) at Arlington, Virginia, and Elwell obtained permission to install a 30-kW Poulsen radio at the same site in December 1912. The Navy then sent cruisers across the Atlantic Ocean to make comparative tests. Impressed by the results, the Navy equipped most of their ships with Poulsen radios made in the United States. In 1913 the Navy also established regular night radiotelegraph communication over 7250 km between their radio station at Arlington, and Pearl Harbor, Hawaii, with a repeater station in San Francisco. Two years later, radiotelegraph communication was also possible from Arlington, with a 100-kW station at Darien, Panama. In the following years, Federal

Telegraph installed a radiotelegraph network around the world, including stations at Paris, Rome, Cairo, Mogadishu, Shanghai, Cavite (the Philippines), Guam, and New Zealand with 200- and 300-kW stations. The U.S. Navy bought this radio network as well as all radio patents of Federal Telegraph in 1917, when the United States entered the war with Germany. A Poulsen radio station with the highest antenna output power of 400 kW was installed at Croix d'Hins, near Bordeaux, when U.S. forces landed at France.

Lee de Forest also developed an arc transmitter for radiotelephony using the arc of a spirit burner in a hydrogen atmosphere. He applied his Audion as a detector in the receiver. He founded the *De Forest Radio Telephone Co.* and supplied 28 stations for the Pacific Fleet of the U.S. Navy around 1907. A 10-kW coastal station was installed on the 210-m-high tower of the *Metropolitan Life Insurance Co.* in New York. The British Navy, too, made satisfactory radiotelephone experiments between its battleship *Furious* and its training station at Vernon, over a distance of almost 100 km with this equipment in 1908, using a numeral code to improve the secrecy.

Radiotelephone experiments with arc equipment were also made by the German high-frequency physicist Ernst Ruhmer (1878–1913) over a distance of 3 km in Berlin in 1908. The French Marine lieutenants Colin and Jeance achieved radiotelephony over 240 km, and the American Collins over 130 km between New York and Philadelphia around 1910.

In 1920, a series of experiments with radiotelephony on arc equipment were made at the German station Königs-Wuesterhausen with a 4-kW arc transmitter operating between 80 and 110 MHz. The radiotelephone transmission was received at Moscow on February 2; at Sweden on February 25; in Luxembourg, the Netherlands, and at Yugoslavia on December 23; and in England on December 31.

17.1.4 Frequency Alternator Radio Transmitter

Unfortunately, the Poulsen convertor arc radio produced numerous undesired harmonics, which hampered other radio transmission, and the transmitter efficiency was quite low. A better solution came from the United States, where at the end of the nineteenth century Reginald Aubry Fessenden took up Nicola Tesla's idea of the Tesla transformer (Section 12.3). After a few years of working for Edison, Reginald Aubry Fessenden (1866–1932), born at East Bolton, Quebec, Canada, became a professor of electrotechnique at Purdue University and lectured on Hertzian waves at the University of Pittsburgh and later at Western University of Pennsylvania in Allegheny City. In 1898 he constructed a 15-kHz ac generator, called a *frequency alternator*, which enabled him to make radiotelephone experiments in December 1900. On September 28, 1901, he received the world's first patent for radiotelephony; for "the improvement of equipment for the wireless transmission of electromagnetic waves and for the improvement of the transmission and reception of words and other audible signs." He also invented an electrolytic device for detection of electromagnetic waves, which he called a *barretter*. With his patents he founded the *National Electric Signaling Company* based at Washington in 1902. Technical problems motivated him to get the *General Electric Company* (GE) involved in the production of his products, and the General Electric Signaling Co. was founded. This company produced the first pair of frequency alternators, which achieved radiotelephony over a distance of 40 km in 1904. One year later, Fessenden invented the heterodyne principle, which substantially improved the selectivity of receivers and which later, as a superheterodyne circuit, became part of all radio receivers.

In the meantime, GE had employed a young Swedish immigrant, Ernst Frederik Werner Alexanderson (1878–1875), born in Uppsala, who after having studied under Slaby at the Technical University in Berlin came to the United States in 1901. Alexanderson was given the task of improving Fessenden's frequency alternator. He first constructed a 75-kHz machine and then a 100-kHz frequency alternator,[11] which on Christmas Eve in 1908 transmitted *Silent Night, Holy Night*, played on the violin by Fessenden, and was heard at a distance of 320 km from the transmitter at Brant Rock, Massachusetts. A second transmission on the following New Year's Day, under even better atmospheric conditions, was heard in the West Indies.

The frequency alternator developed by Fessenden and Alexanderson became widely used for radiotelephony. A permanent radiotelephone connection over 320 km between Brant Rock, Massachusetts, and Long Island was installed in 1907. The equipment had a typical output power of 200 kW at a frequency of 12 kHz (25,000 m), 75 kW at 25 kHz, and 50 kW at 50 kHz.

At the same time, the German engineer Rudolf Goldschmidt developed a *frequency alternator* in combination with a static frequency multiplier based on a reflection principle, which enabled the generation of relatively high frequencies at a multiple of the rotation speed and a high power output. A transmitter output power of typically 100 kW at a frequency of 20 kHz was achieved with 3000 rotations per minute. Such transmitters, with 100-kW antenna power manufactured by *C. Lorenz AG* and the *Hochfrequenz-Maschinen-Aktiengesellschaft für drahtlose Telegraphie* (Homages), took radiotelegraphy over a distance of 6500 km between stations at Tuckerton, New Jersey, and Eilvese near Hannover, Germany, with greetings between President Wilson and Kaiser Wilhelm II on June 20, 1914. Radiotelephone experiments were also made between the two stations, resulting in a few understandable messages. It provided Germany's only overseas contact during World War I, together with the spark radio at Nauen.[12]

At about the same time, Telefunken developed a frequency alternator with external frequency multiplication in premagnetized transformers which achieved 400 kW over the range 17 to 50 kHz. A 100-kW station with this equipment was installed at Funabashi, Japan, just before the beginning of World War I. The German station at Nauen was also equipped with two Telefunken frequency alternators with a power of 150 and 400 kW in 1916. A huge antenna was supported by two 250-m masts, three 150-m masts, and seven 210-m masts. After World War I, Telefunken supplied 400-kW frequency alternators for the stations at Kootwijk, the Netherlands, and Torre Nova at Rome, and a 150-kW station for Prado del Rey, Madrid, in 1923. The last transmitter of this type was also installed in Japan, at Nagoya, in 1928 with an output power of 600 kW.

Frequency alternators developed by the Belgian engineer Marius Latour and manufactured by the *Compagnie Générale Radioélectrique* were used in France, Czechoslovakia, Italy, and Romania. In Russia, the engineer Valentin Petrovitch

---

[11] Alexanderson constructed a generator with 600 poles and a rotating speed of 332 rotations per second (19,920 rpm), which thus generated a current with a frequency of $600 \div 2 \times 332 = 99,600$ Hz.

[12] The German submarine cables were seized and integrated in the British submarine cable network upon the outbreak of war.

TABLE 17.1   Radio Equipment in Operation, 1913

Country	Land	Ships	Total	Marconi	Telefunken
Belgium	1	19	20	20	—
France	36	228	264	—	—
Germany	25	521	546	1	545
Great Britain	143	1347	1490	1317	—
Italy	27	159	186	186	—
Japan	7	32	39	—	—
The Netherlands	6	97	103	52	37
Spain	8	58	66	58	8
Sweden	5	49	54	10	44
United States	189	789	978	—	—
Uruguay	3	6	9	—	9
Total	450	3305	3755	1644	643

Source: Data from *Journal Télégraphique* 1914, 1915.

Wologdin (1881–1953) constructed a frequency alternator in 1922. Marconi produced frequency alternators under license of Goldschmidt.

Frequency alternator radios operating at very long waves between 25,000 and 6000 m required huge antennas with 100- to 250-m mast heights and areas of a few square kilometers; furthermore, it required a large power plant, which limited its use to fixed long-distance communications.

At the beginning of World War I, about 60% of all radiotelegraph equipment in operation worldwide was manufactured by Marconi and Telefunken; 25 other companies shared the rest. Some of these companies, not yet mentioned in this chapter, were the Japanese company Teishinsho, the Compagnie Russe des Télégraphes et des Téléphones Sans Fil, the British Lodge Muirhead Syndicate, the French Compagnie Générale Radiotélégraphique, and the Société Française Radiotéléctrique. A summary of the approximate worldwide distribution of radio equipment in 1913 is given in Table 17.1.

17.1.5   Electronic Radio Equipment

The discovery of the triode in 1907 brought a radical change in radio system development. The triode, which amplified the received antenna signal by at least a factor of 10, replaced the various forms of coherers and detectors in the radio receiving equipment. In 1912, Lee de Forest and Alexander Meissner discovered that triodes could be used not only to detect and amplify signals but also to generate continuous electromagnetic waves which could be used for radiotelephony. Experiments were made with transmitters using a triode in an oscillator circuit to generate the required frequency and using in the transmitter output stage a number of triodes in parallel to obtain the desired output power. A first experiment using triodes for radiotelephony was made on June 21, 1913 between stations at Berlin and Nauen in Germany. In the United States in 1915, AT&T's chief engineer John J. Carty spoke from Montauk Point, Long Island, to Wilmington, Delaware, and St. Simons Island, Georgia. The first transatlantic radiotelephone experiment was made in the same year with transmission from the U.S. Naval station in Arlington, Virginia, to a receiver on the Eiffel Tower at Paris. The transmitter produced by AT&T used 500 vacuum tubes, which generated an output power of between only 2 and 3 kW. After weeks of experimenting, on October 21 the engineers on the Eiffel Tower heard the words "And now, good night, Shreeve." Colonel Shreeve, together with Austin Curtis, were the Western Electric engineers in Paris. A year later a radiotelephone message was conveyed to an aircraft flying near Brooklands airfield in England.

World War I held up further experiments for commercial application of radio equipment with vacuum tubes. After the war and taking advantage of progress in the production of more reliable vacuum tubes for higher frequencies and higher output power, and the development of transmission of a single sideband,[13] radiotelephone experiments were once more made between the United States and Europe in 1923. About 60 people, including Marconi, gathered at London on January 15, listened for two hours to messages spoken by H. B. Thayer, president of AT&T, and other officials of AT&T from their offices at 195 Broadway, New York City. This time, instead of a homodyne receiver and nondirectional antennas as used in 1915, a double-demodulation (superheterodyne), single-sideband receiver and complex, highly directive antenna arrays were used. The one-way system operated at 57 kHz with a transmitter on an RCA station at Rocky Point in New York and a receiver at Western Electric factory at New Southgate, England. The transmitter generated 200 kW of output power in three stages using twenty 10-kW water-cooled vacuum tubes.

Experimental two-way radiotelephony between the United States and the U.K. took place on March 7, 1926, with the transmitter in the United States sending at 57 kHz and the transmitter in the U.K. at 52 kHz. Commercial service started on January 7, 1927 (Section 17.4).

17.1.6 Shortwave Transmission

Marconi and radio amateurs discovered around the 1920s that long-distance radio transmission could be achieved at much shorter wavelengths and with less power than used so far. At the first U.S. National Radio Conference in 1922, the radio amateurs had been given exclusive use of a frequency band around 2 MHz, as this band was considered to be useless for long-distance radio communication. Surprisingly, the U.S. amateur Fred H. Schnell and the French amateur Leon Deloy achieved a radio communication at 2.7 MHz across the Atlantic Ocean in November 1923. The radio amateurs were primarily ex-military radio operators who made a hobby out of their military experience. Very soon they found that on those frequencies, they could communicate worldwide. It was understood that the ionized layers predicted by Heaviside and Kennelly in 1902, called *Kennelly–Heaviside layers*, actually reflected shortwaves a few times between those layers and Earth,

---

[13] Radio signals which are amplitude-modulated by a message radiate a carrier and two sidebands: an upper sideband located above the carrier frequency and a lower sideband located below the carrier. The carrier transmits no intelligence, but the complete message is transmitted in duplicate since each sideband contains the complete message. By eliminating one of the sidebands and the carrier, thus by single-sideband transmission, the message can be transmitted at much lower power and using a much narrower frequency band.

albeit in a rather irregular way, depending on various factors, such as radio frequency, radiation angle, daytime, weather, season, and sunspot activity. The shortwaves could be directed with suitable small antennas and thus required far less power than was needed for long and very long waves. A most efficient directive antenna with reflectors and directors was developed by a Japanese engineer, Hidetsugu Yagi (1886–1976), together with a colleague, Shintaro Uda, called the *Yagi antenna*, well known throughout the world for their use as TV antennas. Yagi, who received postgraduate training in the United States, Britain, and Germany, proposed his device in a paper published in 1928 entitled "Beam Transmission of Ultra Short Waves."

Marconi summarized the advantages of shortwave radio transmission in his statement that "20 kW applied to the antenna gives the same results at the receiving end as 20,000 kW under the old system." He had been experimenting with shortwaves for the Italian Navy from 1916, when he proposed the shortwaves to limit the propagation to quasi-optical ranges, thus preventing eavesdropping by an enemy beyond the horizon. His experiments showed that this assumption was right only as far as distances up to about 300 km were involved. Surprisingly, at distances beyond 1000 km, reception sometimes suddenly became possible. In 1919 he achieved radiotelephony in the 15-m band (20 MHz) over a distance of 125 km. In 1923, on his private steam yacht *Elettra*, he spanned the 4300 km between the U.K. and the Cape Verde Islands with radiotelephone equipment operating at 3 MHz (97 m) with only 1 kW transmitting power. On May 30, 1924, he transmitted intelligible speech for the first time in history between Caernarvon, North Wales, and Sydney, Australia, operating at 3.3 MHz (92 m) with 28 kW of transmitting power generated in 24 vacuum tubes (without forced cooling) and using "beam" antennas. One month later, the GPO contracted with Marconi to build shortwave stations for communicating from Rugby with Canada, India, South Africa, and Australia. In addition to this network for the GPO, the Marconi Company created its own network, with a shortwave transmitter station at Bodmin, a receiver station at Bridgewater, and a dozen stations worldwide.

In Germany, commercial shortwave operation began in 1924 with an 800-W transmitter at Nauen. Radiotelegraphy operating at 4.3 MHz (70 m) over 12,000 km was achieved with Buenos Aires, Argentina, in the same year. Radiotelephony using a 20-kW transmitter operating on 14 and 30 m at Nauen communicated with about 300 shortwave stations worldwide. At that time, water-cooled shortwave transmitter vacuum tubes were available with a typical output power of 7 kW for radiotelephony and up to 20 kW for radiotelegraphy. A 500-kW vacuum tube for low-frequency transmission was available in 1931.

17.2 MARITIME RADIO

As a means of ending isolation for those at sea, Marconi created the *Marconi International Marine Communication Company* in 1900. Beginning in May 1901, the company erected coastal stations in Britain, Ireland, Italy, Belgium, Canada, and Newfoundland and equipped large ocean steamers with its equipment. A ship's radio equipment was leased from Marconi, together with a Marconi radio telegrapher, who had strict orders to communicate with Marconi stations only. Next, Marconi made an exclusive deal with the world's major insurers, *Lloyd's of London*. This deal implied that any ship had to be equipped with a Marconi radio if it were to take advantage of the worldwide network of marine intelligence that centered on Lloyd's, and be insured by Lloyd's. These restrictions were lifted in 1906, as described in Chapter 23. The first vessel with a radio onboard is thought to be the U.S. liner *St. Paul*, which in November 1899, eastbound to Southampton, received a wireless message over a distance of 105 km from the Isle of Wight. The next vessel to have a ship radio, one year later, was the German liner *Kaiser Wilhelm der Grosse*.

Maritime radiotelegraphy very early proved its usefulness for help on the sea when on January 23, 1903,[14] the Russian icebreaker *Yermak* in the Baltic Sea received an emergency call by radio and rescued a group of 50 fishermen from an ice floe near Hogland Island. In 1909, also on January 23, the Italian steamer *Florida*, with 830 emigrants aboard, westward bound to the United States, in deep fog some 300 km east off the U.S. coast, was struck by the U.S. ship *Republic*. A wireless distress signal was sent immediately by the *Republic*, received by a U.S. coastal station, and rebroadcast to the steamer *Baltic*, which was guided by wireless from the *Republic* on a six-hour odyssey through heavy fog to the two sinking ships. All 1700 people on the two ships were saved. The entire world had known of and participated in this major tragedy at sea: Without radio, no help could have been summoned, nor would anyone have known about the disaster. Even more publicity was given to the arrest of a notorious British murderer, Dr. Henry Crippen, who after having poisoned his wife, embarked at Antwerp on the Canadian Pacific liner *Montrose* with his secretary, who was disguised as a boy. The captain got suspicious about their behavior and informed Scotland Yard, which set out in a much faster ship, the *Laurentic*, and arrested the two upon their arrival in Canada in July 1910.

Two years later, the "Queen of the Ocean," the White Star liner *Titanic*, on its maiden voyage from Southampton to New York, would have arrived safely in the United States if four serious attempts to warn the ship of nearby ice fields had not been ignored by the chief radio operator J. G. Phillips and his superiors. This is documented clearly in the protocol of the *Titanic Disaster Hearings* before a *Subcommittee of the Committee on Commerce United States Senate*, from April 19 to May 25, 1912.

On April 14, 1912, the day of the *Titanic* disaster, ice warnings were telegraphed to the ship by the British steamers *Baltic* and *California* and by the German steamer *Amerika*. The first warning came from the *Baltic* around midday, indicating the position of a large ice field with some icebergs on the route of the *Titanic*. The *California*, controlled by the same company as the *Titanic*, informed the *Titanic* at 5:35 P.M. (New York time) about three large icebergs north of the route of the *Titanic*. A few hours later, the *Amerika* also informed the *Titanic* of two large icebergs. The fourth warning came again from the small freighter *California*, when it was about 30 km from the *Titanic* and had to stop because of ice. The captain requested his radio operator, Cyril Evans, once more to send an ice warning to the *Titanic*. At 9:05 P.M., Evans, who knew Phillips personally, sent him a message

---

[14] This date is given in ITU's publication *Great Discoveries Telecommunications*, and it gives as reference an earlier ITU publication, *From Semaphore to Satellite*, where it is stated that this happened on January 26, 1900. The German magazine *Post und Telekommunikations Geschichte* Vol. 2. 1998, gives February 6, 1900 as the date of rescue. It is highly unlikely, however, that fishing boats and icebreakers were equipped with radios in 1900. Even 1903 would be surprisingly early.

informing that the *California* had stopped because of ice. This message interrupted a communication between the *Titanic* and the radio station at Cape Race at Newfoundland. This annoyed Phillips, who told Evans brusquely "to shut up" and keep out of the conversation. Evans waited for about half an hour and then switched off his radio and went to bed.

At 10:13 P.M. (11:46 P.M. local time), the *Titanic*, almost at full speed, made its fatal collision with an iceberg. Beginning at 10:25 P.M., Phillips sent an emergency signal "CQD. Struck iceberg, come to our assistance at once. Position: Lat. 41.46 N; Long. 50.14 W." This signal was first heard by Harold Thomas Cottam, the radio operator of the Cunard liner *Carpathia* at 10:45 P.M., just before he started to remove his headset and go to bed. The captain of the *Carpathia* decided immediately to go to help the *Titanic*. It took the *Carpathia* three and a half hours to cross the almost 100 km to the *Titanic*. Upon arrival, only 706 persons could be picked up from the lifeboats; the *Titanic* had already disappeared completely and 1517 persons, including J. G. Phillips, had died in the ice-cold water.

Officers of the *California* saw white rockets on the horizon that night. They started sending Morse-coded light signals, but as they received no answer, they woke their radio operator, Evans, at 3:30 A.M. and asked him to investigate the significance of the rockets. Within a few minutes, Evans received information from the German steamer *Frankfurt* that the *Titanic* had gone under. Two hours later the *California* arrived at the site of the disaster, shortly after the last shipwrecked persons had boarded the *Carpathia*. Upon the arrival of the *Carpathia* at New York, when the rescued passengers noticed the presence of Marconi, they thanked him, saying "Ti dobbiano la vita" (thanks to you, we are alive). According to the protocol of the Titanic Disaster Hearings, the *California* could have arrived four hours earlier and picked up all the *Titanic* passengers if the initial radio conversation between Phillip and Evans had been more civilized. Moreover, the disaster would not have happened if the captain and his officers had respected the radio warnings properly and reduced the speed of the *Titanic* in time.

The International Bureau of Berne reported that by 1912 there were 2752 ship stations, of which 1964 were open for public service. Those ship stations were served by 479 coastal stations (about 30% Marconi and 22% Telefunken) in 60 countries, of which 327 stations were for public use. Figure 17.7 shows a typical ship station.

After World War I, radio became widely used on ocean ships, as electronic vacuum tubes substantially improved the quality, reliability, receiver sensitivity, and transmitter output. In 1920 the number of ship stations had increased to 12,622, served by 937 coastal stations. The number of fixed nonmaritime stations was only 95, a clear indication of the prevalence of maritime radio over other uses of radio. Figure 17.8 shows one of the oldest coastal stations, Norddeich Radio on the German North Sea coast, which started operation in November 1905 (for local service and beginning July 1, 1907 for international service) and closed on January 31, 1999.

Maritime radio communication so far was limited to communication between ships and communication between ships and shore stations, without through-connections to landline telephone systems. First trials for the development of a radiotelephone system capable of enabling Bell Telephone System service in the United States to be extended to include ships at sea were made from 1919 to 1922. Two land stations were established: at Deal Beach, New Jersey, and Green Harbor, Massachusetts, and a field experimental station was located in between, at Cliffwood,

Figure 17.7 Radiotelegraphy ship station, around 1912. (© Siemens-Archiv, Munich.)

New Jersey. Two ships operating between Boston and Baltimore, the SS *Gloucester* and the SS *Ontario*, and the ocean liner SS *America* were equipped with 1-kW, long-wave radios operating between 700 and 840 kHz. Daytime ranges up to 400 km, contrasted with occasional communication over a few thousand kilometers at night. The experiments culminated in telephone conversations between a ship in the Atlantic Ocean via the transcontinental telephone line to Los Angeles/Long Beach and then via radio to Catalina Island in the Pacific Ocean. This was demonstrated to delegates of the Preliminary International Communications Conference on October 21, 1920. The trials were successful from a technical standpoint, but adverse postwar economic conditions delayed commercial use.

The development of shortwave radio systems substantially reduced the cost of radio stations, and successful trials with such equipment operating on 4.5 MHz (66 m) were made in 1925 between New York and Bermuda. Public telephone service to ships on the Atlantic Ocean via shortwave radio was then opened on December 8, 1929. Within one year this service from both the U.S. and British coasts was available on the British White Star liners, the *Olympic* and the *Majestic*, and on the

Figure 17.8 Coastal station, Norddeich radio, 1955. (Scanned from *Post- und Telekommunikationsgeschichte*, Vol. 2, 1998, cover page.)

American liners *Leviathan, Olympic, Majestic*, and *Homeric*. The system operated in the band from 3 to 17 MHz, and the radio stations had 500-W transmitters, with 15-kW transmitters used at the coastal stations. Reliable communication with surface waves was achieved for distances up to about 500 km. Less reliable communication using reflected sky waves started at a distance of about 2 km between the ship and the coastal station. Figure 17.9 shows the relation between the operating frequency and the distance achieved during day, night, summer, and winter.

17.3 MOBILE RADIO

In addition to marine and intercontinental radio, Marconi, in 1901, also developed the first mobile car radio. A steam-driven wagon was equipped with a transmitter, a receiver, and a cylindrical antenna about 5 m high mounted on the roof. The high chimneylike vertical antenna could be brought into a lower horizontal position if circumstances required (Figure 17.10).

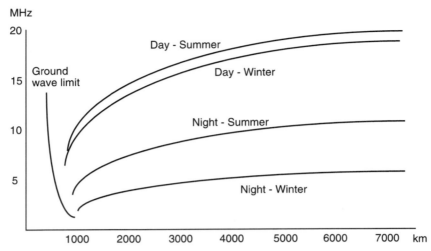

Figure 17.9   Distance–frequency characteristic of the transatlantic radiotelephone operation. (After *Bell System Technical Journal*, July 1930, p. 411.)

The U.S. *Army Signal Corps* installed radio equipment on horse carriages in 1909. It took about 1 minute to erect the several-meter-high antenna. Military mobile and transportable radiotelegraph equipment was used widely during World War I.

The first mobile radiotelephone service on land reportedly was set up by the Detroit Police Department in 1921 using a frequency close to 2 MHz. In 1923 the British police used a "radio car" to handle traffic at the annual Derby horse race. The next year the police in New York began to equip their motorcycles with radio-telegraph sets. The U.S. Army Signal Corps used radiotelephone sets on their air-planes beginning in the mid-1920s. *Imperial Airways* in London were first to equip planes with commercial Marconi radios in 1935. In 1926, radiotelephone service became available for first-class railway passengers on the Berlin–Hamburg route (Figure 17.11).

By the late 1920s, large, expensive two-way car radios came on the market for public use in the United States. Paul V. Gavin, owner of the *Galvin Manufacturing Corporation* in Chicago, founded in 1928, made devices that enabled battery-powered radios to use the electric facility's lines. He challenged his engineers to design a simpler car radio for the mass market. They succeeded in 1931. To celebrate their success, Galvin linked "motion" and "radio" and gave the radio the trade name *Motorola*. In 1947 that name was also given to the company. Soon, U.S. cars began to be equipped with Motorola two-way radios. Motorola launched the first portable radiotelephone in 1943, the Handy Walkie, with a weight of 16 kg. How-ever, commercial service available to the general public in the United States did not begin until 1946. The city of St. Louis, Missouri, was the first to offer Domestic Public Land Mobile Radio Service (DPLMRS, as classified by the FCC), operating three channels in the 150-MHz band. The equipment was developed and the service introduced by the Bell System. One year later, a "highway system" began operation along highways between New York and Boston, with equipment operating between 35 and 44 MHz.

Figure 17.10   First mobile radio, 1901. (Scanned with permission of the ITU from Anthony R. Michaelis, *From Semaphore to Satellite*, International Telecommunication Union, Geneva, 1965, p. 133.)

17.4   INTERCONTINENTAL RADIOTELEPHONY

Intercontinental radiotelephony started on January 7, 1927, between New York and London. Separate routes were applied for the eastbound and westbound transmission (Figure 17.12). Transmitting power of about 200 kW was produced by 35 large 10-kW water-cooled vacuum tubes operating at 7 kV. The frequencies of 58.5 to 61.5 kHz, corresponding to a long wavelength of about 5000 m, needed antennas of that length. At Rocky Point, six towers, each 120 m high, supported the 5-km-long transmitting antenna. At Cupar the receiving antenna consisted of two parallel pole lines each at a length of 5 km. A third pole line connected the two parallel lines, which were separated at a distance of about 3.2 km. For interconnecting the radio circuits with telephone lines on a four-wire basis, and to prevent voice-frequency singing through residual imbalances, special *vodas* (voice-operated device, anti-singing) were used. To improve the signal-to-noise ratio, *companders* (compressor–expanders) were introduced which raised the amplitude of the weaker parts of speech previous to transmitting, depressed the raised parts to their proper value, and reduced the radio noise after reception.

Figure 17.11 Radiotelephony of the German railways in 1926. (Courtesy of Museum für Kommunikation, Frankfurt, Germany.)

Shortly after this transatlantic radiotelephone service was begun with long-wave transmission, three shortwave radiotelephone systems were installed and began service in June 1928, June 1929, and December 1929. Each system was adjustable to operation on 19 MHz (16 m), 14 MHz (21 m), and 9 MHz (33 m), to enable adaptation of the frequency to time of day and season. The longest antenna used for these systems, installed at Lawrenceville, New Jersey, had a length of about 1.6 km and was supported by 21 towers, each at a height of 55 m.

Observations over the first years of operation showed that the long-wave system was available for commercial operation during about 80% of the time, with major disturbances from lightning during the summer months. The availability of the shortwave systems was 64%.

Radiotelephone calls were limited to a maximum of 12 minutes. The New York–London rate was initially $75 for 3 minutes plus $25 per additional minute, but was soon reduced to $45 and $15 per minute, and to $21 ($7 per minute) in 1936. Whereas two operators in London and two in New York could handle the calls, a total of about 40 persons were required to operate the four long-wave and the two shortwave stations. In the first year an average of only seven calls per day were established, originating almost equally in the United States and in the U.K., but this increased to about 50 in 1929.[15]

Radiotelephony was established from the United States to Hawaii on December 23, 1927, and to Belgium in 1928. The next year, operation was begun to Holland, Germany, Sweden, France, Denmark, Norway, Switzerland, Spain, Austria, Hungary, and Czechoslovakia. South America and Australia followed in 1930, South Africa in 1932, and Japan and Java (Indonesia) in 1934. Eventually, 240 radio cir-

[15] Differences in the accents of English speakers in London and in New York sometimes made it easier for British operators to understand the English spoken by telephone operators in Holland than that spoken by operators in New York City.

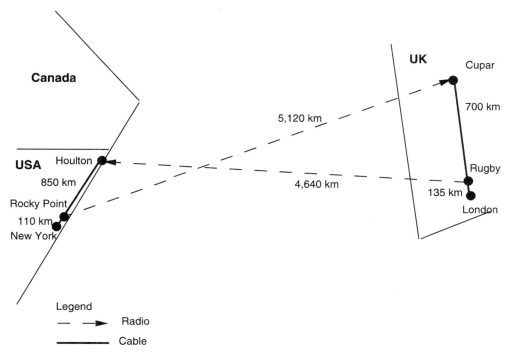

Figure 17.12   First intercontinental radiotelephone connection, 1927. (After *Bell System Technical Journal*, Vol. VII, 1928, p. 169.)

cuits connected the United States with 140 countries. A radiotelephone long-distance record was made in 1930 with a radio communication over 23,000 km between stations on Java, Indonesia, and in Argentina via Berlin. A first around-the-world radiotelephone conversation starting and terminating in New York took place on April 25, 1935.

One of the largest shortwave stations for intercontinental telephony was installed in France at Saint-Assise, southeast of Paris, in 1929. The 39-m-tall antenna was composed of two layers supported by two masts 75 m apart. It operated on 15, 55, and 24.5 m and communicated with the United States, Africa, and Southeast Asia within an 11,000-km range.

17.5   RCA AND C&W CREATED TO BEAT MARCONI

Marconi was the catalyst, and the big loser, in the establishment of two major telecommunications companies: the *Radio Corporation of America* (RCA) in 1919 and *Cable & Wireless* (C&W) in 1928.

17.5.1   Radio Corporation of America

Marconi, in 1919, once more endeavoring to establish his delayed "Wireless Girdle Round the World," approached General Electric with the intention of buying 24 frequency-alternator radio stations. Franklin D. Roosevelt, then acting secretary of the U.S. Navy, feared that this would result in a worldwide British monopoly on radio transmission, in addition to the existing British monopoly on submarine communication. He informed U.S. President Wilson, then at the peace conference in Paris,[16] who instructed the Navy to prevent the sale to Marconi and to establish an American-owned radio company. General Electric, eager not to lose the business, bought the Marconi Wireless and Telegraph Company of America and established the *Radio Corporation of America* (RCA) as the successor of that company on November 21, 1919. To give RCA access to all U.S. radio patents, a complex set of agreements was made between the U.S. Navy, General Electric, AT&T, Westinghouse, the United Fruit Company, and RCA. AT&T held patents on the vacuum-tube technology of de Forest; Westinghouse had the patents on Edwin Armstrong's feedback oscillator and heterodyne technology; the U.S. Navy controlled the arc technology of Poulsen and Elwell; and the United Fruit Company operated a large radiotelegraphy network for communication between its plantations, ships, and offices. In the same year, RCA negotiated a Four Power Pact with Marconi, the French *Compagnie Générale de Télégraphie*, and the German company Telefunken. The four companies settled the first international radio cartel, in which the world was divided into six territories:

1. *RCA territory:* the United States and the Philippines
2. *Marconi territory:* the U.K. and the Commonwealth
3. *RCA for internal and Marconi for external communications:* Canada, the Caribbean, and Guyana
4. *RCA for internal and the others for external communications:* Latin America apart from Argentina, the Caribbean, and Guyana
5. *Limited competition:* China
6. *Open competition:* the rest of the world

Under this agreement, each company was to have exclusive rights to use of the other companies' patents within its respective territories. The agreement was to run until January 1, 1945. RCA took over from Marconi the radiotelegraph services from the United States with the U.K., Hawaii, and Japan, and established new services to France, Germany, and Norway. It started national radiotelegraph service on March 1, 1920. The first South American service started on January 1924, to Buenos Aires, Argentina, with shortwave equipment. By that time, radio broadcasting had begun, and RCA, unlike Marconi without a global empire to serve, turned its interest to this new field.

17.5.2 Cable & Wireless

At the end of the 1920s, Great Britain still dominated world communications by submarine cable, with 450,000 km of cable compared with 270,000 km owned by U.S. companies, 37,000 km owned by French companies, and 30,000 km by all

---

[16] Marconi participated as a plenipotentiary delegate to the peace conference in Paris, in which capacity he signed the peace treaties with Austria and Bulgaria.

others. However, the British submarine cable companies experienced a significant loss of revenue due to the success of Marconi shortwave radio operated by the GPO inside the British Empire (the "Imperial Chain") and by Marconi Wireless outside the Empire. Moreover, they had reason to be concerned about the rapidly growing role of U.S. telecommunications companies, especially of IT&T, which in 1927 bought All America Cables, which had 50,000 km of submarine cable to the West Indies and to Central and South America. Discussions on the British challenge in telecommunications were widely covered by the *New York Times* and other newspapers. To examine the situation and find means of counterattacking, the *Imperial Wireless & Cable Conference* was held in London in 1928, with the participation of the major British radio, telephone, and cable companies, which were united in a loose alliance known as the Electra House Group. A decision was taken that both cable and radio operations should be merged into a common system. *Cable & Wireless* was formed as the holding company for a new communications company called *Imperial and International Communications Ltd.* (I&IC). I&IC was the merger of the leading cable company, the Eastern Telegraph Company (Section 8.7.3), and Marconi Wireless. In 1934 the name of I&IC was changed to *Cable & Wireless Ltd.* and the holding company was renamed *Cable & Wireless (Holdings) Ltd.* Cable & Wireless became a public company on January 1, 1947. In 1950, C&W employed a staff of 800 in the U.K. and 8200 overseas. By 1954 it operated 240,000 km of submarine cable and 320,000 km of radio circuits. Currently, C&W has evolved from a worldwide telecommunications operator to a global carrier focused on the business market. It has four main operating units: global markets, network operations and carrier services, Cable & Wireless United States, and Cable & Wireless IDC, which is responsible for the C&W activities in Japan and East Asia.

For Marconi, this merger was the end of his international radio activities.[17] He returned to Italy, where he got involved in politics and supported Mussolini, who appointed him president of the National Council of Research in 1928 and president of the Royal Academy of Italy in 1930. He was made a marchese and nominated to the Italian Senate in 1929. As his last involvement in radio, he presented to Pope Pius XI a radiotelephone connection between Vatican City and the summer residence of the Pope at Castel Gondolfo, a distance of about 20 km, in 1932. This was a novelty, as it presented the first practical application of a radio frequency at 600 MHz.

REFERENCES

Books

Bertho Lavenir, Catherine, *Great Discoveries: Telecommunications*, Romain Pages Editions, 1991.

Deloraine, Maurice, *When Telecom and ITT Were Young*, Lehigh Books, New York, 1976.

Garratt, G. R. M., *The Early History of Radio from Faraday to Marconi*, Institution of Electrical Engineers, London, 1994.

[17]The Marconi operation companies became part of C&W, while the Marconi production facilities became part of General Electric of Coventry (GEC). It was not until 1999 that Marconi Communications was again created as a transmission manufacturing company in the U.K.

Godwin, Mary, *Global from the Start: A Short History of Cable & Wireless*, Cable & Wireless, London, 1994. [Leaflet]

Gööck, Roland, *Die großen Erfindungen Radio: Fernsehen Computer*, Sigloch Edition, Künzelsau, Germany, 1988.

Hugill, Peter J., *Global Communications since 1884: Geopolitics and Technology*, Johns Hopkins University Press, Baltimore, 1999.

Kuntz, Tom, *The Titanic Disaster Hearings*, Simon & Schuster, New York, 1998.

Michaelis, Anthony R., *From Semaphore to Satellite*, International Telecommunication Union, Geneva, 1965.

Nesper, Eugen, *Die drahtlose Telegraphie*, Verlag von Julius Springer, Berlin, 1905.

Oslin, George P., *The Story of Telecommunications*, Mercer University Press, Macon, GA, 1992.

Polleit, Reinhard, *Die Geschichte der drahtlosen Telegrafie*, published by the author at Neustadt am Rübenberge, 1979.

Articles

Anon., Die weitere Entwicklung der drahtlosen Telegraphie in Deutschland, *Archiv für Post und Telegraphie*, Vol. 7, July 1919, pp. 245–251.

Arnold, H. D., and Lloyd Espenschied, Transatlantic radio telephony, *Bell System Technical Journal*, Vol. 2, 1923, pp. 116–145.

Beyer, Klaus, Hertz und die Anfänge der Funkentelegrafie, *Archiv für deutsche Postgeschichte*, Vol. 2, 1988, pp. 155–168.

Blackwell, O. B., Transatlantic telephony: the technical problem, *Bell System Technical Journal*, Vol. 7, 1928, pp. 168–194.

Bown, Ralph, Transoceanic radio telephone development, *Bell System Technical Journal*, Vol. 16, 1937, pp. 560–567.

Jentsch, Otto, Die erste deutsche Funkentelegraphenanlage für den allgemeinen Verkehr, *Archiv für Post und Telegraphie*, Vol. 14, July 1900, pp. 573–582.

Jentsch, Otto, Die Fortschritte der Funkentelegraphie, *Archiv für Post und Telegraphie*, Vol. 8, April 1901, pp. 25–422.

Jentsch, Otto, Die neuere Entwicklung der drahtlosen Telegraphie und Telephonie, *Archiv für Post und Telegraphie*, Vol. 18, October 1914, pp. 557–584.

Leclerc, Herbert, Von Apia bis Yap, Ehemalige deutsche Postanstalten in der Südsee, *Archiv für deutsche Postgeschichte*, Vol. 21, 1988, pp. 7–32.

Nichols, H. W., and Lloyd Espenschied, Radio extension of the telephone system to ships at sea, *Bell System Technical Journal*, Vol. 3, 1924, pp. 141–186.

Pieper Hans, Die englischen Bemühungen vor dem Ersten Weltkrieg um ein weltweites Kabel- und Funkmonopol, *Archiv für deutsche Postgeschichte*, Vol. 1, 1975, pp. 79–95.

Thurn, H., Drahtlose Telephonie, *Archiv für Post und Telegraphie*, Vol. 12, June 1910, pp. 287–299.

von Ellisen, Hans-Joachim, Drahtlose Telegrafie mit gedämpften Wellen, *Archiv für deutsche Postgeschichte*, Vol. 2, 1993, pp. 25–52.

von Kniestedt, Joachim, 100 Jahre Funkpatent von Ferdinand Braun, *Post- und Telekommunikationsgeschichte*, Vol. 1, 1999, pp. 47–49.

Wilson, William, and Lloyd Espenschied, Radio telephone service to ships at sea, *Bell System Technical Journal*, Vol. 10, July 1930, pp. 407–428.

Young, W. R., Advanced mobile phone service: introduction, background, and objectives, *Bell System Technical Journal*, Vol. 48, January 1979, pp. 1–14.

Internet

www.britannica.com, Development of radio technology, by Reginald Leslie Smith-Rose.

www.britannica.com, Marconi, by Reginald Leslie Smith-Rose.

www.cwhistory.com, History of C&W, by Mary Godwin, C&W Curator, Porthcurno, Penzance, U.K.

18

PHOTOTELEGRAPHY

18.1 KOPIERTELEGRAPH OF GUSTAV GRZANNA

The German engineer Gustav Grzanna was the first person to use a method of scanning the X/Y axes, combined with the use of photographic paper in the receiver, in 1901. The position of the stylet on the X/Y axes in the transmitter was coded such that each position corresponded to a certain strength of the line current. Instead of a stylet, a little mirror attached to a magnet needle was used in the receiver. Two electromagnets, one for the X and the other for the Y axis, controlled the movement of the magnetic needle and thus of the mirror in such a way that a light ray directed toward the mirror and reflected on the photographic paper wrote the received message on the photographic paper. Grzanna called his device a *Kopiertelegraph* (German for "copying telegraph"). He founded the company Kopiertelegraph in Dresden in 1901 but stopped production in 1905 when a compatriot, Arthur Korn, introduced a more efficient product.

18.2 TELAUTOGRAPH OF ARTHUR KORN

As photography became popular at the beginning of the twentieth century, the German physicist Arthur Korn (1870–1945) developed a facsimile machine suitable for transmission of photographs. He was the first to introduce a facsimile machine with optical scanning in the transmitter and photographic reproduction in the receiver, in 1902. Korn used a selenium cell inside a cylinder made of glass at the transmitting end. The light, which scanned a transparent film on the outside of the cylinder, was

The Worldwide History of Telecommunications, By Anton A. Huurdeman
ISBN 0-471-20505-2   Copyright © 2003 John Wiley & Sons, Inc.

converted by this selenium cell into a current which varied according to the intensity of the light. Photographic paper was placed on a drum inside a dark container on the receiving side. The drum revolved synchronously with the glass cylinder at the transmitting end. A light ray was directed to the photographic paper in the dark container through an opening which varied in size as a function of the line current.

Korn's Telautograph was first tested in 1904 on a Munich–Nuremberg–Munich loop. The transmission time for a photo of Prince Regent Luitpold took 42 minutes. Korn improved his device by adding a second selenium cell, combined with a compensation method on the transmitting side. On the receiving side he replaced the adjustable opening of the dark container of the receiver by a fixed opening and directed the light ray through this opening via the mirror of a much more sensitive galvanometer. With those improvements the transmission time of a 13- by 18-cm photo could be reduced to 12 minutes, or even 6 minutes at a lesser resolution. Regular picture transmission between Munich and Berlin started on April 16, 1907.

The new Telautograph drew international interest from the press. The French newspaper *L'Illustration* bought Korn's Telautograph in November 1906 and obtained a monopoly for operating the machine in France. The next year, the *Daily Mirror* installed the Telautograph at their offices in London and Manchester and opened a line with Paris in 1908. This service became popular when a picture taken of a jewel thief in Paris, and published the next day in the *Daily Mirror*, was used to identify and arrest that person at a hotel in London. The newspapers *Politiken* in Copenhagen and *Dagens Nyheter* at Stockholm used Telautographs around 1908. World War I ended the picture service.

The German company *Telefunken* made the first experimental radio transmission with the Telautograph in 1910. An experimental link was installed in 1917 between Berlin and Constantinople (now Istanbul). A most impressive demonstration of the performance of Korn's Telautograph was given on May 6, 1922: A picture of Pope Pius XI was sent via cable from Rome to Berlin, then in about 40 minutes by radio from Berlin to Otter Cliffs on the Atlantic coast of Maine, and published in the New York newspaper the *World*.

Another German company, *C. Lorenz*, together with Arthur Korn, developed a Telautograph for wireless transmission in 1927 called the *Lorenz–Korn picture telegraph* (Figure 18.1). This telegraph was first used for a network of Prussian police in Berlin in 1928. The pictures were scanned at a speed of 2400 picture points per second. The resulting signal was modulated on a 4-kHz carrier and connected with a radio station via a cable 15 to 20 km long. The radio stations were located outside the towns to reduce electric radiation disturbances (electro smog) which was already a problem at that time. The radio equipment was remotely controlled and operated at a frequency of about 270 kHz with a transmitter output of 5 kW. The transmission speed was about 2 minutes for a 13- by 18-cm photo.

18.3   TELEGRAPHOSCOPE OF EDOUARD BELIN

In 1906 the French scientist Edouard Belin (1876–1963) developed a telephoto machine which he called the *telegraphoscope*. The first telephoto transmission was made from Paris to Lyon and Bordeaux back to Paris in 1907. Korn and Belin rivals

Figure 18.1   Lorenz–Korn picture telegraph. (Scanned from *50 Jahre Lorenz, 1880–1930*, C. Lorenz Aktiengesellschaft, Berlin, p. 320.)

for the newly emerging press reporting market. Eventually, Belin won in 1913 with a portable version of his telegraphoscope, called the *Belinograph*. It was smaller than a typewriter, easy to operate, and capable of being connected to an ordinary telephone. The first transatlantic radio Belinogram was sent from Malmaison, Paris, to Annapolis, Maryland, in 1921. The Belinograph was adopted in Britain in 1928 and was used almost exclusively by European news media during the 1930s and 1940s.

18.4   SIEMENS–KAROLUS–TELEFUNKEN PICTURE TRANSMISSION SYSTEM

A quantum leap in image transmission was made in the mid-1920s when the relatively insensitive selenium cells could be replaced by photoelectric cells, especially in a version known as a *Kerr cell*. The German physicist August Karolus (1893–1972) developed a Kerr cell with inertless light modulation called the *Karolus cell*, for which he obtained German patent 571,720 on June 27, 1924. With the Karolus cell as a light-intensity sensor, Karolus developed[1] a facsimile machine with excellent image resolution and a high transmission speed of typically 22 seconds for a standard 9- by 12-cm photo (Figure 18.2).

Successful trials were made in 1925 on a telephone line between Berlin and Leipzig, and the same distance of 150 km was covered by radio in March 1926. Regular picture transmission began on December 1, 1927 between Berlin and Vienna and in 1930 between Berlin and London and Frankfurt and London. Also in 1930, the German patent office issued the first patent for "a machine for the electrical transmission of characters," developed by Rudolf Hell (1901–2002[2]). Hell produced the

[1] The development was in cooperation with the German Imperial Post and the companies Siemens and Telefunken, hence the name Siemens–Karolus–Telefunken Picture Transmission System.

[2] Rudolf Hell celebrated his hundredth birthday on December 19, 2001, in the town of Kiel, where he had made his major developments. He died on March 11, 2002.

Figure 18.2 Siemens–Karolus–Telefunken picture telegraph, 1927. (© Siemens-Archiv, Munich.)

characters with a coding of 7×7 points. Picture transmission between Berlin and Tokyo began on March 20, 1940.

18.5 FACSIMILE MACHINES OF AT&T AND WESTERN UNION

Both AT&T and Western Union developed facsimile machines to be used on long-distance telephone lines. Western Union introduced a facsimile system named *Telepix* in 1924, which was used by 20 newspapers over 13,000 km of leased lines. The police also used the service and in July 1924 transmitted from New York to Chicago the first fingerprint of a criminal who was arrested in Chicago and could be identified by his fingerprint on file in New York. The service was costly and slow, however, and was discontinued after one year.

AT&T developed a more elaborate and efficient facsimile machine named *Telephoto* in 1925. Much similar to the approach of Karolus in Germany, AT&T employed transparent cylindrical drums that were driven by motors synchronized between transmitter and receiver. At the transmitter a positive transparent print was placed on the drum and was scanned by a photoelectric cell. The output of the photocell modulated a 1800-Hz line carrier signal. At the receiver the line signal controlled the intensity of a narrowly focused light beam that progressively illuminated a photographic paper. The transmission time of a standard 13- by 18-cm photo with a resolution of 40 lines/cm took about 7 minutes. The Telephoto service was introduced between New York, Chicago, and San Francisco in 1925 and a few years later was extended to Boston, Atlanta, Cleveland, St. Louis, and Los Angeles. A portable transmitter could be connected on the network anywhere between those cities. It had an impressive start when, on March 4, pictures of the inauguration of President Coolidge taken in the morning in Washington were published in the afternoon papers in New York, Chicago, and San Francisco.

Radio picture service between the United States and Great Britain started in 1926 and with other European countries and with South America in 1933. Transatlantic picture service via cable started in 1939. It took 20 minutes to transmit a standard 13- by 18-cm photo.

A major step forward was made in 1934 when engineer Raleigh G. Wise of Western Union invented electrosensitive dry recording paper, named *Teledeltos*. This was a black, carbon-impregnated paper, coated initially with vermilion mercuric sulfide, but later with a lead thiosulfide. With this paper, typewritten, handwritten, or graphics could be handled without an elaborate photographic process. The message to be sent was wrapped around a drum that moved under a photoelectric cell. In the receiver a stylus moving on a page of Teledeltos paper decomposed and burned away minute portions of the coating of the paper, in response to the output of the photocell in the transmitter. Western Union introduced this machine on November 14, 1935 on the New York–Buffalo line and one year later between New York and Chicago. Desktop facsimile to speed customers' messages to its offices and intercity facsimile services were introduced in 1935. By 1939, this facsimile service was available from coast to coast. The Teledeltos facsimile machine was still large and expensive. Automatic facsimile machines resembling mailboxes were used in hotels and public places as well as at the World's Fairs in New York and San Francisco in 1939. In 1941, Western Union installed a 33,000-mile printing telegraph network for the Civil Aeronautics Authority, linking 180 airports, 400 Weather Bureau offices, and various military locations.

18.6 PHOTOGRAPH TRANSMISSION EQUIPMENT IN JAPAN

Photograph transmission equipment developed in Japan came just in time to transmit photographs of the coronation of Emperor Hirohito in 1928. Public phototelegraph services were introduced in 1930. Later, in 1936, successful experiments were made in transmitting wireless photographs between Tokyo and Berlin, London, and San Francisco. Pictures shown in Japanese newspapers of Japanese athletes at the Olympic games in Berlin in 1936 were very popular.

REFERENCES

Books

Gööck, Roland, *Die großen Erfindungen Nachrichtentechnik Elektronik*, Sigloch Edition, Künzelsau, Germany, 1988.

Oslin, George P., *The Story of Telecommunications*, Mercer University Press, Macon, GA, 1992.

Articles

Barnekow, Rolf, and Manfred Bernhardt, Die Vorläufer der Telefaxgeräte, *Post- und Telekommunikationsgeschichte*, Vol. 1, 1995, pp. 57–62.

Ives, H. E., et al., The transmission of pictures over telephone lines, *Bell System Technical Journal*, Vol. 4, April 1925, pp. 187–214.

19

TELEPRINTERS

After telegraphy and telephony came the teleprinter, another form of instantaneous telecommunication in the 1920s. Teleprinters became commonplace in the offices of companies and governmental organizations. Beginning in 1962 with the addition of switching to this new form of telecommunication, *telex* (abbreviated from "teleprinter" and "ex-change") networks, were developed, so company headquarters could communicate with their branch offices and with other companies equipped with teleprinters. Initially, modified manual telephone switchboards were used for switching, but very soon, special telex switching equipment was used in the telex networks. In Europe, where language differences made written communications attractive and more urgent than in the United States, automatic switching exchanges for teleprinters were developed employing the same switching principles and devices as used for telephony. This service started in Germany in 1932. Teleprinter service before World War II was limited to a few countries, such as the United States, Great Britain, Belgium, Denmark, Germany, The Netherlands, and Japan.

19.1 TELEPRINTER DEVELOPMENT IN THE UNITED STATES

In 1867, Christopher L. Sholes invented the typewriter, which beyond use in business and government offices, soon also found use in telegraph offices. The telegrapher listened to the Morse sounder and wrote the received messages directly in plain language with the typewriter. This service was such a success that in 1873, Sholes sold the manufacturing rights of his typewriter to firearm, sewing machine, and farm

The Worldwide History of Telecommunications, By Anton A. Huurdeman
ISBN 0-471-20505-2   Copyright © 2003 John Wiley & Sons, Inc.

toolmakers *E. Remington & Sons*, which then began industrial production of type-writers.

With direct-letter-printing telegraphs, the message was usually printed at both the transmitting and receiving ends in a continuous straight line on a paper tape. The tape was then cut into strips of equal length, which were pasted on a telegram form. Direct-letter-printing telegraphs required very accurate means to maintain synchronism between the transmitting and receiving apparatus. To overcome this disadvantage, two major improvements were made: typewriter-type keyboards and start–stop signals for each character. Donald Murray, a farmer from New Zealand who had worked in a newspaper printing shop, made the first improvement. He became a journalist and occupied himself with telegraphy. In 1901, he constructed a keyboard layout similar to that of a typewriter. This keyboard was connected with a perforator that produced a tape in which the code of each character was perforated transversely. The tape was placed in a transmitter which produced the line code. Similarly, on the receiving side a perforated tape was produced that could be used for transmission onward or for printing the message in plain language on a paper tape.

The major improvement came from effective cooperation between Charles L. Krum and his son, Howard Krum. Charles Krum was vice-president of the *Western Cold Storage Company* when in 1902 a young engineer named Frank Pearne was given the opportunity to use the factory facilities for his plan to develop a direct keyboard teleprinter. After one year of unsuccessful experimenting, Pearne lost interest. Charles Krum became interested, however, and continued the experiments successfully. He filed patents for a type-bar page printer in 1903 (U.S. patent 888,335, issued May 19, 1908), and for a type-wheel printing telegraph in 1904 (U.S. patent 862,402, issued August 6, 1907). In 1906, Howard Krum graduated as an electrical engineer and joined his father in teleprinter development. Father and son together modified several kinds of commercial typewriters to perform the duty of handling telegraph signals, but none met their requirements, so they developed a high-precision typewriter suitable for telegraph operation. By 1908 they could test an experimental model on the telegraph lines of the *Chicago & Alton Railroad*. Commercial operation started in 1910 by the *Postal Telegraph Company* between New York and Boston.

Synchronizing remained a problem until Howard Krum worked out a start–stop method.[1] A start signal was transmitted immediately preceding the code of each character, similarly, a stop signal was transmitted at the end of the code of each character. The code employed for the characters was a five-unit code similar to that introduced by Baudot in 1874. For the start signal one code element was added, and for the stop signal, one or in some cases two, elements. The resulting overall code was therefore referred to as a *7*-unit code*. In this way the transmitter and receiver resynchronize at the start of each character. The Krums, together with Joy Morton (owner of the *Morton Salt Company*), founded the Morkrum Company and produced the first start–stop *teletypewriter* in 1912 for Western Union. The teletypewriter printed the received signals directly in plain language onto a paper tape at a speed of 40 words per minute, operating without the intermediate use of perforated tape at either end of the system.

---

[1] A patent for this method was filed on May 31, 1910, and U.S. patent 1,286,351 was issued on December 3, 1918.

Associated Press was the next major customer. Morkrum teletypewriters were used from 1915 to deliver news from the AP office in New York to newspapers in New York and nearby towns and in Philadelphia, a job previously done by messenger boys. Within a few years, over 800 newspapers received their news from AP by Morkrum teletypewriters controlled by a single operator in the AP office in New York. Other press associations soon followed the example of AP.

At the same time, Ernst Eduard Kleinschmidt (1875–1977) began developing a teletypewriter. Kleinschmidt, born in Bremen, Germany, went to the United States at the age of 8. Although without much school education, he obtained 118 patents in his 101-year life. He patented first a Morse keyboard transmitter[2] and later a Morse keyboard perforator, which became known as the *Wheatstone perforator*. In 1916 he filed an application for a type-bar page printer.[3] Shortly after Morkrum obtained their patent for the start–stop method, Kleinschmidt filed (on May 1, 1919) an application entitled "Method of and Apparatus for Operating Printing Telegraphs," including an improved start–stop method, for which U.S. patent 1,463,136 was issued on July 24, 1924. Teletypewriters based on this patent were delivered to Western Union in the early 1920s.

Instead of wasting energy and money in patent battles on the start–stop method, Kleinschmidt and the owners of the Morkrum Company decided to merge the companies into the Morkrum–Kleinschmidt Company in 1924. The new company combined the best features of both printers into a new type-wheel printer, for which Kleinschmidt, Howard Krum, and Sterling Morton together obtained U.S. patent 1,994,164. This printer had the alphabet spaced around the rim of a wheel mounted on a shaft attached to a gear wheel. Each time the armature of a magnet was attracted by the line codes, the type wheel revolved until the letter desired faced a ribbon of paper. Another magnet pushed the paper against the type with an inked ribbon between and thus printed the letter.

In 1925, the company name was changed to Teletype Corporation, which became a worldwide-known brand name for a highly reliable teleprinter. Figure 19.1 shows a version manufactured under license in Germany in 1927.

AT&T inaugurated a *Teletypewriter Exchange Service* called TWX on November 21, 1931 for their 16,000 teletypes in operation. One year earlier, on October 1, 1930, AT&T made Teletype Corporation a subsidiary of Western Electric. The foreign rights of teletype were sold to IT&T. Further thousands of teletypes were installed at the premises of private companies and banks and newspapers. Manual central switching exchanges were established through which a subscriber could communicate by teletype with any other subscriber in the United States. To make a call, the customer looked up the number in the nationwide TWX directory and called the operator to be connected with the desired party. Once connected, the two subscribers could type their messages and replies. On December 31, 1932, Western Union also started a manually switched *Timed Wire Service* for their 18,200 teletypes in operation, whereby a telegraph line was placed at the disposal of a customer but on a one-way basis only. By 1938, over 160 teletype exchanges served about 120,000 subscribers.

---

[2] U.S. patent 964,371, filed February 7, 1895, issued January 11, 1910.

[3] U.S. patent 1,448,750, issued March 20, 1923.

Figure 19.1   Teletypewriter. (Courtesy of Alcatel SEL.)

19.2   TELEPRINTER DEVELOPMENT IN GREAT BRITAIN

The development of teleprinter equipment in Great Britain is strongly connected with Frederick George Creed. Born in Mill Village, Nova Scotia, in 1871, Creed started his career as a check boy for Western Union at nearby Canso. There he taught himself on cable and landline telegraphy. He then worked for the *Central and South American Telegraph and Cable Company* in Peru and Chile. Tired of having to use hand-operated Morse keys and Wheatstone tape perforators, he conceived the idea of a typewriter–style machine that would enable complete Morse code signals to be punched in a tape simply by operating the corresponding character keys. He went to a suburb of Glasgow in Scotland, where in a garden shed, he changed an old typewriter into a keyboard perforator. Compressed air was used to punch the holes. He also constructed a receiving perforator (re-perforator) and a printer. The re-perforator recorded the incoming signals into a perforated tape identical with the transmitting tape. The printer decoded this tape into plain language on ordinary paper. Thus was born the *Creed High Speed Automatic Printing System*.

Although Lord Kelvin told him that "there is no future in that idea," Creed managed to obtain an order for 12 machines from the GPO in 1902. He opened a small factory in Glasgow in 1904 and moved five years later with six of his mechanics to Selsdon Road, South Croydon. In 1915, he settled at the present site of Telegraph House in East Croydon.

A big boost came for the Creed machine when in 1912 the *Daily Mail* adopted the Creed system to transmit daily the entire contents of its newspaper from London to Manchester for simultaneous publication. In 1913, the first experiments were made in high-speed telegraphy by wireless transmission with radio equipment installed at the Croydon factory and Creed's home about 5 km away. However, World War I diverted their activities to military equipment.

In the 1920s, Donald Murray made a valuable contribution to telegraphy by introducing the *Murray multiplex system*, a five-unit telegraph code. Murray applied a rationalized allocation of the combinations of the five bits to the characters of the

Figure 19.2   World's first automatic teleprinter exchange, Berlin, 1932. (© Siemens-Archiv, Munich.)

alphabet on a frequency-of-occurrence basis, thereby reducing the wear in the teleprinter. Murray's multiplex system and other telegraph patents were acquired by Creed in 1925 and used for a new teleprinter, Model 3. This teleprinter printed the messages directly onto gummed paper tape at a speed of 65.3 words per minute. It was the first Creed teleprinter to go into mass production. Many thousands were sold worldwide in the years 1927–1942.

The world's first network of teleprinter machines was installed in the U.K. The *Press Association* in London installed a private news network around 1920 using several hundred Creed teleprinters to serve newspapers in Bath, Bristol, Cardiff, Exeter, Glasgow, Leeds, Manchester, Newport, Plymouth, and Swindon. It served practically every morning daily in the U.K. and for many years was the world's largest single private teleprinter network. Other newspapers followed, as well as telegraph administrations and companies in Australia, Denmark, India, South Africa, and Sweden.

In July 1928, Creed & Company became part of IT&T. Frederick Creed, wealthy after having sold his company, retired in 1930 and turned his inventive mind to other, albeit less successful projects, such as a midocean "Sea Drome" and an "unsinkable" boat. He died at his home in Croydon at the age of 86.

By 1931, when teleprinter operation was limited primarily to press and railway applications, GPO introduced a person-to-person public teleprinter service to open telegraphy to business and industrial users. Creed & Co. thereupon developed its Model 7 teleprinter. The Model 7 was a revolutionary machine in those days, with many new features, such as interchangeability between page and tape writing, ribbon

Figure 19.3   Siemens teleprinter, 1936. (© Siemens-Archiv, Munich.)

inking, answer-back device, and the use of ball bearings on all high-speed shafts with a lubrication system that permitted 100 hours of continuous unattended operation. Over 80,000 units of the Model 7 in different versions were manufactured over an almost 50-year period.

By 1927, the GPO adopted a uniform transmission system for telegraphy and teleprinter based on voice-frequency transmission on the existing telephone network.[4] In 1936, GPO started international teleprinter operation with the Netherlands and one year later with Germany.[5]

[4] Initially, a 300-Hz signal, later a 1500-Hz signal, and eventually, 18 telegraph channels were allocated within one telephone channel.

[5] Since Germany had a separate teleprinter network, the messages sent from England had to be demodulated in Amsterdam and sent to Germany as normal telegraph signals.

19.3   TELEPRINTER DEVELOPMENT IN GERMANY

At the beginning of the twentieth century, the gradual penetration of telephony kept the total amount of telegraph apparatus in service almost constant. Furthermore, the Morse sounder replaced the Morse writing telegraph in many applications. At the beginning of World War I, the number of Morse writing telegraphs had been reduced to about 9700. The number of Morse sounders had increased to 4200 and some 1250 Hughes telegraphs were in operation. After the war and a long period of dismantling and reconstruction, for lack of a national product, the German Imperial Post and Telegraph Administration decided to test foreign teletypewriters. Long trial operations started on June 5, 1926, on the Berlin–Chemnitz and Nuremberg–Munich lines with teletypewriters from Morkrum–Kleinschmidt. In the same year C. Lorenz acquired patent licenses from that company and Siemens & Halske developed their own teleprinter. Based on the positive experience on those trials, in 1932 the administration decided to install a national public automatic teleprinter network (in German, *Fernschreibnetz*), which began service on October 16, 1933. The first automatic teleprinter offices were opened in Berlin (Figure 19.2) with 13 subscribers and in Hamburg with eight subscribers. The world's first international teleprinter operation began in 1934 from Germany with The Netherlands and Switzerland and in 1936 with Belgium, Denmark, and Great Britain.

The transmission of teleprinter signals was made on lines separate from the telephone network. Whereas the administration installed and operated the infrastructure and took care of the maintenance of the teleprinter machines, the subscribers had to buy their teleprinter machines directly from the two manufacturers, C. Lorenz and Siemens & Halske. Figure 19.3 shows a teleprinter produced by Siemens and Figure 19.4 a teleprinter produced by C. Lorenz. The German teleprinter network had 100 subscribers in 1935, about 1500 at the beginning of World War II, and 3000 before the network collapsed at the end of the war.

Figure 19.4   Teleprinter LO-15 of C. Lorenz. (Courtesy of Alcatel SEL.)

19.4 TELEPRINTER DEVELOPMENT IN JAPAN

Various types of teleprinters able to print Japanese characters were tested beginning in 1922. U.S.-made teleprinters printing Japanese character were introduced in June 1927 on the Tokyo–Osaka line. Production of teleprinters in Japan began in 1937.

REFERENCES

Books

Gööck, Roland, *Die großen Erfindungen Nachrichtentechnik Elektronik*, Sigloch Edition, Künzelsau, Germany, 1988.

Michaelis, Anthony R., *From Semaphore to Satellite*, International Telecommunication Union, Geneva, 1965.

Oslin, George P., *The Story of Telecommunications*, Mercer University Press, Macon, GA, 1992.

Das deutsche Telegraphen-, Fernsprech- und Funkwesen, 1899–1924, Reichsdruckerei, Berlin, 1925.

50 Jahre Lorenz, 1880–1930, C. Lorenz Aktiengesellschaft, Berlin.

150 Years of Siemens: The Company from 1847 to 1997, Siemens, Munich, 1997.

Articles

Barnekow, Rolf, and Manfred Bernhardt, Die Vorläufer der Telefaxgeräte, *Post- und Telekommunikationsgeschichte*, Vol. 1, 1995, pp. 57–62.

Bernhardt, Manfred, Entwicklungsgeschichte der elektrischen Telegrafie bis zur Einführung des öffentlichen Fernschreibdienstes in Deutschland (Kurzfassung), *Archiv für deutsche Postgeschichte*, Vol. 1, 1992, pp. 53–60.

Wenger, P.-A., The future also has a past: the telefax, a young 150-year old service, *Telecommunication Journal*, Vol. 56, No. 12, 1989, pp. 777–782.

Internet

To HTML from *The Early History of Data Networks*, by Gerard J. Holzmann and Björn Pehrson.

http://Japan.park.org/Japan/NTT/Museum, History of telegraph.

www.rtty.com, Creed and Company Limited, The first 50 years, by Alan G. Hobbs; latest revision, January 28, 1998.

www.massis.lcs.mit.edu, Telecom Digest, A brief history of the Morkrum Company, manuscript from Howard L. Krum, 1925.

www.massis.lcs.mit.edu, Telecom Digest, History of teletypewriter development, by R. A. Nelson, Teletype Corporation, October 1963.

20

COPPER-LINE TRANSMISSION

20.1 TELEGRAPHY TRANSMISSION ON COPPER LINES

At the beginning of the twentieth century, a worldwide electrical telegraphy network existed, connected by copper lines with a total length of about 1,800,000 km. Land-line telegraphy, however, stagnated, as telephony became the major form of telecommunications. The development of transmission of telegraphy at the beginning of the twentieth century therefore concerned mainly submarine cable. Telephony on submarine cable or on radio was not yet possible. Relay repeaters, which extended the distance of landline telegraph line performance, were not yet available for submarine cable. The siphon recorder of William Thomson, with minor improvements, still determined a low transmission speed of 25 words/min (125 letters/min). The invention of electronic-tube-operated signal-shaping amplifiers in the 1910s increased the speed to 40 words/min, and 2000 words/min was achieved with permalloy cable loading in the 1920s.

This worldwide submarine cable telegraphy network was dominated by British private enterprises, which owned almost 70% of the network and supplied almost 90% of the cable. In total, about 300,000 km of submarine cable belonged to private companies, and only 40,000 km was state owned. The state-owned submarine cable concerned mainly coastal cables or direct connections between two nearby countries where international agreements could easily be made. Strong efforts were made by U.S., French, and German companies to reduce British domination. As a first success, the *German–Atlantic Telegraph Company* on September 1, 1900 opened a direct cable between Germany and the United States, albeit via the Azores, because landing rights in Great Britain were denied despite 10 years of negotiating. The cable was then the longest transatlantic cable. It had a length of 3389 km between Emden,

The Worldwide History of Telecommunications, By Anton A. Huurdeman
ISBN 0-471-20505-2   Copyright © 2003 John Wiley & Sons, Inc.

Germany, via Vigo, in the north of Spain, to Horta on Fayal, Azores, and 4528 km between Horta and New York, thus a total of 7917 km. The maximum depth of the route was 5670 m between Horta and New York and 4850 m between Vigo and Horta. Two years later the same company laid a second cable on the same route. Horta became an international Atlantic submarine cable center where telegraph operators of various companies exchanged messages received for onward transmission at Trinity House (Trinity reflecting the cooperation of three nations: the United States, England, and Germany) until this function was taken over by automatic repeaters in the 1920s. Currently, about 15 transatlantic cables still land at the Azores. Similarly, the South Atlantic islands Cape Verde, Ascension, and St. Helena connected submarine cables between Europe, Africa, and South America.

France laid additional cables to its African colonies with landings at Morocco, Algiers, and Dakar, Senegal. From Dakar, coastal cables were laid up to Cape Town, and a transatlantic cable to Pernambuco in Brazil. To reach their colony of Indochina without using a British cable, a French cable was laid to Copenhagen, where it was connected with the network of the Great Northern Telegraph Company.

The United States and Great Britain laid their first cables across the Pacific in 1902. Both cables were manufactured by the *British Telegraph Construction and Maintenance Company*. The American cable went from San Francisco via Hawaii to Guam, and from there, to Manila in the Philippines (since 1898 a U.S. colony), where the service began in July 1903. The cable was laid and operated by the *Commercial Pacific Cable Company*. Many years later it was revealed that even this cable was not U.S.-owned at all; 50% was owned secretly by the British Eastern Telegraph Co. and 25% by the Danish Great Northern Telegraph Co.

The British cable was operated by the *Pacific Cable Board*, which was co-owned by the governments of Great Britain ($\frac{5}{18}$), Canada ($\frac{5}{18}$), and New-South Wales, Victoria, Queensland, and New Zealand, each with $\frac{1}{9}$. The major protagonist of this cable project was Sir Sandford Fleming,[1] who since 1880 had promoted a direct connection from Canada through the Pacific Ocean to Australia without passing through London. The cable went from Vancouver via the Pacific Ocean islands of Fanning, Christmas, Penrhyn, and Suwarrow (which for this purpose became British islands!), and Fiji to the Australian island Norfolk, where a branching was made with one cable to Australia and another to New Zealand. Cable laying had begun on March 13, 1902, and commercial operation began on December 8, 1902. The network was completed with a direct cable between Australia (at Bondi Bay near Sydney) and New Zealand (at Muriway Creek). This *All-British Cable* was extended around the world with cable from Perth, Australia, via the Indian Ocean islands Cocos, Rodriquez, and Mauritius to Durban, South Africa, from there via landline to Cape Town, then passing the Atlantic Ocean islands of Ascension, Cape Verde, and Azores, and terminating at Porthcurno, Cornwall, in Britain. With a cable between the Cocos Islands and Batavia (now Jakarta, Indonesia) the All-British Cable was also connected with an existing British cable that ran via India and the

---

[1] Sir Sandford Fleming (1827–1915), born in Scotland, went to Canada in 1845, where he became the driving force behind the Canadian Pacific Railway. He is the founder of standard world time, with 24 equal 15°, 2-hour time zones starting at Greenwich, England, at the zero meridian, which was adopted at the *Prime Meridian Conference* in 1884.

Figure 20.1   German cable ship *Stephan* at Yap, 1905. (Scanned from *Archiv für deutsche Postgeschichte*, Vol. 1, 1982, p. 28.) See insert for a color representation of this figure.

Mediterranean to Britain, and with another British cable via Hong Kong to Shanghai and Nagasaki.

Like Fayal in the Atlantic Ocean, Guam became an important crossing point for submarine cable between the countries with a shore on the Pacific Ocean. Japan opened telegraph service with the United States on August 1, 1907, via the islands of Bonin and Guam. Another cable connected Guam with a submarine cable network of the *Deutsch-Niederländische Telegraphengesellschaft AG* (German–Dutch Telegraph Company) in 1905. This network served the Dutch Indies and the German colonies, consisting of some of the Caroline, Marshall, Samoa, and Solomon islands south of Guam, such as Nauru, Jaluit, Yap, and Samoa, and part of New Guinea.[2] The company laid a 3035-km cable from Guam via Yap to Menado, Celebes (now Manado, Sulawesi). The cable conductor consisted out of seven-stranded copper wires, each with a diameter of 0.86 mm. From Yap a cable 3295 km long was laid to Shanghai, where it was connected with the network of the Great Northern Telegraph Co. The conductor was made out of seven 0.71-mm copper wires stranded around a 2.13-mm copper wire. This cable had to pass the Ryukya Rift at a depth of 7460 m, then the largest depth passed by submarine cable. Figure 20.1 shows the German cable ship *Stephan*[3] arriving at the island of Yap (now belonging to the Federal States of Micronesia).

The period between 1900 and 1914 was one of intensive submarine cable laying. A

---

[2] These territories came under German protection on October 12, 1899 and under U.S., British, and Australian protection after World War I.

[3] The *Stephan* could take a cable load of 4500 tons and was built in 1902 at Stettin, Germany (now Szczecin, Poland).

TABLE 20.1   Submarine Cable Ships Fleet at the End of 1906

Country	Number of Ships	Names of Ships
State-Owned Cable Ships		
Canada	2	*Tyrian, Lady Laurier*
China	1	*Fee Cheu*
France	3	*Ampère, Charente, Diolibah*
Germany	2	*Großherzog von Oldenburg, Stephan*
Great Britain	2	*Monarch, Alert*
Italy	1	*Citta di Milano*
Japan	1	*Okinawa Maru*
The Netherlands	1	*Telegraaf* (prior to 1905, German cable ship *Von Podbielski*, launched in Glasgow in 1899)
New Zealand	1	*Tutanekai*
United States	1	*Burnside*
Total	15	
Cable Ships of Private Companies		
Denmark	3	*H. C. Oersted, Store Nordiske, Pacific*
France	3	*François Arago, Contre-Amiral Caubet, Pouyer-Quertier*
Great Britain	25	*Anglia, Amber, Britannia, Cambria, Colonia, Electra, John Pender, Duplex, Levant I, Levant II, Sherard Osborn, Recorder, Patrol, Magnet, Buccaneer, Dacia, Silvertown, Pattrick Stewart. Iris, Faraday, Retriever, Henry Holmes, Minia, Relay, Viking*
Mexico	1	*Mexican*
United States	5	*Norseman, Norse, Cormorant, Mackay–Bennett, Restorer*
Total	37	
Total worldwide	52	

Source: Data from *Archiv für Post und Telegraphie,* 1907.

total of 52 cable ships were in use laying new or repairing existing submarine cable by 1907. The oldest, the *Dacia*, owned by the *Indian Rubber Company*, was launched in 1867. A summary of those ships is given in Table 20.1. Those ships laid almost 200,000 km of submarine cable in the period from 1900 until the beginning of World War I. Table 20.2 gives a summary of the total submarine cable lengths in operation by the end of 1913.

Deployment of submarine cable was interrupted by World War I and did not really recover after the war. The world's large networks had been laid, colonial expansion came to a stop, economical recovery went slowly, and radio presented a workable alternative. Also, the performance of telegraphy on submarine cable was still delicate and affected by high signal distortion, low sensitivity of the siphon recorder, and interference from external sources along the cable routes. Electromechanical signal magnifiers, developed in the 1910s, increased the prevailing trans-

TABLE 20.2   Submarine Telegraph Cable in Operation in 1913

Continent and Country	Number of Cables	Length (km)	Continent and Country	Number of Cables	Length (km)
State-Owned Cables					
The Americas					
Canada	2	740	Venezuela	7	1,125
United States	13	3,980	Brazil	35	88
Mexico	2	735	Argentina	27	135
Bahamas	1	395	Uruguay	5	18
			Total Americas	92	7,215
Europe					
Austria	50	780	The Netherlands	49	480
Belgium	6	190	Norway	770	2,600
Denmark	148	850	Portugal	6	225
France	77	21,045	Russia	32	1,370
Germany	100	5,245	Spain	24	5,800
Great Britain	223	5,050	Sweden	106	555
Greece	53	115	Switzerland	3	25
Italy	59	3,025	Turkey	25	685
			Total Europe	1,731	48,025
Africa					
Senegal	1	5	Portuguese colonies	2	50
French colonies	2	2	Total Africa	5	57
Asia					
British India	13	3,720	Persia	1	28
French Indochina	1	1,432	Russia	1	10
Japan	180	9,115	Siam	2	23
Malacca	1	2	Total Asia	199	14,330
Oceania					
Australia	40	1,020	New Zealand	31	670
Dutch Indies	18	5,700	Pacific Cable Board	5	14,540
New Caledonia	1	2	The Philippines	26	1,915
			Total Oceania	121	23,847
			Total state-owned	2,148	93,475

mission speed of 25 words per minute (125 letters per minute) only slightly, to 40 words per minute. Most of those magnifiers used a sensitive moving-coil galvanometer, which moved a device a small distance to control a much greater power. For instance, in the selenium amplifier a beam of light from a galvanometer mirror moved over a group of selenium cells which through their varying resistance produced a larger signal in an external circuit powered by a local battery. Heurtley developed an even more sensitive device using two heated wires in a Wheatstone

TABLE 20.2 *(Continued)*

Continent and Country	Number of Cables	Length (km)	Continent and Country	Number of Cables	Length (km)
Submarine Cables of the Private Companies					
British Companies					
Eastern Telegraph	105	79,990	Indo-European Tel.	4	355
Eastern Extension Australasia and China Tel.	36	47,100	Anglo-American Tel.	15	24,110
			Direct United States Cable	3	5,885
Eastern and South African Tel.	17	19,500	Halifax and Bermudas Cable	1	1,580
West African Tel.	8	2,725	Cuba Submarine Tel.	10	2,120
African Direct Tel.	9	5,610	West India and Panama Tel.	22	8,080
West Coast of America Tel.	7	3,670	River Plate Tel.	4	405
Western Tel.	30	44,215	South American Cable	4	5,155
Europe and Azores Tel.	2	1,960	Total British companies	285	256,125
Direct Spanish Tel.	4	1,320			
U.S. Companies					
Western Union Tel.	9	13,610	Central and South American Tel.	24	20,645
Commercial Cable	15	30,780	U.S. and Haiti Tel. and Cable	1	2,580
Commercial Pacific Cable	6	18,570	Mexican Tel.	5	5,235
Commercial Pacific Cable Co. of Cuba	1	2,385	Total U.S. companies	61	93,805
German Companies					
German–Atlantic Tel.	5	17,730	German–South American Tel.	5	3,640
Dutch–German Tel.	3	6,335	East-European Tel.	1	343
			Total German companies	14	38,048
Compañia Telegràfico del Plata (Argentina)	1	52			
Great Nordic Tel. Co.	28	16,665			
Française des Cables Tel.	24	21,205			
Total private companies	413	425,2900			
Total worldwide	2,561	519,375			

Source: Data from *Archiv für Post und Telegraphie*, 1914.

circuit which changed its balance when the receiving signal moved one of the wires slightly out of a constant heat stream.

As explained in Section 20.4, the transmission on open-wire lines could be improved substantially by adding inductance to the lines. Searching for such an solution on submarine cable, the AT&T engineers Oliver E. Buckley, H. D. Arnold, and G. W. Elmen discovered around 1915 that a thin tape of annealed nickel–iron alloy called *permalloy* wrapped around a copper conductor dramatically improved performance, so that transmission speeds over 2000 letters/min could be achieved. In October 1923 Western Union and Western Electric made the first successful experiment with a permalloy-loaded cable 220 km long laid in a loop from the south shore of Bermuda. The British Telegraph, Construction & Maintenance Company Ltd. manufactured the cable with permalloy tape supplied by Western Electric Company. Upon this success, a permalloy-loaded cable was laid between New York and Horta and put into operation in September 1924 by Western Union. Two years later a cable with permalloy loading was laid between Horta and Emden, Germany, thus completing a U.S.–German link. In the same year a permalloy cable was laid from New York via Bay Roberts, Newfoundland, to Penzance, U.K. A transmission speed of 2500 words/min was obtained on both cables. The combined traffic capacity of the two cables was about as great as the capacity of all 16 nonpermalloy cables in service at that time across the Atlantic Ocean. The new permalloy cable could be used to its full transmission capacity thanks to new high-speed siphon recorders and vacuum-tube-operated signal-shaping amplifiers. By 1928, about 28,000 km of permalloy cable crossed the Atlantic and Pacific Oceans, which represented 5% of the total worldwide submarine cable length.

Another important achievement was the development in 1938 by the British *Imperial Chemical Industries* (ICI) of a substitute for gutta-percha called *polyethylene*. Polyethylene presented a unique combination of electrical and mechanical characteristics and could be extruded on the same machines as those used for gutta-percha, which eventually, it replaced completely. It was also the base for coaxial cable with wideband characteristics, which appeared in the late 1930s.

20.2 TELEPHONY TRANSMISSION ON COPPER LINES

With the introduction of telephony, mostly overhead copper[4] wire lines (o/w lines) were used to connect subscribers with telephone exchanges and for connection between telephone exchanges. In the beginning a single wire was used for those telephone lines, and Earth was used as the return circuit, just as for telegraph circuits. But houses, factories, and streetcars, also grounded their electrical circuits using Earth as the return circuit, and with the increase in telegraph and telephone penetration, a huge amount of static and noise on the ground circuit substantially disturbed the telegraphy and, even more, the telephony service. On July 19, 1881, A. Graham Bell obtained a patent for a metallic circuit which used two wires to com-

---

[4] Instead of conductors made of copper, initially, iron or steel was generally used because of its higher strength, albeit with a conductivity only 10% that of copper. Around 1880, bronze (an alloy of copper, tin, lead, iron, and phosphorus or silicon) was commonly used because of its much higher mechanical strength than that of copper and a conductivity only 10% below that of copper.

plete the electrical circuit, avoiding the ground altogether and thus eliminating grounding disturbances. The metallic circuit was introduced commercially in October of that year on a telephone circuit between Boston and Providence. It reduced noise greatly over those 45 miles and heralded the beginning of long-distance service. Using the metallic circuit on o/w lines, which have a relatively large distance between the two wires, and air as the dielectric resulted in a relative low capacitance and dielectric loss, so that telephony could be transmitted over distances of up to about 1000 km, albeit under good weather conditions[5] and with the use of sensitive telephones. The longest o/w line, which was installed between New York and Chicago in 1893, had a length of 1500 km and used one pair of copper wires with a diameter of 4 mm.

Introduction of metallic circuits on subscriber access lines took many years because of the additional cost. At the end of the nineteenth century, the metallic circuit was used in the United States for less than 50% of the telephone access lines. With the rapid increase in telephone penetration over longer distances at the beginning of the twentieth century, however, the metallic circuit became a necessity. This led, however, to an overcrowding of o/w lines carried on poles through the streets of cities with a very high concentration near the exchanges, which was ugly as well as causing substantial crosstalk, frequent short circuits, and line interruptions. Figure 20.2 shows a special tower structure that was required to enter a few thousand o/w lines to an exchange in Frankfurt, Germany around 1900. Figure 20.3 shows the accumulation of open wires in a U.S. street in 1911.

This accumulation of o/w lines and the occasional total interruptions of the telephone service due to severe climatic conditions with hoarfrost and snowstorms led to the introduction of telephone cable. On cable, however, with a much higher capacitance due to the smaller distance between the wires and the higher dielectric loss in the insulation, the attenuation was much higher than on o/w lines, and telephony messages beyond 30 km became very difficult to understand. As a consequence, the introduction of cable was limited to urban telephone access networks. A first improvement came at the beginning of the twentieth century when the gutta-percha insulation of the copper conductors was replaced by thin-paper-tape insulation, which reduced the capacitance substantially. A seamless lead sheath hermetically sealed the cable core to prevent moisture ingression.

In Europe, the first submarine telephone cable, with a length of 40 km, was laid in 1890. The four-wire submarine cable was manufactured by Siemens Brothers Telegraph Works and laid between St. Margarets, England, and Sangatte, France, by the cable ship *Monarch* with an 89-person crew. A 135-km two-pair 4-mm o/w line connected London with St. Margarets; a 325-km one-pair 5-mm o/w line connected Paris with Sangatte. Gower–Bell telephones were used in England and d'Arsonval telephones in France. Two public telephone cells each were installed in Paris and in London. Telephone subscribers in Paris could use the international telephone service from their telephone sets, whereas in London the private *National Telephone Company* was not connected with this service and persons interested needed an additional line to the GPO. Initially, one cable pair only was used; one year later the second

---

[5] The attenuation of an o/w line typically increases by 60 to 70% due to rain or 200 to 400% due to snow and hoarfrost, while thunderstorms are dangerous for o/w lines and for subscribers connected by such lines.

Figure 20.2   Open-wire support and access tower on the roof of a telephone exchange. (Courtesy of Museum für Kommunikation, Frankfurt, Germany.)

pair was used, too. In 1897, two additional cables were laid each with two pairs only. In 1903 a submarine cable with a length of 88 km was laid between Dover, England and La Panne, Belgium. The telephone intelligibility was very low and traffic was limited to direct connections between London and Brussels.

20.3   PHANTOM CIRCUITS

The use of a pair for the transmission of each telephone or telegraph channel either on o/w or cable was not very economical. The American engineer Frank Jacob proposed in 1882 a method whereby three telephone circuits could be derived from two pairs of wires. The additional circuit, the phantom circuit, was obtained by using the two wires of each pair in parallel as one side of the phantom line circuit (Figure 20.4). John J. Carty made the first application in 1886. Efficient balanced transformers were needed, however, for the end connections, and complex transpo-

Figure 20.3 Open wires in a U.S. street, 1911. (Courtesy of Museum für Kommunikation, Frankfurt, Germany.)

sition arrangements of the wires were required to achieve a minimum of crosstalk between the three circuits. It therefore took until the beginning of the twentieth century until phantom circuits were widely applied and a 30% reduction of number of required pairs was achieved. A further economy was obtained by applying a phantom circuit over two other phantom circuits, thus operating seven circuits on four pairs of wires.

20.4 PUPIN COILS

At the beginning of the twentieth century, transmission of telephony was possible up to a distance of about 30 km on cable and about 2000 km on o/w lines. For tele-

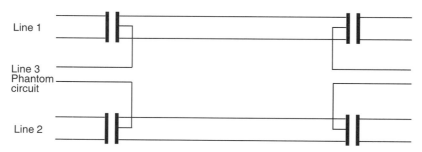

Figure 20.4 Phantom circuit arrangement.

phone transmission over larger distances the capacitance between a pair of wires in a cable or an o/w line needed to be balanced by additional inductance.[6] For this purpose, the second outstanding invention for line transmission, after Wheatstone's relay, came from Michael Idvorsky Pupin (1858–1935). Pupin, born in Serbia, came to the United States at the age of 15 with "five cents in his pockets." Within 10 years he graduated from Columbia University, then went to Germany, where he attended lectures of Helmholtz and Kirchhoff (just after Heinrich Hertz had left Helmholtz) and obtained a doctor of physics degree in Berlin. Back in the United States he became a professor at Columbia University in New York, where he made investigations on long-line transmission and worked out the theoretical basis for the determination of the inductance to be added to telephone lines to extend their range significantly. In 1899, he published a study entitled "Propagation of Long Electrical Waves" in which he proposed to improve telegraph lines for the transmission of telephony with the introduction of coils at regular distances. He published a more specific paper in 1900 entitled "Wave Transmission over Non-uniform Cables and Long Distance Lines," in which he claimed that with special designed coils, later called *Pupin coils*, at regular intervals, telephony on underground cable could be extended by a few hundred kilometers and on o/w by several thousands of kilometers without an increase in conductor diameter.[7] His study culminated in a precise mathematical formula for the design of loading coils for which he submitted patent claims in the same year.[8] Two patents were granted on June 19, 1900 under the numbers 652,230 and 652,231. AT&T immediately bought these patents for the sum of $455,000, surely the highest price ever paid for a formula! In fact, although George Cambell and John Stone of AT&T had been working for several years on a solution to the capacitance problem on their long telephone lines, they had missed the opportunity to patent their findings, so that AT&T was forced to buy Pupin's patents. Some 20 years later, AT&T claimed that with the Pupin coils they had saved over $100 million that otherwise would have been required for different cable and additional amplifiers. AT&T installed the first Pupin coils on a 17-km cable between New York and Newark, New Jersey, in 1902. Other cables followed rapidly, so that in 1906 loaded cables were installed, for instance, between New York and New Haven (127 km) and between New York and Philadelphia (140 km). The cables, with No. 14 AWG (American Wire Gauge) 1.6-mm conductors, were loaded with 250-mH coils at 1-mile (1850 m) intervals. The longest loaded cable (before the introduction of telephone amplifiers) was laid between Boston and Washington via New York, a distance of 724 km. Loading on phantom cable circuits was first used in 1910 on a cable between Boston and Neponset.

On the European continent, Pupin experienced substantial difficulty in showing that his patent application was not just derived from the respective theories of Oliver

---

[6] The necessity of a balance of inductance and capacity had already been recognized by Oliver Heaviside in 1873 and published between 1885 and 1887 in the British weekly the *Electrician*, in which he stated: "Without sufficient inductance, permitting energy to be stored in the magnetic field of the line, efficient transmission would not be possible and much of the energy of the signal would be transformed into heat."

[7] Routes with Pupin coils were implemented up to around 700 km on cable and up to 2000 km on o/w.

[8] The idea of improving the transmission signals over a line by adding distributed inductance to it had been presented by Oliver Heaviside in 1887 and 1893 with papers in the *Electrician*. Coil loading of lines was also proposed by S. P. Thompson and patented in the U.K. in 1891 and in the United States in 1896. However, Pupin was first to define the exact criteria for the design of loading coils.

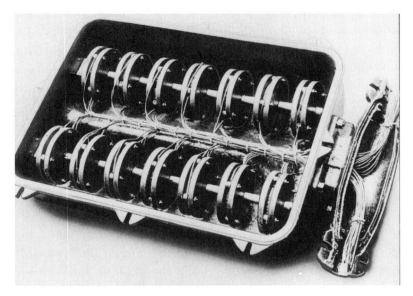

Figure 20.5   First Pupin coils used in Germany, 1902. (© Siemens-Archiv, Munich.)

Heaviside in the U.K. and Vaschy in France. However, Siemens, in cooperation with Pupin, immediately developed loading coils based on Pupin's theory. The coils were installed on 14 pairs of a 28-pair 32-km telephone cable between Berlin and Potsdam (Figure 20.5). An official test was made in March 1902 in which 25 experts from Germany and abroad could be convinced about the improvement with the new coils. After a three-year patent suit in Germany, whereby Pupin improved his patent with the support of Siemens & Halske, the patent was granted on February 4, 1904, and bought immediately by Siemens.[9] Initially, the Pupin coil was used primarily on o/w lines. The longest o/w line in Europe, 1350 km long, was installed between Berlin, Frankfurt, Basel, and Milan.[10] The line, completed in 1914, used two 4.5-mm hard-drawn copper conductors. It passed through the Simplon tunnel via a Krarup cable (Section 20.5) with a length of 22 km, which had been laid in 1906.

The first European project involving underground cable laid over a large distance using Pupin coils has begun in 1912 in Germany for the so-called *Rheinland cable*. The cable covered a distance of 600 km between Berlin, Magdeburg, Hannover, Dortmund, and Cologne. It was laid in concrete cable ducts with four conduits, thus providing three conduits for future cables (Figure 20.6). The first section, a 52-pair cable between Berlin and Magdeburg, was put into operation on November 11, 1913. Hannover was connected to the cable in August 1914. World War I delayed further work until 1921, at which time a 142-pair cable (Figure 20.7) was laid between Hannover and Cologne. Figure 20.8 shows the placing of a Pupin coil in a cable duct of that cable. The cable construction was optimized for the transmission

---

[9] Pupin was given one-third of the profit made by Siemens on those coils.

[10] Substantial opposition to this line was experienced in Switzerland, where the 15-m-wide obstacle-free trail for the o/w line was considered to be an irresponsible impairment of nature, with subsequent danger of falling rocks.

Figure 20.6   Cable duct for Europe's first long-distance telephone cable. (© Siemens-Archiv, Munich.)

of telephony by twisting conductors to pairs. Two different approaches to the twisting of pairs to a quad (four conductors for one physical circuit) were made, called the Dieselhorst–Martin quad (DM quad, used mainly in Germany) and the star quad. In a *DM quad* two conductors were twisted to form a pair, and two such pairs were twisted with different twist length into a quad. In a *star quad*, four conductors were twisted simultaneously into a quad. Coverage of telephony on such telephone cable with a conductor diameter of 1.5 mm was typically 40 km without using Pupin coils and 220 km with Pupin coils. For a conductor diameter of 2.0 mm those values were 60 and 320 km. Pupin coils in paper-tape-insulated underwater cable were first used in 1906 on a 12-km seven-pair cable laid in Lake Constance at a depth of 250 m between Friedrichshafen, Germany, and Romanshorn, Switzerland. Flexible cable sleeves were used to accommodate the 22 Pupin coils. Those Pupin coils, combined with 1.5-mm copper conductors, enabled telephone service between Zurich, Stuttgart, and Munich. The cable was inaugurated in the presence of Count Zeppelin and served about 40 years. The first submarine cable, 37.1 km long, with Pupin coils was installed between St. Margaret, U.K. and Sangatte, France, in May 1910.[11] The Pupin coils for the two pairs were integrated in the cable, which increased the diameter of the cable from about 5 cm to 10 cm, over a length of 1 m. Similar cables were laid one year later from St. Margaret to La Panne, Belgium, with a length of 89 km, and in 1913, with a length of 165 km, to the Netherlands. Submarine telephone cable using Pupin coils with 24 pairs for the transmission of 12 telephone channels were

[11] This cable was laid in parallel with and in addition to the first three unloaded submarine telephone cables laid between Great Britain and France in 1891 and 1897.

Figure 20.7   The 142-pair Rheinland cable. (© Siemens-Archiv, Munich.)

also installed over a distance of 47 km between Gledser in Denmark and Warne-
münde in Germany in 1926 and over a distance of 117 km between Malmö, Sweden
and Stralsund, Germany, in 1927.

20.5   KRARUP CABLE

A Danish telegraph engineer, Carl Emil Krarup (1872–1909), found another solution
to increase the inductance of cable. Krarup, born in Copenhagen, studied civil
engineering and was put in charge of public works in Copenhagen until he joined the
Danish Telegraph Administration in 1898. In 1901 the administration sent him to the
University of Würzburg, Germany, where he made experiments on copper conductors bandaged with iron wires.[12] Back in Copenhagen in 1902, he wrote a paper
on the inductance of electrical lines for which he received a second prize from the
University of Copenhagen. Krarup, now an telegraph engineer, proved his theory in
the same year when an underwater cable was planned between Helsingör in Denmark and Helsingborg in Sweden. He proposed to distribute inductance evenly along
the cable by winding thin iron wires, typically 0.2 to 0.3 mm in diameter, around the
entire length of each of the copper conductors. The 5-km cable was laid and performed well. One year later, two submarine Krarup telephone cables were laid. A

---

[12] Continuous loading by means of a longitudinally discontinuous layer of iron covering the conductor was
proposed by J. S. Stone of AT&T in 1897 and patented in U.S. patent 578,275. The German professor
Breisig suggested the use of an open helix of iron wound around the conductor in 1901, whereas Krarup
used a closed spiral so that the adjacent turns were in contact.

Figure 20.8 Placing of a Pupin coil in a cable duct, 1912. (Courtesy of Museum für Kommunikation, Frankfurt, Germany.)

two-pair cable 20 km long was laid through the Fehmarnbelt connecting Germany with Denmark. A second Krarup cable with two pairs was laid on the same route in 1907. A cable of remarkable length 75.5 km connected the isle of Helgoland with the German mainland at Duhnen near Cuxhaven. It used two pairs for two telephone channels and one pair for telegraphy. The copper conductors had a diameter of 4 mm, which made it the longest submarine telephone cable without application of telephone amplifiers. The production of Krarup cable was relatively expensive and the amount of inductance added was low,[13] so that it found limited application and was later replaced by permalloy cable (described in Section 20.1), which provided a much higher inductance.

20.6 TELEPHONE AMPLIFIERS

Amplification of telephone signals became possible when Lee de Forest and Robert von Lieben invented the triode in 1906. Similarly, as telegraph signals on long-distance transmission were regenerated by means of relays, electronic line repeaters could be applied to amplify the speech band for long-distance transmission of telephony beyond 100 km on pupinized cable or beyond 600 km on o/w lines. In 1912, first experiments with a telephone amplifier took place on a line 1000 km long

[13] Measurements made around 1910 showed that Krarup cable increased the telephony intelligibility by 60%, whereas with Pupin coils an improvement of 250 to 370% was obtained.

between Königsberg (now Kaliningrad, Russia) and Strasbourg via Berlin. Three years later, AT&T took into operation the longest o/w telephone line in the world, 5419 km, between New York and San Francisco, on January 15, 1915. Construction of the line started in 1911 with the New York–Denver section. The line was constructed with two pairs of hard-drawn copper wires 4.2 mm in diameter, which were supported by 130,000 poles. Pupin coils with an inductance of 250 mH were placed at intervals of 13 km. A phantom circuit was applied on the two pairs so that three telephone circuits could be used. Three telephone amplifiers were used in 1915 and eight in 1918. In 1920, all Pupin coils were removed and 12 improved telephone amplifiers installed, which resulted in an increase of bandwidth from 1500 Hz to 3000 Hz. The telephone charges amounted to $20.70 for 3 minutes.

Open-wire lines remained very popular in the United States for long-distance communication. By 1930 over 5.5 million kilometers of open wires crossed the United States and another 185,000 km was added annually. At the same time, 34,000 km of cable was installed on long-distance routes, of which about 50% was aerial cable. The telephone subscribers were connected with their local exchange almost exclusively (94%) with cable. Typically, 1800 pairs of 28 AWG (diameter 0.32 mm) copper conductors were accommodated in this cable with an outside diameter of 6.7 cm.

After World War I, the Rheinland cable (cited in Section 20.4) was equipped with telephone amplifiers at intervals of 75 km on four-wire circuits with 0.9-mm conductors and 150 km on two-wire circuits with 1.4-mm conductors. It was extended into a national network with a cable length of over 7000 km in 1927 and 16,000 km in 1939. It became the nucleus of a European telephone network that served international telephony between surrounding European countries. Table 20.3 shows the major lines between European capitals that could be operated through this cable network around 1930.

Cable became the major transmission medium for telephony in Europe. In Germany, for instance, in 1922, only 25% of the long-distance lines were on cable; this figure increased to 50% in 1925 and to 84% by the end of 1939. At that time, 89% of the local telephone lines in Germany were also on cable. In 1939, the German transmission network consisted of 28 million kilometers of cable, of which 3000 km was coaxial cable and 2.4 million kilometers was on open-wire lines.

Telephone amplifiers were also used for submarine cable. On December 1, 1919, operation began on a 120.5-km-long cable between Stralsund, Germany and Trelleborg, Sweden, then the longest submarine telephone cable. The cable had two pairs,

TABLE 20.3 Telephone Lines between European Capitals

Route	Length (km)	Route	Length (km)
London–Stockholm	2456	Paris–Prague	1219
Paris–Stockholm	2362	Amsterdam–Vienna	1163
Geneva–Stockholm	2249	Brussels–Vienna	1092
Amsterdam–Stockholm	1985	Prague–Zurich	1031
London–Vienna	1711	Amsterdam–Zurich	985
Paris–Vienna	1335		

which in a phantom circuit were used for three telephone channels. Telephone amplifiers were used in Stralsund and in Malmö (30 km north of Trelleborg). Another remarkable early submarine telephone cable began operation on April 11, 1921 between Havana, Cuba, and Key West, Florida, by the *Cuban–American Telephone and Telegraph Company*. This company was established in 1917 by the powerful AT&T and the newcomer IT&T to construct and operate a cable between the United States and Cuba. World War I delayed the project until 1921, when three one-pair cables each 190 km long were laid through the Straits of Florida at a depth of 1860 m. The Telegraph Cable Company of the U.K. manufactured the cables in accordance with a design by the British cable expert Sir William Slingo. Each cable consisted of one Krarup-type conductor, consisting of a copper wire loaded with a wrapping of fine iron wire insulated by gutta-percha compound. The return conductor was a heavy copper tape wrapped outside that insulation, thus in contact with the seawater. Each cable transmitted one telephone channel and four telegraph channels, three above and one below the voice band. Telephone amplifiers were provided at both terminals. A fourth cable was added in 1931. Another impressive achievement was construction of the trans-Andean telephone line, connecting Argentina and Chile between Buenos Aires and Santiago in 1928.

A project for the implementation of a transatlantic telephone cable was also begun in the early 1930s. AT&T and the GPO planned to lay a continuously loaded cable for a single telephone circuit without using submerged telephone repeaters. The cable was to provide a single telephone circuit between Newfoundland and Ireland. Four layers of Perminvar[14] tape were planned to be used for loading. Cable manufacturing had already started in Germany when the project was dropped due to the economic depression and heavy competition from shortwave radio.

20.7   ANALOG MULTIPLEXING

In the 1870s, A. Graham Bell experimented with transmission of six to eight telegraph signals of different frequencies simultaneously on a single line. Further experiments were made by Ernest Mercadier (1836–1910), who proposed in 1889 to operate eight telegraph signals on different frequencies on a single line between Paris and Lyon. He intended to use eight different tuning forks on the transmitting end and *monotelephones*, each operating selectively at a specific frequency, on the receiving end. In 1891, two Frenchmen, Maurice Hutin and Maurice Leblanc, invented the use of electrical resonance circuits instead of mechanical resonance devices for selecting carrier frequencies for the transmission of telephone signals. Demonstrations of such carrier techniques, combining a number of telephone channels on a common circuit, were conducted by the German high-frequency physicist Ernst Ruhmer in 1908 and in 1910 by George O. Squier, a major general in the U.S. Signal Corps. With the prevailing state of electrical engineering, however, commercial realization of the transmission of telephone signals on carrier frequencies could not be achieved.

Analog multiplex systems, also called *carrier frequency systems*, could be developed with the advent of electronic valves, the realization of filters and resonance circuits, and the subsequent possibility of electronic generation of specific frequencies. In multiplex or carrier frequency equipment, a number of telegraph and/or telephone

---

[14] Perminvar is a ferromagnetic alloy similar to permalloy, but consisting of 45% nickel, 30% ferrum, and 25% cobalt.

channels are each modulated on a different *carrier frequency*, so that they can be transmitted on a common physical circuit without mutual interference.

The first experiments were made in Germany around 1913 in the laboratories of Telefunken and the Imperial Post and Telegraph Administration, which resulted primarily in a military application. In the United States, AT&T made the first tests with a laboratory model on an o/w line between South Bend, Indiana, and Toledo, Ohio, in 1914. Around 1918 it developed two commercial carrier-frequency telephone systems: a four-channel *carrier-suppressed system*, called *type A*, and a three-channel *carrier-transmitted system*, called *type B*. At the same time, multiplexing was developed with four telegraph channels on top of a telephone channel, or 28 telegraph channels instead of one telephone channel. To enable multiplexing of telegraph channels, the direct-current telegraph signals were replaced by signals on different frequencies within the voice-frequency band of 300 to 3400 Hz. Similar multiplex systems were also introduced on high-voltage power lines.

AT&T installed the first multiplex system for the simultaneous transmission of four telephone channels, together with one nonmultiplexed telephone channel on one pair of an existing o/w line between Baltimore and Pittsburgh in 1918. Within five years AT&T installed telephone carrier equipment for 27,200 channel-kilometers on o/w lines 7300 km long, and telegraph carrier equipment for 155,000 channel-kilometers on o/w lines 13,700 km long. In 1928, over 445,000 km of telephone channels was in operation on A and B systems. In the meantime, progress in electronic circuitry resulting in stable oscillators and amplifiers enabled development of a long-haul three-channel *system C*, which, like type A, was carrier suppressed but in addition used single-sideband transmission and flexible carrier frequency spacing, which did not need synchronization between transmitting and receiving stations. The three channels were arranged in two bands (for the two directions) between 6 and 28 kHz. Repeaters were required at intervals of 250 to 550 km, depending on line quality and climatic conditions. This allowed use on coast-to-coast lines. Figure 20.9 shows a typical arrangement in which, by means of highpass/lowpass filters, three multiplexed telephone channels can be connected to a line that operates one non-

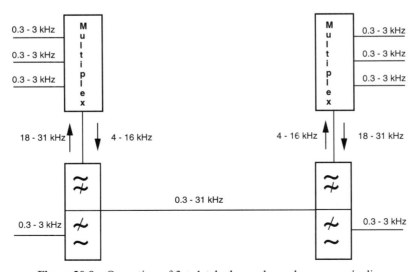

Figure 20.9   Operation of 3 + 1 telephone channels on one o/w line.

multiplexed telephone channel. The C system, one of the most successful, was first used on a line between Pittsburgh and St. Louis in 1924. About 800,000-km o/w lines were equipped with system C by 1950 on lines with individual lengths between 120 and 3200 km. The last system was removed from service in 1980. By 1938, when 100,000 km of lines was equipped with system C, a 12-channel *J system* was developed for operation in parallel with system C on a single o/w pair. Thus, system J increased the transmission capacity of one o/w pair to $12 + 3 + 1 = 16$ telephone channels within a band of 0.3 to 140 kHz. System J was first applied on the Oklahoma–Albuquerque section of the transcontinental line in 1939.

Multiplex on o/w lines was also applied at an early stage in Norway. The geographic and climatic conditions in this country are very difficult for cable, so that most long-distance lines used o/w line. Even this was problematic, due to the high disturbances caused by the northern lights, which could be avoided only by carefully drilled pairs. In 1925, the first three-channel multiplex system was installed on such an o/w line between Lilleström, Trontheim, and Fauske. The world's longest o/w line, 2000 km long, operated with a three-channel multiplex system, was opened between Oslo and Vardø at 71° north latitude on July 1, 1935. Four years later, this "world's longest" rating went to an 8715-km-long o/w line operated with a three-channel multiplex between Moscow and Khabarovsk in southeastern Siberia.

In Australia, five-channel multiplex systems operating on o/w lines were installed between New South Wales and Victoria in 1925, between West and East Australia in 1931, and on a 4000-km line between Sydney and Perth in 1934.

In Asia, a first three-channel carrier frequency system on an o/w line 205 km long was installed in Malaysia between Kuala Lumpur and Ipoh in 1930. One year later followed a carrier frequency system on an o/w line 395 km long between Kuala Lumpur and Singapore for the operation of three telephone channels and four telegraph channels.

Multiplex equipment for operation on cable—which has much higher attenuation than o/w lines and thus requires much shorter repeater spacing—could be developed after the principle of *negative feedback* was discovered at Bell Labs in 1927, and thus the design and operation of stable repeaters became possible. Simultaneously, cable construction was improved to enable the transmission of wider frequency bands than required for telephony. Basically, careful twisting and application of better insulation material resulted in symmetrically balanced pairs. This cable was called *symmetrical* or *transmission cable*.

AT&T developed a nine-channel system for operation on quads (two pairs of conductors), which was first used on an experimental link in 1933. A 68-pair AWG 16 (1.3 mm diameter) cable 25 miles (46.3 km) long was installed in the ducts of the New York–Chicago route in such a manner that both ends terminated in a long-lines repeater station at Morristown, New Jersey. Repeaters at 25-mile intervals were connected with the 68 pairs at Morristown to form an 850-mile (1575 km) four-wire circuit. The prevailing economical conditions prevented commercial implementation. In 1937, a *K system* was developed for the transmission of 12 channels in the frequency band 12 to 60 kHz on cable. Up to 200 amplifiers in tandem were used at 27-km intervals.

The world's first long-distance cable carrying 12 multiplexed telephone channels was put into operation in the U.K. on a 170-km link between Bristol and Plymouth in 1936.

A remarkable achievement was the construction by NEC of a six-channel carrier-telephone cable over 3000 km between Tokyo and Shengyang (then called Mukden) in China in 1939. From Tokyo, the cable ran west through Japan's mainland to Fukuoka, there crossed the Korean Strait by submarine cable to Pusan, ran north through Korea up to Sinuiju, crossed the Yalu River, and reached Mukden via Dandong and Fengcheng. At that time this was the world's longest carrier telephone cable in service. The route of this line is shown in Figure 20.10, which also shows the first submarine telegraph cable in 1871, the first o/w telephone line in 1888, and the first optical fiber cable in that region in 1985.

The first submarine cable adopted for carrier telephone transmission was laid across the 75-km stretch between the Californian coast and Santa Catalina Island in 1923. A first submarine cable with 30 pairs, each operating a 1 + 2 multiplex system, was laid in 1932 between St. Margaret, U.K., and La Panne, Belgium. In 1939, St. Margaret was connected with Sangatte, France, by a 36-km-long cable with 1.3-mm copper conductors. This was the first cable that operated a 12-channel multiplex system in both transmission directions within the same cable. It was manufactured by the French companies Lignes Télégraphiques et Téléphoniques and Câbles de Lyon (now belonging to Alcatel). Operation of 12-channel links between London and Calais began in December 1939.

The first submarine cable using submerged telephone repeaters was laid in 1943 over a distance of 275 km in the U.K. between Anglesey, Wales, and the Isle of Man. Another submarine telephone cable with submerged repeaters was laid between Lowestoft, U.K., and Borkum, Germany, in 1946. These repeaters were designed for operation in shallow water. Repeaters designed for deep water were first used for a pair of cables 220 km long laid in 1950 between Key West, Florida, and Havana, Cuba, with a depth of about 1700 m. Multiplex equipment was used for the transmission of 24 telephone channels on a separate cable with three carrier repeaters for each transmission directions.

After World War II, analog multiplex systems appeared for transmission of 24, 60, and 120 channels on symmetrical cable and radio relay successively, followed by 300, 960, 1260, 1800, 2700, and even 10,800 channels in the early 1970s operating on coaxial cable and up to 2700 channels on radio-relay systems. How such a multitude of telephone channels can be multiplexed with the application of separated basic groups and only a small number of different carrier frequencies is shown in Figure 20.11 and described briefly in Technology Box 20.1.

20.8 DIGITAL MULTIPLEXING

The idea of digitizing speech was first conceived by a British radio engineer, Alec H. Reeves (1902–1971).[15] Reeves, born in the U.K., studied electrical engineering at the

---

[15] Reeves lived in two worlds: He was a brilliant engineer and practiced spiritualism. He claimed to maintain contact with his friend and adviser Michael Faraday, the great electrical pioneer of the nineteenthth century. After he had observed in 1931 that radio-relay signals reflected from the white cliffs of Dover (thus contributing to the conception of radar), he became more and more convinced that people in other universes were trying to communicate with us and that their signals could be translated into Morse code. Reeves, a pacifist, refused to develop systems for weapons but cooperated on the development of defensive devices such as radar.

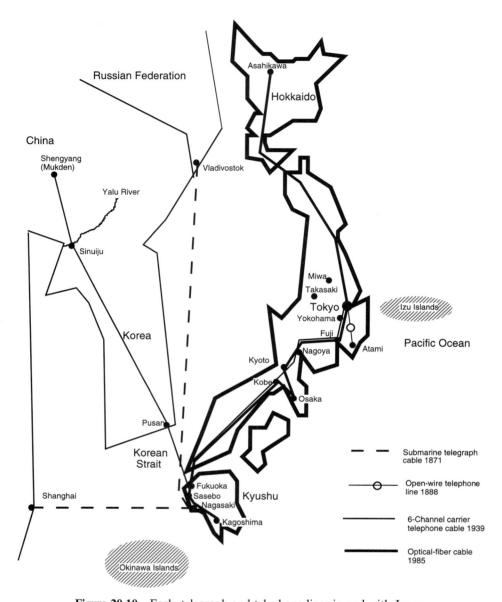

Figure 20.10 Early telegraph and telephone lines in and with Japan.

City and Guilds Engineering College in London. He joined *International Western Electric*, London, in 1922 and moved to Paris in 1927 to work at IT&T's newly created research center, *Les Laboratoires Standards* [later named *Laboratoire Central des Télécommunications* (LCT)], where he first worked on the technology of radio relay and radar. While investigating various methods of overcoming the transmission problems of noise, distortion, and crosstalk, he also reflected on the attempts of Philipp Reis, Bourseul, and Bell to "telegraph speech" and came to the conclusion

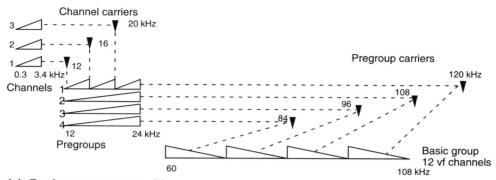

(a) Basic group composition

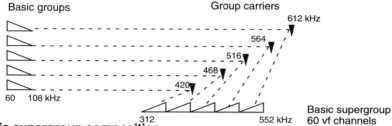

(b) Basic supergroup composition

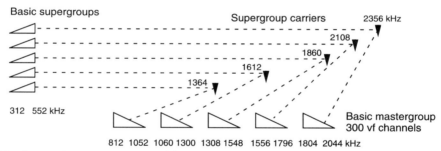

(c) Basic mastergroup composition

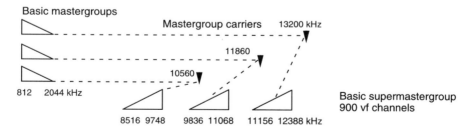

(d) Basic supermastergroup composition

Figure 20.11   Configuration of multiplex groups. (From A. A. Huurdeman, *Guide to Telecommunications Transmission Systems*, Artech House, Norwood, MA, 1997, Figure 2.2; with permission of Artech House Books.)

TECHNOLOGY BOX 20.1

Analog Multiplexing

In analog multiplexing, several voice-frequency (VF) channels, which are transmitted via common transmission media and come from an exchange or in the access network from the PABX of large subscribers, are connected in parallel to multiplex channel modulation equipment. The modulation equipment first limits the bandwidth of each VF channel to 4 kHz for the accommodation of a 300- to 3400-Hz speech band and a signaling channel. The signaling channel, either *inband* and thus at a frequency within the speech band, or *outband*, for example, 3850 or 3825 Hz, is added in the exchange to each speech channel for transmission of the relevant dialing, calling, and switching criteria, thus creating the VF channel. With this channel as a starting point, historically two different multiplexing technologies emerged: pregroup translation, which was standardized by CCITT, and single-channel translation, which was used in North America and in part of Asia.

The *pregroup translation technology* combines three VF channels by modulating each VF channel on a high-frequency (HF) carrier which is spaced 4 kHz from the next carrier in the same pregroup. For the VF-channel translation, single-sideband modulation with surpressed carrier is used so that each VF channel carried occupies only 4 kHz instead of 8 kHz. The carriers and the lower sidebands are filtered out, resulting in a 12- to 24-kHz pregroup band. Four such pregroups are then modulated on four HF carriers spaced 12 kHz apart, resulting in a 60- to 108-kHz basic group. Pregroup translation has been dictated by the engineering that prevailed at the time when the original equipment was designed, especially in view of the industrial production of crystals for exact frequency generating and effective filters for specific frequency bands. To accommodate 12 VF channels within one group by means of the intermediate modulation of three VF channels in one pregroup, only three channel carriers and four pregroup carriers are required instead of the 12 carriers required for *single-channel translation*. Single-channel translation saves one translation stage, however, and thus creates less modulation noise and signal distortion.

Starting again with the 12-channel *basic group*, two different higher translation schemes emerged, CCITT-standardized analog multiplex and Bell analog multiplex.

In *CCITT-standardized analog multiplex*, CCITT has defined *basic supergroups* for 60 channels, *basic mastergroups* for 300 channels, and *basic supermastergroups* for 900 channels. The translation of 300 VF channels in one basic mastergroup thus requires three channel carriers, four pregroup carriers, five group carriers, and five supergroup carriers, a total of only 17 carriers instead of 300. Similarly, for the translation of 10,800 channels into one multiplex system, in addition to the 17 carriers above, another three mastergroup carriers and 12 supermastergroup carriers are required: a total of 32 carriers.

In the system used by Bell, instead of a 300-channel mastergroup, a lower and an upper mastergroup for each 600 VF channels and a supermastergroup for 3600 VF channels are used.

Source: Adapted from A. A. Huurdeman, *Guide to Telecommunications Transmission Systems*, Artech House, Norwood, MA, 1997; with permission of Artech House Books.

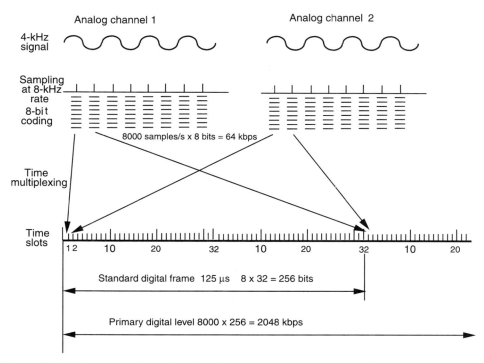

Figure 20.12   The conversion of analog channels into a 2-Mbps digital signal. (From A. A. Huurdeman, *Guide to Telecommunications Transmission Systems*, Artech House, Norwood, MA, 1997, Figure 2.7; with permission of Artech House Books.)

that coding speech in such a way as to transmit it in telegraphic code as a series of digits could replace analog speech signals. He worked out the principles of *pulse code modulation* (PCM) and applied for his first patent in France in 1938 (852,183), a second in the U.K. in 1939, and a third in the United States, where it was filed as U.S. patent 2,272,070 on February 3, 1942. At the outbreak of World War II, Reeves returned to the U.K., and from there he went to the United States, where he developed a 24-channel PCM system for the U.S. Army produced at Bell Labs. However, PCM required complex electronic circuitry that was difficult to realize economically with electronic vacuum tubes. It therefore took almost 25 years, until reliable and low-cost transistors were available, before PCM could be implemented on a commercial scale. The principle of PCM is shown in Figure 20.12 and described in Technology Box 20.2.

20.9   COAXIAL CABLE

The transmission of multiplexed telephone channels on o/w lines becomes problematic with frequencies beyond 150 kHz, due to crosstalk and interference from long-wave broadcast transmitters. Transmission of 12 channels in two-wire operation or 24 channels in four-wire operation is therefore the upper limit for reliable o/w oper-

---

TECHNOLOGY BOX 20.2

Principle of Pulse Code Modulation

Transmission digitalization starts with the conversion of analog telephone signals into a digital format. An analog signal can be converted into a digital signal of equal quality if the analog signal is sampled at a rate that corresponds to at least twice the signal's maximum frequency. Analog VF channels which are limited to the 300- to 3400-Hz band are therefore sampled at an internationally agreed rate of 8 kHz. Each time the analog signal is sampled, the result (the measured value of the signal at the sampling moment) is encoded using an 8-bit (= one octet or one byte) code. Because sampling proceeds at a rate of 8 kHz (8000 samples per second), and each sample is coded with 8 bits, the transmission speed of a digitized VF speech channel is 8000 samples/s $\times$ 8 bits = 64,000 bps or 64 kbps.

This analog-to-digital (A/D) conversion is carried out by pulse-code modulation (PCM) equipment, which time-multiplexes a number of digitized VF channels into a standard digital frame, similar to analog multiplex, where 12 VF channels are frequency-multiplexed to form a basic group. Unfortunately, as with the analog multiplex, two different PCM systems, and consequently, two (in fact, three) different digital multiplex hierarchies, have been developed. A 30-channel PCM system standardized by CCIT has found worldwide application. A 24-channel system developed in the Bell Labs a few years before development of the 30-channel system in Europe is used in North America and, slightly modified, in Japan and Korea.

After A/D conversion, in line with CCITT recommendation G.701, 30 VF channels plus one channel for signaling the 30 VF channels and one channel for frame synchronization, for maintenance, and for performance monitoring—a total of 32 channels—are time-multiplexed in a standard PCM digital frame called the *primary digital frame*. The transmission speed of this primary digital frame is 32 $\times$ 64 kbps = 2048 kbps. This transmission speed is usually referred to as the 2-Mbps primary level, first-order PCM, CEPT-1, or E1 standard.

The conversion from analog to digital described above changes the continuous analog signal to a pulse-type signal which at the receiving end is again converted to a continuous analog signal. The 24-channel T1 system has a primary level of 1544 kbps.

Source: Adapted from A. A. Huurdeman, *Guide to Telecommunications Transmission Systems*, Artech House, Norwood, MA, 1997; with permission of Artech House Books.

---

ation. In symmetrical cable with *unshielded twisted pairs* (UTPs) and paper isolation, the upper limit is around 250 kHz (equal to 120 channels in four-wire operation) or 550 kHz (equal to 120 channels in two-wire operation) with styroflex-isolated UTPs. Those limitations and the introduction of television in the 1930s presented a challenge for development of a transmission medium with a much wider bandwidth.

A solution was found in the coaxial construction of cable conductor pairs with a copper inner conductor inside a cylindrical copper tube that acted as an outer conductor.[16] In coaxial cable a high-frequency transmission circuit is formed between the inner surface of the outer conductor and the outer surface of the inner conductor. By virtue of the skin effect, the outer conductor serves both as a conductor and as a shield that provides protection against outside interference, this protection being more effective the higher the frequency. The inner conductor is supported within the tube in such a way that the intervening dielectric is mainly gaseous (either air or pressurized gas), thus minimizing dielectric losses.

Coaxial cable was first patented for use as an antenna feeder in the U.K. by C. S. Franklin under British patent 284,005 of January 17, 1928. In Germany, the Norddeutsche Seekabelwerke (North-German Submarine Cable Factory) and Siemens & Halske jointly developed and manufactured a coaxial cable for telephony and television transmission. A first link with this coaxial cable was installed in September 1934 on an 11.5-km route in Berlin between the Reichspostzentralamt (Imperial Central Post Office) at Berlin–Tempelhof and a television laboratory in Berlin–Witzleben. In addition to a TV channel, the cable carried 200 multiplexed telephone channels. The diameter of the inner conductor was 5 mm and of the outer conductor was 18.5 mm (Figure 20.13). The world's first commercial service on long-distance coaxial cable was opened in 1936 between Berlin and Leipzig. Repeaters for 200-channel telephone operation were placed at 36-km intervals and for the TV signal at 18-km intervals. In 1938, the system was extended to Nuremberg and Munich and in 1939 to Hamburg and Frankfurt.

In the U.K. the first coaxial cable for the transmission of 40 telephone channels was installed between London and Birmingham in 1936. The route was extended to Manchester and via Leeds to Newcastle in 1939. Experimental transmissions with 600 telephone channels were made on those routes, too. A coaxial submarine cable used for telephony was laid in 1937 between the U.K. and the Netherlands.

In the *Bell System Technical Journal*, Volume 13, in 1934, Bell Labs presented a detailed description of *wideband transmission over coaxial lines*, which had been tested in an 8-km loop at Phoenixville, Pasadena. Despite the positive results of the test, it was stated in the paper that "the practical introduction is not immediately contemplated and in any event, will necessarily be a very gradual process." Contrary to this cautious statement, the first coaxial cable was laid three years later between New York and Philadelphia for the transmission of 240 telephone channels. The experimental route, 150 km long, used 10 two-way repeaters at 10-mile (15-km) intervals. The diameter of the inner conductor was 1.8 mm (AWG 13) and that of the outer conductor was 6.8 mm. The repeaters included pilot-regulated amplification to overcome the about $\pm 7\%$ attenuation change of the aerial cable due to temperature variations from day to night and summer to winter temperatures. One year later, another experimental link was installed between New York and Princeton, with a transmission capacity of 480 telephone channels and repeaters at 5-mile intervals. Based on the positive results, construction of a national coaxial cable network for the transmission of 480 telephone channels or 240 channels and a TV signal began in 1939. In 1941, 600-channel systems called *L1* were installed between the major metropolitan areas, and by 1948 a complete transcontinental L1 coaxial cable network

[16] A British physician, Lord John William Strutt Rayleigh (1842–1919), had described the principle of coaxial cable operation in 1897.

Figure 20.13   First coaxial cable. (© Siemens-Archiv, Munich.)

was in operation. The first TV use of the L1 system was to transmit an Army–Navy game celebrating the end of World War II in 1945.

In France, installation of a coaxial cable for the transmission of 600 telephone channels between Paris–Bordeaux and Toulouse started in 1939, but the war stopped the project.

In Australia, the world's first submarine coaxial cable was laid between Apollo Bay, Melbourne, and Stanley in Tasmania in 1936. The 300-km cable passed Kings Island midway, where a repeater was installed. The cable transmitted one 8.5-kHz broadcast channel and seven telephone channels, with a bandwidth of 3 kHz each.

AT&T once more started a transatlantic telephone cable project in the early 1940s. Twelve multiplexed telephone channels were to be operated on a coaxial cable with underwater vacuum-tube repeaters at 90-km intervals, but World War II stopped the project.

The first submarine coaxial cable with an underwater repeater was laid in the U.K. between Anglesey and the Isle of Man in 1943. After the war, the first submarine coaxial cable was laid between the U.K. (Lowestoft) and Germany (island of Borkum). A submarine repeater was added in 1946, which increased the capacity to five telephone channels. The first submarine coaxial cable in deep water for the transmission of 24 multiplexed telephone channels using submarine repeaters in flexible repeater housings was laid through the Straits of Florida between Key West and Havana in 1950.

REFERENCES

Books

Feyerabend, E., *50 Jahre Fernsprecher in Deutschland, 1877–1927*, Reichspostministerium, Berlin, 1927.

Gööck, Roland, *Die großen Erfindungen Radio: Fernsehen Computer*, Sigloch Edition, Künzelsau, Germany, 1988.

Horstmann, Erwin, *75 Jahre Fernsprecher in Deutschland, 1877–1952*, Bundesministerium für das Post- und Fernmeldewesen, Bonn, Germany, 1952.

Hugill, Peter J., *Global Communications since 1884: Geopolitics and Technology*, Johns Hopkins University Press, Baltimore, 1999.

Kobayashi, Koji, *Computers and Communications*, MIT Press, Cambridge, MA, 1986.

Oslin, George P., *The Story of Telecommunications*, Mercer University Press, Macon, GA, 1992.

Seeleman, Claus, *Das Post- und Fernmeldewesen in China*, Gerlach-Verlach, Munich, 1992.

Siemens, Georg, *Der Weg der Elektrotechnik Geschichte des Hauses Siemens*, Vols. 1 and 2, Verlag Karl Alber, Freiburg, Germany, 1961.

Articles

Affel, H. A., et al., Carrier systems on long distance telephone lines, *Bell System Technical Journal*, Vol. 8, 1928, pp. 564–570.

Anon., Die Kabelflotte der Welt, *Archiv für das Post- und Fernmeldewesen*, No. 19, 1907, pp. 310–313.

Anon., Das deutsch-niederländische Kabelnetz in Ostasien, *Archiv für das Post- und Fernmeldewesen*, No. 19, 1907, pp. 577–590.

Basse, Gerhard, Deutschlands Fernsprechverkehr mit dem Ausland, *Archiv für das Post- und Fernmeldewesen*, No. 5, 1963, pp. 281–588.

Basse, Gerhard, Die Verbreitung des Fernsprechers in Europe und Nordamerika, *Archiv für deutsche Postgeschichte*, Vol. 1, 1977, pp. 58–103.

Buckley, Oliver E., The loaded submarine telegraph cable, *Bell System Technical Journal*, Vol. 4, 1925, pp. 355–374.

Buckley, Oliver E., High-speed ocean cable telegraphy, *Bell System Technical Journal*, Vol. 7, 1928, pp. 225–267.

Clark, A. B., and B. W. Kendall, Carrier in cable, *Bell System Technical Journal*, Vol. 12, 1933, pp. 251–254.

Curtis, Austen M., The application of vacuum tube amplifiers to submarine telegraph cables, *Bell System Technical Journal*, Vol. 6, 1927, pp. 425–441.

Espenschied, L., and M. E. Strieby, Systems for wide-band transmission over coaxial lines, *Bell System Technical Journal*, Vol. 13, 1934, pp. 654–679.

Fromageot, A., France–England submarine cable (1939) and Paris–Calais cable, *Electrical Communication*, Vol. 24, No. 1, 1947, pp. 24–39.

Gherardi, Bancroft, and F. B. Jewett, Telephone communication system of the United States, *Bell System Technical Journal*, Vol. 9, March 1930, pp. 1–100.

Green, C. W., and E. I. Green, A carrier telephone system for toll cables, *Bell System Technical Journal*, Vol. 17, 1938, pp. 80–84.

Kendall, B. W., and H. A. Affel, A twelve-channel carrier telephone system for open-wire lines, *Bell System Technical Journal*, Vol. 18, 1939, pp. 119–142.

Körber, Heinz, and Karlheinz Heyer, Geschichte der Leitungs- und Übertragungstechnik, *Archiv für deutsche Postgeschichte*, part I: Vol. 1, 1983, pp. 130–146, part II: Vol. 2, 1983, pp. 112–139.

Leclerc, Herbert, Von Apia bis Yap Ehemalige deutsche Postanstalten in der Südsee, *Archiv für deutsche Postgeschichte*, Vol. 1, 1982, pp. 7–32.

Meyer, W., Das allbritische Telegraphenkabel durch den Stillen Ozean, *Archiv für das Post- und Fernmeldewesen*, No. 17, 1903, pp. 517–520.

Nance, H. H., Some very long telephone circuits of the Bell System, *Bell System Technical Journal*, Vol. 3, 1924, pp. 495–507.

Pfitzner, Das deutsch–amerikanische Kabel, *Archiv für das Post- und Fernmeldewesen*, No. 18, 1900, pp. 701–709.

Pilliod, James J., Philadelphia–Pittsburgh section of the New York–Chicago cable, *Bell System Technical Journal*, Vol. 2, 1923, pp. 60–87.

Pilliod, J. J., Transcontinental telephone lines, *Bell System Technical Journal*, Vol. 18, 1939, pp. 235–254.

Röscher, Max, Das Weltkabelnetz, *Archiv für das Post- und Fernmeldewesen*, No. 12, 1914, pp. 373–389.

Strieby, M. E., Coaxial cable system for television transmission, *Bell System Technical Journal*, Vol. 18, 1939, pp. 438–457.

von Luers, Schnellbetrieb auf langen Unterseekabeln, *Archiv für das Post- und Fernmeldewesen*, No. 3, 1912, pp. 65–70.

von Wittiber, Das neue englisch–fransösische Fernsprechkabel, *Archiv für das Post- und Fernmeldewesen*, No. 22, 1910, pp. 179–183.

21

RADIO-RELAY TRANSMISSION

21.1 EVOLUTION LEADING TO RADIO-RELAY TRANSMISSION

Radio systems operating on long and short waves made possible transmission over long distances without the use of repeater stations, albeit with a capacity of only a single or a few telegraphy or telephony channels per HF carrier and with extremely variable quality, with fading and static, depending on atmospheric, ionospheric, and sunspot conditions. To obtain radio transmission at a quality comparable with that of line transmission, much shorter waves were generated which behaved like light, and a new technology, now called *radio-relay technology*, was developed in the 1930s. A concise description of this technology is given in Technology Box 21.1, and Figure 21.1 shows a typical radio-relay route.

It was found that quasi-optical waves with wavelengths in the range of centimeters, corresponding to frequencies above 1.5 GHz, could be focused by parabolic antennas into very narrow rays and transmitted in a direct line of sight to a corresponding station. At that station, called a *relay* or *repeater station*, the signal could be received, amplified, and retransmitted to the next station. Each repeater station could restore the signal received close to its original quality. In this way it became possible to cover hundreds, even thousands of kilometers, depending on the topography, with repeaters spaced typically at 50 km.

Generation of such quasi-optical waves was not, however, possible with the HF generator developed by Alexander Meissner in 1914. Meissner's HF generator could generate and amplify frequencies up to 10 MHz (30 m wavelength) by using a grid to control the density of the beam of electrons traveling from cathode to anode. However, this density control is less effective if the transit time of the electrons from cathode to anode is an appreciable fraction of a cycle of the frequency generated, which is the case with very high frequencies. Separation of the cathode and anode

The Worldwide History of Telecommunications, By Anton A. Huurdeman
ISBN 0-471-20505-2   Copyright © 2003 John Wiley & Sons, Inc.

TECHNOLOGY BOX 21.1

Radio-Relay Technology

Radio-relay transmission is the technology of generating, modulating, amplifying, and directing very high frequencies through the atmosphere for subsequent selective receiving, amplifying, and demodulating. In other words, radio-relay technology essentially shapes modulated, very short electromagnetic waves into rays or beams for propagation at the velocity of light through the atmosphere in a specific direction where it will be received practically undisturbed by other waves that might be in the air simultaneously. The frequency of the waves used in radio-relay transmission range from 300 MHz to about 100 GHz. They behave, as Hertz confirmed, like light and can thus only be propagated in an almost straight line and lose their intensity as a function of the distance covered. Consequently, radio-relay transmission is limited in two ways: to the direct line of sight between stations and by attenuation in the atmosphere.

As a consequence of these limitations and depending on the actual topology between the terminal stations of a link, a number of repeater stations might be required. A repeater station amplifies the weak radio-frequency signals received from the transmitters of adjacent stations and, if necessary, and the signal is digital, regenerates the signal and retransmits it to the adjacent stations. In addition to performing the same functions as a repeater, double-terminal stations give access to the information carried on the link so that information can be extracted and added. Usually, a few radio-relay channels, called *operating channels*, each carrying a separate wideband signal, operate in parallel via a common antenna. The operating channels are usually protected against equipment and/or propagation failures by a common protection channel. Beyond standard electronics, radio-relay technology concerns primarily modulation, radiation, propagation, and protection switching.

In contrast to transmission via copper or optical fiber cable, which have homogeneous characteristics, radio relay uses the atmosphere as a transmission medium, which is by nature nonhomogeneous. To overcome the disadvantages of this nonhomogeneity, beyond careful planning of the path between the terminal stations, further specific radio-relay techniques are required, such as space and frequency diversity, signal equalization, cross-polarized transmission on a common frequency, interference cancellation, and error correction.

Source: Adapted from A. A. Huurdeman, *Radio-Relay Systems*, Artech House, Norwood, MA, 1995; with permission of Artech House Books.

could not be made much shorter to reduce the transit time. The German physicist Heinrich Georg Barkhausen (1881–1956), a professor at the *Institute für Schwachstromtechnik* (Institute for Low-Voltage Current Technology)[1] in Dresden, found a solution in 1920. He proposed that electronic valves for the generation and amplifi-

---

[1] The Institute für Schwachstromtechnik, now the Institute of Telecommunications, was the first institute for telecommunications in Germany, founded by Barkhausen in 1911.

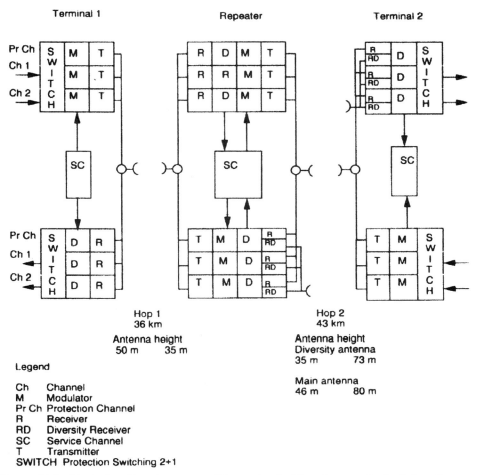

Figure 21.1 Typical radio-relay route. (From A. A. Huurdeman, *Radio-Relay Systems*, Artech House, Norwood, MA, 1995, Figure 4.2; with permission of Artech House Books.)

cation of frequencies beyond 10 MHz should control the velocity of the beam rather than its density. To influence the transit time appreciably, the distance between cathode and anode needed to be increased, which was technically easier than making it smaller. In velocity-modulated tubes, basically, the speed of the direct-current electron beam is modulated by the radio-frequency (RF) input signal. Based on this principle, three types of *velocity-modulated tubes* (also known as *transit-time tubes*) were developed: the klystron, the magnetron, and the traveling-wave tube (TWT). A more detailed explanation of this rather complex principle of velocity modulation is given in Technology Box 21.2 and illustrated in Figure 21.2. The first velocity-modulated tube, the klystron, was developed around 1930; the magnetron followed in 1934 and the TWT in the late 1940s.

The advent of the klystron was a prerequisite for the birth of radio-relay transmission. Four years later, *on January 26, 1934, the radio-relay era started* with the opening of radio-relay service between France and the U.K.

TECHNOLOGY BOX 21.2

Velocity-Modulated Tubes

In velocity-modulated tubes, the speed of the direct-current electron beam is modulated by the radio-frequency (RF) input signal. To make this modulation effective, the speed of the electron beam and the speed of the electromagnetic field caused by the RF input signal should be of the same magnitude. In a klystron, this is achieved by an electron beam first passing through a cavity resonator, where the electrons are accelerated or retarded depending on the RF input signal. In the following long drift space, the electrons accelerated during one half-cycle (of the RF input frequency) catch up with the electrons decelerated in the preceding half-cycle, resulting in a local density increase of the electrons in synchronism with the RF input signal. This effect is called *bunching*, and the first cavity is thus the *bunching cavity*. The next resonant cavity, known as the *catching cavity*, catches the variations in density of the direct-current electron beam by inducing an electromagnetic wave at the frequency of the RF input signal, but substantially amplified as a result of the energy of the electron beam. If the klystron is used as an oscillator rather than as an amplifier, a feedback loop is provided between the two cavities.

In a magnetron a permanent magnet directs the electron beam into a number of resonant cavities in the anode, which is arranged around a central cylindrical cathode. The RF input signal applied to the cavities progresses with the electron beam, clockwise to the output. Interaction between the input signal and the electron beam results in an output wave with the frequency of the input signal but with a much higher energy. The power output of a magnetron is essentially independent of the RF input signal. Because the magnetron operates as a saturated amplifier, it is not suitable for amplifying amplitude-modulated signals, but it is very useful for microwave ovens.

The traveling-wave tube (TWT) has far outlived the klystron and magnetron. The TWT combines the best characteristics of both. The RF input signal is applied to a helix circuit, which is dimensioned so that the propagation velocity is reduced to about the speed of the electron beam passing along the axis of the helix. On its longitudinal path almost from the cathode to the anode, the velocity of the electron beam is adjusted to the axial phase velocity of the input signal. Some electrons are accelerated and others decelerated, resulting in a progressive rearrangement of the electrons in phase with the input wave. Moreover, the modulated electron beam in turn induces additional waves on the helix. This process of mutual interaction continues along the length of the helix, with the net result that the electron stream takes the direct-current energy to the circuit as RF energy, thus amplifying the wave. TWTs have the advantage of a larger bandwidth than magnetrons and klystrons. Although they have been replaced in radio-relay equipment by all-solid-state devices, they are still used in satellites.

Source: Adapted from A. A. Huurdeman, *Radio-Relay Systems*, Artech House, Norwood, MA, 1995; with permission of Artech House Books.

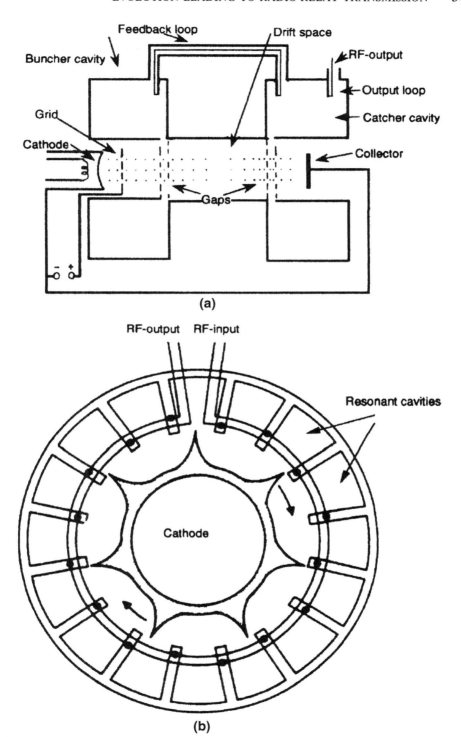

Figure 21.2 Velocity-modulated tubes: (*a*) klystron; (*b*) magnetron; (*c*) TWT. (From A. A. Huurdeman, *Radio-Relay Systems*, Artech House, Norwood, MA, 1995, Figure 1.3a–c; with permission of Artech House Books.)

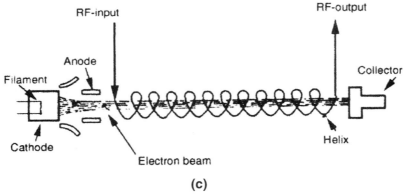

(c)

Figure 21.2 *(Continued)*

21.2   WORLD'S FIRST RADIO-RELAY LINK

The decisive breakthrough for radio-relay transmission was made in France. André Clavier and his team, working under the direction of Maurice Deloraine (born in Paris in 1898) at the *Laboratoire Central des Télécommunications* (LCT, then an affiliate of IT&T) in Paris, conducted experiments to generate, transmit, and receive very short waves with a wavelength of 17 cm (1.7 GHz). Preliminary trials in 1931 (Figure 21.3) demonstrated that it would be possible to construct a radio-relay link across the English Channel, selecting practically the same two sites as Louis Blériot

Figure 21.3   World's first experimental radio-relay terminals at both sides of the English Channel, 1931: left, at Calais; right, at Dover. (Scanned with permission of the ITU from Anthony R. Michaelis, *From Semaphore to Satellite*, International Telecommunication Union, Geneva, 1965, p. 163.)

(1872–1936) had used for his historic crossing in an aircraft on July 25, 1909. Telephone and telegraph messages were exchanged using parabolic antennas with a diameter of 3 m. During these successful experiments, Clavier observed that ships crossing the microwave beam, which ran close to the water in the middle of the English Channel, interfered with transmission in such a way that the size of the ships could be established from the waves reflected. The effect was investigated further in 1935 by *Compagnie Générale de Transmission Sans Fil* (CSF; now Alcatel Telspace) in experiments between two ocean vessels and by Telefunken in Germany using 600-MHz equipment. These experiments resulted in the development of *moving-target indication* (MTI) technology, later called *radar* (radio detection and ranging). The immediate consequence of Clavier's successful experiment, however, was not the development of radar but the development and production of commercial radio-relay equipment. This, the world's first commercial radio-relay equipment, was amplitude-modulated (AM) using a klystron producing 1-W RF output power and operating at 1.7 GHz, manufactured by the then IT&T affiliated companies *Standard Telephones and Cables* (STC, now part of Northern Telecom) in the U.K. and *Le Matériel Téléphonique* (LMT, now integrated into Alcatel Telspace) in France.

The system was put into commercial operation between airports at Lympne in England and St. Inglevert in France at a distance of 56 km across the English Channel. *The date of putting this link into operation, January 26, 1934, marks the beginning of the radio-relay transmission era*, This was 40 years after Heinrich Hertz died and 20 years after Meissner invented the HF generator.

French and English civil aviation authorities used the world's first commercial radio-relay link to coordinate air traffic between Paris and London. It carried one telephone and one telegraph channel simultaneously. World War II interrupted operation in 1940.

Various names have been given to this new technology, as noted in Technology Box 21.3. In this book, in compliance with ITU terminology, the term *radio relay* is used.

21.3   INITIAL RADIO-RELAY SYSTEMS

The second commercial radio-relay system, manufactured by STC, was installed in the U.K. between Stranraer in Scotland and Belfast, Northern Ireland, crossing the North Channel with a distance of 65 km. With the simultaneous transmission of nine-telephony channels, it was the first multichannel radio-relay system, The GPO inaugurated this 65-MHz amplitude-modulated radio-relay system in 1937. Reportedly, there were some problems in preventing intermodulation between channels, and speech could suddenly sound like a mixture of Scottish and Irish bagpipes! In 1942, an eight-channel pulse-modulated (PM) system for military operation was designed in the U.K. using a magnetron operating at 5 GHz.

In Germany, the IT&T affiliated company C. Lorenz began experiments with radio-relay equipment in 1931. In 1932 the company installed its first experimental radio-relay link over 60 km between the Ullstein tower at Berlin–Tempelhof and Fürstenwalde, using equipment operating at 500 MHz and an output power of 0.1 W. An additional 10 sets of this equipment were then tested by the army. The equipment reliability was not satisfactory, and a more reliable transportable two-

---

TECHNOLOGY BOX 21.3

Alternative Names for Radio Relay

André Clavier called the first beam radio system across the English Channel *micro-ray radio* in 1931. In the same year the term *quasi-optical waves* was coined for waves below 10 m. McGrath, an official of *Postal Telegraph* in Chicago, coined the term *microwaves* for waves 30 cm long, and in 1932 this term appeared for the first time in a publication of the *Proceedings of the Institute of Radio Engineers*. In France the term *liaison faisceau hertzien* (literally, Hertzian beam link) commemorates the discovery made by Heinrich Hertz. In Hertz's own language (German) the term is *Richtfunk* (literally, directional radio), which emphasizes the directional character of the transmission. The Italian name *ponte radio* signifies the bridge between two points made by radio. In Spanish, the name *microondas*, like *microwaves*, refers to the very short length of the waves. In English, the term *microwave radio relay* not only draws attention to the wavelengths employed but also hints at the use of repeater stations to cover long distances.

The terms *radio relay* or simply *microwave* have been used for many years, but since microwaves are also used for radar, satellites, and industrial, medical, and scientific (IMS) applications, as well as in ovens for cooking, the terms *microwave radio* and *radio relay* are now generally used in the English language for transmission applications. CCIR recommended the term *radio relay* and defined *radio-relay systems* as "radio communication systems in fixed service operating at frequencies above 30 MHz, which use tropospheric propagation, and which normally include one or more intermediate stations."

Source: Adapted from A. A. Huurdeman, *Radio-Relay Systems*, Artech House, Norwood, MA, 1995; with permission of Artech House Books.

---

channel system also operating at 500 MHz was developed and manufactured by C. Lorenz and Telefunken in 1937. Telefunken had installed their first experimental radio-relay link in 1935 between Groß-Ziethen and Wiesenburg near Berlin. Both companies improved the transmission quality substantially a few years later by applying frequency modulation (FM), which made the traffic signals almost immune to static and fading and required less bandwidth. Frequency modulation was developed and patented by the American Major Edwin Howard Armstrong (1890–1954) in 1933.[2] Telefunken produced a single-channel FM 500-MHz system in 1936, and C. Lorenz began development in 1939 and production of the world's first 10-channel FM 1.3-GHz equipment (Figure 21.4) in 1940. Both types of equipment were

---

[2] Edwin Howard Armstrong developed frequency modulation to improve radio broadcasting. RCA and the FCC, however, initially opposed the introduction of FM. When Armstrong set up his own FM broadcasting station, RCA also introduced FM (e.g., in 1940 transmitting from the Empire State Building tower), but refused to pay royalties to Armstrong. Long and costly patent suits followed. On February 1, 1954, Armstrong wrote a farewell letter, put on his overcoat, hat, and gloves, and stepped out of a thirteenth-floor window.

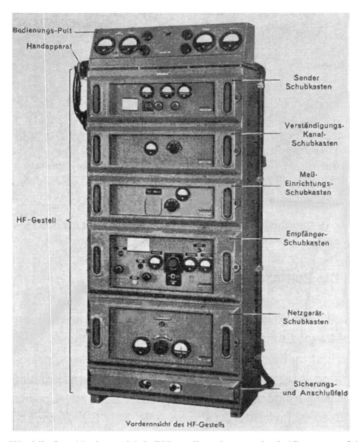

Figure 21.4 World's first 10-channel 1.3-GHz radio-relay terminal. (Courtesy of Alcatel SEL.)

equipped with a magnetron in the RF output stage, providing 1 W at 500 MHz and 0.5 W at 1.3 GHz. They were produced in large quantities, albeit to be used for a huge wartime network, with 2500 stations stretching from North Cape via Sicily to North Africa and from the French Atlantic coast via Greece to Crete and the Black Sea, a total link length of some 50,000 km. The Mediterranean Sea was crossed between the isle of Creta and Derna, Libya, covering a distance of 350 km with 45-MHz equipment. Telecommunication was achieved up to distances of around 5000 km.

The single-channel Telefunken equipment was replaced in 1942 by 10-channel equipment. C. Lorenz enhanced their 10-channel system in 1944 by using an improved *Heilscher generator*,[3] which produced an RF output of 10 W.

In France, an experimental pulse-modulated 600-MHz radio-relay link was installed near Lyon between Saint–Genis-Laval and le Col de la Faucille in 1935. A

---

[3] A Heilscher generator was a klystron with only one resonant cavity for the input and output signals, connected with the drift space by an input gap for electron beam modulation and an output gap for energy transfer from the electron beam to the (input) wave in the common cavity.

TECHNOLOGY BOX 21.4

First Fresnel Zone

The RF energy transmitted by a radio-relay transmitter reaches the receiver in two ways: directly along a straight line and indirectly by reflection. The reflected ray arrives with a phase delay at the receiving antenna, composed of a 180° phase shift at the reflection point plus the phase shift caused by the path length difference. With a path difference of half a wavelength, corresponding to a 180° phase shift, the direct and reflected rays arrive in phase and are thus added. However, with a path difference of a full wavelength and a resulting phase shift of 360° + 180° (reflection), the two rays cancel one another. The reflection points of all (desired) rays with a constant path length of half a wavelength delay are described by an ellipsoid called the *first Fresnel zone*.

The French physicist Jean Fresnel (1788–1827) defined such an ellipsoid for optical radiation. The radius of the Fresnel ellipsoid at any point is defined by the formula $r = \sqrt{D_1 D_2 \lambda}/(D_1 + D_2)$.

This formula implies that for a given link, at longer wavelength, thus at lower frequency, the antennas need to be higher. Consequently, the use of higher frequencies would seem advantageous to reduce tower- and height-associated costs. On the other hand, the signal attenuation increases with the frequency. As a compromise, therefore, frequency bands between 4 and 7 GHz are preferred for high-capacity systems on long-distance backbone routes with typical hop lengths of 50 km, with frequency bands below 4 GHz used primarily for low- and medium-capacity systems for regional and rural applications. The frequencies above 10 GHz are used mainly for urban systems and for backbone systems in densely populated areas, with consequent short hop lengths of about 25 km. The frequency bands above 20 GHz, which experience very high attenuation, are particularly useful for radio-relay access links from public to cellular networks and for interconnecting cellular base stations with hop lengths between 5 and 10 km. This applies even more for frequencies above 60 GHz, where high oxygen absorption limits the propagation to a few hundred meters, which make those frequencies very useful for wireless access systems or *radio-in-the-loop systems* (RITLs).

Source: Adapted from A. A. Huurdeman, *Radio-Relay Systems*, Artech House, Norwood, MA, 1995; with permission of Artech House Books.

12-channel FM radio-relay system operating at about 3 GHz was developed and an experimental link was installed between two stations in Paris in 1944. With the same equipment the first commercial link was put into operation between Paris (at the telephone exchange Vaugirard) and Montmorency in 1946. The world's first radio-relay link, which connected continental France with Corsica and covered 205 km, went into operation in August 1947. Despite the distance, there was almost a direct line of sight, as the station on the mainland was located on a 700-m-high mountain near Grasse and the station on Corsica was on a hill near Calenzana. The 12-channel

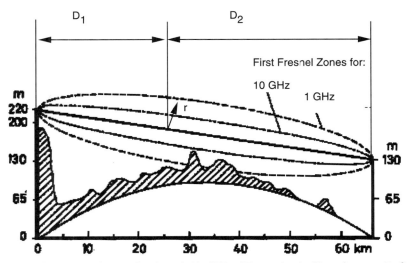

Figure 21.5   First Fresnel zone for 1 and 10 GHz. (From A. A. Huurdeman, *Radio-Relay Systems*, Artech House, Norwood, MA, 1995, Figure 2.5; with permission of Artech House Books.)

100-MHz equipment was supplied by the *Societé Française Radioeléctrique* (SFR, now part of Alcatel Telspace) and later extended to 24 channels.

In Spain, a radio-relay route was installed connecting Barcelona with the island of Majorca and thence with the island of Minorca in 1935. This little published but remarkable radio-relay route was the world's first to include an over-the-horizon link of about 170 km and the first radio-relay repeater station. The equipment was developed by the IT&T engineers Maurice Deloraine (described in his fascinating book *When Telecom and ITT Were Young*) and Alec H. Reeves (the inventor of PCM). Civil war, followed by the Franco dictatorship, diverted attention from this technical achievement.

In Italy, experiments were made with six- and nine-channel AM equipment in 1939, and the resulting equipment was used during World War II.

In Japan, experimental six-channel radio-relay equipment manufactured by NEC was installed across the Tsugaru Strait in 1939.

In the United States, in the early 1930s, Bell Labs began investigations into ultra-shortwave transmission at frequencies above 30 MHz. At that time, Bell Labs and IT&T still freely exchanged technical know-how based on a historical agreement made in 1925, when AT&T agreed to limit its activities to North America, with IT&T to be active in the rest of the world. Bell Labs started a two-year propagation study in 1938 over a 62-km experimental link between Beer's Hill and Lebanon in New Jersey with equipment operating at 80 and 160 MHz. The propagation test also included trials over different land paths and over water, including a 110-km path over water between Atlantic Highlands, New Jersey, and East Moriches on Long Island along the Atlantic coast, and a 60-km path without a direct line of sight. Bell Labs installed another experimental link in 1941 with 160-MHz AM equipment crossing the Chesapeake Bay between Cape Charles and Norfolk, Virginia. The tests were continued after the war on a 65-km link between New York City and Neshanic, New Jersey, with frequencies between 700 MHz and 24 GHz. Those tests provided

Figure 21.6 Radio-relay terminal with 120-channel 2-GHz equipment, 1955. (Courtesy of Alcatel SEL.)

valuable information about the relation between frequency, ground reflection, and fading. A major finding, since then strictly observed in radio-relay planning, was that a wave at those frequencies can be transmitted between two stations with a minimum of loss if an opening defined as the *first Fresnel zone* is free of obstacles. An explanation of the Fresnel zone and its relation to radio-relay system planning is given in Technology Box 21.4 and illustrated in Figure 21.5.

After the war, AT&T and IT&T took advantage of the technological impetus that came from the tremendous wartime radar development effort and made a big jump to higher frequencies, which provided greater transmission capacities. The result was a 4-GHz FM system, put into operation in 1947 between New York and Boston as the first commercial radio-relay network in the United States. It consisted of 10 stations carrying television from New York to Boston with 480-channel multiplexed telephony in both directions.

IT&T developed a radio-relay system using pulse modulation (PM), which appeared to be very advantageous for cost-effective small-capacity (24-channel) systems. The basic IT&T design was soon taken over by its affiliates, STC, Lorenz, FACE, and NEC, and even by competitors in Europe and Japan.

In Australia, Standard Telephones and Cables Pty (at that time an IT&T affiliate, now Alcatel Australia) produced three-channel 40-MHz radio-relay equipment that was first used in 1943 as a backup route for the 150-km Bass Strait submarine cable laid in 1936 connecting Tasmania with Australia. At that time, this was the world's longest submarine telephone cable and longest radio-relay hop.

In Germany in 1945–1946, the U.S. occupation forces installed a 1200-km radio-relay link from Bremen via Frankfurt (station on the Feldberg) to Munich using 500- and 600-MHz equipment. On August 1, 1947 this network was officially ceded to the German administration.

Radio-relay transmission became widely used after World War II for the reestablishment of destroyed networks. Practical use was made of the technological progress of the last few years. Instead of reconstructing the o/w lines, more rapidly installable radio-relay equipment was used, partially modified from military equipment.

In Europe around 1950, radio-relay equipment went into production, operating mainly at 2 GHz for the transmission of 24-, 60-, and 120-channel equipment and for TV (Figure 21.6).

Radio-relay transmission enjoyed a strong impetus from the emergence of television and found worldwide application in the second half of the twentieth century, as described in Chapter 25.

REFERENCES

Books

Brémenson, C., *Connaissance des liaisons hertziennes*, Alcatel Radio Space Defense, Paris, 1990.

Carl, Helmut, *Radio-Relay Systems*, Macdonald, London, 1966.

Deloraine, Maurice, *When Telecom and ITT Were Young*, Lehigh Books, New York, 1976.

Huurdeman, Anton A., *Radio-Relay Systems*, Artech House, Norwood, MA, 1995.

Libois, L. J., *Faisceaux hertziens et systèmes de modulation*, Éditions Chiron, Paris, 1958.

Michaelis, Anthony R., *From Semaphore to Satellite*, International Telecommunication Union, Geneva, 1965.

Oslin, George P., *The Story of Telecommunications*, Mercer University Press, Macon, GA, 1992.

Articles

Budischin, F., Rückschau über den Ausbau des Richtfunks im Fernmeldenetz der Deutschen Bundespost bis zum Jahre 1968, *Archiv für das Post- und Fernmeldewesen*, January 1974, pp. 3–97.

Fox, Jerome, et al., in Proceedings of the symposium on modern advances in microwave techniques, New York, November 8–10, 1954, *Microwave Research Institute Symposia Series*, Vol. IV, Polytechnic Press of the Polytechnic Institute of Brooklyn, New York, 1955.

Friis, H. T., Microwave repeater research, *Bell System Technical Journal*, Vol. 27, April 1948, pp. 183–194.

22

CRYPTOGRAPHY

The information transmitted through optical telegraph systems was basically secret-coded information that could only be understood by those who were in the possession of the code translation books. With Chappe's system, only the operators at the terminal stations were in possession of the elaborate codebooks. The operators on the other stations only knew a few operational codes and repeated the message signals received without knowing their contents. The people who saw the moving arms had no idea of the information being transmitted.

With the advent of electrical telegraphy, everybody could read the transmitted messages directly on the receivers of the needle telegraphs or on the Morse receivers if one knew the nonsecret dash–dot code. Moreover, the open-wire lines connecting telegraph stations could easily be tapped and connected with a telegraph receiver by unauthorized persons. Radiotelegraphy made unauthorized listening to radio messages even easier. The enormous success of electrical telegraphy therefore very soon made secret coding a necessity. Basically, two types of coding are used:

1. *Substitution coding*, whereby letters, words, expressions, and complete phrases are replaced by other letters or by artificial or nonrelated words or numbers from a special codebook before being sent on a telegraph line. On the receiving end the same codebook has to be used for manual decoding of the text received.

2. *Transposition coding*, whereby the text is jumbled into a disarranged order following a predeterminable code key so that a disarranged text, unintelligible to outsiders, is sent on the telegraph line. At the receiving end a similar device needs to be set to apply the same code key in order to produce the original uncoded text.

The Worldwide History of Telecommunications, By Anton A. Huurdeman
ISBN 0-471-20505-2   Copyright © 2003 John Wiley & Sons, Inc.

The substitution coding of letters can be mono- or polyalphabetic. The Italian architect, painter, writer, and philosopher Leon Battista Alberti (1404–1472) first used monoalphabetic coding in 1470. He introduced an encryption disk with an adjustable ring changing the sequence of the letters. Another Italian, Giovanni Battista Della Porta (1535–1615), philosopher, astrologer, and mathematician, in 1563 introduced polyalphabetic coding by adjusting the ring of the encryption disk during a message according to a prearranged plan. In 1891 the French military cryptologist Etienne Bazieres (1846–1931) introduced a polyalphabetic coding device in which separate sequences of letters were used successively on a number of separate encryption disks. Thomas Jefferson (1743–1826), the third U.S. president, had suggested this procedure around 1800.

22.1 MANUAL CODING

The first known effort to improve secrecy on electrical telegraph lines was made by Francis O. J. Smith in 1847, when he published *The Secret Corresponding Vocabulary: Adapted for Use to Morse's Electro-magnetic Telegraph*. In the same year, Henry Rogers, who had been Morse's telegraph operator in 1845, published *The Telegraph Dictionary and Seaman's Signal Book, Adapted to Signals by Flags or Other Semaphore; and Arranged for Secret Correspondence, through Morse's Electro-magnetic Telegraph*. Those and other secret codes that soon followed were used widely, so that by 1854, one in eight of the telegrams between New York and New Orleans passed in code.

After successful laying of transatlantic telegraph cable in 1866, a further impetus was given to nonsecret commercial coding to decrease the cost of telegrams. In 1874 the first edition of the long-lived *The ABC Code* appeared, compiled by William Clausen Tue, a shipping manager later elected a Fellow of the Royal Geographical Society. Another famous commercial code published in 1877 was *The Merchants Code*, with 15,000 dictionary words compiled by John Charles Hartfield.

The cable companies charged for code words as if they were plain language, limiting both plain and code words to a maximum of seven syllables. This gave rise to the creation of very long artificial words with 10 to 20 letters combined into seven syllables. In 1875, at its conference in St. Petersburg, the ITU specified that extra-European telegrams should have a maximum length of 10 letters. This did not stop the use of words that were difficult to understand. Four years later, therefore, the ITU stipulated: "Code-language telegrams can contain only words belonging to the German, English, Spanish, French, Italian, Dutch, Portuguese, or Latin languages." The United States, not being a member of the ITU, further allowed the use of "any pronounceable group of letters" as a single word. In 1903 in London the ITU accepted that practice provided that the artificial words could be pronounced in any of the eight approved languages and would not be more than 10 letters long. Within one year, *Whitelaw's Telegraph Cyphers: 400 Millions of Pronounceable Words* appeared in England. The book consisted of 20,000 code words of five letters each. As any word could be combined with any other in a pronounceable 10-letter word, a total of $20,000 \times 20,000 = 400,000,000$ words were created, which met the recommendation of the ITU but caused further mistakes and confusion. One year later, Ernest Lungley Bentley (1860–1939) founded a code company and in 1906 published

Bentley's Complete Phrase Code, consisting of clearly distinguishable five-letter codes. It soon sold some 100,000 copies and became the most widely used commercial code.

In a further effort to standardize coding, the ITU set up a *Code Control Committee*. This committee was active between 1908 and 1913. Strict censorship during World War I prevented the public use of codes. After the war the ITU set up another *Committee for the Study of Code Languages*, which produced both a majority report, in which 14 countries allowed only five-letter codes to be used, and a minority report, by Great Britain, which wanted to retain 10-letter codes. The current *Telegraph Regulations* allow artificial words up to five letters. The introduction of the five-letter code substantially reduced the cost of telegrams and reduced mistakes, so that within the next 10 years, almost all other dictionary-word types of code disappeared.

22.2   AUTOMATIC CODING

The first example of mechanical coding of a telegraph text was given by Wheatstone in 1867 at the World Exhibition in Paris. He demonstrated his *cryptograph*, consisting of two concentric disks: an outer disk with the letters of the alphabet in the usual order, and an inner disk with the letters of the alphabet in random order. This was a solution similar to the one that Leon Battista Alberti had used 400 years earlier. The two disks were connected by a toothed wheel and had a common pointer. Coding was obtained by turning the pointer to the desired letter of the plain text on the outer disk, the inner disk turned with the pointer, placing the letter to be used for transition of the letter from the plain word opposite the original letter. Together with Earl Playfair of St. Andrews, Wheatstone developed an improved version of his first crypthograph, the *Playfair cipher*, which still was used in World War II.

The American Eduard Hugo Hebern (1869–1952) took a first step toward automatic ciphering in 1915. Hebern connected two electrical typewriters in such a way that the type bars of the second typewriter could be variably connected with the type bars of the first writer. A message written in plain language on the first typewriter appeared coded on the second typewriter. Heber established a factory for the production of his cryptographic typewriters in Oakland, California, in 1921. Apparently, he sold only 11 machines and went bankrupt in 1926. At the beginning of World War II, however, production of improved versions began under military control.

In October 1919, the Dutchman Hugo Alexander Koch obtained a patent for an automatic ciphering device using rotating drums. Koch did not bring his device into actual use despite much experimenting. The German engineer Albert Scherbius obtained the patent rights from Koch and used it to construct the most complex mechanical–electrical ciphering machine produced in the twentieth century and named it *Enigma* (the Greek word for "riddle"). The *Scherbiusschen Chiffriermaschinen AG* in Berlin produced the Enigma in various versions, but limited quantities from 1923 onward. The Enigma used a normal typewriter keypad for input of the plain text, and the letters substituted could be read on a lamp panel. Initially, with three different drums, each with 26 alphabets, 17,576 ciphering possibilities were given. The location of the drums could be interchanged, which increased the possible variations to $6 \times 17,576 = 105,456$. Figure 22.1 shows one of the early versions of Enigma.

Figure 22.1 Enigma. (Scanned with the permission of the Museum für Kommunikation Frankfurt, Germany, from Klaus Beyer et al., *Streng Geheim*, Museumsstiftung Post und Tele-kommunikation, Germany, 1999, p. 105.)

Typical of most pioneers, Scherbius was not very successful with Enigma. Only a few units were sold to the German army, which in the Peace Treaty of Versailles of June 28, 1919 was limited to 100,000 soldiers. Scherbius died in an accident in 1931 and his company was dissolved in 1934. One year later, the *Chiffriermaschinen-Gesellschaft Heimsoeth und Rinke* revived the activities and delivered over 100,000 units for the military forces in Germany, Italy, Spain, and Japan.

Despite the high number of ciphering possibilities and the complexity of Enigma, in 1933 a group of Polish mathematicians succeeded in deciphering a radio message sent by German Marines. They secretly had an Enigma machine in their possession for a few days, which enabled them to construct an even more complicated machine called a *bomba*. For each position of each of the three drums used in Enigma they used a separate bomba. From an intercepted message, each bomba could then detect the key applied to the corresponding drum. A bomba operated like a reversed Enigma; the coded text went through all possible combinations until it reproduced the original German text. The Polish intelligence lost their lead, however, when in 1938 two additional drums were added to Enigma, resulting in over 11 million variations. Changing the external wiring between the drums increased the coding variations to over 150×10^{12}.

Enigma was a pure ciphering machine used mainly for ciphering of messages before they were sent by radio. With the advent in the late 1920s of the teletypewriter (teleprinter), which uses 5-bit coding, it was a natural step to encipher this 5-bit teleprinter coding. Within a few years, a *secret teleprinter* appeared which included its

Figure 22.2   Siemens's secret teleprinter. (© Siemens-Archiv, Munich.)

own ciphering devices (Figure 22.2). Plaintext could be entered in the usual way on the teleprinter keyboard. A ciphering device with a complex arrangement of up to 10 drums changed the polarity of the five-digit teleprinter code in an irregular way before being sent out on the fixed teleprinter line. Similar to Enigma, the secret teleprinter could be used for both ciphering and deciphering.[1] Secret teleprinters were used initially mainly for diplomatic and commercial applications. The German military forces used extensively modified versions manufactured by Siemens & Halske, Telefunken, and C. Lorenz shortly before and during World War II on fixed lines and on radio-relay links. Enigma was used primarily to cipher radio messages.

In an effort to decipher German military messages, the British Secret Intelligence Service (SIS) established a top-secret high-security department called the *British Government's Code and Cipher School* (GC&CC) at Bletchley Park, an attractive nineteenth-century mansion 80 km north of London At its peak, some 10,000 persons—mathematicians, physicists, military experts, philologists, chess players, and "native speakers"—worked to decipher daily up to 2000 radio messages intercepted from the German forces. Initially, following the Polish example, electro-mechanical calculating machines were developed for each drum and its various positions. In January 1940, their first calculating machine, which at Bletchley Park was called a *bombe*, could be put into operation. From May 1940, the Bletchley Park team was in a position to decipher the radio traffic of the German Air Force. The German Marines used an even more secure Enigma with an additional drum and other refinements. From the end of 1940, however, the messages from the Ger-

---

[1] This teleprinter was developed by August Jipp (1896–1977) and Ehrhard Rossberg (1904–1989) at the Siemens factory in Berlin. Patents were obtained on July 18, 1930 (German patent 615,016) and June 6, 1933 (U.S. patent 1,912,983).

man Marines could also be deciphered. As proof of their efficiency, on August 8, 1940, the Bletchley Park team sent to Winston Churchill the following deciphered text: "vonreichsmarschallgoeringanalleeinheitenderluftflottendreiundzwanzigundfu-enfoperationadlerinkuerzesterzeitwerdensiediebritischeairforcevomhimmelfegenheil-hitler" (from Reichsmarshall Goering to all units of the Airforce 23 and 5. Operation Adler. Within a shortest possible time you will sweep the British Air Force down from heaven. Heil Hitler!). Further intercepted messages revealed important details of the planned attack. Thus Churchill was informed in advance about the German air attacks, which started on August 13. Rather than the British Air Force, it was the German Luftwaffe that was finally "swept from heaven," losing some 2000 airplanes.

To further improve the efficiency of the British SIS and to attempt to decipher the messages sent by the German secret teleprinters, an effort was made to replace the electromechanical bombes by quicker electronic devices. The first prototype, called *Heath Robinson*, managed to read 2000 symbols per second. In 1942, a development team was brought together under the direction of Britain's leading mathematician, Alan M. Turing (1912–1954); a team of engineers from the research station of the GPO at Dollis Hill, headed by Thomas H. Flowers; and the mathematician Maxwell H. A. Newman. Working under high pressure, they managed within one year to develop the world's first electronic computing machine, called *Colossus*. This cryptoanalytic calculating machine was equipped with 1500 electronic tubes. It failed, however, to decipher the messages of the secret teleprinters.

The existence and apparent successes of the Bletchley Park establishment were not revealed until 1974, when a book entitled *The Ultra Secret* was written by ex–Group Captain Frederick William Winterbotham, after the British SIS had declassified the Bletchley Park activities. Despite the complexity of Enigma, the people at Bletchley Park had managed to decipher many of the secret German messages and so could prevent much war damage in Britain and improve the effectiveness of the Allied invasion in Normandy in 1944. Alan M. Turing, who should have been given much honor for having saved many lives, sadly committed suicide in 1954 when he came under suspicion of homosexual offenses. After the war, Colossus became the starting point from which digital electronic telephone switching was developed in the U.K. (Section 29.4).

Another ciphering device that made history originated from a Swedish development in 1919. The Swedish textile engineer Arvid Gerhard Damm, working with the company *Aktiebologat Cryptograph* (founded in 1916), used his experience gained with Jacquard weaving machines for construction of an automatic ciphering machine. He applied for a patent in 1919 for his *rotory system*. This company almost went bankrupt, but in 1921, Boris Hagelin (1892–1983) joined the company, bringing international experience and support from the Swedish Nobel dynasty. Hagelin improved the cryptograph and in 1925 succeeded in getting the Swedish Army to use his Swedish product, the new prototype B-21, instead of the German Enigma. Arvid G. Damm died in 1927 and Hagelin became the owner of Aktiebologat Cryptograph, later called AB Cryptoteknik. The B-21 had a lamp field similar to that in Enigma. In a new compact version, the C-35, the lamp field was replaced by a printer, which produced the ciphered text at a speed of three letters per second. To improve the operating comfort, the C-35 was connected to an electric typewriter, which the U.S. company Remington had just introduced. The C-35, as small as a telephone, became very successful. More than 5000 units were sold, mainly for the French secret service, la Deuxième Bureau. In 1940, when the German Army invaded Norway, Hagelin took the manufacturing information and 50 units of his latest model to the United States, where he convinced the army of the superior quality of the C-35. In 1941, production of a modified version, type M-209, began in the Corona typewriter factory at Groton, New York. More than 140,000 units were produced during World War II, and Hagelin became the first cryptograph pioneer to make a fortune.

In September 1944, Hagelin returned to Sweden with the intention of starting production of a secret teleprinter of his own design for commercial use. However, prevailing Swedish law stipulated that technologies with military interest be offered to the government before commercial production could be considered. In the interest of producing cryptographic equipment for commercial applications, Hagelin left for Switzerland, where in 1948 he founded *Crypto AG* in Zug. A product of this company made history during the Cold War when a ciphering unit, the TC-52, was used for the red telephone line between the White House in Washington, DC, and the Kremlin in Moscow.

REFERENCES

Books

Beyer, Klaus, et al., *Streng Geheim*, Museumsstiftung Post und Telekommunication, Germany, 1999.

Kahn, David, *The Codebreakers*, Weidenfeld and Nicolson, London, 1967.

Articles

Mache, Wolfgang, Der Siemens-Geheimschreiber: ein Beitrag zur Geschichte der Telekommunikation 1992: 60 Jahre Schlüsselfernschreibmaschine, *Post- und Telekommunikationsgeschichte*, Vol. 2, 1992, pp. 85–94.

23

INTERNATIONAL COOPERATION

Reflecting the progress in development of telegraph apparatus, the International Telegraph Union, at their administrative conference in London in 1903, advised relegating the Morse telegraph from international lines (as recommended in 1865) to lines of modest activity. The Hughes equipment was recommended for more active lines, and those that handle more than 500 telegrams per day were advised to use the Baudot system, or similar improvements.

The *Marconi International Marine Communications Company* built its own radiotelegraph stations on land, strategically located along the sea trade routes and placed its own operators onboard ships fitted with equipment leased from the Marconi company. Within a few years of starting in 1901, Marconi stations were opened in Belgium, Britain, Canada, Ireland, Italy, and Newfoundland. Marconi instructed his operators to exchange wireless signals with any other wireless station only if that station had Marconi equipment. This restriction soon caused a serious incident. In 1902, Prince Heinrich of Prussia (1862–1929), brother of Wilhelm II (1859–1941), the king of Prussia and emperor of the German empire, while crossing the Atlantic Ocean on his way back from a visit to the United States, wanted to send a courtesy telegram to President Theodore Roosevelt (1858–1919) but was refused service because the radio on the ship was not made by Marconi. To prevent Marconi from monopolizing the radiotelegraph network, and to open radiotelegraphy to all ships in distress, Wilhelm II initiated an international conference in 1903 that took place in Berlin with the participation of delegates from Austria–Hungary, Britain, France, Germany, Italy, Russia, Spain, and the United States. One of the chief Russian delegates was Alexander Popoff, later honored by the government of the USSR as "the Russian inventor of radio." The conference resulted in a protocol—signed by all participants apart from those of Britain and Italy—stating: "Coast stations should

The Worldwide History of Telecommunications, By Anton A. Huurdeman
ISBN 0-471-20505-2 Copyright © 2003 John Wiley & Sons, Inc.

be bound to receive and transmit telegrams originating from or destined for ships at sea without distinction as to the system of radio used by the latter."

Three years later, again upon invitation of Wilhelm II, an international radio conference took place in Berlin. The German government had prepared a complete draft "Convention and Radio Regulations," modeled closely on the Convention of the International Telegraph Union signed in St. Petersburg in 1875. This conference was a big success. The delegates from all 29 participating countries, including Britain and Italy, signed the final convention and the annexed radio regulations. Three major conditions agreed upon were an obligation (1) to connect coastal stations to the international telegraph service, (2) to give absolute priority to all distress messages, and (3) to avoid radio interference as much as possible.

The conference also decided that the Bureau of the International Telegraph Union at Berne should act as central administrative organ of the Radiotelegraph Conference. The conference allocated the high-frequency frequencies 500 and 1000 kHz for public communication in the maritime service. Frequencies below 188 kHz were reserved for long-distance communications by coastal stations, whereas frequencies between 188 and 500 kHz were reserved for nonpublic services (military and naval services). The bureau in Berne was also assigned the task of acting as a central frequency registration office to safeguard proper coordination of the allocation of frequency spectrum for radio services. The frequency spectrum was then already recognized to be a valuable resource, and like other resources, of limited availability. It was agreed that all stations should register with the bureau their major station details, such as frequencies used, hours of operation, call signs, and radio system used.

Prior to the conference in Berlin, two different radio-emergency call signals were in use. The signal "CQ" was the general call signal at the time, and to distinguish its urgency in case of distress, the letter "D" was added (popularly interpreted as "Come Quick, Danger"). German ships used the distress call "SOE" and proposed this at the conference for international acceptance. However, as the "E" in Morse code is a single dot, that was not considered satisfactory, so the rhythmic signal "SOS" (dot–dot–dot, dash–dash–dash, dot–dot–dot, popularly interpreted as "Save Our Souls") was generally accepted.

In 1912, the International Radiotelegraph Conference met again, in London, As this was three months after the *Titanic* disaster, it was called the Titanic Conference. It was only then that the Marconi Company instructed their operators "to communicate with all other ships, regardless of the system adopted by the latter." Three new radio services were officially included in the radio regulations: radio beacons, time signals, and weather reports. Radio beacons were allowed to use any frequency above 2 MHz, whereas for time signals and weather reports, any frequency below 188 kHz was allowed.

The first international telecommunications conference after World War I was the administrative conference at Paris in 1925. The most important achievement of that conference was the organizing of two technical consultative committees: the *Comité Consultatif International de Télégraphy*, abbreviated CCIT and also called International Telegraph Consultative Committee, and the *Comité Consultatif International de Communications Téléphoniques à Grande Distance*, also called the International Telephone Consultative Committee (CCIF). The purpose of both committees was to study the technical issues of telephony and telegraphy and to work out recommendations concerning the operation and, specifically, the interfaces between various parts of the telephone and telegraph networks. The recommendations were published by the bureau at Berne and have since constituted the internationally accepted standards for telecommunications equipment and transmission lines.

The International Radio Conference came together again in 1927, this time in Washington, DC. In addition to 80 countries, 64 private companies (broadcasting organizations, radio operators, and manufacturers) were represented. A noteworthy decision was taken as follows: "French is the official language of the Conference. Nevertheless, since the presiding administration has so requested, and as an exceptional measure, English may be used. Delegations are recommended to use this privilege with discretion."

The allocation of frequencies was extended at the conference to the entire range from 10 to 60,000 kHz. It was decided that the old type of radio, with a spark transmitter occupying a wide bandwidth, should no longer be installed and should not be used after January 1, 1930, unless the radio-frequency output power was below 300 W.

The most important decision of the conference in Washington was to set up a radio consultative committee called the *Comité Consultatif International Technique des Communications Radioélectriques* [International Radio Consultative Committee (CCIR)]. This new Committee held its first meeting in Den Haag in the Netherlands, in September 1929.

The next conference took place in 1932 in Madrid. Both the 13th International Telegraph Conference and the 3rd International Radio Conference met there simultaneously and decided to merge the two organizations into a single union: the *International Telecommunication Union* (ITU). Delegates of 80 countries signed the new convention.

The Madrid conference also bore fruit on the American continent. In Havana in 1937, the *First Inter-American Radio Conference* met and the 16 American states represented set up an *Inter-American Radio Office* and allocated frequencies in three different zones of the Americas. This conference also made the recommendation to the ITU that the frequency allocation table should be extended beyond 30 MHz to 300 MHz, a recommendation adopted by the ITU at their conference in 1938 in Cairo.

Again a war interrupted the valuable international telecommunications cooperation. Radio silence for amateurs was to be observed for the next six years, as radio was monopolized for military communications during World War II.

PART V

PERIOD FROM 1950 TO 2000

24

EVOLUTION OF TELECOMMUNICATIONS FROM 1950 TO 2000

This part begins with the situation five years after the end of World War II. That war not only took the lives of millions of human beings but also heavily destroyed the telecommunications networks in many European and Asian countries. An impression of the extent of destruction can be obtained from information released by the French government in 1946. Over 90,000 km of open-wire lines were down (the copper being used for ammunition), 30 cities had their underground cable network cut, 110 telegraph offices and 60 relay stations lay in ruins, tens of thousands of telephone sets had to be replaced, and 50 submarine cables had been cut. Those figures, representing about 90% of the total network, also give an indication of the already high penetration of telecommunications by the beginning of World War II.

Thanks largely to the immediate and effective Marshall Plan Aid[1] granted by the U.S. government for the reconstruction of Europe's destroyed infrastructure, the telecommunications networks were rapidly brought back to at least provisional operation. The introduction of semiconductor technology and the digitalization of at first transmission, and later switching, spurred an enormous progress in reestablishing national networks for telephony, telegraphy, and telex. Moreover, national and international networks were established with new technologies such as satellite networks, data networks, intelligent networks, cellular networks, integrated services digital networks, and the Internet.

In the second half of this period, a worldwide wave of deregulation, privatization, and liberalization brought competition into telecommunications. This resulted in a thorough restructuring of telecommunications operators, and the total number of

---

[1] After the war the U.S. Congress passed the European Recovery Program, as aid program to rebuild Europe proposed by General G. C. Marshall and called the *Marshall Plan*. In addition to loans, the plan provided merchandise, food, raw materials, and technical assistance totaling $13 billion. Marshall received the Nobel Peace Prize in 1953 for this contribution.

The Worldwide History of Telecommunications, By Anton A. Huurdeman
ISBN 0-471-20505-2   Copyright © 2003 John Wiley & Sons, Inc.

telephone subscribers, which after 100 years of telephony had reached 350 million, increased to almost 1 billion by the end of the century.

The end of this period is characterized by an evolution from narrowband, circuit-switched, state-owned telecommunications networks to broadband, packet-switched, private telecommunications networks. Simultaneously, the price for telecommunications services went down significantly, and the previously impersonal telecommunications subscriber became more and more a valued customer of competitively offered telecommunications services.

24.1 THE SEMICONDUCTOR ERA

The electrical conductivity of materials ranges in an order of magnitude of 10^{23} between the best conductors, made of copper, and the worst conductors, made of pure glass. The name *semiconductor* applies to a normally solid material in crystal form whose electrical conductivity lies roughly in the middle of the aforementioned range and that varies as a function of: temperature; degree of chemical purity; radiation with light rays, ultraviolet rays, or x-rays; and the electrical potential applied.

In the 1930s, research began at the Bell Labs toward the goal of replacing electromechanical relays—used extensively in switching equipment—by a suitable semiconductor device. World War II interrupted the research, but soon after the war a three-person team resumed the investigations: John Bardeen (1908–1991), Walter Brittain (1902–1987) and William Shockley (1910–1989). They started a series of experiments with the aim of finding a solid material arrangement that could replace not only relays but also vacuum tubes. At issue was the fact that current flow takes place at the point of contact between a semiconductor and a metal. Their aim was to determine whether or not the electric field set up by the current at the point of contact could be made to control the current flowing through the slab of a semiconductor. If so, a small signal at the point of contact would cause a large current to flow through the material, and this would effect amplification, similar to Lee de Forest's triode.

A tiny piece of germanium was placed on top of a piece of metal, and two very thin gold leaves, separated almost invisibly from each other, pressed on the germanium. On December 23, 1947, the device was shown to work—a little change of current through the metal plate indeed caused a big change in current flow through the piece of germanium. One of their colleagues, John R. Pierce, suggested the name *transistor* for this device, from "transfer resistor": where *transist* conveys the idea of a gain resulting from an intensity amplification [*trans*-(res)*ist*ance], and "*-or*" relates it to the varistor and the thermistor.

In June 1948, after patent application, the invention of the transistor was made public. The transistor was presented as a device that could perform the same functions as those performed by a vacuum tube but was smaller, more reliable, and required much less power. As a protection against the prevailing accusations of monopoly, AT&T adopted a liberal patent policy. For $25,000 a company—whether American or foreign—could have access to the transistor patents. The production of germanium transistors was begun in 1951 by Western Electric. The first commercial application came in 1952 with the production of transistorized hearing aids manufactured by the U.S. company Raytheon. Two years later, Texas Instruments brought the first transistorized portable radio on the market. A further impetus and worldwide attention was given in 1956 when Shockley, Brittain, and Bardeen were given the Nobel Prize in Physics for the invention of the transistor. By 1959, worldwide sales of transistors overtook those of vacuum tubes for the first time. It was in Japan where the transition from electronic tubes to transistors was first realized. Using the Bell patents, Japanese manufacturers started mass production of transistors for applications not only in consumer products but very soon also for telecommunications equipment. In 1958, NEC built the first fully enclosed clean plant for manufacturing semiconductor devices. Transistor radios made in Japan soon could be heard in every corner of the world, and Japanese all-solid-state telecommunications equipment gained significant market share worldwide.

In 1963, a British company introduced the first calculator using transistors. It was the size of a cash register. Texas Instruments introduced a scientific transistorized table calculator in 1967. Four years later the transistorized pocket calculator, again predominantly made in Japan, began its worldwide penetration.

From then onward, the development of many types of transistors and other solid-state components, their manufacture, and their application in all electronic domains constituted a real technological revolution. Now, 50 years later, some scientists do not hesitate to rank the invention of the transistor at the same level of human innovation as the invention of the wheel or the control of fire.

The major milestones in the evolution of transistors are shown in Table 24.1. All-solid-state technology significantly reduced power consumption (easily by 80%), heat dissipation, and equipment size, and subsequently reduced the cost and dimensions of the primary power supply, cooling, and equipment accommodation.

Using photolithography and micrometallurgy, it soon became possible to accommodate active and passive elements together with their connections on a single automatically produced microcircuit called a *chip*. The major step toward this semiconductor technology was made in late 1958, when Jack Kilby of Texas Instruments developed an assembly technique that came to be known as an *integrated circuit*. In this circuit, passive elements, resistors and capacitors, were built into the same silicon chip as the transistors, with the metal connections between these components running across the surface of the chip.

TABLE 24.1   Major Milestones of the Evolution of Transistors

Year	Achievement
1948	Point-contact transistor
1950	Single-crystal germanium
1951	Grown junction transistor
1952	Alloy junction transistor
	Single-crystal silicon
1955	Diffused-base transistor
1958	Field-effect junction transistor
1960	Planar transistor
	MOS transistor
	Epitaxial transistor
1962	Silicon field-effect transistor

TABLE 24.2 Evolution of Circuit Integration

Type	Grade of Integration
IC	Integrated circuit
MSIC	Medium-scale integrated circuit
LSIC	Large-scale integrated circuit
VLSIC	Very large scale integrated circuit
MMIC	Monolithic microwave integrated circuit
ASIC	Application-specific integrated circuit

A few months after Kilby demonstrated his invention, another engineer, Robert N. Noyce, came up with the same idea. Together with Gordon E. Moore, he founded the *Intel Corporation*. The next major step in semiconductor technology was made in 1971, when Intel designed an integrated circuit named 4004, known as a *microprocessor*. An arithmetic–logic unit (ALU) and its control unit were combined on a single large-scale-integrated 12-mm chip so that logical functions could be performed by this *large-scale-integrated circuit* (LSIC). With microprocessors it became possible to let equipment perform logical programs, depending on specific variations, without requiring the permanent assistance of a human operator.[2] With steadily increased refinement of the material used and greatly improved production methods, the evolution of circuit integration technology progressed from 1960 to 1980 as summarized in Table 24.2. Whereas the IC 4004 combined one transistor with four passive elements, current 64-Mbps chips contain some 70 million transistors. Simultaneously, the failure rate of chips has been reduced from 20% initially to nearly zero today. The 1-Gbps chip available at the end of the twentieth century accommodated over 1 billion transistors, which corresponds to the information on 30,000 typewritten size DIN-A4 pages.

24.2 DIGITALIZATION

(*Chronological continuation of Section 20.8*) Digitalization of transmission and switching equipment, conceived by Alec H. Reeves before World War II, could be realized in the 1960s when transistors and ICs became available. Digitalization of transmission equipment began with the development of the American 24-channel T1 PCM system and a 30-channel E1 PCM system in Europe. With those digital systems the capacity of existing telephone cables could be increased by 24 and 30 channels, respectively, per copper pair. Based on the 24-channel system, two different *plesiochronous*[3] *digital multiplex hierarchies* (PDH) evolved in North America and in

---

[2] In 1965, Moore proclaimed *Moore's law*: "Integrated circuits double their complexity every year [later adjusted to every 18 months]."

[3] The term *plesiochronous* is derived from the Greek expression *plesio*, which stands for "almost but not exactly" because the plesiochronous digital multiplex hierarchy is almost, but not exactly, synchronous in order to cope with small deviations of the synchronization equipment on various corresponding multiplex stations. Each higher level in the hierarchy has a bit rate four times that of the lower level plus a few stuffed bits. For instance, level $2 = 4 \times 2048 + 256 = 8448$ kbps (8.448 Mbps), and level $3 = 4 \times 8.448 + 0.576 = 34.368$ Mbps.

Japan and Korea; a CCITT-recommended hierarchy evolved from the 30-channel system in the 1970s. In the meantime, the digitalization of switching began also (Chapter 29).

The introduction of transmission via optical fibers increased the transmission capacity from the order of Mbps on radio-relay channels and coaxial cable pairs to the order of Gbps, and at the end of the twentieth century even to Tbps on single optical fibers. For those optical fibers and with the progress made in synchronization, in the 1980s Bell Labs developed a new fully synchronous digital multiplex hierarchy called *Sonet* (Synchronous Optical Network). The Sonet concept was still based on the 24-channel system; therefore, at a conference in Seoul, Korea, in February 1988, CCITT adopted a new *synchronous digital hierarchy* (SDH) based on the 30-channel system, compatible with Sonet from a level of 155 Mbps.

Within 40 years digital transmission made such enormous progress that the transmission capacity of a single transatlantic cable increased from 36 telephone channels with coaxial cable TAT-1 (Section 26.2.1) in 1956 to the equivalent of 58,060,800 telephone channels with optical fiber cable Flag Atlantic-1 (Section 28.3.4) in 1999. This increase in transmission capacity resulted in a substantial decrease in the price of transmission networks. On average, each fourfold increase in transmission capacity increased the cost of equipment just two-and-a-half times and thus led to a 40% price decrease.

24.3   NEW TELECOMMUNICATIONS NETWORKS

A great improvement in international telecommunications was made in 1965 as the first commercial telecommunications satellite, *Early Bird* (*Intelsat I*), started operation of 240 telephone channels and one TV channel across the Atlantic Ocean. The *Intelsat II* and *III* satellites were launched in 1968 and 1969, covering the Pacific Ocean and Indian Ocean regions, so that worldwide satellite service could begin in 1969.

Less then 200 years after military requirements created telecommunications, military requirements created new applications that adapted to civil service and have revolutionized worldwide telecommunications. In the early 1970s, the government of the USSR claimed to have developed rockets that with atomic warheads could be launched from Siberia, and after a few seconds' journey, could destroy North American towns. Alarmed by this threat, the U.S. government conceived a national *star wars* defensive network known as the *strategic defense initiative* (SDI). A grid of satellites circling the globe at an altitude of about 1000 km were immediately to detect any missile, wherever launched, determine its speed and direction, and transmit that information to a fail-safe terrestrial computer network in the United States. The computer network was to determine the exact time to launch a counterattack missile and guide that missile successfully to intercept and destroy the foreign missile before arriving above North American territory. A fail-safe transmission network had to interconnect the satellites and connect the satellites with the terrestrial computer network. To make that transmission network insensitive to jamming and other deliberate disturbances, including cosmic radiation, a new transmission mode was conceived which applied spread-spectrum frequency hopping with coded signals.

A huge program was begun to conceive, develop, test, and implement the highly complex star wars network elements. The first element to be realized, in 1971, was a terrestrial computer network under the name *Arpanet*. With the Cold War coming to an end in the late 1980s, star wars did not happen and SDI was canceled.[4] Very soon, efforts were made to modify the star wars elements into the following commercial telecommunications systems:

- Arpanet became the Internet (Section 34.4), currently the fastest-growing telecommunications service, with over 360 million users.
- Frequency hopping and spread-spectrum coded signal transmission helped to make cellular radio (Chapter 32) the second fastest-growing telecommunications service, with 740 million subscribers by year-end 2000.
- The spectacular idea of a grid of satellites, now not to observe missiles but to serve as cellular radio base stations in space, was realized in the Iridium system, which on November 1, 1998, with 66 satellites in *low earth orbit* (LEO), began the era of *global mobile personal communication by satellite* (Section 27.4.5).

---

[4] SDI was revitalized on a reduced scale in early 1999 as the *national missile defense* (NMD) program.

25

RADIO-RELAY NETWORKS

25.1 TECHNOLOGICAL DEVELOPMENT OF RADIO-RELAY SYSTEMS

Radio-relay transmission received a big impetus from the emergence of television around 1950. The TV studios had to be connected with the TV transmitters dispersed throughout a country. The transmission of video signals, with a typical bandwidth of 4 MHz for a black-and-white signal and 6 MHz for a color signal, was not possible on symmetrical pair cable and was very expensive on coaxial cable. With radio-relay transmission, the carrier frequency, in the range 2 to 20 GHz, was sufficiently high to enable modulation of the broadband TV signals. The first national radio-relay link for TV transmission began operation in the United States in 1949 the inauguration of President Harry S Truman was broadcast. A challenge for an international radio-relay network covering seven European countries was given with the coronation of Queen Elizabeth II of Britain on June 2, 1953. In a special effort, CCIR and CCITT worked out their recommendations for TV transmission, including a vital one for *air interface* of radio-relay signals across national borders. Experimental exchange of images over the English Channel took place on August 23, 1950. The coronation of the Queen could be watched live on TV in seven European countries, thanks to an international chain of radio-relay stations. Even then, CCIR recommended that cable and radio-relay systems be treated as transmission systems of equal standing, and that the objective should be to achieve for radio-relay transmission the performance recommended by CCITT for international telephone circuits on metallic conductors.[1] This was a courageous objective in 1951, as it was by no means certain that

---

[1] CCIR Recommendation 40, issued in 1951, stated: "Between fixed points, telephone communication should be effected wherever possible by means of metallic conductors or radio links using frequencies above 40 MHz, where this can be realized, and the objective should be to obtain the transmission performance recommended by the CCITT for international telephone circuits on metallic conductors."

The Worldwide History of Telecommunications, By Anton A. Huurdeman
ISBN 0-471-20505-2   Copyright © 2003 John Wiley & Sons, Inc.

Figure 25.1   Typical radio-relay repeater station in Thailand. (Courtesy of Gerd Lupke.) See insert for a color representation of this figure.

it could be met within the foreseeable future. However, in just five years the radio-relay engineers were able to meet this ambitious objective, and radio-relay transmission became the standard solution for applications where the terrain or required speed of implementation limited the use of cable. Radio-relay stations are, therefore, frequently found at very inaccessible sites: in dense jungle, on the tops of hills, and on mountains. Figure 25.1 shows, as a typical example, a radio-relay repeater station on a mountain in Thailand. The state of the art of radio-relay equipment in the mid-1950s is illustrated in Figure 25.2, which shows a terminal with a transmission capacity of 960 telephone channels and one TV signal and a common standby channel operating in the 4-GHz band.

25.1.1   All-Solid-State Radio-Relay Systems

The production of all-solid-state radio-relay equipment became technically and economically possible with the advent of microwave semiconductors around 1960. Gunn and IMPATT diodes could be used for very high frequencies, and the *field-effect transistor* (FET) could replace traveling-wave tubes in RF power output stages.

All-solid-state technology substantially reduced power consumption (by about 80%), and equipment size (by about 50%), and subsequently reduced the cost and dimensions of the primary power supply (diesel generators, rectifiers, and batteries) and equipment accommodation (shelters or prefabricated buildings for repeater stations at isolated sites). As a result, radio-relay transmission became an increasingly flexible, economical, and reliable means of transmission for low, medium, and high transmission capacities. Soon many telecommunication administrations in Africa, Asia, and Latin America could afford to construct reliable national transmission networks, thanks to the highly competitive all-solid-state radio-relay equipment.

Within 40 years, radio-relay transmission became most effective for regional and national distribution of TV programs and for high-capacity backbone systems with

Figure 25.2   Radio-relay terminal for the transmission of 960 telephone channels and one TV channel, 1957. (Courtesy of Alcatel SEL.)

small- and medium-capacity access routes for public telecommunication networks as well as for "linear" networks alongside oil and gas pipelines, electricity distribution lines, railways, and highways.

The transmission capacity of radio-relay systems gradually increased from 60, 120, and 300 channels to 960, 1260, and 1800 telephone channels per radio-frequency (RF) channel in the 1960s, and even to 2700 channels (1963 in Italy) and 3600 channels (1977 in Japan). Figure 25.3 shows an all-solid-state radio-relay repeater operating at 6.2 GHz as produced in 1980 for the transmission of 1800 telephone channels per RF channel in a 3 + 1 configuration (three RF operating channels and one RF standby channel).

In the late 1970s, the availability of strictly linear RF amplifiers enabled the development of single-sideband, suppressed-carrier, amplitude-modulation (SSB-AM) radio-relay equipment with a transmission capacity of 5400 channels in Japan and 6000 channels in the United States.[2] Technically, this was a great achievement. Operationally, however, the requirement for such very high capacity analog systems suddenly faded away as telephone exchanges were being digitized and radio-relay systems were needed for the transmission of telephone and data signals in a digital mode.

25.1.2   Digital Radio-Relay Systems

Early in the 1970s, the availability of low-cost semiconductor technology opened a new epoch in electronics, leading to another breakthrough in communications. Integrated semiconductor technology revolutionized progress in computers and tele-

---

[2] This information appeared in CCIR Report 781.

Figure 25.3   An 1800-channel 6.2-GHz radio-relay repeater in $3 + 1$ configuration, 1980. (Courtesy of Alcatel SEL.)

communications equipment. Whereas in telephony the human voice can be understood even if slightly distorted on its long analog transmission route from subscriber to subscriber, computers need fault-free interconnections to prevent disaster. Consequently, to meet the communication needs of computer users and because of the increase in the number of digital exchanges, the transmission network gradually had to be converted from analog to digital. Optical fiber was not yet available for transmission, while digital transmission on coaxial cable required repeaters at extremely short intervals, thus making it uneconomical, at least on most new routes.

Digital transmission has a major advantage over analog: that the signal received can be regenerated into its original shape irrespective of the actual signal-to-noise ratio (SNR) up to a certain limit, called the *threshold*. Whereas with analog transmission the SNR gets worse after each reamplification or demodulation at a repeater station, with digital transmission the signal received can be restored to its original shape independent of the number of cascaded repeaters. Digital radio-relay transmission is thus also possible at frequencies beyond 10 GHz, which cannot be used for analog transmission because of the very short hops and large number of reamplifications required for a given distance.

Analog radio-relay transmission proved itself in long routes for trunk networks, where many centers have to be linked across areas that are difficult to cable. Today, with digital transmission making it feasible to use the less crowded higher-frequency bands, radio-relay transmission has become equally suitable for short hops, where users are close together but separated by densely packed streets and buildings, as is the case in major towns and cities all over the world.

Digital radio-relay transmission thus became the solution not only for public networks but also for private networks. Banks, insurance companies, travel agencies, brokers, transport companies, and others all started using computers, and high-quality medium and high-capacity lines were needed to interconnect the computers in their branch offices with headquarters. In many cases the public network could not provide the required data lines quickly enough, so many companies installed their own private digital radio-relay networks. Digital radio-relay equipment appeared on the market rapidly in the 1970s and gradually stopped the production of analog equipment.

The first digital radio-relay system went into operation in Japan in 1969. The equipment operated in the 2-GHz band with a transmission capacity of 17 Mbps, corresponding to 240 telephone channels. The first systems were used primarily between digital telephone exchanges and for connecting digital telephone exchanges with existing analog backbone networks. At digital telephone exchanges, 1.5- or 2-Mbps digital data streams were modulated onto the RF carrier of radio-relay equipment without the need for costly analog frequency-division multiplexing (FDM). Digital-to-analog conversion at stations where the digital access routes interfaced with an analog backbone route was done by using a transmultiplexer to convert, for example, 2×2 Mbps data streams into a 60-channel analog super group, and vice versa.

In the 1970s, analog-to-digital (A/D) conversion twice on a route could be eliminated by introducing equipment for inserting digital data streams in addition to the analog baseband on existing analog radio-relay systems. The names *DUV*, *DAV*, and *DAVID* given to this digital insertion equipment stood for:

- *DUV:* data under voice (e.g., a 2-Mbps stream under a 1800-channel baseband)
- *DAV:* data above voice (e.g., a 1.5-Mbps stream above a 1800-channel baseband)
- *DAVID:* data above video (e.g., a 2-Mbps stream above a baseband with a TV signal)

Soon, however, this type of equipment was no longer required either, as high-capacity digital radio-relay equipment became available in the 1980s for the transmission of 34 and 140 Mbps in Europe and 45, 90, 135, 180, and 270 Mbps in the United States, Canada, and Japan. Since then, digital radio-relay systems operating in the 2-, 6-, 7-, 11-, 13-, 15-, 18-, 23-, 38-, 42-, 50-, and 60-GHz bands have been introduced worldwide. Figure 25.4 shows a digital radio-relay terminal complete with antenna for outdoor installation, equipped with two transmitters and two receivers operating at 18 GHz for transmission capacities of 8, 34, 140, and 155 Mbps. By that time, worldwide some 40,000 radio-relay transmitter/receivers were manufactured annually and radio-relay transmission covered 20 to 50% of telephony in national networks and almost 100% of TV transmission.

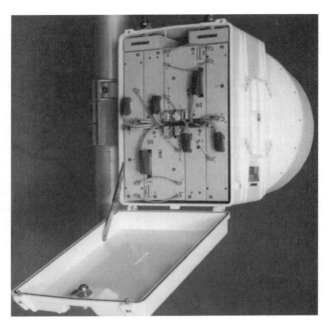

Figure 25.4 Outdoor 18-GHz digital radio-relay terminal, 1985. (Courtesy of Alcatel SEL.)

25.1.3 Radio-Relay Systems for the Synchronous Digital Hierarchy

Successful implementation of transmission on optical fiber cable in the 1980s and the rapid increase in transmission capacity and repeater spans (jumping from a few kilometers on coaxial cable systems to over 100 km on optical fiber cable), and especially the introduction of the *synchronous digital hierarchy* (SDH), presented a new challenge to radio-relay engineering. As radio-relay technology matured and the frequency spectrum available became increasingly congested, the future did not look bright for the survival of radio-relay transmission. Radio-relay engineers took up the challenge and developed new technologies:

- Cross-polarization interference canceling techniques to enable simultaneous use of horizontal and vertical polarizations of the same RF carrier frequency, thereby doubling the transmission capacity
- Modulation schemes with higher-frequency spectrum efficiencies that can accommodate higher capacities within a given bandwidth[3]
- Multicarrier transmission with two or four RF carriers per transmitter/receiver instead of just one, thereby increasing the transmission capacity two- or even fourfold

---

[3] The state-of-the-art digital radio-relay modulation in the late 1980s was 64 QAM (quadrature amplitude modulation) with a spectral efficiency of 6 bps per hertz. To accommodate SDH into the existing frequency plans, a combination of QAM and trellis coding was introduced [e.g., 64 TCM (64-state trellis-code modulation)] with a spectral efficiency of 5.5 bps per hertz and substantially improved error correction, allowing lower receiver reception thresholds.

As a result of these new technologies, radio-relay systems were introduced in the 1990s for the transmission of SDH at 155-Mbps, 2 × 155-Mbps, and 622-Mbps levels.

25.1.4 Transhorizon Radio-Relay Systems

Marconi had observed in 1932 that VHF signals could be received far beyond the horizon. IT&T installed the first VHF transhorizon system between Barcelona and the island of Majorca in 1935 (see Section 21.3). Military transhorizon transmission was used during World War II in Newfoundland, Canada, and Alaska.

A remarkable commercial transhorizon radio-relay link was installed in Germany in 1948 over a distance of 196 km when the Soviet occupation forces suddenly created a border around the Soviet-occupied zone and blockaded Berlin from the rest of Germany. Immediately, radio-relay equipment was manufactured from VHF mobile radio equipment by the company C. Lorenz and installed in West Berlin–Wannsee on a high tower and on the 799-m Torfhaus mountain just behind the newly formed Iron Curtain. The eight-channel equipment, operating in the 41- to 68-MHz band, provided an output of 100 W, which was soon increased to 1000 W. In 1950 the transmission capacity was increased to three RF channels, each operating 15 telephone channels. Uninterrupted telephone and telex traffic with isolated West Berlin was maintained over this link for many years.

Together with Bell Labs, the Massachusetts Institute of Technology (MIT) investigated the phenomenon of transhorizon transmission in the early 1950s. Propagation experiments were made in the arctic region with powerful signals directed at the troposphere and picked up by special ground receivers at distances as great as several hundred kilometers. The investigations resulted in empirical planning data for commercial transhorizon systems operating between 30 MHz and 5 GHz. Two commercial systems were installed in 1957 based on this experience: one system, installed and operated by AT&T and IT&T, covered 298 km between Florida and Cuba, and a second system, between Italy and Spain, bridged 385 km between the Mediterranean Sea islands of Sardinia and Minorca. In Japan a little later, a 340-km system was installed between Kagoshima and Amami Oshima.

The world's largest commercial transhorizon network was installed in the early 1970s in Brazil. This network, installed by the French company Thomson CSF in the states of Mato Grosso and Amazonas, covered 3600 km of very inaccessible terrain between the towns of Campo Grande and Manaus. With hop lengths up to 400 km, only 11 hops were required. The network transmitted 120 channels on 900 MHz with an RF output power between 10 and 1000 W using antennas with diameters of 9 to 27 m. Space diversity was used on most hops. Part of the network was still in operation 25 years later.

Frequencies below 1 GHz were used for hops beyond 400 km. Hops between 200 and 400 km were preferably operated at 2 GHz, whereas below 200 km operation up to 5 GHz was possible with relatively small antennas. Transhorizon communication met the requirements for special applications, where lack of infrastructure made cable transmission practically impossible and radio-relay line-of-sight transmission too expensive. The satellite transmission in the 1970s, and the worldwide development of transportation infrastructures, gradually phased out transhorizon transmission.

25.2   RADIO-RELAY SYSTEMS WORLDWIDE

25.2.1   Radio-Relay Systems in North America

In the United States, 32 TV stations in the regions of New York, Boston, Philadel-phia, Baltimore, and Washington broadcast the inauguration of President Truman (1884–1972) in 1949. The TV stations were interconnected by a combined coaxial cable and 4-GHz radio-relay network operated by AT&T. In the same year, another link for television transmission was installed between New York and Philadelphia. In September 1951, New York was linked with San Francisco using 107 stations for the transmission of 600 telephone channels or one television signal.[4] In January 1953, some 75 million people watched on TV the inauguration of President Eisenhower (1890–1969). The ceremony on TV was distributed throughout the United States from New York to San Francisco via a 12,000-km radio-relay network serving over 100 TV broadcast transmitters. By the end of 1958, AT&T operated 110,000 km of TV circuits and 32,000,000 km of telephone circuits on its radio-relay network. In that year it also completed a radio-relay network across Canada.

Western Union also installed a radio-relay network in a triangle between New York, Philadelphia, Washington, and Pittsburgh in 1948. The network was extended to Chicago in 1959 and up to San Francisco in 1964, when it had a total length of 11,000 km. The network was later expanded to 13,500 km and connected with the Canadian Pacific network, which covered 5000 km between Montreal and Vancouver.

Radio-relay systems were used extensively for the transmission of telephony and carried the major portion of long-distance traffic in the United States and Canada until the massive deployment of optical fiber cable in the 1990s. Even then, in 1992, 620-Mbps radio-relay equipment was installed instead of cable in large mountainous areas of Canada.

25.2.2   Radio-Relay Systems in Latin America

In Latin America, Thomson CSF (later integrated into Alcatel Telspace) installed one of the first radio-relay routes in Argentina in 1960. The 1800-km-long route between Buenos Aires and Campo Duran used some 160 transmitters and receivers operating at 2 GHz and offered a capacity of 120 channels. In Mexico, Standard Elektrik Lorenz AG (SEL) constructed a 2000-km national radio-relay network connecting major towns in 1962. A 4-GHz system provided 960 telephone channels and TV. The project was executed on a turnkey basis, including survey, roads, buildings, towers, and power plant. Air interface was provided with the network of AT&T in the United States. A repeater station where 16 RF channels converged was at 3000-m altitude on Mount Culiacan (Figure 25.5).

In 1968, on the occasion of the Olympic games in Mexico, SEL extended the net-work on a turnkey basis by an additional 6000 km. The 6.2-GHz radio-relay network, which provided 1800 telephone channels and TV transmission, included international air interfaces with the United States and Guatemala.

---

[4] These stations were equipped with the first of a long series of T systems, such as the famous TD-l (T for transmission, D for 4 GHz, H in TH for 6 GHz, and TJ and TL for the 11-GHz bands).

Figure 25.5   Radio-relay repeater station on Mount Culiacan, Mexico. (Courtesy of Alcatel SEL.)

The world's first commonly planned multinational radio-relay network is the Comtelca network (Comisión de Telecomunicaciones Central America) in Central America. This network, installed by NEC in the early 1970s on a turnkey basis with 960-channel 4- and 6-GHz equipment, connects via 25 repeater stations the capital cities of five Central American countries: Guatemala, El Salvador, Honduras, Nicaragua, and Costa Rica. Figure 25.6 shows one of the stations used for this network in El Salvador. The network was operated initially with one operating channel and one standby channel, and in line with the increased need for communication, extended to three operating channels and one common standby channel. This analog network was upgraded in the mid-1990s with digital equipment with a transmission capacity of 140 Mbps. It was extended to 35 stations and a total network length of 1510 km under a contract with Alcatel Standard Electrica of Spain.

Another remarkable international radio-relay network was installed in the Caribbean in 1989. Five long over-water hops with lengths between 46 and 92 km and a total length of 345 km connect the islands of the Lesser Antilles: Montserrat, Guadeloupe, Morne la Treille, Dominica, Martinique, and St. Lucia. The equipment, with a transmission capacity of 140 Mbps operating at 7 GHz, was supplied by Alcatel ATFH and operated by C&W and France Telecom. Although all hops were on line-of-sight conditions, space diversity had to be used on all five hops

Figure 25.6   Typical radio-relay repeater station in Central America: the Las Pavas station in El Salvador, 1968.

to overcome the very difficult propagation conditions above the sea in the tropical climate.

In South America, traditionally a radio-relay subcontinent, radio-relay networks were erected in all countries. In 1970, Thomson CSF installed in Argentina an interesting radio-relay route using underground stations. It connected the cities of Cordoba and Mendoza, with a link passing through the Pampa de las Salinas. The all-solid-state 600-channel radio-relay equipment operating in the 7-GHz band was accommodated at the repeater stations in buried cylindrical containers to avoid the expense of buildings and to protect the equipment against the extreme day/night temperature differences. Almost maintenance-free turbo generators, using a high-efficiency closed-circuit vapor system, provided a no-break power supply at those stations. The generators, mounted on the tower foundation with an output between 400 and 2000 W, were designed for a life expectancy of 50,000 hours.

In 1993, Siemens supplied 155-Mbps equipment for some 7000 km of radio-relay routes in Argentina, Brazil, and Colombia, in addition to optical fiber cable transmission equipment for only 2500 km. This included the first Latin American nationwide SDH radio-relay network in Colombia, with 4300 km of backbone routes.

Figure 25.7 World's second-highest broadband radio-relay repeater, in Bolivia. (Courtesy of Siemens.)

The following three examples of record radio-relay links installed by Siemens give good proof of the survival of radio-relay transmission despite optical fiber, thanks to its suitability for use in adverse geographical and climatic conditions. In 1990, the world's highest broadband radio-relay repeater station was installed in Bolivia on the 4760-m Mount Luribay (Figure 25.7). This repeater station is part of a 140-Mbps, 4.7-GHz route between La Paz and Santa Cruz, which includes an exceptionally long direct line-of-sight hop of 151 km between two stations at heights of 4000 and 1000 m.

Mount Luribay lost its ranking as the highest station one year later when Siemens installed a radio-relay repeater station on Cerro Esperanza in north Argentina at a height of 4902 m. The third example is a 160-km hop also operating at 4.7 GHz, with a capacity of 140 Mbps, installed in Mexico in 1995. The hop is across the Gulf of California between San Lucas, at a height of 575 m, near Guaymas on the mainland, and Vigia, at 430 m above sea level, on the Baja California peninsula. The hop was part of a 635-km route which continued from Vigia to La Paz, the capital of Baja California. This hop was particularly difficult because of fading due to atmospheric layers above the water surface created by the unfavorable climatic conditions. To meet the extremely problematic propagation conditions, an unusual technical solution was implemented: a combination of space diversity with four antennas in each direction, and frequency diversity with two frequencies on each antenna. Large parabolic antennas are used with a diameter of 4.6 m instead of, usually, 3 m, and a frequency spacing of 160 MHz instead of 80 MHz. Thus at any time, the best signal can be selected out of eight signals received and the transmission quality required is met with a complex solution which under the prevailing geographic conditions is more economical than an optical fiber cable.

25.2.3 Radio-Relay Systems in Europe

In 1950, the U.K. GPO brought into operation the first radio-relay link for TV transmission in Europe between London and Birmingham, operating at 1 GHz. Two years later followed a 395-km link for TV transmission between Manchester and Edinburgh and from there to Kirk O' Shotts. The equipment, manufactured by STC and operating at 4 GHz, was the first to use a TWT in the transmitter output stage. A few years earlier (in 1948), the company had introduced the first portable radio-relay equipment, first used in 1950 by the BBC for live outside broadcasting of a boat race.

In France, a first radio-relay link for TV transmission came into operation in 1951 between Paris and Lille. The equipment was manufactured by Thomson–Houston and operated at 1 GHz. Two years later, a 4-GHz system called GDH 101, developed jointly by the French PTT and CSF (Companie Générale de Téléphone Sans Fil), went into operation between Paris, Lille, and Strasbourg. Within five years a national radio-relay network 10,000 km long was installed with this equipment connecting Paris, also with lines terminating at Lyon, Marseille–Cannes, Bordeaux, and Rennes–Nantes.

A second over-water line-of-sight link with Corsica between Grasse and La Punta at a distance of 250 km was installed in 1953 with three RF channels operating between 50 and 80 MHz, each carrying 24 telephone channels. The system was extended in 1956 to Algiers via two repeater stations on the isle of Sardinia, with equipment operating at 400 MHz manufactured by SFR (Societé Française Radioeléctrique).

In Italy the companies Magneto Marelli and FACE produced equipment for the first radio-relay link for TV transmission which came into operation in 1952, just in time to participate in the Eurovision TV broadcasting of the coronation of Queen Elizabeth II of Britain. This historic television transmission to 60 million spectators in the U.K., Belgium, The Netherlands, France, West Germany, Switzerland, and Italy, on June 2, 1953 marked a major step in the development of radio-relay equipment in Europe. Within 10 years, a radio-relay network for TV transmission connected all major Western European cities in an area linking Dublin, Helsinki, Palermo (Sicily), and Madrid.

In Germany, all radio-relay development was stopped in 1945, but around 1950, the companies C. Lorenz [from 1958 Standard Elektrik Lorenz AG (SEL) and from 1992 Alcatel SEL], Siemens & Halske (from 1966 Siemens AG), and Telefunken[5] began again with development and production of radio-relay equipment with frequency modulation for 24-, 60-, 120-, and 240-channel telephony transmission and TV transmission, and for pulse modulation for 24- and 60-channel telephony transmission operating in the 2-GHz band. A TV-transmission system on the route Hamburg–Hannover–Cologne–Frankfurt–Stuttgart–Munich was completed in 1954 but used successfully in 1953 during the first Eurovision transmission. The same companies extended their production program to 960, 1800, and 2700 channels and TV-transmission systems operating in the 4-, 6-, 7-, and 11-GHz bands and in the 1980s to a complete range of digital radio-relay equipment. Change of ownership and rationalization stopped the production of radio-relay equipment in the 1990s at Alcatel SEL and Telefunken.

The world's longest radio-relay link was taken into operation in Russia under extremely difficult climatic and logistic conditions in 1996. The 7800-km link from Moscow via Samara to the city of Novosibirsk, in the middle of Siberia, built by Siemens, and from that city to Chabarovsk, in eastern Siberia, built by NEC, was

---

[5] Ownership of Telefunken went from AEG to Bosch in 1984 and to Marconi in 2000.

completed in only 18 months. The radio-relay link has a transmission capacity of 155 Mbps on each of the six RF channels and the two standby channels. It operates primarily on 6.7 GHz, but some hops have to use other frequencies between 3.4 and 6.2 GHz. The route continues from Chabarovsk to Nakhodka via an existing terrestrial optical fiber cable and from Nakhodka via two submarine cables to Jóetsu in Japan and to Pusan in South Korea.

25.2.4   Radio-Relay Systems in Asia

The first high-capacity radio-relay system in Southeast Asia was installed in 1954 by STC between Kuala Lumpur and Singapore and subsequently extended to George Town on the island of Pinang. It used 600-channel 4-GHz equipment. Some 30 years later the telecommunications administration in Malaysia signed a contract with SEL for the largest national radio-relay network ever put out to international tender thus far. This huge turnkey project included the supply of 2500 transmitters and receivers, partly in analog but mainly in digital systems. It was necessary to survey around 3000 km of radio-relay routes, primarily in dense jungle, to determine the best location for several hundred new stations. The associated access roads, buildings, towers, and no-break power supplies were all part of the contract. A typical station of that project is shown in Figure 25.8.

In the same region, in Papua New Guinea, the world's first low-power-consumption high-capacity (960 channels) radio-relay equipment was taken into operation in 1970 by Telettra (now Alcatel Italia) within the scope of a 3000-km backbone network. The equipment was installed at several difficult-to-access sites in mountainous jungle areas, with helicopters used to transport the equipment and prefabricated shelters. The system, called the IR-20 RF repeater, applied RF amplification and optimized circuitry using only 10 active components for operation in a $1 + 1$ cold-standby mode. The power consumption of a repeater station was reduced to 18 W from about 300 W, which was typical at that time. For this extremely low power consumption, initially dry nickel–cadmium batteries were used, which were replaced every six months by helicopter visits. A few years later the dry batteries were replaced by solar cells, which were still in use 20 years later.

Figure 25.9 illustrates another example of the use of solar cells for a radio-relay repeater station in the desert of the Sultanate of Oman, erected by Siemens in 1984. The solar array, with a peak capacity of 2.2 kW, provides the power supply for the radio-relay equipment. The radio-relay equipment, with 500 W continuous power consumption, is accommodated in a shelter with a passive cooling system that accumulates the heat dissipation of the equipment during the day and radiates this heat via an outside condenser during the night.

In Japan the first radio-relay system for TV transmission went into service at the beginning of the 1950s using equipment imported from STC. Around the same time, in parallel with European companies and taking advantage of its affiliation with IT&T, NEC began its development of a pulse-modulated system, which was first used by an electricity distribution company in April 1953. The first radio-relay system, with a capacity of 360 channels operating at 4 GHz, produced in Japan by NEC based on IT&T TD-2 know-how, went into operation on the 470-km Tokyo–Osaka route in 1954. Radio-relay networks were also installed in, India, Indonesia, Thailand, the Philippines, Taiwan, and other Asian countries.

Figure 25.8   Radio-relay station in Malaysia, 1988. (Courtesy of Alcatel SEL.)

25.2.5   Radio-Relay Systems in Australia

In Australia, the first radio-relay link was constructed in 1943 as a backup route for the 150-km Bass Strait submarine cable connecting Tasmania with Australia (see Section 21.3). The first broadband radio-relay link in Australia was installed in 1959 over a distance of about 140 km between Melbourne and Bendigo in the state of Victoria. Since then, Australia, with its vast inhabited areas and scattered population, became the most radio-relay-minded continent. The first major radio-relay link in the world to be powered exclusively by solar energy was opened in 1979 at a length of about 500 km between Alice Springs and Tennant Creek, locations that were previously stations of the overland telegraph line (see Section 8.6.1). An even longer solar-powered radio-relay route, 1600 km long, was constructed along the coast of the Indian Ocean between Port Hedland and Kununurra. Yet another unusual energy source is being used for a radio-relay link along the more than 1000-km-long route of a gas pipeline from the Moomba gas fields in South Australia and the coastal regions of New South Wales. The radio-relay stations of this link are

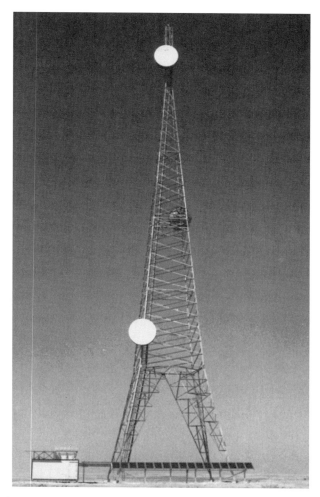

Figure 25.9   Solar-powered radio-relay repeater station in the Sultanate of Oman. (Courtesy of Siemens.)

powered by generators using gas from the pipeline. Wind generators are also used as a power supply, for instance on King Island in the Bass Strait and on some stations of the east–west radio-relay route between Townsville and Tennant Creek.

Toward the end of the twentieth century, by June 1996 the incumbent operator Telstra operated radio-relay links that spanned more than 200,000 km, compared with some 44,000 km of optical fiber cable and 14,000 km of coaxial cable.

25.2.6   Radio-Relay Systems in Africa

One of the first radio-relay links in Africa was a 60-channel 2-GHz system installed in 1951 in Morocco on a route between an electricity plant in Afourer and a barrage in Bin el Quidane, by SFR of France. It achieved good results by using five passive

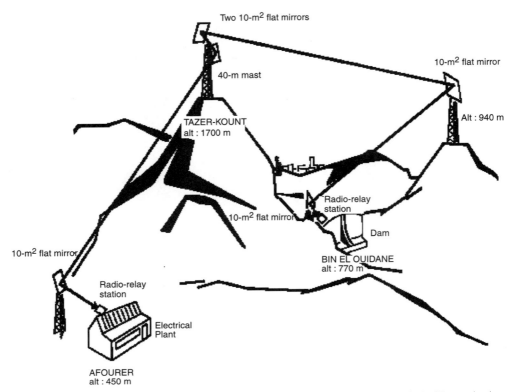

Figure 25.10 One of the first radio-relay networks in Africa, in Morocco, 1951. (From A. A. Huurdeman, *Radio-Relay Systems*, Artech House, Norwood, MA, 1995, Figure 1.7; with permission of Artech House Books.)

reflectors to overcome geographical obstacles in the Atlas Mountains (Figure 25.10). Three decades later, in 1980, the African and Asian continents were linked by a record 355-km hop over the Red Sea between the Sudan and Saudi Arabia. This 960-channel 2-GHz equipment was supplied by Telettra (now Alcatel Italy). On both the African and Asian sides, repeater stations were built on tops of mountains: Jebel Erba (2179 m) and Jabel Dakka (2572 m), respectively. To obtain a suitable space-diversity performance improvement, 120-m-high towers had to be built at both stations. A feature of the relay stations was that instead of just one antenna per direction, a pair of antennas was used for each of the two directions, vertically spaced, forming an antireflection system to neutralize the effects of rays reflected from the Red Sea. Both repeater stations were fully powered by solar cells with batteries for seven days of no sunshine, as well as a diesel generator in the event of a total failure of the solar generator and the impossibility of reaching the station to clear the fault before the storage battery was depleted.

Radio-relay communication brought some improvement to the notoriously underdeveloped telecommunications in Africa. National radio-relay networks of significance were installed mainly in South Africa (Figure 25.11), Egypt, Morocco, and

Figure 25.11   Radio-relay repeater station in South Africa. (Courtesy of Gerd Lupke.)

in most western and eastern African countries. In the 1970s an effort was made to connect the major African countries by a regional radio-relay backbone called *Panaftel*. Lack of infrastructure and poor maintenance, as well as a lack of traffic (in Africa most international traffic is intercontinental traffic), seriously limited the success of Panaftel.

25.3 WIRELESS ACCESS SYSTEMS

Telephone penetration in the developing world is largely hindered by the high cost to implement and maintain the access network infrastructure by means of a pair of copper wires for each subscriber. Cable laying is expensive, o/w overhead lines are vulnerable, and both need time-consuming right-of-way clarifications. The advent of cost-effective integrated circuits (Section 24.1) made it economically possible in the 1980s to use radio-relay transmission in *multiple access* (MA) and *point-to-multipoint* (P-MP) modes of operation in the underserved vast rural areas in both developed and developing countries.

For decades, most rural areas, if served by telecommunications at all, were provided with open-wire lines or, even less reliable, with HF radio. The cost of the civil works involved for the installation of cable systems make them too expensive to serve a few scattered faraway subscribers. Cost-effective digital P-MP radio-relay transmission equipment, however, offered an economical solution for these underprivileged areas. In a P-MP network, a central radio-relay station connected (e.g., by a conventional radio-relay link) with a telephone exchange at the fringe of the PSTN radiates its RF energy via an omnidirectional or sectional antenna to a number of radio-relay "outstations" within its line of sight. Outstations even farther away, out of sight of the central station up to a typical distance of 500 km, can still be served by means of cascaded repeater stations. Figure 25.12 shows a P-MP network developed in the 1980s which can serve up to 960 subscribers scattered in an area within a radius of 500 km around the central station. Each outstation can serve 80 subscribers within a maximum distance of 10 km. Up to 14 repeaters can be cascaded to cover the full service area. As with a PABX, the 960 subscribers can be interconnected within the network, and up to 60 subscribers simultaneously can be connected with the PSTN. Public telephone boxes can also be connected to an outstation. The entire network can be centrally supervised by an operation and maintenance system (OMS) connected to the exchange interface part (EIP) of the central station.

The first P-MP radio-relay network with six subscriber stations was manufactured by the Canadian company SR Telecom and installed in 1977 in an isolated fishing village near the landing points of the first transatlantic telegraph and telephone cables in Newfoundland, Canada, a country in which a large part of the population lives in rural areas. SR Telecom installed P-MP radio-relay networks in various countries the largest network for over 50,000 subscribers was installed in Saudi Arabia.

The world's highest P-MP repeater station was installed in 1989 in Bolivia on Cerro Mercedes at an altitude of 5400 m, using solar cells for the power supply. Another repeater station of the same P-MP network, shown in Figure 25.13, was installed on an approximately 4100-m unnamed mountain, which was then officially named Cerro Panait, to honor the radio-relay survey engineer who selected this mountain as an excellent site for a P-MP repeater station.

Thanks to P-MP radio-relay transmission, thousands of inhabitants of rural and isolated areas in over 100 countries are no longer isolated but can communicate with the rest of the world at a price comparable with that for urban subscribers. The success of P-MP systems encouraged the manufacturers of cellular radio systems to extend cellular radio operation to fixed subscribers. Initially, fixed cellular terminals were given to potential subscribers who were not yet connected to the PSTN and

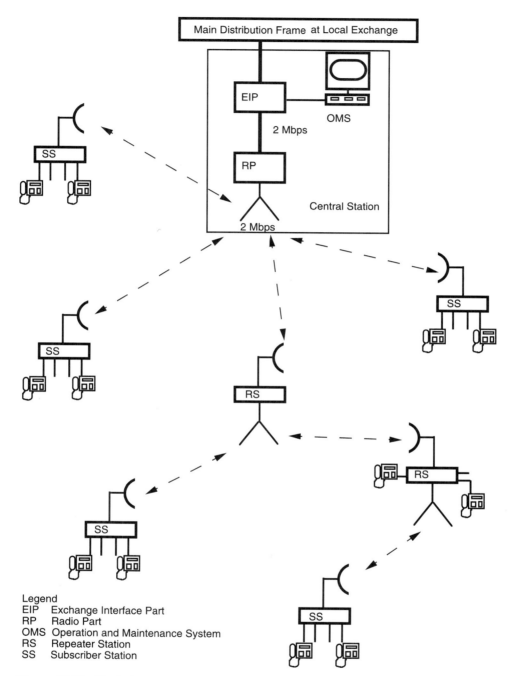

Figure 25.12 Typical P-MP network. (From A. A. Huurdeman, *Radio-Relay Systems*, Artech House, Norwood, MA, 1995, Figure 1.12; with permission of Artech House Books.)

Figure 25.13   P-MP repeater station on Cerro Panait in Bolivia, the world's first mountain named in honor of a radio-relay engineer. (Courtesy of Alcatel SEL.)

happened to live within the service area of a cellular system. In this way, due to the high system capacity of cellular networks, suddenly, thousands of would-be subscribers could be connected with the PSTN. Fixed cellular subscribers do not need the highly complex facilities required for mobile subscribers such as automatic roaming and cell hand-over, as a fixed subscriber does not move from cell to cell or change location. Battery-operated handheld phones are not required, as conventional telephone sets can de connected to the fixed radio terminal. The operating procedures and signaling tones are the same as in the PSTN. Subsequently, stripped-down cellular systems appeared in the early 1990s to serve high densely populated areas so far deprived of telephony. A typical solution is NEC's minicellular system, which serves about 5000 fixed subscribers in a cluster of seven cells, each with a radius in the range of 20 km. In additional to the fixed operation, subscribers can also use their cordless telephones outside their home or office within their cell. The prefix *mini* does not refer to cell size (small cellular cells being called *minicells* and *microcells*) but to the small service area of the stripped-down cellular system. This system was first installed in Malaysia in 1991. In a similar way, some 50,000 subscribers were connected to the PSTN in fixed cellular networks and P-MP networks in the previous German Democratic Republic within one year in the early 1990s.

In the meantime, radio systems appeared that were specially developed for the access network in urban areas. These are now mostly called *wireless local loop* (WLL) systems, or similar to optical fiber access systems, *radio in the loop* (RITL).[6] Whereas the P-MP systems operate primarily in the 1.5-, 2.5-, and 2.7-GHz frequency bands, WLL systems operate in the bands 3.5, 10.5, 24, 26–28, 38, 40–42, and 50 GHz. P-MP systems offer only telephone and fax services. The WLL systems

---

[6] WLL systems are also named *broadband fixed wireless access* (BFWA), *wireless in the local loop* (WiLL), *wireless access* (WA), *fixed radio access* (FRA), *fixed wireless access* (FWA), *local multipoint distribution service* (LMDS), *multichannel multipoint distribution service* (MMDS), *last-mile radio*, or *radio in the local loop* (RLL).

are available in various capacities for telephone, fax, and data services up to 155 Mbps, which, in terms of capacity, places WLL between xDSL and optical fiber (both covered in Chapter 28).

In Africa, some progress is made with the installation of WLL systems, primarily in South Africa, Senegal, Côte d'Ivoire, Guinea. Ghana, Kenya, Nigeria, and Uganda, where at the end of 1998, slightly over 200,000 subscribers, thanks to WLL, got a solution to their years-long place on a waiting list for a fixed-line telephone connection.

In Indonesia a large rural project was started in 1997 to connect 280,000 business and residential subscribers to a digital WLL network in Jakarta and West Java called FONET (flexible overlay network).

In the U.K. the national operator Ionica (then third after British Telecom and Mercury) started commercial operation in May 1996, aiming to serve residential and small business customers all over the country using fixed radio access. The Ionica customers were connected by radio operating at 3.5 GHz via a nearby base stations with the next local exchange, thus bypassing the outside line plant. Initially, voice, fax, and low-speed data were offered; operation in the 10.5-GHz band was planned to enable broadband operation. Alas, Ionica is already history. After serving only 62,000 subscribers instead of the expected about 200,000, Ionica went bankrupt in October 1998.

The systems PHS, and especially DECT, described in Sections 32.4.2 and 32.4.3, are also used as RITL. Those systems are in mass production, which enabled a significant price decrease in RITL applications. Thus a significantly larger number of persons could afford to buy a RITL instead of waiting one to five or even more years for a wireline telephone. The telephone can be bought in a department store[7] and used immediately once a private operator or even the national telecom administration installs a small RITL base station in their residential area. Without need to dig up roads to lay cables, WLL systems can be set up within hours rather than months. In Spain, as an example, this became reality for over 200,000 subscribers in 1994, and in Mexico for some 1000 subscribers in 1995, all of whom had been waiting for years in vain for a wireline connection. According to studies of the *Strategis Group* and *Pyramid Research*, almost 10 million persons had a WLL connection by the end of the twentieth century, of which about 40% used a downgraded cellular system, 25% used DECT or PHS, 1.7% used a satellite system, and the remaining 13.3% were connected to the PSTN via a P-MP or other wireless system. The geographical distribution was 40% in Asia, 35% in the Americas, 15% in Europe, and 10% in Africa.

Most of those systems operate in the higher-frequency bands, where neither operation licenses nor frequency coordination are required. This may, however, lead to interference with other services in urban areas. An effective way to get around this negative effect was developed at the end of the century with smart antennas which operated multiple narrow beams controlled by a special *spectral reuse and filtering technology* (SeRFiT) using digital processing techniques to create clear signals. By focusing narrow beams of RF energy, the same frequencies can be reused many times within a network.

As the latest broadband fixed wireless access development at the end of the twentieth century, very short range broadband radio-relay equipment operating in the 60-

---

[7] The Spanish department store chain *El Cortes Ingles*, for instance, sells WLL equipment through 60 stores around Spain as a package of WLL and computer called a *broadband office suite*.

GHz band came on the market. This GigaLink equipment operates in the 60-GHz oxygen band, where up to 98% of the RF energy is absorbed by the oxygen content of the atmosphere, which limits transmission to a maximum of about 2 km. Within this short range the transmission quality is practically independent of rain and fog. The high bandwidth enables application of simple *direct digital modulation* (DDM) of signals with a capacity of 100 to 622 Mbps. The U.S. manufacturer of this equipment, *Harmonix*, claims that up to 100,000 subscriber units can be deployed in a 10-km area without causing mutual interference. As a first major application, several buildings in downtown Tokyo were connected by such a system with a transmission capacity of 622 Mbps.

A different wireless solution for last-mile access was introduced at the end of the twentieth century with laser technology called *free space optics* (FSO). With FSO the light of a laser is encoded with a digital signal and beamed through free space toward a sensitive optical receiver within line of sight. FSO is derived from free-space infrared transmission technology, developed in the 1960s for missile guidance. Commercial FSO systems came on the market in 1995 with a transmission speed of 2 Mbps over a distance of 6 km. Compared with the WLL solutions, FSO has a much lower cost; it does not require spectrum licensing or planning permission and can have very high transmission capacities. A major drawback is its vulnerability to fog, especially frozen fog. Trials made in Moscow and Seattle, both cities with frequent heavy fog, however, resulted in an availability of over 99.9%, with 1.6-Mbps transmission over 1.6 km. Another drawback is scintillation caused in the atmosphere by hot weather, a phenomenon already observed by George Everest in the 1820s in India. Wind and building sway also affect performance unless automatic tracking keeps the beams aligned. A further consideration is eye safety. The systems operate on wavelengths of 1550 and 800 ns, which can cause damage to the retina in case of unprotected presence near the transmitter. On the other hand, FSO does not produce electro smog and therefore might achieve a higher acceptance than that for radio systems. Its ease of installation will also support a broad application: The equipment can be plugged directly into any office power outlet and communicate through a window or from a rooftop. As a typical example, a local area network was recently installed with FSO systems operating at a speed of 155 Mbps at the Smithsonian Institution in Washington, DC, connecting all museums on the national Mall and the offices in nearby L'Enfant Plaza.

At the end of the twentieth century, FSO systems were available for transmission speeds up to 620 Mbps over distances up to about 6 km. A few thousand terminals are in operation in over 40 countries. A major pioneering company is PAV Data Systems in the U.K. Some FSO startup companies made cooperation agreements with the largest telecommunications equipment manufacturers, such as Nortel with San Diego–based AirFiber, and Lucent with Seattle-based TeraBeam Networks. The Canadian startup fSONA (free-space optical networking architecture) based its technology on research made in the U.K. at the R&D center of BT[8] at Martlesham Heath.[9]

---

[8] The name of the incumbent operator in the U.K. changed from General Post Office (GPO) to British Post Office (BPO) in 1968, to British Telecom in 1980, and BT in 1991.

[9] Systems for transmission of 2.5 Gbps, even four times 10 Gbps in DWDM technology (see Section 28.1) over 5 km, are announced for release in the early years of the twenty-first century.

25.4 RADIO-RELAY TOWERS AND AESTHETICS

Radio-relay towers have to accommodate antennas at such a height that a free line of sight is given between the corresponding stations. Consequently, a radio-relay tower usually is an intrusion in the landscape which poses a big architectural challenge, especially in urban areas. Unlike high-tension transmission lines, which always disturb the landscape, radio-relay towers quite often provide an aesthetic landmark. Even the most basic tower, shown in Figure 25.14, matched with the surrounding forest and presented in its environment a sign of progress. It was erected temporarily in Guatemala in 1968 to enable TV broadcasting in that country of the Olympic Games in Mexico. Most radio-relay towers are self-supporting towers constructed from hot-dip-galvanized angle steel, which certainly at first view hardly create an aesthetic impression. Still, many such towers became landmarks, similar to the Eiffel Tower, which was highly criticized in the beginning for disturbing the skyline of Paris

Figure 25.14 Provisional radio-relay tower in Guatemala, 1968. (Courtesy of Gerd Lupke.)

Figure 25.15   Inside view of a self-supporting radio-relay tower in Malaysia, 1975. (Courtesy of Alcatel SEL.) See insert for a color representation of this figure.

but became the world's most visited tower. Those strictly functional constructions can give an aesthetic filigree impression, as shown in Figure 25.15.

A good combination of old and new technology is shown in Figure 25.16. This ancient windmill in Portugal became a radio-relay station in 1973 to link the landing point of the SAT 1 cable (South Atlantic telephone cable No. 1) at Sesimbra, with landing points of TAT 5 (transatlantic telephone cable No. 5) at Conil and a Mediterranean cable at Estapona in Spain. An example of a radio-relay tower with a Chinese touch is given in Figure 25.17, which shows a tower on top of a high building in Taipei, Taiwan-China.

A new type of tower made of reinforced concrete appeared in the 1950s which made it easier to meet aesthetic requirements. Those towers were used for TV broadcasting and for radio-relay stations at nodal points with one or more antenna platform(s) for antennas in various directions. The first such tower for TV broadcasting, including a panoramic restaurant and an observation platform designed by

Figure 25.16 Ancient windmill used as radio-relay station at Sesimbra, Portugal, 1973. (Courtesy of Alcatel SEL.)

Fritz Leonhardt, was inaugurated in 1956 at Stuttgart, Germany. It was a model for TV-broadcasting towers erected in many cities. This tower still had a modest height of 217 m. In the meantime, the world's highest TV tower, 553 m, was opened in Toronto, Canada, in July 1978. The Telmex tower in Mexico City, constructed in the 1950s (Figure 25.18), is an example of an early large concrete tower for radio-relay application.

In the U.K., many towers are given planning consent only if disguised: for example, as trees if in the countryside. In Germany, a country with one of the world's most meshed radio-relay networks, requiring many antennas per tower, the Bundespost started in 1951 with the erection of reinforced concrete towers with several antenna platforms. The towers have a maximum diameter of 8 m and a height of at least 45 m. The equipment rooms are inside the tower near the antenna platforms. About 300 such towers were erected all over Germany. Figure 25.19 shows the highest of these towers (331 m), which was built at Frankfurt. The lower platform, with a diameter of 57 m, is the largest tower-antenna platform in the world.

In the nineteenth century, high chimneys in the landscape were appreciated as symbols of industrial progress. Similarly, in the twentieth century, telecommunications towers became visible signs of the information society.

Figure 25.17   Radio-relay tower in Taipei, Taiwan-China, 1973. (Courtesy of Alcatel SEL.)

Figure 25.18   Telmex radio-relay tower in Mexico City, 1960. (Courtesy of Gerd Lupke.)

Figure 25.19 Highest telecommunications tower in Germany, at Frankfurt, with the world's largest antenna platform.

REFERENCES

Books

Carl, Helmut, *Radio-Relay Systems*, Macdonald, London, 1966.

Deloraine, Maurice, *When Telecom and ITT Were Young*, Lehigh Books, New York, 1976.

Huurdeman, Anton A., *Radio-Relay Systems*, Artech House, Norwood, MA, 1995.

Libois, L. J., *Faisceaux hertziens et systèmes de modulation*, Éditions Chiron, Paris, 1958.

Michaelis, Anthony R., *From Semaphore to Satellite*, International Telecommunication Union, Geneva, 1965.

Oslin, George P., *The Story of Telecommunications*, Mercer University Press, Macon, GA, 1992.

Articles

Anon., The phone goes Bush, extract from *Australian Geographic*, No. 29, January–March 1993.

Anon., Optical wireless beams in on last-mile market, *FibreSystems Europe*, January 2001, pp. 29–32.

Anon., Untapped unlicensed band: a blessing and a curse, *Telecommunications Americas*, April 2001, pp. 11–12.

Anon., 60-GHz band: fiber's new alternative, *Telecommunications Americas*, August 2001, pp. 16–18.

Bonin, J.-P., et al., 140 Mbit/s inter-island microwave system in the Caribbean, *Commutation & Transmission*, No. 2, 1990, pp. 25–34.

Buckley, Sean, Free space optics: the missing link, *Americas Telecommunications*, Vol. 35, No. 10, October 2001, pp. 26–33.

Budischin, F., Rückschau über den Ausbau des Richtfunks im Fernmeldenetz der Deutschen Bundespost bis zum Jahre 1968, *Archiv für das Post- und Fernmeldewesen*, January 1974, pp. 3–97.

Channing, Ian, It's WLL, Jim, but not as we know it, *Mobile Communications International*, November 1999, pp. 89–90.

Drechsler, Bettina, Kommunikationstechnisches Bindeglied für russische Föderationsrepubliken, *Telcom Report* (Siemens), Vol. 18, November–December 1995, pp. 302–305.

Friis, H. T., Microwave repeater research, *Bell System Technical Journal*, Vol. 27, April 1948, pp. 183–194.

Häfner, Andreas, A giant leap over the Gulf of California, *Telcom Report International* (Siemens), Vol. 18, November–December 1995, pp. 16–18.

Kenward, Michael, Free-space lasers light urban broadband links, *FibreSystems Europe*, May 2001, pp. 63–66.

Oguchi, Bun-ichoi, Microwave radio system, *Telecommunication Journal*, Vol. 45, No. 6, 1978, pp. 323–330.

Telstra, *The Story of Long Distance Communications*, Information kit 4.

Teutschbein, Werner, Fernmeldetürme Symbole des Kommunikationszeitalters, *Archiv für deutsche Postgeschichte*, Vol. 1, 1989, pp. 109–122.

26

COAXIAL CABLE TRANSMISSION

26.1 TERRESTRIAL COAXIAL CABLE

In the early 1950s, three copper-wire transmission media were commonly used: open wire, symmetrical cable, and coaxial cable. Open-wire systems with a transmission capacity of 3, 4, 10, or 12 channels were used primarily to connect small telephone exchanges in sparsely populated areas. Symmetrical cable was used in urban regions where transmission capacities of 24, 60, 120, or 240 channels were required. Coaxial cable came into use for the transmission of 300 and 600 telephone channels and for the transmission of TV signals. In the United States the L3 system was introduced in 1953. It had a transmission capacity of 1860 channels or 600 channels and a 4.1-MHz bandwidth for the 525-line NTSC (National Television Standard Committee) signal. It was designed for transmission of telephony over 1000 miles and of TV over 4000 miles.

Coaxial cable with a 2.6-mm-diameter inner conductor and a 9.5-mm-diameter outer conductor was standardized by CCITT as *standard tube coaxial cable 2.6/9.5*, also called *standard CCI-tube* (Figure 26.1).

In France a coaxial cable with a 1.2-mm-diameter inner conductor and a 4.4-mm-diameter outer conductor was developed which was standardized at the CCITT plenary meeting at New Delhi in 1960 as *small-diameter coaxial cable 1.2/4.4*, also called *pencil gauge* or *small tube coaxial cable*. Coaxial cable transmission systems were developed in the 1960–1970s for the transmission of 300, 1260, 2700, 3600, and 10,800 telephone channels on a pair of standard CCI tube cables, and for transmission of 300, 600, 960, 1260, and 2700 telephone channels on a pair of small-tube coaxial cable.

The transmission of such large numbers of channels required bandwidths of 1.3 MHz for 300 channels, up to 60 MHz for 10,800 channels. To compensate for the

The Worldwide History of Telecommunications, By Anton A. Huurdeman
ISBN 0-471-20505-2   Copyright © 2003 John Wiley & Sons, Inc.

Figure 26.1 Coaxial cable with 12 standard CCI tubes, 1970. (Courtesy of Museum für Kommunikation, Frankfurt, Germany.)

high attenuation of coaxial cable at those bandwidths, numerous repeaters were required at intervals between 1.5 km for a 10,800-channel system and 9.5 km for a 1260-channel system on a standard CCI tube and between 2 km for a 2700-channel system and 8 km for a 300-channel system on a small-tube coaxial cable. Figure 26.2 shows the relation between transmission capacity and repeater spacing for the three copper-wire media.

With the availability of reliable semiconductors that could be used in buried repeaters, coaxial cable transmission became the preferred solution for long-distance transmission. Whereas repeaters for the symmetrical cable, and usually also for o/w systems, were accommodated in buildings, repeaters for coaxial cable systems were buried in special underground containers. Power for the repeaters was supplied from terminal stations via the inner conductors of two coaxial pairs. Typically, up to 20 repeaters of a 960-channel system (Figure 26.3) could be fed from one terminal.

The world's longest terrestrial coaxial cable route, with a length of about 8500 km running from Moscow via Chabarovsk and Vladivostok to Nakhodka, was opened in 1966. A submarine cable laid between Nakhodka and Nautsu in Japan made it possible to set up direct intercontinental circuits between Europe and Japan in 1969.

A special version of coaxial cable called *Cloax* was developed in the Bell Labs in the late 1960s. Cloax was developed for two reasons: to improve the mechanical and electrical characteristics of coaxial cable, and to reduce the amount of copper in view of worldwide copper shortage. A thin copper skin was laminated to a tinned-steel

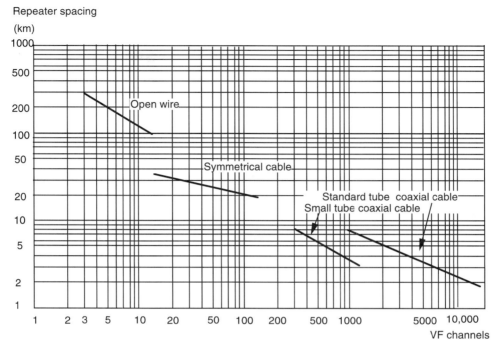

Repeater spacing

(km)

Figure 26.2   Repeater spacing versus transmission capacity. (From A. A. Huurdeman, *Guide to Telecommunications Transmission Systems*, Artech House, Norwood, MA, 1997, Figure 3.2; with permission of Artech House Books.)

sheath with a copolymer adhesive. The entire laminate was corrugated to obtain higher flexibility and to add crush resistance.

With the introduction of digitalization of switching and transmission, digital transmission systems for operation on coaxial cable gradually replaced analog transmission systems in the 1970–1980s. Such digital systems, with a transmission capacity of 34 Mbps, were operated on small-tube coaxial cable with repeater spacing of 4-km and 140-Mbps systems with 2-km repeater spacing. Systems with a capacity of 565 Mbps were operated on standard coaxial cables with a repeater spacing of 1.5 km.

Coaxial cable was often considered to be superior to radio-relay transmission. However, both domains were used: coaxial cable for long-distance high-capacity transmission lines in moderate geographical environments, and radio-relay transmission for applications where the terrain or other conditions limited the use of cable.

26.2   SUBMARINE COAXIAL CABLE

26.2.1   Transatlantic Coaxial Telephone Cables

The planning and implementation of the first transatlantic telephone cable presented various interesting technological challenges. Actually, the planning dates back to the

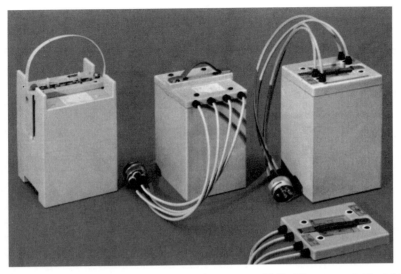

Figure 26.3   Power-fed coaxial underground repeaters, 1965. (Courtesy of Alcatel SEL.)

early 1930s (see Section 20.6). After World War II, electronic technology had advanced to a point where serious consideration could be given to a wideband system with numerous long-life repeaters laid on the bottom of the ocean, powered by current supplied over the cable from sources on shore. In 1952, planning coordination was began between AT&T, the British GPO, and the Canadian Overseas Telecommunications Corporation. AT&T and the GPO had gained experience with multiplexed and repeatered telephone submarine cable operation, for example, between the United States and Cuba in deep water and between the U.K. and France in shallow water. Both sides had ample experience with submarine cable—not so much, however, with transocean repeaters.

Locating, raising, and replacing a defective submarine repeater can easily cost over $1 million (today's submarine repeater price amounts to around $750,000). Consequently, those repeaters were meant to lie on the bottom of the ocean without failure for at least 20 years. To obtain this result, it was necessary either to use a minimum number of components in the repeaters, all of the utmost reliability, or to provide duplicate components to take over in case of failure. Chances for successful deep-sea cable laying were considered greater the smaller the repeaters, which spoke against the duplication solution. AT&T had good experience with a flexible type of repeater housing mounted in continuation with the cable so that it could pass around the submarine cable-laying gear without requiring the ship to be stopped each time a repeater was laid, as used to be the practice.

Out of those considerations grew the approach to using a minimum of components for the repeaters and thus using one-way repeaters—instead of two-way repeaters with their associated directional filtering devices—and consequently, to apply four-wire operation with two cables running in parallel. To further assure the long life of the repeaters, a conservative design approach was adopted: using electronic tubes with characteristics typical of the late 1930s rather than more advanced tubes or even transistors.[1]

For the short, shallow-water part of the transatlantic cable between Newfoundland and Nova Scotia, a more advanced approach was taken with the latest design electronic tubes, enabling a wider bandwidth and thus two-wire operation with one cable, thus gaining experience on the newest technology.

The transmission capacity of the coaxial cable was determined by the lower and upper frequency of the repeater, which with conservative design was between 20 and 170 kHz. A band of about 150 kHz was thus available, enabling the accommodation of three groups of 12 channels.

For the power feeding of submerged repeaters spaced at 60-km intervals, a 1950-V dc source was used, with half of the power supplied at each end of the cable at opposite potential: 1950 V positive with respect to ground at one end and 1950 V negative at the other end. This placed the maximum potential and risk on the repeaters near the shore ends, while the repeaters in the middle of the cable, in deep water, had potentials near ground.

The cable was named *TAT-1* (transatlantic telephone cable No. 1). Figure 26.4 is a schematic of the route, with an indication of transmission capacity and distances. The route constituted a combination of almost all transmission media available at the time. Describing the route from east to west, it started with a 24-channel carrier system on symmetrical-pair cable between London and Glasgow and a 36-channel carrier on coaxial cable between Glasgow and the U.K. landing point, Oban. In the Atlantic Ocean, two parallel coaxial submarine cables were used, whereas a single coaxial submarine cable was laid in the Gulf of St. Lawrence between Clarenville in Newfoundland and Sydney Mines in Nova Scotia. From Sydney Mines a radio-relay system followed to Spruce Lake, New Brunswick, where the 36 telephone channels were divided into six telephone channels to Canada and 29 telephone channels to the United States. The thirty-sixth channel was used for 12 voice-frequency telegraph channels, of which six were routed to Canada and six to the United States. From Spruce Lake one radio-relay link went to St John, Canada, and another to Portland, Maine. From Portland a symmetrical-pair cable with a 12-channel carrier system went to West Haven and a coaxial cable ended the U.S. route in White Plains, New York. The Canadian route used a 12-channel carrier on an open-wire line from St. John to Quebec and on a symmetrical cable between Quebec and Montreal.

TAT-1 began operations on September 25, 1956. Since 1927 the transatlantic telephone service had been by single-channel HF radio. The 36 more reliable telephone channels now available via cable were considered by traffic forecasters to be sufficient for the next 20 to 30 years. The HF-radio stations were therefore closed down. The superior performance of the submarine cable, which offered domestic quality on international calls instead of noisy and unreliable radio calls, evoked such a demand that within two months the radio stations had to be opened again to cope with the increase in transatlantic telephone traffic: an experience of overcautious forecasting to be repeated many times. TAT-1 carried 10 million calls during its 22-year life span. Transatlantic telephone traffic grew at an annual rate of 20%. The capacity of TAT-1 was raised to 48 channels in 1959 by lowering the channel band-

---

[1] Transistors were first used in submarine repeaters in 1964 in a submarine telephone cable laid between the U.K. and Belgium.

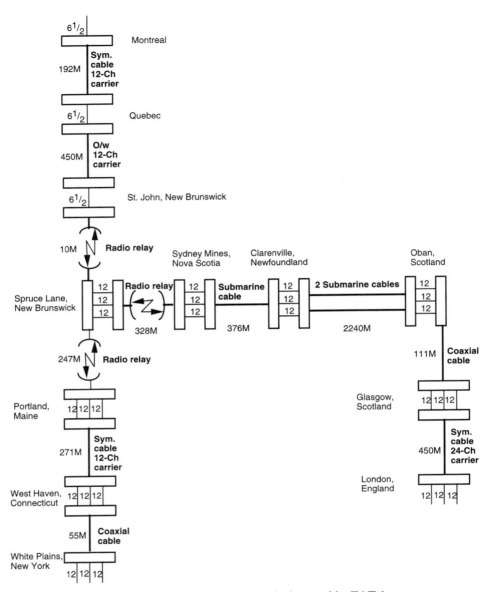

Figure 26.4   First transatlantic telephone cable, TAT-1.

width from 4 kHz to 3 kHz. In the same year, *TAT-2* was laid between Canada and France, also with a capacity of 48 telephone channels with a bandwidth of 3 kHz. This transmission capacity was doubled in 1960 by means of *time assignment speech interpolation* (TASI) equipment. TASI, developed by AT&T, is based on the fact that in a disciplined call, one partner only is talking; thus each direction of a speech channel is occupied only half of the time and thus (with complex electronic equipment) can be used for another call. *TAT-3*, laid down in 1963, was the first bidirec-

Figure 26.5 Cable ship *Neptun* in Cape Kennedy Harbour, 1963. (Courtesy of Museum für Kommunikation, Frankfurt, Germany.)

tional transatlantic cable. It had a capacity of 138 (non-TASI) channels. Two years later, *TAT 4* was laid down between the United States and France with 138 channels. An improved *lightweight* type of coaxial cable was used. Instead of external armoring of the cable, the inner conductor was made of a copper tube welded around a 41-wire steel rope, which provided the mechanical strength of the cable. The outer conductor, a thin copper tube, was protected by a sheath of high-density polyethylene without armoring. Solid low-density polyethylene separated the conductors. The cable was laid by AT&T's cable ship *Long Lines*, which was the largest in the world. Figure 26.5 shows a contemporary cable ship.

The new cable, produced with great accuracy and combined with a fully transistorized system, enabled the transmission of 845 non-TASI channels on *TAT-5*, laid down in 1970 between the United States and Spain. *TAT-6*, laid down between the United States and France in 1975, had a capacity of 4000 telephone channels. In 1983, *TAT-7* was laid down between the United States and the U.K. with a capacity of 4200 channels (without considering TASI). It was the last coaxial transatlantic cable. Since then, a dozen optical fiber cables have been laid. Table 26.1 summarizes the major details of TAT-1 to TAT-7. Figure 26.6 gives an impression of the short time intervals at which the new cables were laid and the significant reduction in annual operating cost per speech channel achieved with each new cable. TAT-1 was taken out of operation in 1978 after 22 years of service and TAT-2 in 1983 after 23

TABLE 26.1   Transatlantic Coaxial Telephone Cables

Cable	Year	Route	Distance (km)	Number of Channels at 3 kHz	Repeater Spacing (km)	Power Feeding	
						kV	mA
TAT-1	1956	Canada–U.K.	4390	48	82	1.94	325
TAT-2	1959	Canada–France	4090	48	82	2.14	225
TAT-3	1963	U.S.–U.K.	6515	138	42	5.0	370
TAT-4	1965	U.S.–France	6665	138	42	5.1	370
TAT-5	1970	U.S.–Spain	6410	845	22	2.78	137
TAT-6	1976	U.S.–France	6290	4000	11	3.5	657
TAT-7	1983	U.S.–U.K.	6070	4200	11	5.15	657

years of service. TAT-3 to TAT-7 were still in operation at the end of the twentieth century. Submarine cable repair requiring special cable ships was expensive, time consuming, and resulted in a high traffic loss. Fortunately, however, the reliability was so high that the de facto standard of submarine cable reliability was often expressed as two or fewer ship-repair operations in a 25-year period.

26.2.2   Worldwide Submarine Coaxial Telephone Cables

The success of the TAT cables provoked worldwide activity in submarine telephone cable laying. Britain made strong efforts to regain the submarine cable hegemony. C&W commissioned three new cable ships, and the British Commonwealth Conference in London planned an all-British submarine telephone network in 1958. As part of this network the Canada transatlantic telephone cable No. 1 (Cantat-1) between Newfoundland, Canada and Oban, Scotland opened in 1961. This cable carried 80 telephone channels. One year later this network was extended via a radio-relay route

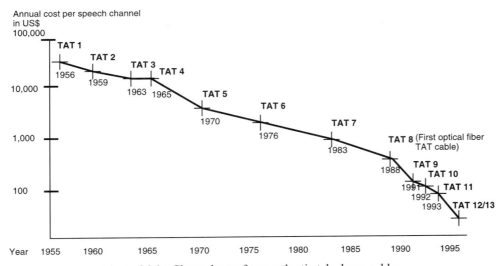

Figure 26.6 Chronology of transatlantic telephone cables.

through Canada and by a Commonwealth-Pacific cable, called *Compac*, from Vancouver via Hawaii and Fiji to New Zealand, and Australia. The Compac cable had a length of 12,450 km and a capacity of 82 channels; it was decommissioned in 1985. In 1962, another transatlantic 24-channel cable called *Icecan* was laid connecting Canada and Iceland, with a landing at Frederiksdal, Greenland, and a 26-channel cable called *Scotice* between Iceland and Scotland.

A transpacific cable, *Transpac 1*, was laid in 1964 connecting the United States via Hawaii with Guam and from Guam with Japan, the Philippines, Hong Kong, and Vietnam. Transpac 1 had a capacity of 142 channels. *Transpac 2* in 1975, with a capacity of 845 channels, went from Guam to Taiwan, Korea, Hong Kong, and Singapore. *Transpac 3*, in 1986, had a capacity of 3780 channels.

A Southeast Asia Communication submarine cable called *Seacom 1*, laid in 1965, connected Singapore, Hong Kong, and the Philippines via Guam and Papua New Guinea with Australia. In 1974, *Cantat-2* was laid with a capacity of 1840 channels. A cable called *Tasman*, laid in the Tasman Sea in 1976, connected Australia and New Zealand with 480 channels. One year later, those two countries were connected with Canada by a cable called *Anzcan*, with a capacity of 1380 channels.

From the European continent in the meantime, in 1973, South Atlantic telegraph cable No. 1 (*SAT 1*) was laid between South Africa and Portugal. The landing point at Sesimbra in Portugal was connected via a radio-relay link (see Figure 25.16) with the landing point of TAT-4 at Conil in Spain and a Mediterranean cable at Estapona in Spain. Another South Atlantic system, *Atlantis*, connected Portugal via Senegal with Recife, Brazil. It was laid in 1982 with a capacity of 2580 channels. Various other submarine telephone cables were laid from Europe to Africa and South America and between North and South America.

A total of 300,000 km of submarine telephone cables had been laid in 1985. The Indian Ocean, however, was still the only sea in the world without submarine telephone cable. This changed one year later when the world's longest submarine coaxial

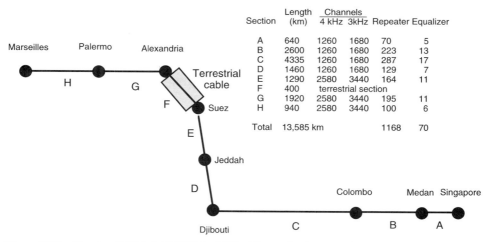

Figure 26.7 Route of SEA-ME-WE 1, 1986. (From A. A. Huurdeman, *Guide to Telecommunications Transmission Systems*, Artech House, Norwood, MA, 1997, Figure 3.12; with permission of Artech House Books.)

cable for telephony was laid to connect Europe with the Middle East and Southeast Asia. The cable, called *SEA-ME-WE 1* (Southeast Asia, Middle East, Western Europe No. 1), 13,585 km long, connected eight countries: Singapore, Indonesia, Sri Lanka, Djibouti, Saudi Arabia, Egypt, Italy, and France (Figure 26.7). The section passing through Egypt is constructed with terrestrial coaxial cable. The cable, laid in shallow water as usual, is armored with steel wires as protection against fishing boats and trawlers. About 85% of the length is deepwater cable, which is unarmored. The maximum depth of the route is 4500 m between Medan and Colombo. Two multiplex systems operate on the cable: a 1260-channel system with a bandwidth of 12 MHz on sections A to D, and a 2580-channel system with a bandwidth of 25 MHz on sections E to H (these capacities refer to 4-kHz channels). By the use of 3-kHz channels, the capacities increase to 1680/3440 channels, whereas with TASI and 4-kHz channels, the capacities are 2500/5000 channels. Power feeding is with a constant current of 545 mA and a maximum voltage of 6500 V. The repeaters, located in shallow water, are equipped with temperature-regulated amplifiers to compensate for the temperature changes. Temperature-regulated amplifiers are not required in the deep-sea water, where the temperature variation does not exceed 0.5°C, which corresponds to an attenuation change of ±0.12%. The cable project was implemented within two years in cooperation of the companies NEC (section B), STC (section E), and Submarcom (now Alcatel), which built the other six sections.

REFERENCES

Books

Hugill, Peter J., *Global Communications since 1844: Geopolitics and Technology*, Johns Hopkins University Press, Baltimore, 1999.

Huurdeman, Anton A., *Guide to Telecommunications Transmission Systems*, Artech House, Norwood, MA, 1997.

Articles

Anon., Cable transmission characteristics, Vol. 4, No. 2, February 1955, pp. 981–1035 in selected articles from *The Lenkurt Demodulator* (Lenkurt Electric Co.), 1971.

Mottram, E. T., et al., Transatlantic telephone cable system: planning and over-all performance, *Bell System Technical Journal*, Vol. 36, 1957, pp. 7–141.

Paul, D. K., Communications via undersea cables: present and future, *SPIE Fiber Optics*, Vol. 599, 1985, pp. 36–50.

27

SATELLITE TRANSMISSION

27.1 EVOLUTION LEADING TO SATELLITE TRANSMISSION

Satellite transmission presents the youngest version of radio transmission. The bright but initially almost unnoticed idea of using an in-space Earth-orbiting satellite as a relay for communication between stations on Earth came from the British science-fiction writer Arthur C. Clarke (1917–), born at Minehead, Somerset, who wrote a prophetic article entitled "Extra-terrestrial Relays: Can Rocket Stations Give World-wide Radio Coverage?" in the technical magazine *Wireless World* (Vol. 51, October 1945).[1] In that article Clarke proposed peaceful use of the World War II V-2 rocket, in a combination of rocketry and radio-relay engineering, to operate relay stations in space.

The V-2 was developed and produced in the period 1936–1944 by Wernher von Braun (1912–1977) with the help of over 10,000 people at the Rocket Center of the German Army and Air Forces at Peenemünde, on the Baltic Sea isle of Usedom. The peace treaty of Versailles in 1919 powerfully limited the scope of weapons permitted in Germany; it did not mention rockets, however. The German army therefore took advantage of the remarkable achievements of a small group of rocketry pioneers in Germany, including Hermann Ganswindt, Helmut Gröttrup, Walther Hohmann, Willy Ley, Rudolf Nebel, Herman Oberth, Klaus Riedel, Max Valier, Johannes Winkler, and the student Wernher von Braun. Beginning in 1929, the German Army

---

[1] Arthur C. Clarke, later Sir Arthur, also called the "grandfather of satellites," wrote that article in May 1945 when he was a flight lieutenant in the Royal Air Force and instructor on ground-controlled approach radar. He was paid £15 for the article! Clarke is now living in Sri Lanka, where he is the proprietor of *Underwater Adventures* and writes science-fiction books and articles. Clarke was a fellow of King's College, London; he received a Marconi International Fellowship and the gold medal of the Franklin Institute. The International Astronomical Union has named the geostationary orbit the *Clarke Orbit*.

The Worldwide History of Telecommunications, By Anton A. Huurdeman
ISBN 0-471-20505-2   Copyright © 2003 John Wiley & Sons, Inc.

took control of all rocketry activities in Germany and established its own rocket development center, first in Berlin–Kummersdorf and in 1936 at Peenemünde. The first successful experimental flight of the V-2 took place on October 3, 1942. This development had been possible thanks to the theoretical and practical work of rocketry pioneers from Austria, Czechoslovakia, France, Italy, Russia, and the United States, as summarized in the next section.

27.1.1 Rocketry Pioneers

In 1881, Hermann Ganswindt (1856–1934) presented a technical proposal with a detailed description of a space vehicle for two persons. This was the first proposal that applied Newton's law of "action equals reaction" by using the backward force of dynamite explosions as the motive force for a space vehicle.

Konstantin Eduardovich Ziolkovsky (1857–1935), born in Ijewskoje, Rjasan, near Moscow, lost his hearing through an attack of scarlet fever at the age of 10, his mother died three years later and his father, a forestry expert, was rarely at home. To overcome his loneliness and social isolation, he started reading books on physics and mathematics and developed a strong interest in space exploration. He published the first calculations on rocket propulsion in 1903. The world's first rocket society, called OIMS (after the Russian for "society for the exploration of interplanetary communication"), was founded in Moscow in 1924 but existed for only one year. An international rocket exhibition took place in Moscow in 1927. In the same year, Friedrich Arturowitsch Zander (1887–1933) made a proposal for a space vehicle. He showed a model of it two years later in Moscow. Nikolai A. Rynin (1877–1942) published a nine-volume *Encyclopaedia about Space Traveling* (in Russian) in the years 1928–1932. A new rocket society was founded in 1931, called GIRD (Gruppa Isutschenija Reaktiwnogo Dwishenija; group for research of reaction motors), as a department of the Osoaviachim (Obschestvo Sodeistvija Oborone, Aviazionnomu I Chimitscheskomu Stroitelstvu; society for the support of defense and aerochemical development). GIRD achieved international recognition but very soon became a military department and was integrated into the Research Institute for Jet Propulsion (RN II) under Minister of Armament Michael Tuchatschewski.[2] A series of experimental rockets, ORM (Opitnij Raketnij Motor, experimental rocket engine) 1 to 102, were developed, which reached a maximum altitude of 4500 m on August 17, 1933 and of 12 km two years later.

Beginning in 1916 in France, Henri Melot experimented with the use of rockets for airplanes.

In 1919, Robert H. Goddard (1882–1945) published a classic paper on rocket propulsion and in 1930 was the first person to launch a small liquid fuel–propelled rocket into the atmosphere.

Herman Oberth (1894–1989), a German physicist of Austrian–Hungarian origin born in Hermannstadt (now Sibiu, Romania), wrote the classic book *Die Rakete zu den Planetenräumen* (The Rockets to the Planets) in 1923. In 1925, Walther Hohmann (1880–1945) published *Die Erreichbarkeit der Himmelskörper* (The Possibility

---

[2] Michael Tuchatschewski was an active promoter of rocketry, but on June 10, 1937, in one of Stalin's reckless purges, he was imprisoned and later killed, together with 35,000 officers.

of Reaching the Celestial Bodies), in which he laid the mathematical foundation for much of rocket technology, including detailed calculations about the launching of rockets.

In 1928, Guido Pirquet proposed establishing a space station at an altitude of 12,760 km called *Cosmonautic Paradoxon*, from which space vehicles would be launched to our moon and to other planets.

Willy Ley (1906–1969) was the first journalist to specialize in space technology. In 1926 he published his book *Die Fahrt ins Weltall* (Journey into Space), in which in easily understandable language he covered a wide range of space technology and described various conditions for human survival in space. In 1927 he was a cofounder and later vice-president of the German *Verein für Raumschiffahrt* (VfR; astronautics society) and its monthly magazine *Die Rakete* (The Rocket) in 1927. Ley emigrated to the United States in 1935, where he continued to give lectures and to write numerous articles and books on space technology. Most remarkable was Ley's book *The Conquest of Space*, published in 1950, with drawings by the American artist Chesley Bonnell. Ley, a strong promoter of the Apollo space project, died June 24, 1969, less than one month before Neil Armstrong and Edwin Aldrin landed on the moon on July 21.

Maximilian Valier (1895–1930), a German engineer of Austrian origin, born in Bozen (then an Austrian town, now Italian) gave lectures on rocketry in various European countries. He was the initiator and cofounder of the VfR. Valier was the first pioneer to die at the site of a rocketry accident, the explosion of his "rocket-auto" on May 17, 1930.

General Arturo Grocco made experiments with solid-fuel rocket engines in Italy in 1927.

Ludwig Otschenaschek (1872–1949) launched solid-fuel rockets up to a height of 1500 m in 1930. He also used a two-stage rocket.

Robert Esnault-Pelterie (1881–1957) held a remarkable first lecture on space travel for the Societé Physique at Paris in 1912. In 1929 he created a new scientific name for space technology with his book title *Astronautique.*

In 1931, Major J. I. Barré, inspired by Esnault-Pelterie, began investigating various fuel combinations for rocket engines. He constructed the first French liquid-fuel rocket, a 100-kg device that reached an altitude of 15 km in a 60-km trajectory in 1941.

In 1932, Eugen Sänger (1905–1964) began a wide range of experiments on jet propulsion for rockets and stratospheric airplanes. One year later he published the world's first textbook on rocket propulsion technology, entitled *Raketenflugtechnik*.

In 1930, G. Edward Pendray, together with David Lasser and others, mainly science-fiction authors, founded a U.S. counterpart of the German VfR, initially called the *American Interplanetary Society* (AIS) and later the *American Rocket Society* (ARS). The AIS issued a magazine, initially called *Bulletin*, then *Astronautics*, and finally, *Jet Propulsion*. David Lasser published the first book on rocketry in the English language under the title *The Conquest of Space* in 1931.

Philip Ellaby Cleator founded the British Interplanetary Society (BIS) in 1933. Experiments with rockets were forbidden in the U.K. by an antiexplosives law dating to the eighteenth century. Rocketry research in the U.K., was therefore limited to theoretical investigations. The *BIS Journal* (now the *Journal of the British Inter-*

planetary Society) was first published in 1934. Cleator published the first book in the U.K. on aeronautics, *Rockets through Space*, in 1937. The BIS made extensive investigations and preparations for a vehicle to be launched from Titicacasee, Peru/ Bolivia, destined to land on the moon, but World War II stopped the project.

Sergeji Koroljow (1906–1966), cofounder of GIRD, published a book about stratospheric rocketry in 1934. After World War II he was responsible for the reconstruction in the USSR of the V-2 and was the chief constructor of *Sputnik* in 1957.

Most of the above-mentioned experts practiced the free international exchange of ideas. For example, the Austrian Guido Pirquet, the Russian Nikolai A. Rynin, and the French Robert Esnault-Pelterie were members of both the BIS and the VfR. This came to an end, however, when rocketry research came under military control.

27.1.2 Passive Satellites

Before active satellites could be launched, the moon was used as a passive repeater for the transmission of signals over a long distance. Experiments with the reflection of radio signals for radar and communication purposes were made repeatedly in the late 1940s and early 1950s. With those experiments, at least one of the questions raised by Clarke (at which frequencies radio waves would pass through the ionosphere) could be answered: Radio waves passed best through the ionosphere at frequencies in the range 1 to 10 GHz. In this range, also called the *microwave window*, there is a minimum of cosmic noise and a minimum of attenuation of electromagnetic waves. In July 1954, the U.S. Navy transmitted the first voice messages via the moon back to Earth, and two years later a permanent moon relay service was established between Washington and Hawaii. Two steerable radio telescopes with a diameter of 25 m were used at either end. This circuit was in operation successfully until 1962, albeit only when the moon was above the horizon at both ends.

Instead of the moon, balloons were used as passive repeaters. The National Advisory Committee for Aeronautics (since 1958 the *National Aeronautics and Space Administration*, NASA), in cooperation with Bell Telephone Laboratories and Jet Propulsion Laboratories, in 1956 began development of an aluminized plastic balloon with a diameter of 30 m and a weight of 50 kg. The balloon, *Echo I*, was launched on August 12, 1960 by a Delta rocket to an altitude of 1600 km, where it expanded to its full size. In the same month, communication via this balloon was made across the United States between Goldstone, California, and Holmdel, New Jersey. Antennas with an aperture of 400 square feet (37 m^2) were used at those sites. Two-way telephony was possible via *Echo I* at a frequency of 136 MHz. Even more impressive was the transatlantic communication via *Echo I* made in the same month with France. *Echo II* was launched on January 25, 1964. It had a diameter of 41 m, a weight of 258 kg, an apogee of 1300 km, and a perigee of 1260 km.

27.1.3 Postwar Rocket Development in the United States

Wernher von Braun came to the United States as a voluntary "prisoner of peace" after the war in September 1945. An additional 126 experts from Peenemünde also became prisoners of peace rather than aiding rocket development in the USSR. The move to the United States, hailed as the biggest transfer of technology and brainpower ever made after a war, was organized by the U.S. Army in an action called

Overcast and later, Paperclip.[3] Wernher von Braun wrote a letter to the U.S. Navy Bureau of Aeronautics in July 1945 in which he proposed to use rockets for the launching of satellites for measurement and research purposes. Upon arrival in the United States he certainly would have liked to take up immediately Clarke's idea of developing together with the U.S. rocketry pioneers an adequate rocket for launching telecommunication satellites. Instead, the German experts had to assist in the Hermes rocket project[4] of the U.S. Army at the newly erected rocket development center at Fort Bliss, near El Paso, Texas. Parts for about 100 V-2s and 40 tons of paper on development and production of the V-2 had been seized in Germany. The German experts were of invaluable help in using that material and information for Hermes and following projects. About 70 V-2s were reconstructed and fired from nearby White Sands in New Mexico from April 16, 1946 until 1952. On August 22, 1951 a V-2 reached an altitude of 214 km.

As a further result of this cooperation, a two-stage rocket was lifted successfully to an altitude of 403 km on February 24, 1949. The rocket, called *Bumper*, used a V-2 as the first stage and a *WAC Corporal* as the second stage. The WAC (Women's Auxiliary Corps) Corporal was the first genuine U.S. rocket developed by the U.S. Army in cooperation with the *California Institute of Technology* in 1945. The WAC Corporal was first launched on September 26, 1945 at White Sands and reached an altitude of 70 km. In total, eight *Bumper* rockets were launched. *Bumper 7* and *8* were the first rockets fired at the new rocket-launching center at Cape Canaveral, Florida. *Bumber 7*, launched on July 29, 1950, reached a record altitude of 403 km.

In 1950, the U.S. Army moved its rocket research center from Fort Bliss to the Ordnance Missile Command at the Redstone arsenal in Huntsville, Alabama, where Wernher von Braun became the head of the development department for guided missiles. By 1962 he had a staff of 6000 persons, including his early mentor, Herman Oberth. In Huntsville, the V-2 technology was used for the development of a series of *Redstone* and *Jupiter* rockets, which, within eight years, resulted in the successful launch of America's first satellite, the *Explorer*, on January 31, 1958.

27.1.4 Postwar Rocket Development in the USSR

Most of the experts from Peenemünde who were not selected to work in the United States accepted work at the newly created Russian rocket production institute RABE at Bleicherode near Nordhausen in the Russian-occupied zone of Germany. By the middle of 1946, over 5000 persons worked in RABE under the management of the Russian rocket expert Major Boris Tschertok of the Russian Rocket Research Institute NII-I (in Moscow) and Helmut Gröttrup, the German ex-Peenemünde expert for missile guidance. Suddenly, on October 21, 1946 at 3 o'clock in the morning a secretly selected group of 175 of the German experts, including Gröttrup, were forced to prepare themselves within two hours with their families for transportation to the USSR in an action called Ossoaviachim, in honor of the prewar Russian research group of that name. In the next days all RABE material was transported to

---

[3] The name *Paperclip* referred to the procedure applied for selection of the German experts. A paperclip was placed on the cards of those experts who were to be invited to come to the United States.

[4] The objective of the Hermes project was to develop a suitable ballistic missile for carrying nuclear weapons.

the USSR. The German experts enabled the Russians to launch a V-2 from the newly established rocket center Kapustin Jar 200 km east of Stalingrad (now Volgograd) on October 18, 1947. The V-2 reached a distance of 350 km. After this achievement, the German experts were systematically excluded from Russian rocketry research. In fact, most of them were moved to the Black Sea, where they could enjoy the pleasant climate, albeit in complete isolation. They returned to Germany between December 1951 and November 1953.

27.1.5 *Sputnik*, the First Satellite

On April 5, 1950, James Van Allen, professor at the University of Iowa, proposed using a satellite to make observations from space during the International Geophysical Year (IGY), which was to take place from July 1, 1957 to December 31, 1958, a period with an expected maximum of solar radiation.[5] During a subsequent IGY-preparatory meeting in Rome on October 4, 1954, it was officially agreed to attempt to include an Earth satellite. The obvious choice was to ask the U.S. authorities. The U.S. government agreed, but valuable time was lost in a competency struggle between the U.S. Army with their *Orbiter* (supported by von Braun) and the U.S. Navy with their *Vanguard*, which was finally favored. On September 29, 1955, President Eisenhower's press secretary announced that within the scope of the IGY, the United States would launch a satellite that scientists of all nations could use to enhance scientific progress.

The world was greatly surprised, however, when on October 4, 1957, a Russian *Sputnik* sent its mysterious "bleep-bleep" from space! *Sputnik I*,[6] a 58-cm ball weighing 80 kg, transmitted space telemetry information on 20 and 40 MHz for 21 days. It moved around the Earth in 96.2 minutes on a 229/946-km elliptic orbit. After 57 days it reentered the atmosphere and burned up. *Sputnik* shocked American scientists, engineers, and politicians, and the U.S. Congress quickly voted funds for satellite programs. Thanks to the Cold War, the strategic importance of communication satellites was recognized. The U.S. Navy decided immediately to upgrade a *Vanguard* test flight planned for October 26, 1957 to a live flight with a small, experimental, 50-cm satellite. The launch had to be delayed a few days, and then on December 6, after a 1.25-m flight, the rocket exploded and a bleeping satellite was found on a nearby beach. In the meantime, to make the situation even more dramatic, *Sputnik II*, with the dog Laika on board, had been in orbit since November 3. *Sputnik II* weighed 508 kg and was still connected with a central part of the rocket, which weighed 3000 kg. The perigee was 224 km and the apogee 1661 km. After the failure of the rocket of the U.S. Marines, the U.S. Army got a chance to prepare a crash program to launch a newly conceived *Explorer* satellite with a modified *Jupiter C* three-stage rocket, in a tight race with the Marines, who prepared a new *Vanguard* launch for January 26, 1958. Again that *Vanguard* start had to be delayed to Febru-

---

[5] In continuation of the two *International Polar Years* (IPYs), in 1882–1883 and 1932–1933, in which 12 and 49 nations, respectively, had participated, a third IPY was proposed which should include observation stations in space. Considering that the observations this time should not concentrate on the poles but be really global, an IGY instead of an IPY was then organized with 67 participating countries.

[6] Konstantin E. Zaiolkowski created the word *Sputnik* (signifying "fellow traveler") for an artificial earth satellite in his book *Dreams about Heaven and Earth, and the Effects of Universal Gravity*, published in 1895.

ary 5, when it actually made a short 500-m flight but then exploded. Fortunately, the Army launched its *Explorer* successfully on January 31.

Explorer 1, a 2-m-long 14-kg cylinder, orbiting on a 360/2500-km ellipse, was a big success; it transmitted telemetry information for nearly five months. Evaluation of his information led to the most important discovery of the IGY: the existence of *Van Allen radiation belts* around Earth. It was discovered that high-energy particles such as protons, electrons, and heavy ions, which are emitted from the sun, are trapped by the magnetic field of Earth in specific "belts" and can cause substantial damage to electronic circuitry.[7]

The first successful launch of a *Vanguard* took place on March 17, 1958. It reached a perigee of 651 km and an apogee of 3960 km. It was the first satellite that used solar cells and mercury batteries. The batteries failed after three months, but the transmitter and receiver of the satellite continued to operate from the solar cells until 1965. Solar cells were also used in *Sputnik III*, which was launched on May 15, 1958.

27.1.6   First Communication Satellites

The U.S. Air Force launched the first communication satellite *Score* (signal communication orbit repeater experiment) successfully on December 18, 1958. *Score* was a delayed-repeating satellite, which received signals from Earth stations at a frequency of 150 MHz, stored them on tape, and later retransmitted them. *Score* retransmitted the 1958 Christmas message of President Eisenhower to the world. The 68-kg *Score* was placed in a low Earth orbit (LEO) on a 182/1048-km ellipse. The significance of different satellite orbits in relation to the Van Allen belts and the time of satellite visibility on Earth is shown in Figure 27.1 and explained briefly in Technology Box 27.1.

Score had two transmitter/receiver units and a tape recorder that were still battery powered. The battery was discharged after 12 days and the transmission stopped. A second delayed-repeating satellite called *Courier I-B* was launched on October 4, 1960 in a 586/767-km ellipse. Although it used solar cells to recharge the batteries, operation stopped after only 17 days.

In continuation of successful experiments with the passive satellite *Echo I* in 1960, and with the availability of reliable transistors, solar cells, traveling-wave tubes, microwave resonators, and low-noise receiver circuitry, AT&T and Bell Labs decided to proceed with the development of an active satellite called *Telstar*. The major objectives were to look for the unexpected; to establish transatlantic communication; to evaluate multichannel two-way transmission of telephony, data, and TV; to gain experience in tracking a large ground station antenna to the satellite; to gain numerical knowledge of the Van Allen belts; and to evaluate the effects of cosmic radiation.

---

[7] The radiation effects on satellites are twofold. The first occurs gradually, as the electronic equipment slowly degrades over time, either by damage to the silicon lattice within the components or by building up charges in the semiconductors which eventually disable the circuits. The second effect is sudden, in which high-energy particles induce either single-event latchup (SEL) or single-event upset (SEU). SEL is a catastrophic event where a single particle can "latch" a device to the point where it no longer can function without removing power or, in the extreme case, destroying the device. An SEU is not permanently damaging, but it can "flip bits," change the state of a memory device from a "1" to a "0," or vice versa. This effect, if unmitigated, can upset spacecraft processors. (From *Iridium Today*, Spring 1997.)

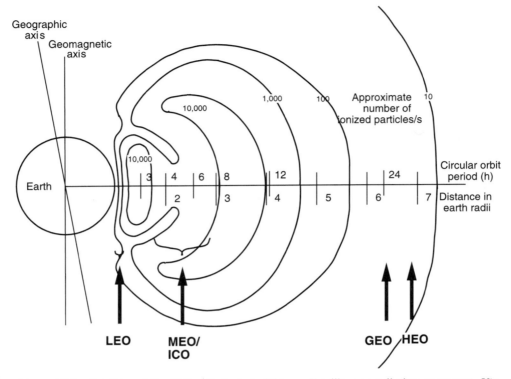

Figure 27.1 Satellite orbits. (After Robert A. Nelson, Satellite constellation geometry, *Via Satellite*, March 1995, p. 112. Originally from A. A. Huurdeman, *Guide to Telecommunications Transmission Systems*, Artech House, Norwood, MA, 1997, Figure 7.4; with permission of Artech House Books.)

Previous research had indicated that the preferred frequency for communication between satellite and Earth stations should be in the range 1 to 10 GHz. The frequencies in this range, however, were already allocated worldwide to various other services. After considerable study and consultation with the ITU and others, it was decided that sharing frequencies with radio-relay systems could easily prevent interference with other services. The 4-GHz band was selected for the down direction, from satellite to Earth station, so as to minimize the deleterious effects of rain on the signal received. The 6-GHz band was selected for the up direction. All electronic devices were all-solid-state, with the exception of transmitter power stages, for which traveling-wave tubes were used with a guaranteed life span of five years. The power supply was critical. Three alternatives were considered: miniature nuclear reactors, isotope power supply, or solar cells. It was decided that solar cells would present the best solution, the major uncertainty being its resistance against cosmic radiation and the effect of large day/night temperature differences. About 30% of the surface of *Telstar* was covered with a total of 3600 solar cells. A film of synthetic sapphire covered the cells as protection against cosmic radiation. The efficiency of those cells was about 10% and the capacity was not sufficient for continuous feeding of the electronic equipment. A 26-V nickel–cadmium storage battery was added to carry

TECHNOLOGY BOX 27.1

Satellite Orbits

A satellite orbit is the lifetime location of a satellite in space. A satellite remains in orbit as long as its centrifugal force is in balance with the gravitational attraction of Earth, the sun, the moon, and other cosmic influences. The rotation time, or *orbit period T* (in hours) for which a satellite for a given height h remains balanced in orbit is in accordance with *Kepler's third law*, given by the formula $T = 2\pi\sqrt{(h + R)^3/GM}$, where R is the Earth's radius = 6378.155 km, G the gravitational constant = 6.67×10^{-11} N · m^2/kg^2, and M the Earth's mass = 5.95×10^{24} kg.

The relation between the distance from Earth's center and the circular orbit period T is indicated in Figure 27.1 and summarized as follows:

Altitude (km)	Orbit Period	Rotations per Day (Approx.)
150	1 h 27 min	16
500	1 h 34 min	15
800	1 h 41 min	14
1,500	1 h 56 min	12
4,000	2 h 55 min	8
10,000	5 h 47 min	4
20,000	11 h 50 min	2
36,000	24 h 07 min	1

(A satellite reaches exact geosynchronism at an altitude of 35,786 km and an orbital velocity of 3.075 km/s.)

In accordance with Kepler's first law, a satellite orbit is an ellipse with the center of Earth at one focus. The point at which the satellite is closest to Earth is called the *perigee*, while the point at which the satellite is farthest away from Earth is called the *apogee*. The most frequently applied circular orbit is a special case of the elliptical orbit with both foci coinciding with the center of Earth. Kepler's second law says that the line joining the satellite with the center of Earth sweeps over equal areas in equal time intervals (thus achieving the highest speed in apogee). The point where this line meets the equator is called the *subsatellite point*. The distance between the subsatellite point and the satellite, thus the altitude above the mean equatorial radius of Earth, is a major characteristic of the satellite, determining the coverage area, the orbiting period, and thus the time of satellite visibility, the signal propagation delay, the path attenuation, and the influence of the two Van Allen belts.

Four distinct altitude ranges are used for telecommunications satellites, moving in the same direction as the Earth's rotation, in orbits as follows:

1. *Low Earth orbit* (LEO): between 500 and 2000 km
2. *Medium Earth orbit* (MEO): between 5000 and 15,000 km; also called *intermediate circular orbit* (ICO)
3. *Geostationary Earth orbit* (GEO): at 35,786 km; also called *Clarke orbit*
4. *Highly elliptical orbit* (HEO): with an apogee that may be beyond GEO

Source: Adapted from A. A. Huurdeman, *Guide to Telecommunications Transmission Systems*, Artech House, Norwood, MA, 1997; with permission of Artech House Books.

Figure 27.2   Model of *Telstar I*. (Courtesy of Museum für Kommunikation, Frankfurt, Germany.)

the peak load and to permit operation during sun eclipses. Because of those power limitations, most electronic equipment was turned on and off by radio commands from Earth stations. Another important parameter was the orbit position. Geostationarity, although preferred for worldwide coverage, was not possible with the Delta rocket used to launch the satellite. To meet the conditions of transatlantic communication, a 960/6140-km elliptical LEO was chosen with an orbital period of 158 minutes, providing almost half an hour of common visibility per day in the United States and France and in the United States and the U.K., respectively. *Telstar* was constructed as an 87-cm sphere weighing 80 kg. A model of *Telstar I* is shown in Figure 27.2.

Telstar I was launched by NASA from Cape Canaveral with a Delta rocket in the early morning of July 10, 1962. Speech and TV was transmitted between Washington and Andover via *Telstar*. Great was the surprise when the first signals of those transmissions were received on the European continent on July 11, 1962 at the Pleumeur–Bodou Earth station in France. One day later, TV signals were transmitted to the United States from the European Earth stations Pleumeur–Bodou and Goonhilly Downs, Cornwall, U.K. After 10 days of experiments followed the official introduction of transatlantic TV transmission via satellite on July 23 with a speech of President Kennedy, a baseball game, and the sound of Big Ben. ***Thus the era of commercial space telecommunications began on July 23, 1962.***

The Earth stations in the United States and in France, manufactured by AT&T, got large horn-reflector antennas with a horn diameter of 20 m and a reflector length

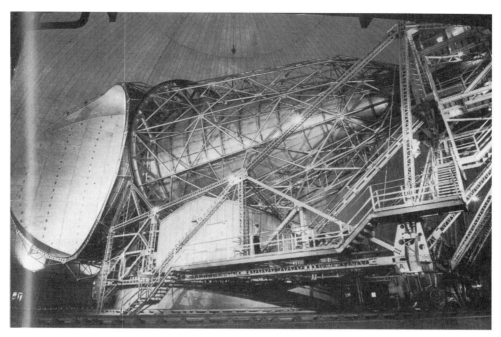

Figure 27.3   Construction of the first Earth station antennas. (Scanned with permission of the ITU from Catherine Bertho Lavenir, *Great Discoveries: Telecommunications*, International Telecommunication Union, Geneva, 1990, p. 85.)

of 54 m (Figure 27.3). The antenna, with a weight of 380 tons, could be moved in azimuth and elevation to track the satellite at an accuracy of $\frac{1}{20}$ of a degree. A huge Dacron and synthetic rubber radome with a diameter of 64.5 m protected the antenna, including the electronic equipment, against humidity, wind, and rapid temperature changes. At Goonhilly Downs a large steerable parabolic antenna was used with a diameter of 32 m manufactured by Marconi. This antenna type became the standard for Earth stations in the next few years, with a diameter from 32 m down to 2 m. In 1964, the next Earth station (Figure 27.4), manufactured by Siemens, was installed at Raisting, Germany, near the area where in 1801 the world's first private optical telegraph system went into operation.

The worldwide deployment of Earth stations in the first 25 years is summarized in Table 27.1. Figure 27.5 shows the unusually shaped antenna of a provisional installation in 1968 at Si Racha, Thailand, manufactured by GT&E.

Telstar was used for experimental telephone, image, and TV transmission, including color TV transmission of a surgical operation. Telemetry information received from *Telstar* indicated that the density of electrons at high energy was much higher than had been anticipated. As a consequence, *Telstar I* interrupted operation after its 1242nd orbit on November 23. An exciting procedure was begun to achieve remote repairwork in space. Laboratory tests pointed to certain transistors as being the most likely sources of trouble. Special codes were devised to take advantage of certain circuit features that would permit bypassing these particular transistors. On December 20, the satellite received successfully one of these modified codes. In

Figure 27.4   Earth station at Raisting, Germany. (Courtesy of Museum für Kommunikation, Frankfurt, Germany.)

subsequent operations, all voltages were removed from the command decoders, and as had been predicted, this action allowed recovery of the transistors. *Telstar* was "repaired" so that transmission restarted on January 3, 1963. On February 14, operation of the command system degraded again. It took longer and longer to respond to the normal command codes. By February 20 it no longer responded to the normal codes but responded to the modified codes. The next day, however, one such command was misinterpreted and a relay was operated that disconnected most

TABLE 27.1   Worldwide Deployment of Earth Stations

Service	Year					
	1965	1970	1975	1980	1985	1990
Public service						
Intelsat	3	45	113	228	347	502
Eutelsat	—	—	—	—	15	78
Arabsat	—	—	—	—	17	24
European Broadcast Union	—	—	—	—	6	15
Domestic satellite	—	—	11	190	1,057	1,880
Business						
Medium stations	—	—	—	5	168	540
VSATs	—	—	—	2,330	6,970	60,900
TVRO (excluding DTH)	—	—	—	240	400	12,600
Inmarsat						
Coast stations	—	—	—	2	16	47
Total deployment	3	45	123	2,995	8,996	76,586

Figure 27.5 Provisional Earth station at Si Racha, Thailand, 1968. (Courtesy of Gerd Lupke.)

of the electronic system from the power supply and interrupted the transmission permanently. With improved cosmic radiation protection, *Telstar II* was launched on May 7 of the same year.

27.2 FIRST SYNCHRONOUS COMMUNICATION SATELLITES

The first synchronous communication satellite, *Syncom I*, in orbit at an altitude of 36,000 km (as proposed by Arthur C. Clarke in 1945), was produced by Hughes Aircraft Company[8] and launched by NASA for the U.S. Army on February 14, 1963. Unfortunately, it went silent 20 seconds after ignition of its apogee motor for synchronous orbit injection. Clarke conceived of the idea of placing telecommunications satellites in orbit at a distance of 36,000 km from Earth,[9] so that the satellites would move synchronously with Earth and thus have a *geostationary* position relative to their Earth stations. He proposed placing three satellites in this orbit equidistantly around Earth to provide worldwide transmission coverage as shown in Figure 27.6. Clarke also envisioned direct broadcasting of TV signals from a satellite to home TV receivers. His paper, written in 1945, however, was seen as science

---

[8] Hughes Aircraft Company, incorporated in 1953, shifted the satellite business to Hughes Space and Communication (HSC), which in October 2000 was acquired by Boeing Satellite Systems. By then Hughes had manufactured 165 satellites.

[9] Clarke clarified that he took the idea of satellites in geostationary orbit from a Slovenian engineer, Herman Potocnic, who mentioned this solution in his book *Das Problem der Befahrung des Weltraums*, which he wrote in 1929 under the alias Herman Noordung. Clarke also made a reference to this book in his article "Extra-terrestrial Relays."

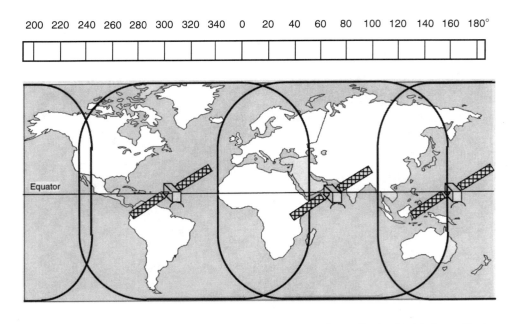

Figure 27.6   Three geostationary satellites provide worldwide coverage. (From A. A. Huur-deman, *Guide to Telecommunications Transmission Systems*, Artech House, Norwood, MA, 1997, Figure 7.6; with permission of Artech House Books.)

fiction and went almost totally unnoticed until his vision became reality with the launching of *Syncom II* on July 26, 1964. *Syncom II* accomplished successful synchronous orbit placement. It was used for telephone, telex, and facsimile transmission between Africa, Europe, and the United States. Operation between the United States and Africa was opened on August 23, 1963 with a telephone conversation between U.S. President John F. Kennedy and the Nigerian Prime Minister, Sir Abubakar Tafawa Balewa. Kennedy repeated the remark "What hath God wrought!" made on May 24, 1844 when Morse opened the electrical telegraph line between Baltimore and Washington.

Syncom III was launched on August 19, 1964, and reached geostationary Earth orbit, giving coverage to the Pacific Ocean region and making it possible to broadcast the opening ceremony of the Olympic games in Tokyo on October 10 live on TV screens in the United States. Live transmission of the games would technically have been possible and took place in Canada during the two weeks of the games, but was prohibited in the United States by NBC's exclusive rights to carry videotapes of the games by plane from Tokyo via Anchorage (for plane refueling) to Seattle. *Syncom III*, which weighed 39 kg, had to be very light and simple to enable launching into GEO, so it could not transmit video and audio simultaneously. Audio from Tokyo was therefore dispatched via coaxial submarine cable to California; at Seattle it was combined with the video and, with commercials added, distributed via AT&T's terrestrial network to NBC's nationwide TV network.[10] This was completely different at the last Olympic games of the twentieth century in Sydney, Australia, in 2000. A network of 35 geostationary satellites transmitted 3400 hours of live video of 28 different sports from 37 venues to 3.7 billion spectators in 220 countries.

The average operational lifetime of geostationary telecommunications satellites increased from a few years in the early 1970s to 15 years at the end of the twentieth century. The total typical weight of a geostationary satellite at launching, including onboard fuel, amounts to 3500 to 4000 kg, of which the payload[11] is at most 10%. The total RF output power is typically 2.5 to 3.5 kW. The typical solar array end-of-life output power of a telecommunications satellite increased from 400 W in 1970, to 1000 W in 1983, to 2500 W in 1990, and to 9000 W in the year 2000. About 250 geostationary telecommunication satellites were in orbit at the end of the twentieth century. Figure 27.7 shows a satellite made ready for launch.

The geostationary-satellite market was shared at the end of the twentieth century by fewer than 10 manufacturers in the following approximate relation[12]: Boeing, 37%; Lockheed Martin, 18%; Space Systems/Loral, 12%; Alcatel, 10%; Astrium,[13] 8%; and others, 15%. The regional distribution of GEO-satellite services was approximately as follows: Asia Pacific, 24%; North America, 22%; Europe, 22%; Atlantic Ocean, 12%; Indian Ocean, Pacific Ocean, and Latin America, each 5%; the Middle East, 4%; and Africa, 1%.

27.3 SATELLITE LAUNCHING

Satellite launching is a crucial action in the creation of a satellite system. Launching into LEO or MEO can be done in one step. To launch a satellite into GEO usually requires three successive orbits (Figure 27.8). The satellite is first transported into a circular parking orbit. Ignition of the rockets in the third stage brings the satellite into an elliptical *geostationary transfer orbit* (GTO), also called the *Hohmann ellipse*, after the inventor of multiorbit launching. When the satellite is at the apogee of the GTO, an apogee kick motor of the satellite is fired to bring the satellite through near-GEO into geostationary orbit.

The vehicles used for satellite launching are divided in two categories: reusable launchers and expendable launcher vehicles (ELVs), which are destroyed in space during the course of their mission.

Reusable launchers are the U.S. Space Shuttle and the Russian Energia. These *space transport systems* launch one or, depending on size, several satellites into a low Earth parking orbit at an altitude of 300 to 500 km. The satellite then must use its

---

[10] The *New York Times* Saturday evening edition of October 10, 1964 reported enthusiastically: "Live television coverage of this morning's opening of the Olympic Games in Tokyo was a superlative of quality, a triumph of electronic technology that was almost breathtaking in its implication for global communications."

[11] The payload or communication subsystem of a satellite is the actual radio-relay station, consisting of antennas, transponders (receiver/downconverter/transmitter), and high-power transmitting amplifier.

[12] The data on worldwide production and service distribution were published in *Via Satellite* in July 2001, in "Via Satellite's Global Satellite Survey," by Rob Fernandez.

[13] Astrium is a new company, established in May 2000 as a merger of the space businesses of Aerospatiale, Matra, DaimlerChrysler Aerospace, and BAE Systems. It has about 7500 employees.

Figure 27.7 Satellite HS 601 HP made ready for launch, 1999. (Courtesy of Boeing Satellite Systems.) See insert for a color representation of this figure.

own propulsion for injection into the GTO, near-GEO, and finally, into its orbital slot.

Expendable launcher vehicles such as the European Ariane, the U.S. Delta and Atlas, the Chinese Long March, and the Japanese H-2 rockets place one or two satellites successively into parking orbit, transfer orbit, and near-GEO. The U.S. Titan and the Russian Proton expendable launchers have an upper stage that is ignited when it is at apogee in the GTO and there injects the satellite into near-GEO.[14]

The launching of satellites requires special launching centers adequately located for minimum launching cost. Currently, four such centers exist. The first was established at Cape Canaveral at 28.5°N in the United States in 1950. The second was established in the USSR at Baikonur Cosmodrome near Leninsk (48°N), Kazakstan, in 1955. The third was established in 1979 almost on the equator at 5.2°N at Kourou in French Guiana. The fourth was established in China in 1985 at the Taiyuan satellite launch center 640 km southwest of Beijing. Japan used a launching site on

---

[14] On November 26, 2001, the upper stage of a Proton ELV failed and an Astra-1K satellite (the world's largest satellite for transmitting 112 TV channels to 90 milllion people) stayed in a 300-km orbit, from which it was brought down and disappeared in the Pacific Ocean on December 10, 2002.

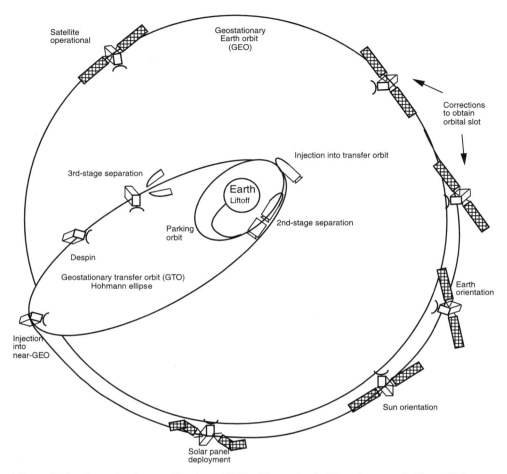

Figure 27.8 Launch of a satellite into GEO. (From A. A. Huurdeman, *Guide to Telecommunications Transmission Systems*, Artech House, Norwood, MA, 1997, Figure 7.8; with permission of Artech House Books.)

Tanegashima Island off the southern coast of Kyushu up to 1992. The Baikonur Cosmodrome Leninsk became a major city, with 150,000 people living in the middle of nowhere. *Sputnik* was launched there in 1957. After the Cold War this military rocket center was converted to a commercial satellite-launching center under the name Khrunichev Space Center.[15] In 1993, the *International Launch Service* (ILS) based at San Diego, California, was formed as a joint venture of Lockheed Martin Corporation[16] with the Russian companies Khrunichev State Research and Production Space Center and RSC Energia. ILS provides commercial satellite launch ser-

---

[15] The Khrunichev Space Center is named after Mikhail Khrunichev, a former minister of aviation. The place started as an automobile factory on the order of Tsar Nicholas II in 1916. The production capacity was leased to Germany in the 1920s for the production of the first metal-covered U-20 airplanes. Proton rockets have been produced there since 1962. The center had a workforce of 18,000 people in 1995.

[16] The U.S. company, Lockheed Martin Corp., was formed in March 1995 as a merger of the companies Martin Marietta and Lockheed.

vices on the U.S. Atlas rocket from Cape Canaveral and on the Russian rocket Proton from the Khrunichev Space Center in Kazakstan. At the end of the twentieth century some 50,000 people lived at the Khrunichev Space Center. The Proton rocket dates back to 1965; since then it has had over 200 launches, with a success rate of 96%. The latest version of Proton, Proton-M/breeze-M, has a launching capacity of 21,000 kg into LEO, or 5500 kg into GTO, or 2920 kg directly into GEO. The latest version of Atlas, the Atlas-3, is capable of delivering payloads of between 3400 and 4060 kg into GTO.

The launching center at Kourou was the first center especially implemented for launching of commercial satellites. This spaceport, located near the equator, allows maximum advantage to be taken of Earth's rotation to impart velocity to the satellite and minimize the energy required for launching and maneuvering into orbit. The center was established by the *European Space Agency* (ESA, formed in 1975) and belongs to *Arianespace*.[17] Arianespace developed and produced a series of launchers called Ariane 1 to 5. Ariane 4 was the most successful commercial satellite launcher, having had 50 consecutive successful launches within two years, by the time the first commercial Ariane 5 launch took place on December 10, 1999. Over 80% of the Ariane 4's launched lifted two satellites with a total weight up to 5000 kg. Ariane 5 was designed to lift two satellites with a total weight of 12,000 kg into GTO.[18] Figure 27.9 shows successive versions of Ariane 4 and Ariane 5.

Commercial satellite launching from the Chinese center at Taiyuan is made by the *China Great Wall Industry Corporation* with their *Long March* rocket. This rocket, in seven different versions, can carry a payload of 2500 to 5000 kg into GTO.

An interesting new concept for launching of satellites was introduced at the end of the twentieth century by the Sea Launch Company, LLC,[19] founded in 1995. Instead of launching from a terrestrial site, Sea Launch Co. launches satellites from a floating platform which is currently located at the equator 154°W on the Pacific Ocean between Hawaii and the Christmas Islands. Sea Launch Co. takes advantage of the maximum possible boost from Earth's rotation by launching from the equator, without requiring a costly infrastructure on a terrestrial site.

The launching center consists of two vessels: the *Odyssey*, a launch platform transformed from a North Sea oil platform; and the command and rocket-assembly ship *Sea Launch Commander*, a modified roll-on, roll-off cargo vessel, providing the launch control facilities and accommodations for 240 crew members, customers, and other visitors. *Odyssey* is the world's largest self-propelled semisubmersile vessel. Before launching, the platform legs are lowered some 20 m into the sea, which sta-

---

[17] Arianespace, based at Evry (near Paris), was formed on March 26, 1980 by 36 leading European manufacturers in the aerospace and electronics industry, together with 13 major European banks and the French National Space Agency (CNES, Centre Nacional d'Etudes Spatiales).

[18] The first "10-ton" Ariane 5 was launched on December 11, 2002, but after 186 seconds in launching mission, the launcher demonstrated erratic behavior, and at 456 sec in mission it was destroyed (including the payload: Eutelsat's *Hot Bird 7* and an experimental French satellite called *Stentor*) at an altitude of 69 km and a distance of 800 km of the coast of French Guiana.

[19] The multinational Sea Launch Company belongs to the American Boeing Commercial Space Co. of Seattle (40%), which provides the spacecraft integration and payload fairings; the Russian RSC-Energia of Moscow (25%), which provides the upper stage of the launch vehicle; the Norwegian Kvaerner Maritime A.S. of Oslo (20%), which provides the vessels; and the Ukrainian KB Yuzhnoye/PO Yuzhmash of Dnepropetrovsk (15%), which provides the first two stages of the launch vehicle.

Figure 27.9 Four successive versions of Ariane 4 and (at the right) Ariane 5. (© Arianespace, 1995, D. Ducros.)

bilizes the launch platform, and a dynamic positioning system maintains the precise coordinates for blast-off. The launching rocket, derived from the Russian Zenith (for the first and second stages) and Proton (for the third stage) rockets, currently has the capacity to place into GTO satellites with a total weight up to 5000 kg. Figure 27.10 shows the Odyssey being prepared for its inaugural launch, and Figure 27.11 shows the inaugural launch on March 27, 1999.

By the end of the twentieth century a total of 376 commercial satellites had been launched into GEO at a rate of 16 in the period 1963–1970, 33 in the period 1971–1980, 111 in the period 1981–1990, and 216 between 1991 and 2000. By that time, 233 of the satellites were still in operation. The total transmission capacity of those satellites is slightly below 640 Gbps,[20] which is the capacity of the transatlantic submarine cable TAT-14, put into operation in March 2001 (Section 28.3.1). Despite the relatively low transmission capacity of satellites, optical fiber and satellite transmission play complementary roles. Whereas fiber offers practically unlimited bandwidth but limited geographical reach, satellites offer limited bandwidth but an essentially limitless reach. It is estimated that almost half of the world's countries are dependent on satellites for international connectivity.

[20] According to a *TeleGeography* report entitled "International Bandwidth 2001."

Figure 27.10   Ocean satellite launching platform, 1999. (Courtesy of Sea Launch.) See insert for a color representation of this figure.

A launch into GEO costs typically $90 million. GEO satellite launching at the end of the twentieth century was roughly as follows[21]: with Ariane, 55%; with Atlas and Delta, 26%; with Long March, 10%; and with Proton, 9%.

27.4 SATELLITE TRANSMISSION SYSTEMS

The first telecommunications satellites were used primarily for long-distance continental and intercontinental broadband transmission, for narrowband long-distance operation as a replacement for HF communication, and for direct broadcast of TV. The introduction of optical fiber with very high transmission capacities at a relatively low cost in the 1970s and 1980s shifted the application of satellite communication toward thin-route point-to-multipoint (P-MP) systems and to TV- and data-distribution systems. The ITU reserved frequency bands for satellite transmission are summarized in Table 27.2.

In addition to global operating satellite systems, regional and domestic satellite systems were implemented. With the evolution toward satellites with higher radiated power and smaller Earth stations, satellite transmission also became possible for moving objects. Global mobile personal communication by satellite systems (GMPCS) introduced in the late 1990s with satellites mainly in LEO but also in GEO and MEO provide satellite service to individual persons in any part of the globe. At the end of the twentieth century, special broadband satellite systems were being implemented to enable direct multimedia communication at business and residential premises.

---

[21] Acording to the German scientific magazine *Bild der Wissenschaft* of February 1999.

Figure 27.11 Sea Launch Co.'s inaugural launch, March 27, 1999. (Courtesy of Sea Launch.)

27.4.1 Global Satellite Systems

The promising prospects of satellite transmission and the obvious necessity to establish agreements primarily with foreign government-owned telecommunications authorities induced the U.S. government to put the subject of communication satellites under government control. The U.S. *Communications Satellite Act*, signed by President Kennedy on August 31, 1962, became the most important communications legislation in the United States since the Communications Act of 1934. The act provoked the foundation of the *Communication Satellite Corporation* (Comsat) on February 1, 1963 with the task of commercial development of communication satellites. Satellite operation, however, should not be a national but a global affair. In accor-

TABLE 27.2 Satellite Transmission Frequency Bands

Band	Frequency (GHz)	Band	Frequency (GHz)
L	0.4–0.46	Ku	12.4–18
	0.6–0.8	K	18–26.5
S	1.5–2.7	Ka	26.5–40
C	3.4–8.4	Q	33–50

dance with Resolution 1721 of the United Nations, therefore, *International Tele-communications by Satellite* (Intelsat), with its headquarters in Washington, DC, was founded in July 1964, Comsat being one of the major consortium members.[22] Intelsat began as an international commercial cooperative of governments and tele-communications organizations with a charter membership of 11 nations (the *signatories*), which together owned a global communications satellite system.

Implementation of the Intelsat network began in April 1965 with the launch of *Intelsat I*, also called *Early Bird*. The satellite, manufactured by Hughes Aircraft Co., was located over the Atlantic Ocean, so that commercial intercontinental telephone and television transmission became possible between the United States and Europe. *Intelsat II* was launched on October 26, 1966 to serve the Pacific Ocean region. It did not reach its exact orbit, and a second satellite was launched on January 11, 1967. *Intelsat III* was launched in 1969 to cover the Indian Ocean region, so that world-wide service could start in January 1969. At the end of the twentieth century, Intelsat had a membership of 144 nations and operated 20 satellites.[23]

A second global satellite system, InterSputnik, was founded at Moscow in November 1971. The signatory states were Afghanistan, Bulgaria, Cuba, Czechoslovakia, East Germany, Hungary, Laos, Mongolia, Poland, Romania, South Yemen, Vietnam, and the USSR. Those then-Communistic countries had not joined the Intelsat system. Although a signatory of Intelsat, Nicaragua also joined Inter-Sputnik. InterSputnik used the Soviet communications satellites *Molniya, Raduga*, and *Gorizont*. Considering that all those countries, and especially the USSR, are located north of the equator, instead of a GEO, *Molniya* uses a highly elliptical orbit (HEO) with an apogee of about 40,000 km over the northern hemisphere, and a perigee of about 1000 km over the southern hemisphere. With this constellation, the orbit period is 12 hours, of which, in compliance with Kepler's second law, about 11 hours is visible over the northern hemisphere. Several satellites are used in the orbital plane for the *Molniya* system, and the Earth stations for this system usually have two antennas, which hand-over between satellites automatically.

27.4.2   Regional Satellite Systems

The major regional operating systems are described briefly.

Arabsat   The *Arab Satellite Communications Organisation*, Arabsat, with head-quarters in Riyadh, Saudi Arabia, was founded in 1976 as a consortium of 21 Arab league states serving the Middle East. Arabsat's first satellite, produced by Aero-spatiale and Ford Aerospace, was launched on February 8, 1985.

---

[22] From the beginning, Comsat possessed the sole rights to sell Intelsat services in the United States; this monopoly ended in 1999. In the same year, Lockheed Martin Corporation acquired Comsat, including its stake in Intelsat, and integrated it into *Lockheed Martin Global Telecommunications* (LMGT). The name *Comsat* was retained for the non-US operations. LMGT announced in December 2001 that it would quit global telecommunications and sell its Comsat department to Telenor of Norway.

[23] Intelsat, too, had to adapt to the changing world of telecommunications. In March 1998 it divested six satellites to a wholly commercial affiliate, called *New Skies*, based in Amsterdam. In a historic meeting on November 17, 2000, the Intelsat Assembly of Parties unanimously approved privatizing Intelsat by July 18, 2001.

Eutelsat   The *European Telecommunications Satellite Organization* (Eutelsat) was created in 1977 under an agreement between the PTTs of the 17 CEPT countries. Eutelsat has its headquarters in Paris and now has nearly 50 signatories. Eutelsat provides fixed and mobile services, initially confined to Europe, with the first five satellites *Eutelsat I-F1* to *5*, launched between 1983 and 1988.[24] A second generation of satellites with a design life of nine years, *Eutelsat II F1* to *5*, was launched between 1990 and 1992, which included coverage of eastern European countries up to Armenia and central Siberia as well as large areas of North Africa and the Middle East. Internet service and direct-to-home TV (DTH-TV) were introduced with the satellites *W1* to *4* launched in the late 1990s. With those satellites the coverage was extended to all of Africa and part of the Indian Ocean. In 1995, launch began of a series of six *Hot Bird* satellites colocated in the same orbital position at 13°E for broadband and TV services. Service with North America via *Eutelsat II-F2* was begun in 2000. In the same year, an "always-on broadband anywhere" service was introduced. At the end of the twentieth century, coverage started on the "Marco Polo route," with a steerable beam up to India and Thailand and a fixed beam on China with a Russian-made satellite, *Sesat* (Siberian–European Satellite).

Rascom   The *Regional African Satellite Communications Organisation* (Rascom) was founded in 1986 with the objective to establish an African satellite communications system. The existing transponders already leased or purchased by African countries from Intelsat were pooled with effect from January 1994. Rascom decided in July 1997 to launch the first African satellite in a build–operate–transfer (BOT) partnership. Under the partnership, Alcatel is to supply and operate a satellite with 36 transponders by 2002 and a second satellite three years later. It is planned to serve 500,000 terminals by a combination of VSATs and wireless local loop systems.

SES   The *Société Européenne des Satellites* (SES), based at Château de Betzdorf, Luxembourg, was founded in 1985 as the first private European satellite operator. In 1989, SES established its ASTRA network with multichannel cross-border DTH-TV broadcasting by satellite. ASTRA introduced the concept of satellite colocating in a common orbital position, at 19.2°E, in 1991. This enabled the reception of various TV channels with a fixed dish with only one single feed. Currently, ASTRA-Net is being introduced, which offers multimedia transmission at a capacity of 2 Mbps at interactive satellite terminals connected with high-speed PCs. At the end of the twentieth century, ASTRA transmitted over 1000 TV and radio channels that reached about 90 million households in Europe.[25]

PanAmSat   *PanAmSat* (PAS), with the satellite *PAS-1* launched in 1988, was the first American privately owned satellite system. It belongs to the American com-

---

[24] Although developed for a design life of seven years, *Eutelsat I-F5* was decommissioned in May 2000 after 12 years of operation. In line with the general practice, it was put into its "graveyard" orbit some 150 km above GEO.

[25] SES has an ownership of 50% in NSAB (Nordic Satellite AB) with the satellites *Sirius 2, 3*, and *W*; of 34.13% in AsiaSat with three satellites; and of 19.99% in StarOne (formerly Embratel Satellite Division, Brazil) with the satellites *Brasilsat A2* and *B1* to *4*. The SES Group claims to reach 79% of the world's population on four continents.

pany Alpha Lyracom. *PAS-1* serves Latin America; *PAS-2*, launched in 1994, serves the Asia-Pacific region; *PAS-3*, launched in 1996, serves the Atlantic Ocean region; *PAS-4*, launched in August 1995, to orbit above the Indian Ocean serving Africa, Europe, the Middle East, and South Asia, enhanced the original Latin American system into the first private global satellite network. PanAmSat increased its fleet of satellites to 21 on April 18, 2000, when its high-powered *Galaxy-IVR* spacecraft was launched by an *Ariane-4* at Kourou. This satellite, covering North America, carries advanced services such as AT&T's Headend in the Sky digital cable service, with 140 digital video channels throughout the United States; the AOL Plus via DirecPC interactive direct-to-consumer PC service offered by Hughes Network Systems; and PanAmSat's own Galaxy-3D service, offering end-to-end digital video, audio, and data transmission in major U.S. cities.

GE Americom   GE Americom, Princeton, New Jersey, was formed in 1998 as a merger of GE Spacenet and Gilat, the world's largest VSAT company. It served mainly North America, but in 2000 it acquired Columbia Communications Corporation, with worldwide coverage.[26]

AsiaSat   AsiaSat, founded in 1988 with headquarters in Hong Kong, was the first privately owned regional satellite operator serving Southeast Asian countries, enabling them to develop their domestic systems for national TV programs, rural telephony, and private networks. The first satellite, *AsiaSat 1*, manufactured by Hughes Aircraft Company, was launched in April 1, 1990 with a northern beam covering China, Korea, and Japan and a southern beam stretching from Southeast Asia to Saudi Arabia. *AsiaSat 2*, launched in November 1995, was manufactured by Lockheed Martin Astro Space for an operational life of 13 years. It serves 53 countries and regions with 66% of the world's population in a triangle formed by eastern Europe, Japan, and Australia. On March 21, 1999, *AsiaSat 3*, with a planned operational life of 15 years, was launched to replace *AsiaSat 1* and to extend coverage to Australasia.

Hispasat   Hispasat, founded by the Spanish government–owned companies Telefonica and Retevision, is basically a domestic system but with additional coverage in most of France, Italy, Portugal, the Canary Islands, and the Spanish-speaking part of Latin America. The first satellite was launched on September 10, 1992, the date of the 500th anniversary of Columbus's first landing in the Americas. Only TVRO service was included to Latin America. The second satellite, launched one year later, also provided voice and TV services with Latin America.

Bolivarsat   Bolivarsat, the latest regional satellite system in Latin America, was formed in 1999 by Andesat S. A. and Alcatel to serve the five Andean Pact countries (Bolivia, Colombia, Ecuador, Peru, and Venezuela) and other countries up to northern Canada, with Internet, broadband, and rural telephony. It replaces a previously planned Simon Bolivar system.

---

[26]GE Americom was acquired 100% by SES in March 2001 and emerged as SES Global S.A., which then controlled roughly 20% of commercial GEO communication.

Tongasat   *Tongasat* is a private system founded in 1990 by Matt Nilson, a U.S. businessman who after retirement settled in the kingdom of Tonga. Located in the South Pacific east of the Fiji Islands, Tonga consists of 169 small islands, of which about 36 are permanently populated by slightly over 100,000 inhabitants. In a highly publicized move, Nilson, obtained approval from the *International Frequency Registration Board* (IFRB, the ITU radio spectrum allocation office; since 1992, the Radio Regulations Board) in 1991 to use six orbital slots over the Asia-Pacific region providing full coverage from Africa to North America. One year later, Tongasat made orbital slots a commodity by authorizing the U.S. company *Unicom Satellite Corporation* to operate a satellite system in two of those slots.

Europe*Star   *Europe*Star Ltd.* is the most recent European regional system. It was founded by Alcatel (51%) and Loral Space and Communications (49%) and launched its first satellite on October 29, 2000. It is part of the Loral Global Alliance, which includes Loral Skynet, Satmex, Skynet do Brasil, and Stellat.

Eurasiasat   *Eurasiasat* as the latest regional company started commercial operation on February 1, 2001. Türk Telekom and Alcatel founded the company in 1996 as a follow-on of the Türksat satellite system. It operates *Türksat-1B* and *1C* and *Eurasiasat-1* and covers central Asia and Europe, mainly for DTH-TV, Internet, and data services.

At the end of the twentieth century there were 46 regional satellite operators: 20 in Asia-Pacific, 9 in Europe, 9 in North America, 5 in Africa and Middle East, and 3 in South America.[27] The ranking of the major global and regional operators based on service revenues in the year 2000 was as follows[28]:

Operator	Revenue (millions of dollars)	Number of Satellites
Intelsat	1097	20
PanAmSat	1024	21
SES Group	735	22
Eutelsat	651	15
Loral	591	7
GE Americom	522	12
Inmarsat	417	9

27.4.3 Domestic Satellite Systems

The world's first domestic communications satellite, *Anik I* (Eskimo term for little brother), was placed into geostationary orbit by Telesat of Canada in 1972. *Anik I* was produced by Hughes Aircraft Co. and launched by NASA. Telesat is a subsidiary of the *Bell Canada Enterprises Inc.* (BCE). At the end of the century, Telesat covered the Americas with 12 satellites through more than 3500 Earth stations.

[27] According to information from "Euroconsult 2001," published in *ITU News*, December 2001.

[28] According to company annual reports as evaluated by *Via Satellite* and published in its issues of September (in "Europe's Pursuit of Internet and Broadband Revenues") and October 2001 (in "Connecting the Four Corners of the World").

The second domestic satellite system was installed by Western Union in the United States. *Westar I*, with a transmission capacity of 600 Mbps, began service on July 15, 1974.

The third domestic satellite system was the Indonesian domestic satellite system *Palapa*. Indonesia, with a population of nearly 200 million living on at least 3000 of the 13,700 islands spread across 7000 km, is a country "that can only be seen from space," thus is predestined for satellite operation.[29] *Palapa 1* was launched in 1976 followed by *Palapa 2* in 1977. Beyond being a domestic system, *Palapa* serves the neighboring countries Brunei, Malaysia, Papua New Guinea, the Philippines, Singapore, Thailand, and now Vietnam.

As a novelty, in Indonesia secondhand satellites have recently been sold and reused. In 1992, when the *Palapa* satellites were replaced by new satellites, the decommissioned satellites were bought for low-cost services in the Pacific Rim by a new satellite operating company, *Pasific Satellite Nusantara* (PSN). For this purpose the decommissioned satellites were moved out of their position at 118°E to 134°E, from which position optimal coverage is provided for the Pacific Rim. Moreover, to extend the useful lifetime of the satellites, they were placed in an onboard fuel-saving inclined orbit.[30]

At the occasion of Indonesia's celebration of 50 years of independence mid-1995, a second domestic operator, *Indostar*, started a satellite network for nationwide digital audio broadcasting and TV beamed directly to the population. Low-cost locally made receivers and analog TV-ROs in the $100 range now bring radio and TV to the 200 million population, of which about one-third could not receive TV and one-sixth were deprived of reliable radio broadcast reception.

Thailand established a domestic satellite system, Thaicom, in 1991. Thaicom is owned by the *Shinawatra Satellite Public Company*, which was founded by Thaksin Shiniwatra, then deputy prime minister. *Thaicom-1* was launched in December 1993 and enabled the introduction of DTH-TV transmission in Thailand. Currently, three *Thaicom* satellites are in GEO for worldwide operation.

Other domestic satellite systems in Asia are *Insat* (India), *Koreasat*, and those operated by Cable and Wireless Hong Kong Telecom (CWHKT) and NTT Satellite Communications of Japan. In Japan the first satellite, *Ohsumi*, was launched in 1970.

Aussat, an Australian domestic satellite operator established in 1981, launched its first satellite in 1985.

In Africa, *Nilesat* operates in Egypt DTH-TV satellites that were launched by *Ariane 4* in 1998 and 2000. *Worldspace* launched its first Afrisat in 1998 to provide

---

[29] According to a legend from the twelfth century, a great Indonesian soldier vowed that he would not eat the delicious *palapa* fruit until the islands of his kingdom were united. The advent of satellite communications made it possible to unite the 13,700 islands, and, appropriately, Indonesia's domestic satellite system was named *Palapa*.

[30] Inclined orbit operation is obtained when the station keeping is scaled back to east–west (longitudinal) maneuvers, which requires typically 5% of the annual fuel consumption, and by eliminating the north–south (latitudinal) maneuvers. This results in a 95% reduction in fuel consumption but might require modifications at Earth stations to track the satellite inclination as well as additional coordination in case of shared radio-relay frequency use.

Inclined orbit operation became popular in the early 1990s when an entire generation of satellites reached its original end-of-life dates. In 1994 a total of 117 satellites were in inclined orbit out of a total of 201 GEO satellites, including 42 commercial communications satellites.

DTH radio service. An *Orion 2* satellite from Loral started international service in 2000.

In Latin America, the first domestic satellite system, named *Morelos*, was installed in Mexico in 1985. A second generation of Mexican satellites, *Solidaridad 1* and *2*, was launched in 1993–1994. In addition to Mexico, the system also serves the Spanish-speaking people in the southern part of the United States, the Caribbean, and Central America. In 1999, *Satmex*, the privatized operator, launched its *Satmex 5*, which serves the Americas with Internet.

In Brazil, *Brasilsat*, in 1985, launched its first two satellites, which were produced by Aerospatiale and Ford Aerospace. Brasilsat's fleet was acquired by MCI in 1999.

In Argentina, the first plans for a government-owned domestic satellite system were drafted in 1969. It took until 1997, however, for the private satellite company *Nahuelsat S.A.*, 28% owned by GE Americom, to launch *Nahuel I* with coverage of the entire Americas.

27.4.4 Mobile Satellite Systems

The first satellite system, with three satellites in GEO, serving ships on the oceans was provided by the U.S. *Marisat*, a joint venture of Comsat with RCA, ITT, and Western Union International. The first commercial maritime telephone call via satellite was made on July 6, 1976 from a seismic ship, the *Deep Sea Explorer*, which was off Madagascar in the Indian Ocean, to its home office in Oklahoma. Around the same time, Intelsat started a Maritime Communications System (MCS). In 1973 the *International Maritime Organisation* (IMO), then known as the *Inter-Governmental Maritime Consultative Organisation* (IMCO), decided to establish an international maritime satellite system. As a result, the *International Maritime Satellite Organisation*, Inmarsat, was founded in July 1979 as an internationally owned cooperative, to provide *mobile* (initially only maritime) worldwide satellite communications. Twenty-six states, including some of the InterSputnik signatories, originally signed the Inmarsat Convention and Operating Agreement, presently consisting of 81 member countries.

Inmarsat began maritime service on February 1, 1982 with three leased *Marisat* satellites launched in 1976. A little later, two *Marecs* satellites, launched in 1982 and 1984, were leased from the European Space Agency, and further capacity was leased from Intelsat. In 1990, Inmarsat launched its first own satellite, *INM2-F*, manufactured by a British Aerospace consortium; it had 250 voice circuits.

Within 10 years, Inmarsat served 14,000 ships and other offshore Earth stations with 0.8- to 1.2-m parabolic antennas, which by means of 20 coast Earth stations in 14 countries were connected with the public-switched telephone network, the telex network, and with packet-switched data networks. In the early 1990s, Inmarsat services also became the standard solution for worldwide land mobile satellite services: for example, for long-distance trucking, fleet control, trains and buses, journalists, explorers, and adventurers. In 1997, the 100,000th Inmarsat terminal was commissioned, and 160,000 terminals were in use at the end of 1999.[31]

---

[31] Although originally constituted for mobile operation, the Inmarsat network is also used for fixed services. As an example, a rural pay phone system was started in India in 1997 whereby 1000 rural communities located more than 25 km from the nearest landline telephone are provided with an Inmarsat-based village public telephone.

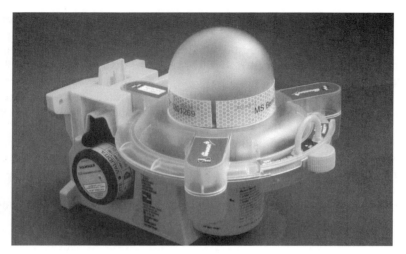

Figure 27.12   Inmarsat E emergency position indication radio beacon. (Courtesy of Navtec, Germany. Photo from Ingo Wenske.)

Inmarsat began aeronautical satellite services, called *Inmarsat Aero*, in 1987, when the first in-flight telephone call was made from a Boeing 747 of Japan Airlines. By the end of the twentieth century, Inmarsat Aero was serving 3000 aircraft from over 80 airlines.

Established to serve the marine community, Inmarsat has since evolved to become the only provider of global satellite communications for commercial and distress and safety applications, at sea, in the air, and on land. Between February 1, 1992 and January 31, 1999, the Global Maritime Distress and Safety System (GMDSS) had been implemented progressively worldwide to replace the Morse SOS distress signal. GMDSS gives ships, wherever they are, the capability to transmit a distress signal to shore-side authorities as well as to ships in the immediate vicinity. Inmarsat was bound by international treaty to develop and maintain this system as part of their public service obligations. The Inmarsat E distress alerting service was introduced for communication with *Emergency Position Indicating Radio Beacons* (EPIRBs), to be used by vessels in trouble. Figure 27.12 shows the world's smallest EPIRB, introduced in 1999. This EPIRB transmits a message automatically when a vessel sinks or the unit is thrown overboard. If time allows, additional preset information about the emergency can be sent by means of remote control of the EPIRB. The beacon includes a GPS (Global Positioning System[32]) receiver for exact location indication.

Inmarsat was the first international treaty organization to be transformed into a commercial company, on April 15, 1999. A residual international cooperative called the *International Mobile Satellite Organisation* (IMSO) remained in place to oversee public service obligations, including the GMDSS service, which will be continued by the new Inmarsat as part of its public obligations.

---

[32] The first global positioning system was developed and implemented jointly by the U.S. Navy and the U.S. Air Force in 1973 to permit the determination of position and time for military troops and guided missiles. It is free for civilian use in a slightly reduced version. Currently, 21 satellites plus three spares are in orbit at an altitude of 20,000 km. A GPS receiver uses those satellites to triangulate its location in latitude, longitude, and altitude, and to calculate its direction and speed.

Around 1985 the Russian Volna network also provided mobile satellite service. In 1991, Alcatel Qualcomm started Euteltracs, a two-way message exchange and position-reporting service for vehicles offered via Eutelsat. Euteltracs is based on the OmniTrac system, which Qualcomm introduced in the United States in 1988. The positioning operates with an accuracy of 300 m for any point in Europe, which makes it suitable for fleet management of road transport. The service extended to maritime fleets in 1998. Within 10 years, Euteltracs was used by 350 vessels and over 20,000 trucks.

By the mid-1990s, additional land mobile satellite services were introduced in Australia, Canada, Mexico, and the United States. All those systems used satellites in GEO.

27.4.5   Global Mobile Personal Communication by Satellite

A concept of direct satellite service to individual persons emerged in the 1980s. At that time fewer than 3 million cellular radio terminals were in operation worldwide, and even optimistic planners did not dare to forecast that this amount would increase beyond 100 million cellular terminals by the end of the twentieth century (in fact, there were 740 million). It was assumed that a maximum of 20% of the world's landmass would be covered by terrestrial cellular radio, serving, at most, 60% of the world's population. A potential market was therefore seen for satellite-based cellular radio for the other 40% of the world's population, of which some 30 to 40 million persons were expected to become subscribers.

Starting in 1985, the ITU-CCIR elaborated a concept for a *Future Public Land Mobile Telecommunications System* (FPLMTS) covering terrestrial cellular radio and mobile satellite operation. The project was taken over by ITU-T and in 1996 separated into its basic components: terrestrial cellular radio into an *International Mobile Telecommunication System* dubbed IMT-2000 (Section 32.5), and the satellite content, which was named *Global Mobile Personal Communication by Satellite* (GMPCS). The GMPCS systems were conceived to interwork with terrestrial cellular radio networks. They should complement the cellular radio systems wherever there is no terrestrial cellular coverage and serve international travelers in regions where the terrestrial cellular networks are not compatible with the traveler's home system. Dual-mode handsets, possibly also in earphone or wristwatch versions, should then automatically select the best available cellular or satellite network. To let it happen, new networks, partly developed for President Ronald Reagan's Star Wars Strategic Defense Initiative (SDI), were modified for personal communications and announced for operation within the next few years under the names *Aries, Celsat, Ellipso, Globalstar, ICO, Iridium, Odyssey, Planet 1, Spaceway,* and *Tritium.* Such mobile satellite systems had been predicted in another little publicized speech that Arthur C. Clarke made at the UN General Assembly in May 1983. Referring to the emerging digital wristwatches, Clarke predicted: "The symbols that flicker across those digital displays now merely give time and date. When the zeroes flash up at the end of the century, they will do far more then that. They will give direct access to most of the human race, through the invisible networks girdling our planet."

Celsat and Tritium disappeared at an early stage. The most impressive of the above-mentioned networks, almost making Clarke's prediction coming true, was the Iridium network.

Iridium   *Iridium* was conceived in 1987 by Raymond J. Leopold, Bary Bartiger, and Ken Peterson of *Motorola Satellite Communications* as a commercial global wireless telecommunications system using handheld telephones. Iridium Inc. was formed by Motorola in 1990 as an international consortium of telecommunications and industrial companies funding and implementing the Iridium system. The name *Iridium* was chosen because the initial system was based on a system consisting of 77 satellites circling the globe, which is similar to the 77 electrons in the element iridium. In 1992 the number of satellites was reduced to 66. Implementation of the Iridium project, at a cost of $5 billion (of which $3.4 billion was for development), became the biggest international telecommunications undertaking of the twentieth century. The board of directors of Iridium comprised 27 international recognized telecommunications pioneers from 14 countries, most of them also representing a major investor in Iridium. Substantial technology pioneering was required, and international cooperation was found with over 25 potential industrial partners from eight countries.

After 10 years of development, preparation, production, and installation in 14 countries, the first satellites were launched on May 5, 1997. A McDonnell Douglas Delta II rocket carrying five Iridium satellites lifted off from Space Launch Complex 2 West at Vandenbergh Air Force Base, near Lompoc, California. The satellites were placed in a circular transfer orbit, from where each satellite was moved to its operational slot by its onboard propulsion subsystem. Within one year, a further eight launches were made with Delta II (40 satellites), and three launches each with a Proton (21 satellites) and the Long March 2C/SD (six satellites), ultimately with 72 satellites in orbit for the constellation of 66, plus six spares. Service was announced to start on September 23, 1998, but it had to be postponed for lack of handsets and the unsatisfactory transmission quality of the system. When service was finally begun on November 1, only 2000 subscribers had been able to acquire a handset. Once additional handsets arrived, interest in the $3000 brick-size handsets had disappeared. Instead of attracting the forecasted 400,000 subscribers in the first operational year, only 10,294 customers had signed up by the end of March 1999, when the international per minute charges were reduced from $7 to $3. On August 13, 1999, with only 55,000 subscribers, Iridium filed for bankruptcy protection and, for lack of a rescue solution, stopped commercial service on March 17, 2000.

During the 10 years of development, Iridium was overtaken by the enormous, largely unexpected penetration of terrestrial cellular radio, with worldwide roaming service, leaving a very small market for the luxury of personal satellite communications service "at any time anywhere in the world." In December 2000, shortly before Motorola would have started a 14-month period of deorbiting the satellites so that they would burn up in Earth's atmosphere, the Iridium system was bought for $25 million. The buyers were a group of investors headed by Ed Staino, former chief executive officer of Iridium (up to April 22, 1999), and Dan Colussy, former president of Pan Am Airlines and Canadian Pacific Airlines. They formed a new company called Iridium Satellite LLC. Operation of the system was contracted to Boeing.

Odyssey   *Odyssey* was proposed by TRW Space & Electronics Group (TRW S&EG, based in Redondo Beach, California) in 1990. TRW and Teleglobe Inc. (based in Montreal, Canada) formed a joint venture in 1994 to develop, construct, launch, and operate a global telecommunication system under the name *Odyssey*. The network was to consist of a constellation of 12 satellites in MEO at an altitude of 10,354 km, providing global dual satellite coverage. The project was abandoned in 1999 due to financial problems and a patent suit regarding the MEO architecture.

Aries *Aries*, a 48-satellite system in LEO proposed in 1992 by *Constellation Communications International* (CCI) of the United States, was recently replaced by a project with eight satellites at an altitude of 2000 km consisting of two parts: an equatorial system with one satellite to be operational in 2002, and a global system with seven satellites to be operational in 2003.

Ellipso *Ellipso*, a consortium led by Ellipsat, founded in 1990, initially proposed a MEO system with 24 satellites in a 3700/740-km elliptical orbit (hence the name *Ellipso*). This was later changed to 17 satellites plus one spare in a highly elliptical orbit with an apogee of 8050 km and a perigee of 633 km. Operation with seven satellites, initially planned to begin in 2001, was delayed.

Globalstar *Globalstar* was founded by the Alcatel-affiliated companies Loral and Qualcomm in 1991. It has 48 satellites in LEO at an altitude of 1410 km (Figure 27.13). Globalstar targets local markets in areas with limited cellular coverage and intends to install numerous fixed pay phone stations. The first satellites (Figure 27.14) were launched from Cape Canaveral on February 14, 1998. Twelve satellites were lost in September 1998 when a launch rocket exploded at the Khrunichev State Space Center, delaying the project by three months. A gradual country-by-country roll-out service began in October 1999. Instead of the 100,000 subscribers envisaged,

Figure 27.13 Globalstar 48-satellite constellation. (Courtesy of Globalstar.)

Figure 27.14   Globalstar satellite. (Courtesy of Globalstar.)

it had only 30,583 subscribers in 102 countries by the end of the twentieth century. Figure 27.15 shows a trimode handset for operation with the Globalstar space network as well as with terrestrial cellular radio systems in the CDMA and AMPS modes (Chapter 32).

ICO   ICO was conceived by Inmarsat in the early 1990s as a system with 12 satellites in two planes, each with five operational and one spare satellite, in an intermediate circular orbit (hence the name ICO) at an altitude of 10,355 km. ICO also filed for bankruptcy protection on September 1, 1999, but the American multibillionaire Craig McCaw, who had refused to rescue Iridium, put together a $1.2 billion rescue package in a new holding company, *ICO-Teledesic Global*, in November 1999.[33]

In addition to the global operating mobile personal satellite systems, regional systems emerged at the end of the twentieth century which operate one or more satellites in GEO, such as:

- *Asia-Pacific-Mobile Telecommunications*, Singapore, since 2000 covering 22 countries in the Asia Pacific and southern Asia region.
- *Garuda, Asia Cellular Satellite International Ltd.*, Indonesia, covering 23 countries in the Asia Pacific and southern Asia region, beginning operation in September 2000.

---

[33] In November 2001, it was agreed to end the merger between Teledesic and ICO Global Ltd. Instead, ICO Global Ltd. became ICO Global Communications Holdings Ltd. Consequently, Teledesic and ICO are again two independent companies.

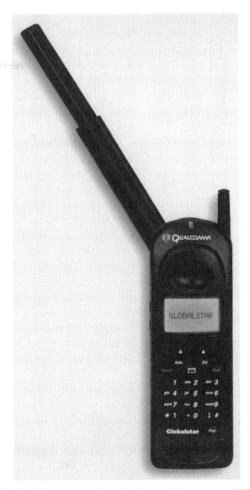

Figure 27.15   Globalstar trimode phone. (Courtesy of Globalstar.)

- *Al-Thuraya Satellite Communications Co.*, United Arab Emirates, covering 99
  countries in the Indian subcontinent, central Asia, the Middle East, central
  Africa, and Europe, beginning at the end of 2001. Dual-band terminals are used
  for this network, which communicates primarily in a terrestrial cellular GSM
  network and in the absence of a GSM signal switches over to satellite service
  automatically. The system includes GPS positioning facilities. A total of three
  satellites will be used. *Thuraya-1* was launched on board a Sea Launch Zenith
  rocket on October 21, 2000.

27.4.6   Multimedia Satellite Systems

The worldwide demand for multimedia services at the end of the twentieth century,
which in many areas of the world could not have been met economically with terrestrial transmission systems, provoked the development of special broadband satel-

lite systems. These new systems will enable direct multimedia communication via satellite at business and residential premises with antennas well below 1 m in diameter. For interactive two-way data transmission as used for multimedia operation, a problem was the signal delay of 500 ms that occurs on the 36,000-km route with satellites in GEO. The first systems were therefore developed for operation in LEO, where the signal delay is on the order of 10 ms only. Special techniques were developed, however, such as proxy caching, which minimize the impact of this delay even for operation in GEO.

The FCC issued licenses in 1997 for the implementation of global broadband "Internet in the sky" networks to the following companies.

Teledesic   *Teledesic* was founded by Craig McCaw and Bill Gates[34] in 1990 as an ambitious super-LEO project with 924 satellites at an altitude of 696 km, providing bandwidth on demand for an envisaged 20 million Internet-using PC owners by about 2002. Teledesic launched an experimental satellite, named *T1*, into LEO on February 25, 1998. This was the world's first commercial non-GEO satellite. The method of launching was also novel. For the first time a Pegasus rocket was carried by an aircraft to a point approximately 13,200 m over open ocean, where it was released and after a 5-second free fall, ignited and within about 10 minutes carried the *T1* into its orbit at an altitude of 565 km. The project was scaled down to 288 satellites operating at an altitude of 1375 km; intersatellite links will be used. A transmission speed of 64 Mbps will be offered for residential customers and up to 622 Mbps for business customers. It is planned to go into broadband service by 2005, under the trademark *Internet in the Sky*.

SkyBridge   *SkyBridge* is a consortium headed by Alcatel and includes Boeing, Loral Space & Communication, Litton Industries, EMS Technologies, Mitsubishi Electric, Sharp, Toshiba, and Qualcomm. It plans to implement a LEO system with 80 (originally, 64) satellites in 20 planes equally inclined by 53° relative to the equator at an altitude of 1469 km. As a "last-mile" project, it will be used for fixed broadband local access offering 20 Mbps reception and 2 Mbps transmission to residential users and a three- to fivefold higher transmission capacity to professional users. About 140 gateways are planned for worldwide coverage, each gateway covering a circular radius of 350 km. Commercial service is announced to begin around the end of 2002.

Astrolink   *Astrolink* was founded in 1999 by a U.S. cable group, Liberty Media, Lockheed Martin Global Telecommunications, and TRW. It is planned as a GEO system with four satellites offering data speeds of 400 kbps, 2 Mbps, and 20 Mbps starting in 2003.

Spaceway   *Spaceway* was founded by Hughes Network Systems in 1999. It plans a GEO system with three satellites (including one spare) above the United States. DirecTV and DirecPC services are planned in cooperation with America OnLine (AOL) by late 2002.

---

[34] Further major investors in Teledesic are Motorola, Boeing, Prince Alwaleed Bin Talal of Saudi Arabia, and the Abu Dahbi Investment Company.

Cyberstar   *Cyberstar*, founded by Loral Orion and Alcatel, is developing a GEO system with four satellites planned for global operation by 2003. Up to 14 gateways will link the system to the Internet. Before launching its own satellites, Cyberstar will offer 1.5 Mbps downlink and 128 kbps uplink service via Intelsat.

Netsat 28   *Netsat 28*, announced in January 2000, when majority-acquired by EMS Technologies of Norcross, Georgia, plans to implement a GEO system initially using one satellite over the United States. Netsat will offer 30 Mbps downlink and 2 Mbps uplink to small businesses and residential users beginning in October 2002. A special antenna design will be applied with 42 spot beams narrowly focused on certain geographic markets covering the largest U.S. cities, and 16 larger beams overlaying the entire country.

Isky   *Isky* plans a GEO system with initially two (later, four to five) satellites serving the Americas. Investors include Liberty Media, TV Guide, and Arianespace. Internet service was announced to begin the third quarter of 2001 but was delayed.

FAISAT   *FAISAT*, founded by the U.S. company *Final Analysis* in 1992, is an example of a little LEO planned to provide narrowband, 300 bps to 300 kbps, global data service with low-cost Internet access and file transfer via e-mail. Operation is planned around 2003 with 32 satellites plus six spares in space in an orbit at 1000 km. Launching is planned in Russia with Cosmos light-class rockets from the cosmodromes at Plesetsk in northern Russia and Kapistin in the south using the Russian launcher OP Polyot. FAISAT launched two experimental satellites with the Russian launcher OP Polyot in 1995 and 1997.

The established operators Intelsat, EutelsaT, PanAmSat, GE Americom, and SES Astra are implementing similar broadband services.

Implementation of the multimedia and GMPCS systems poses the question of where to leave the numerous satellites after their operational life. The operational lifetime depends on the effort required to keep a satellite in its orbit and the corresponding amount of onboard fuel. Current typical operation times are five years in LEO, eight years in MEO, and 10 to 15 years in GEO. So far, the last bit of onboard fuel is used to kick a satellite out of its orbit and deeper into space at the end of its operational life, without caring what might happen there. A satellite kicked out of its LEO, however, can collide with other satellites in LEO, MEO, and GEO. Consequently, they have to be brought down to an altitude of about 250 km, where they will disintegrate in the upper atmosphere (and might create other undesired effects). Letting a satellite come down naturally through the gravitational force of Earth can take hundreds of years before it reaches 250 km and disintegrates: a long time, during which it can collide with many other spacecraft.

The company *Tethers Unlimited Inc.*, Seattle, Washington, is currently developing an ingenious low-cost method, called *Terminator Tether*, of bringing satellites in LEO down quickly to about 250 km at the end of their life. A space tether is a long cable used to couple spacecraft to each other or to other masses. Basically, a Terminator Tether consists of a 5- to 10-km-long conductor of uninsulated aluminum wires

Figure 27.16 Satellite de-orbiting with a Terminator Tether. (Courtesy of Tethers Unlimited.) See insert for a color representation of this figure.

which at the lower side carries an electron emitter (Figure 27.16). At the upper side the tether is connected electrically to a tether control unit in the satellite, and mechanically by means of an insulating section to the satellite body. At the end of the operational life of the satellite, the tether and electron emitter are ejected downward from the satellite and drop away like an anchor. The tether then interacts with the magnetic field of Earth and drags the satellite down. With this device, the de-orbit time from 780 km for an Iridium satellite is calculated to be about three months and from 1469 km for a SkyBridge satellite is about 14 months.

REFERENCES

Books

Balston, D. M., and R. C. V. Macario, *Cellular Radio Systems*, Artech House, Norwood, MA, 1993.

Bergaust, Erik, *Wernher von Braun: Ein unglaubliches Leben*, Econ Verlag, Dusseldorf, Germany, 1976 (translated from *Incredible von Braun*, National Space Institute, Washington, DC, 1975).

Bode, Volkhard, and Gerhard Kaiser, *Raketenspuren, Peenemünde, 1936-1996*, Bechtermünz Verlag/Christoph Links Verlag, Berlin, 1997.

Buedeler, Werner, *Geschichte der Raumfahrt*, Sigloch Edition, Künzelsau, Germany, 1979/1982.

Freeman, Marsha, *Hin zu neuen Welten*, Böttinger Verlag, Wiesbaden, Germany, 1995 (translated from *How We Got to the Moon: The Story of the German Space Pioneers*, Twenty-First Century Science Associates, Washington, DC, 1993).

Gallagher, Brendan, *Never beyond Reach*, Inmarsat, London, 1989.

Huurdeman, Anton A., *Guide to Telecommunications Transmission Systems*, Artech House, Norwood, MA, 1997.

Michaelis, Anthony R., *From Semaphore to Satellite*, International Telecommunication Union, Geneva, 1965.

Oslin, George P., *The Story of Telecommunications*, Mercer University Press, Macon, GA, 1992.

Terrefe Ras–Work, *Tam Tam to Internet*, Mafube Publishing and Marketing, Johannesburg, South Africa, 1998.

Pleumeur-Bodou, Le Centre de Telecommunications par Satellite, Lannion, France, April 1985 [leaflet].

Articles

Anon., The Iridium system joining forces, *Iridium Today*, Fall 1995, pp. 14–24.

Anon., Delhi rural payphones in India, *Via Inmarsat*, April 1999, pp. 35–37.

Anon., The world of satellites: Europe, the Middle East, Africa, Asia, Russia, Latin America, *Via Satellite*, December 1999, pp. 22–86.

Anon., Eutelsat: a new evolution towards global service, supplement to *Via Satellite*, January 2000, pp. 1–16.

Baker, Simon, A question of priorities: Japan's satellite program, *Via Satellite*, January 1992, pp. 32–34.

Bertheux, P., et al., Evolution of geostationary telecommunication satellites, *Alcatel Telecommunications Review*, 4th quarter 2001, pp. 264–269.

Boeke, Cynthia, Via Satellite's 2001 satellite survey, *Via Satellite*, June 2001, pp. 24–30.

Clarke, Arthur C., Extra-terrestrial relays, *Microwave System News & Communications Technology*, Vol. 15, No. 9, August 1985, pp. 59–67 (commemorational repetition of the article originally published in *Wireless World*, Vol. 51, October 1945).

Crawford, A. B., et al., The research background of the Telstar experiment, *Bell System Technical Journal*, Vol. 42, July 1963, pp. 747–764.

Dickieson, A. C., The Telstar experiment, *Bell System Technical Journal*, Vol. 42, July 1963, pp. 739–746.

Fernandez, Rob, Via Satellite's global satellite survey, *Via Satellite*, July 2000, pp. 30–40.

Foley, Theresa, Launch vehicles: taking an industry to the stars, *Via Satellite*, April 2000, pp. 54–60.

Hoth, D. F., et al., The Telstar satellite system, *Bell System Technical Journal*, Vol. 42, July 1963, pp. 765–800.

Hoyt, R., and R. Forward, Performance of the Terminator Tether for autonomous deorbit of LEO spacecraft, presented at the 35th AIAA/ASME/SAE/ASEE Joint Propulsion Conference and Exhibit, June 20–24, 1999, Los Angeles, American Institute of Aeronautics and Astronautics, New York.

Ling, Ruth, The grandfather of satellites, *Via Inmarsat*, October 1997, pp. 20–22.

Marshall, Geoff, and Keith Dyer, Mobile satellite service providers prepare to take on the world, *Mobile Europe*, November 1998, pp. 35–40.

Nagata, H., and D. Wright, A short history of maritime communications, *ITU Telecommunication Journal*, Vol. 57, No. 2, 1990, pp. 117–125.

Nelson, Robert A., Inclined orbit operations, *Via Satellite*, September 1994, pp. 84–91.

Nelson, Robert A., Satellite constellation geometry, *Via Satellite*, March 1995, pp. 110–122.

Ospina, Sylvia, Latin America: a survey of the satellite scene, *Via Satellite*, December 1992, pp. 52–60.

Pritchard, Wilbur L., The history and future of commercial satellite communications, *IEEE Communications*, Vol. 22, No. 5, May 1984, pp. 22–37.

Pritchard, Wilbur L., The development of satellite communications, *Microwaves & RF*, March 1987, pp. 303–312.

Sourisse, P., SkyBridge: global multimedia access, *Alcatel Telecommunications Review*, 3rd quarter 1999, pp. 228–237.

Struzak, Ryszard, Internet in the sky: tests have started, *ITU News*, Vol. 6, 1998, pp. 22–26.

Wold, Robert N., As the world turned: another olympian feat for satellites, *Via Satellite*, November 2000, pp. 38–46.

Internet

www.asiasat.com, company overview.

www.eurastasat.com, company overview.

www.finalanalysis.com, company overview.

www.ses-astra.com, corporate information.

www.skybridgesatellite.com, corporate information.

www.teledesic.com, corporate information.

www.thuraya.com, corporate information.

www.totaltelecom.com.

Hellickson, Amy, Globalstar shares fall 1.7% on handset delays, October 28, 1999.

Naujeer, Husna, Half the world relies on satellite connectivity says report, May 3, 2001.

Wilkinson, Genevieve, Satellite companies face murky fate after Iridium's demise, March 24, 2000.

Young, Anna, Iridium set to rise from the ashes, November 17, 2000.

28

OPTICAL FIBER TRANSMISSION

28.1 EVOLUTION LEADING TO OPTICAL FIBER TRANSMISSION

Optical fiber cable brought a quantum leap in transmission capacity. Whereas coaxial cable and radio-relay systems offered transmission capacities on the order of Mbps, optical fiber cable, introduced in the 1980s, offered transmission capacities on a single optical fiber on the order of Gbps and at the end of the twentieth century even on the order of Tbps. Optical fiber transmission and the introduction of data transmission between computers started an evolution in which the transmission capacity of a telephone line to subscribers was increased from, traditionally, 4 kHz analog, equivalent to 64 kbps digital, for a single telephone channel, to Mbps for residential lines and Gbps for business lines.

The principle of guiding light through a transparent conductor was explained and demonstrated as early as 1870 by the British physicist John Tyndall (1820–1893). Tyndall demonstrated that light could be guided along a curved path in water. Ten years later, in 1880, four years after he invented the telephone, A. G. Bell constructed and patented the Photophone, which he used to transmit speech signals at distances of a few hundred meters. Bell used sound-intensity-modulated sunlight reflected from one mirror to another mirror, detected there with a selenium device. Weather dependence and the insensitivity of selenium photocells hindered practical application.

The invention of the laser (light amplification by stimulated emission of radiation) by Theodore Maiman at the Hughes Research laboratories in Malibu in 1959, by which light could be converted into electricity, and vice versa, inspired research efforts to uncover a low-loss, well-controlled guided optical medium. The decisive breakthrough occurred in 1966 in the U.K. and in Germany. In the U.K., Charles

The Worldwide History of Telecommunications, By Anton A. Huurdeman
ISBN 0-471-20505-2   Copyright © 2003 John Wiley & Sons, Inc.

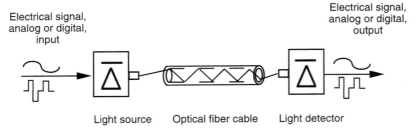

Figure 28.1   Principles of optical transmission. (From A. A. Huurdeman, *Guide to Telecommunications Transmission Systems*, Artech House, Norwood, MA, 1997, Figure 4.1; with permission of Artech House Books.)

K. Kao and George A. Hockham, working under Alec H. Reeves at *Standard Telephone Laboratories* (STL), predicted in the *IEEE Proceedings* of July 1966, under the title *"Dielectric-Fiber Surface Wave-Guides for Optical Frequencies,"* that fibers drawn from extremely pure glass would be an ideal support for the transmission of modulated light waves. The principle of optical fiber transmission as proposed by Kao and Hockham is illustrated in Figure 28.1 and described briefly in Technology Box 28.1. In their publication, Kao and Hockham predicted that a loss below 20 dB/ km should be attainable, although at that time the lowest loss in glass fibers was still on the order of 1000 dB/km. They prepared a patent application about optical fiber transmission, but STL management considered the patent application not to be opportune. In the same year, Kao visited Bell Labs but found there no interest for his ideas. So it happened that in 1966 a German scientist, M. Boerner, while researching for Telefunken, arrived at findings similar to those of Kao and Hockham and obtained the world's first patents for optical fiber transmission. He obtained German patent DBP 1254.513 for a "transmission system using a semiconductor laser, a glass fiber, and a photo detector for transmission of PCM signals." He obtained similar patents in the U.K. and the United States in the same year, but he died soon thereafter.

TECHNOLOGY BOX 28.1

Principle of Optical Transmission

An optical fiber transmission system basically consists of a light source, a cabled optical fiber, and a light detector. The light source is either an inexpensive light-emitting diode (LED) or a complex laser. The light source emits light as a function of an electrical input signal (modulated with the information to be transmitted) at a wavelength suitable for transmission through the optical fiber. At the receiving end the detector converts the light back to the original electrical (modulated) signal. The detector is usually a fast pin-photo diode or an avalanche photo diode.

Source: Adapted from A. A. Huurdeman, *Guide to Telecommunications Transmission Systems*, Artech House, Norwood, MA, 1997; with permission of Artech House Books.

In the same year, the research laboratories of the GPO showed an interest and started optical fiber communications research under F. F. Roberts. William Shaver of the U.S. *Corning Glass Works* visited Roberts and, encouraged by Roberts's progress, initiated a research program at Corning. Robert Maurer took the lead in this research on silica fibers. He concentrated the research on fibers with a core doped with titanium and a pure glass cladding. In 1967, Maurer obtained the assistance of a colleague, Peter Schultz, from the glass chemistry department to help in making pure glass. One year later, Donald Keck joined the team as the first full-time fiber developer at Corning.

In 1968, Kao and Hockham reported having achieved an attenuation of 5 dB/km with special silica samples, and Martin Chow demonstrated the new fiber at a Physical Society exhibition in London in 1969.

In 1970, Maurer, with his team, now including Felix Kapron, achieved the 20-dB goal—in fact, 17 dB/km at 633 nm—with industrially produced fiber with a core of titanium-doped silica glass and a cladding of pure silica glass. Two years later, the team at Corning managed to produce a fiber with a much higher mechanical strength and an attenuation of only 4 dB/km by doping the core with germanium instead of titanium. Reviewing their work some years later, Maurer pointed out that two key decisions led to their success:

1. To use glasses consisting primarily of silica, with oxides added to the extent necessary to increase the refractive index. These glasses have provided the lowest attenuations yet achieved and excellent physical properties.
2. To use vapor-phase processes for manufacturing glass preforms or rods from which the fiber is drawn. Vapor-phase processes have not only provided the necessary purity but in addition have turned out to be surprisingly flexible when adapted to new ways of making the preform. As a result, both ideas have stood the test of time and remain at the heart of today's technology.

The vapor-phase process resulted in the required purification by a factor of 1000 or more with respect to unwanted metal ions. The oxides are added to obtain a distinct difference between core and cladding of the fiber. As promising as these developments were, there were still enormous obstacles to making optical fiber transmission practical. The major issues still unsolved were:

- Could the loss be reduced to the range of a few dB/km to gain a repeater spacing advantage over coaxial cable?
- Could fibers be produced at sufficiently low cost to make optical fiber transmission economically attractive?
- Could fibers be drawn with sufficient strength to survive cabling, installation, and field use?
- Could fiber dimensions and index-of-refraction profiling be controlled well enough to permit satisfactory transmission characteristics and low splice losses?
- Could splicing be performed in a hostile field environment with sufficient alignment accuracy, freedom from contamination, and stability to allow stable system performance?
- Could practical optical fiber connectors be designed and manufactured?

- Could injection laser lifetimes, then still measured in minutes, be extended to the tens of thousands of hours needed for real systems?
- Could laser light be coupled into optical fibers without a large loss?

Affirmative answers to these questions were given within only a few years. The underlying principle of transmission in optical fibers is that light, in an encoded sequence of pulses, is reflected on boundaries of two different optical media, constituting a waveguide for light transmission. The two different media in an optical fiber are the light-conducting cylindrical core of high-grade silica glass surrounded by a concentric sheath (called *cladding*) with a refractive index which is lower than that of the core. The different refractive index is obtained by doping the pure silica glass with boron, which reduces the refractive index, or with germanium, which increases the refractive index.

Basically, three different methods were developed for the production of fiber with core and cladding of different refraction. From the beginning, Corning used an *outside vapor deposition* (OVD) process, whereby fine glass particles were deposited around a glass rod. Bell Labs developed a *modified chemical vapor deposition* (MCVD), whereby the doping is made inside a glass tube, which afterward collapses into a solid glass rod in which the chemicals deposited constitute the fiber core. Application of this process enabled Bell Labs to present at the International Glass Conference held at Kyoto in 1974 an optical fiber with an attenuation of 1.1 dB/ km at a wavelength around 850 nm. In Japan it was observed by the Mizushima Research Laboratory in 1973 that the optimum wavelength would be around 1500 nm. Masaharu Horiguchi of the Optical Components Laboratory of the Electrical Communications Laboratories (ECL) of NTT at Ibaraki[1] and Hiroshi Osanai of Fujikura Cable Works Ltd. produced an optical fiber with an attenuation of 0.47 dB/ km at 1200 nm in 1976, still using the MCVD process. A further major step was made in Japan with the development of a third process for the production of fiber, called *vapor axial deposition* (VAP). In this process the chemical deposition was made in an axial direction with a higher concentration of germanium in the middle while the glass preform was produced, a method said to be similar to the growth of a limestone cave formation. With this complex but superior VAD process, ECL achieved an attenuation of 0.2 dB/km at 1500 nm in 1980.

During this period it became clear that fibers made of silica glass by the prevailing manufacturing processes had three distinct windows of minimum attenuation.[2] These windows are shown in Figure 28.2, and a short explanation is given in Technology Box 28.2.

The first commercial optical fibers produced in the 1970s had a core with a relative large diameter so that the propagation of light through the core proceeded in a multimode fashion, which resulted in a slight distortion of the pulses. An improvement was obtained at the end of the 1970s by grading the refractive index of the core.

---

[1] NTT began systematic research on optical fiber in 1971. Kunio Masuno formed a seven-person team, including Masaharu Horiguchi, which he called his seven samurai of optical devices. Horiguchi and Osanai were awarded a prize by the IEE of the U.K. for the best research paper of the year in 1976.

[2] Currently, fibers have a substantially improved attenuation curve. The curve of reduced water peak fiber, for example, runs almost peak-free from about 0.4 dB at 1250 nm to 0.2 dB at 1570 nm. The theoretical minimum is 0.1 dB/km at 1550 nm.

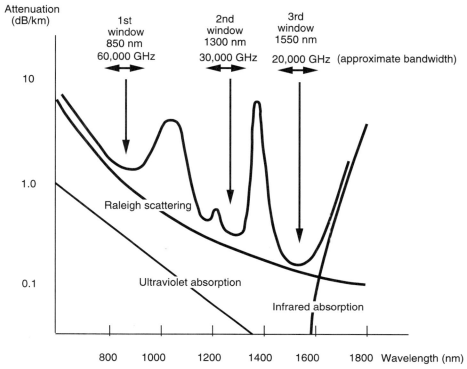

Figure 28.2 Three windows of minimum attenuation for optical fiber transmission. (From A. A. Huurdeman, *Guide to Telecommunications Transmission Systems*, Artech House, Norwood, MA, 1997, Figure 4.3; with permission of Artech House Books.)

A further important improvement followed in the early 1980s such that the core was made so thin that light could travel only in an almost straight line, thus in a single mode. Figure 28.3 illustrates the major characteristics of the three fiber types.

As stated in Technology Box 28.2, optical fiber offers an enormous transmission capacity, far beyond those of coaxial cable, radio-relay, and satellite transmission. Cables made of optical fiber have a significantly smaller diameter then those of coaxial cable, which makes installation easier, faster, and less costly. Despite the brittle nature of the fiber, it is much more elastic than copper. Furthermore, optical fiber is immune to electromagnetic interference and radio-frequency interference from power lines and electrical storms. Optical fiber cable very soon became the exclusive medium for submarine transmission and the standard solution for terrestrial media and high-capacity transmission systems except where adverse terrain, such as jungle, marshes, and rocky areas, had to be crossed and satellite or radio-relay provided less costly solutions.

Multimode fibers were used for low- and medium-capacity systems and for short-haul high-capacity systems. ITU issued the first standard for multimode fibers, No. G.651, in 1980. Single-mode fibers became the exclusive solution for high-capacity long-haul systems. The single-mode fibers were standardized by the ITU in 1984 under the designation G.652. Two improved versions of single-mode fibers were stan-

---

TECHNOLOCY BOX 28.2

Three Windows for Optical Fiber Transmission

Optical fibers made of silica glass have three windows of minimum attenuation between the boundaries of ultraviolet and infrared absorption in the range 800 to 1600 nm. Raleigh scattering, which is a physical material property, determines the theoretical minimum attenuation between the two borders. Glass density changes cause a steady fluctuation of the refractive index that cannot be eliminated and limits the lowest spectral attenuation of the fiber to a value of about 0.13 dB/km at 1550 nm. Further attenuation is caused by material impurities such as hydroxyl (OH) ions that enter the fiber during the production process and evoke peaks of attenuation at 950, 1240, and 1380 nm. As a consequence, three windows of minimum attenuation are located at 850 nm with a bandwidth of 60 THz (1 terahertz = 1000 GHz), at 1300 nm with 30 THz, and at 1550 nm with 20 THz. The total transmission capacity of the three windows thus amounts to 110 THz, which corresponds to some 15 million color TV channels, or 30 billion telephone channels.

Source: Adapted from A. A. Huurdeman, *Guide to Telecommunications Transmission Systems*, Artech House, Norwood, MA, 1997; with permission of Artech House Books.

---

dardized in 1988, G.653 for dispersion-shifted, and G.654 for loss-minimized single-mode fibers. The last fiber has an attenuation of 0.18 dB/km instead of 0.20 dB/km for the standard G.652 single-mode fiber. Two application-specific single-mode fibers were developed and standardized by the ITU in the 1990s: a *nonzero-dispersion fiber*, or ITU G.655 fiber, designed with a carefully controlled chromatic dispersion to minimize pulse spreading on long routes, and a *zero water peak fiber*, or ITU G.652.C fiber, specially designed for a wider usable transmission spectrum from 1280 to 1625 nm, as required for DWDM mode of transmission, discussed below.

By 1992, about 35 million kilometers of cabled optical fibers had been installed worldwide. By that time optical amplifiers had been introduced, which amplified the optical signals directly, without having to convert them into electrical form, thus making it possible to transmit optical signals over almost unlimited distances. Two types of optical amplifiers entered the market: *semiconductor laser amplifier* (SLA) and *erbium-doped fiber amplifier* (EDFA).[3] Both amplifiers use the same physical principle of *stimulated emission* which had been discovered by Einstein. According to this principle, an incident light beam is amplified by stimulated emission in a medium that causes amplification by the injection of energy, also called *energy pumping*. SLAs, which use special pumped lasers and electrical pumping, found limited application because of the far superior performance of EDFAs. In an EDFA the amplifying medium is a short section (typically, a few tens of meters) of optical fiber doped with erbium ions. Erbium (Er^{3+}, number 68 in the periodic system of the elements) is a rare-earth element. Optical energy is injected into this doped fiber section by means

---

[3] Dave Payne at the University of Southampton developed the first EDFA operating at 1550 nm in 1987.

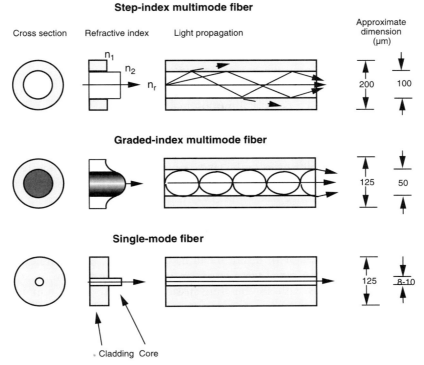

Figure 28.3   Three types of optical fiber. (From A. A. Huurdeman, *Guide to Telecommunications Transmission Systems*, Artech House, Norwood, MA, 1997, Figure 4.11; with permission of Artech House Books.)

of laser diodes, which emit light at a wavelength below the wavelength of the optical signal. For example, a powerful light beam at a wavelength of 980 nm is pumped into the doped fiber, where it excites the erbium ions to a higher energy level, which causes a direct optical amplification of the original signal with a wavelength of 1550 nm. The use of EDFAs extended the range of signal regeneration from about 100 km to about 1000 km.[4]

An EDFA thus amplifies a signal to compensate for the fiber attenuation but not for the signal distortion. The subsequent rediscovery of *Soliton waves* also solved the problem of distortion in the mid-1990s, when over 60 million kilometers of optical fiber had been installed worldwide. *Solitons* are pulses of light that can propagate without changing their shape; thus soliton waves travel practically distortion free.[5]

---

[4] The performance of EDFAs was improved substantially early in 2001 by the use of a cladding–pumping technique. This technique is based on the use of special doped fiber with two concentric cores. One core is dedicated to single-mode amplification of the signal beam, similar to a conventional EDFA. A second, much larger core propagates the various modes of a multimode pump.

[5] The Scottish engineer John Scott (1808–1882) discovered the *soliton phenomenon* in 1834 when on horseback alongside a canal in Edinburgh. He observed a boat being drawn rapidly by a pair of horses, and when the boat suddenly stopped, the bow wave continued its course along the canal for a number of miles, apparently without change of form or diminution of speed.

TECHNOLOGY BOX 28.3

Principle of Soliton Transmission

Soliton is the name for a light pulse that either does not change its shape or changes its shape along the length of a fiber but returns periodically to its original shape. A soliton acts as a perpetual motion machine. This ideal pulse behavior is the result of two detrimental nonlinear effects of optical transmission: the chromatic *group velocity dispersion* (GVD) of the fiber and the *self-phase modulation* (SPM) of pulses caused by the Kerr effect on the refractive index of the fiber core. With SPM the frequencies in the leading half of the pulses are lowered while those in the trailing half are raised; whereas due to GVD above the zero dispersion point, thus in the range where the GVD is negative, the actual group velocity (not the GVD itself) increases with frequency. Hence, the GVD acts contrary to the SPM and, in effect, can correct the SPM. In other words, the self-phase-modulated pulse pulls itself back to its original shape after a certain distance called the *soliton distance*. This distance can be calculated with a complex mathematical formula and can range from less than 1 km to over 100 km on standard single-mode fiber. Thus with soliton transmission, the pulse width remains constant; however, the pulse amplitude remains a function of the fiber attenuation. Hence, optical fiber transmission performance can be optimized by the combined application of soliton transmission and carefully spaced EDFAs.

Source: Adapted from A. A. Huurdeman, *Guide to Telecommunications Transmission Systems*, Artech House, Norwood, MA, 1997; with permission of Artech House Books.

A brief technical explanation of the soliton phenomenon is given in Technology Box 28.3 and illustrated in Figure 28.4. The American physicist Linn Mollenauer of the Bell Telephone Laboratories first demonstrated soliton transmission through 4000 km of single-mode fiber in 1988. Three years later he sent a 5-Gbps signal error-free over 15,000 km of fiber and 2 × 5 Gbps over 10,000 km. In the same year, Masataka Nakazawa of NTT demonstrated soliton transmission through 1 million kilometers of fiber. Two years later, Nakazawa sent soliton signals through 180 million kilometers of fiber and claimed that soliton transmission is limitless.

The combined application of EDFA and soliton technology more than doubled the repeater spans and enabled transmission rates up to about 40 Gbps without signal regeneration up to a distance of about 10,000 km.

Instead of amplifying the optical signal selectively at certain intervals along the line in the short length of erbium-doped fiber in an EDFA, continuous optical amplification along the entire length of an optical fiber line, called *Raman amplification*, was introduced at the end of the twentieth century. With Raman amplification a very strong laser beam at a wavelength slightly shorter than that of the optical signal is pumped into the fiber in the direction opposite to the optical signal. Energy from that beam makes the matrix of the fiber vibrate, which generates nonlinear interactions that cause various scattering processes at different wavelengths. The scattered light at a wavelength different from that of the optical signal is attenuated in the

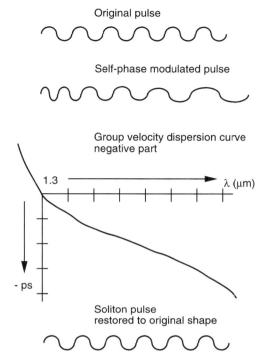

Figure 28.4 Soliton transmission. (From A. A. Huurdeman, *Guide to Telecommunications Transmission Systems*, Artech House, Norwood, MA, 1997, Figure 4.8; with permission of Artech House Books.)

fiber, while scattered light of the same frequency as the optical signal boosts the power of that signal. Basically, with Raman amplification the fiber acts as one long amplifier whereby the energy pumped into the fiber is transferred to the signal. Raman amplification can be used on short cable links instead of EDFAs, or on long-haul cables in order to double the spacing of EDFAs.[6] Whereas EDFAs operate only in the transmission spectrum 1525 to 1620 nm, Raman amplifiers can operate in the range 1300 to 1700 nm.

The worldwide penetration of the Internet and the general increase of data transmission in the last decade of the twentieth century required transmission systems surpassing 40 Gbps. Soliton transmission above 40 Gbps becomes difficult on the fibers available. Similarly, with the present technology, generating transmission speeds becomes increasingly difficult beyond 20 Gbps. A new technology called *wavelength-division multiplexing* (WDM) brought a solution for achieving a higher transmission capacity of optical fibers without increasing the transmission speed beyond practical limits. The simplest form of WDM, which has been in use for several years, is to transmit, for example, one 2.5-Gbps signal through the 1300-nm window and another 2.5-Gbps signal through the 1550-nm window of the same fiber

---

[6] Occasionally, *discrete Raman* is used, in which a special narrow-core fiber is placed within the Raman amplifier.

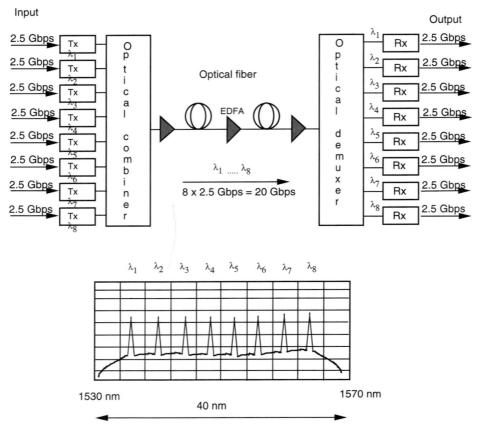

Figure 28.5   WDM with a capacity of 20 Gbps. (From A. A. Huurdeman, *Guide to Tele-communications Transmission Systems*, Artech House, Norwood, MA, 1997, Figure 2.21; with permission of Artech House Books.)

pair, or of the same fiber in the case of bidirectional transmission. The optical signals used on fiber typically have a bandwidth of only a few nanometers, whereas the transmission window at 1550 nm has a bandwidth of approximately 40 nm. Figure 28.5 shows how eight optical signals (usually called *wavelengths* or *colors*), each with a capacity of 2.5 Gbps (and a bandwidth of 0.4 nm), can be transmitted through the 1550-nm window of a single optical fiber, resulting in an aggregate transmission capacity of 8×2.5 Gbps = 20 Gbps. The first WDM systems were used in the standard 1530- to 1570-nm band of the third window. To increase the number of wavelengths that can be accommodated within a window, the bandwidth of both lasers and EDFAs was extended to cover the range 1530 to 1610 nm.[7] This bandwidth of 80 nm can accommodate 200 wavelengths each of 0.4 nm, or with improved systems, 400 wavelengths each of 0.2 nm.

---

[7] The extended band from 1530 to 1610 nm was subdivided into two subbands: a C-band from 1530 to 1570 nm, and an L-band from 1570 to 1610 nm.

To distinguish the operation of multiple wavelengths within one window from parallel operation through two windows, the operation of multiple wavelengths within one window is called *dense WDM* (DWDM). Combined use of DWDM with soliton transmission and EDFAs and Raman amplification, as well as improved fiber, resulted in commercial optical fiber transmission systems operating 80 separate wavelengths each with a capacity of 10 Gbps on one fiber pair by the end of the twentieth century. Repeaterless systems with a capacity of 160 Gbps, upgradable to 320 Gbps, operated over distances up to 300 km.

Experimental record performances reported at that time further demonstrate the enormous progress being made in optical fiber transmission. AT&T, Fujitsu, and NTT reported experimental transmission of signals beyond 1 Tbps at the Optical Fiber Communication Congress at San Jose, California, in 1996. In 1999, the research center of France Telecom, CNET, was the first to demonstrate transmission of 1 Tbps of data across an optical fiber cable 1000 km long using optical amplifiers at 100-km intervals. The U.S.-based international carrier WorldCom, planning to implement a nationwide 40-Gbps-wavelength DWDM system, made field trials in 2000 with the most advanced systems under development. From Siemens AG a DWDM system was tested carrying 80 wavelengths of 40 Gbps each, thus an aggregate capacity of 3.2 Tbps on one fiber pair over a distance of 250 km. Fujitsu demonstrated a system transmitting 176 wavelengths each carrying 10 Gbps, thus an aggregate capacity of 1.760 Tbps on one fiber pair. Nortel demonstrated an 80 × 10 Gbps system: an aggregate capacity of 0.8 Tbps on a single fiber and 1.6 Tbps on two fibers. Alcatel reported an experimental system carrying 32 wavelengths each of 40 Gbps over a distance of 2400 km using Raman amplification without EDFAs and a new reversible fiber arrangement. In 2000, Alcatel demonstrated a laboratory DWDM system with 128 wavelengths with 40-Gbps signals over a single fiber with a length of 300 km, thus a total record capacity of 5.12 Tbps on a single fiber.

NEC and Siemens added a new dimension to optical fiber transmission by sending wavelengths simultaneously in two polarization planes, called *polarization-division multiplexing* (PDM). Siemens, in cooperation with the Deutsche Telekom, made a test at Darmstadt on an optical fiber pair 116 km long with 16 wavelengths each of 40 Gbps on horizon polarization and another 16 wavelengths each of 40 Gbps on vertical polarization, thus an aggregate capacity of 1.28 Tbps on one fiber pair. At the European Conference on Optical Communication in September 2000, NEC reported to have reached 6.4 Tbps over 186 km of fiber using PDM, while Siemens reported to have achieved 7 Tbps with PDM over a distance of 50 km.[8] The Heinrich-Hertz Institute of Berlin University achieved a world record in cooperation with the Deutsche Telekom by transmitting a single 160-Gbps signal (thus without using DWDM or PDM) over a trial link with a length of 116 km without a repeater.

Worldwide deployment of optical fiber experienced an enormous increase in the last years of the twentieth century: 52 million kilometers of fibers was installed in 1998, 70 million kilometers in 1999, and an additional 100 million kilometers in the year 2000. About half of those amounts of fiber were installed in North America, where in the year 2000 the demand for additional fiber was equal to the worldwide

---

[8] In the *Alcatel Telecommunications Review* for the third quarter of 2001, Alcatel reported having achieved a record terrestrial transmission over 100 km of 10 Tbps in a configuration of 128 (channels) × 2 (polarizations) × 40 Gbps (per channel) = 10.240 Tbps.

demand in 1996. The pioneering optical fiber cable company Corning experienced an annual growth rate of 75 to 100% during those three years and became the world's largest manufacturer of optical fiber cable, with a global workforce of 40,000 and factories in the United States, Australia, Germany, the U.K., and Vietnam.

28.2   TERRESTRIAL OPTICAL FIBER CABLE SYSTEMS

Murray Ramsey of STL had the honor of giving Queen Elizabeth the first demonstration of transmission of a digital TV signal over an optical fiber at the Centenary of the Institution of Electrical Engineers (IEE) in London in May 1971. The first commercial optical fiber cable system was put into operation in Germany in 1973: a 2-Mbps system operating on a 24-km link with multimode cable between Frankfurt and Oberursel. AT&T installed a first experimental optical link in Atlanta in 1975. A 144-fiber underground cable was used configured in a loop 14 km long so that the same signal could be sent through the cable many times on adjacent fiber pairs. With a 44.7-Mbps signal, repeater spacing up to 10.9 km could be used with negligible crosstalk and a bit error rate better than 10^{-9}, a promising result better than that ever obtained on coaxial cable.

The first commercial optical fiber link in the U.K. was installed in 1975 for a police network at Dorset when lightning had damaged the existing copper-based network. In the United States, General Telephone and Electronics opened optical fiber transmission with a capacity of 6 Mbps at Long Beach, California, on April 22, 1977. One month later, AT&T operated a 45-Mbps fiber link in Chicago. MCI, in 1982, installed an optical fiber cable between New York and Washington with a capacity of 400 Mbps operating on 1300 nm. In France the first optical fiber cable was installed in 1980 in Paris between the Tuileries and the Philippe–Auguste telephone exchange with a length of 7.5 km and a capacity of 34 Mbps. In Japan a nationwide optical fiber cable network 41,000 km long was installed between 1980 and 1985 running from Asahikawa on the island of Hokkaido in the north to Kagoshima on the island of Kyushu in the south, linking the main cities along the Pacific coast. Fifteen years later a 45,000-km nationwide optical fiber cable network was installed by TransTelecom in Russia along the railway infrastructure of each of Russia's 17 regional railroads. The network connects 974 boroughs in 71 of the 89 regions.

The low attenuation of optical fiber allowed repeater sections exceeding 100 km in length compared with coaxial systems, which required repeaters every 1.5 to 10 km, depending on transmission capacity and type of cable. With this long and flexible repeater spacing the repeaters could be located in telecommunication buildings instead of underground at fixed short intervals along the route. Thus, power feeding of the repeaters via the cable is not required and optical fiber cables can be completely metal-free (thus eliminating lightning damage). The low weight of the optical fiber cable enabled a change in cable laying; instead of pulling the cable through ducts, the fiber cable could be pushed through the duct using compressed air.

In the early 1980s optical fiber transmission systems were used primarily for transmission speeds of 2, 8, and 34 Mbps in short-haul urban applications. Multimode fibers were used operating at 850 nm. Long-haul systems using single-mode fibers operating at 1550 nm with transmission speeds of 140/155 and 565/622 Mbps were introduced in the mid-1980s. Repeaters were flexibly located at 50- to 130-km intervals. Long-haul systems with a capacity of 2.5 Gbps per fiber pair were introduced in the early 1990s, and 10-Gbps systems came into operation at the end of the twentieth century. DWDM systems with four or 16 wavelengths modulated with 2.5 and 10 Gbps were deployed beginning in mid-1990; 32- and 40-wavelength systems came a few years later; and 64-, 80-, and 128-wavelength systems were installed in the year 2000. In metropolitan fiber networks and on some long-distance routes, *optical add/drop multiplexers* (OADMs) were introduced for dropping and reallocation of wavelengths with 2.5- and 10-Gbps optical signals without optical-to-electrical conversion. For example, a DWDM system with 80 wavelengths operated on a standard cable with 48 fibers can thus carry 3840 channels, each with a capacity of 2.5 or 10 Gbps, which can serve a separate customer in a metropolitan network. In the United States the new operator Fiberworks is implementing a huge metropolitan network in major southeastern cities. The cable, with 24 and 144 fibers and a total length of 175,000 km, will bring IP-based traffic to offices in key business districts. The world longest chain of optical fiber rings was installed by AT&T Canada in 1999 from Quebec to Vancouver, with network access points on both sides of the 7000-km Canadian–American border (Figure 28.6).

A revolutionary method of implementing terrestrial optical fiber networks was introduced in France in 1998 using inland waterways as an economical solution for

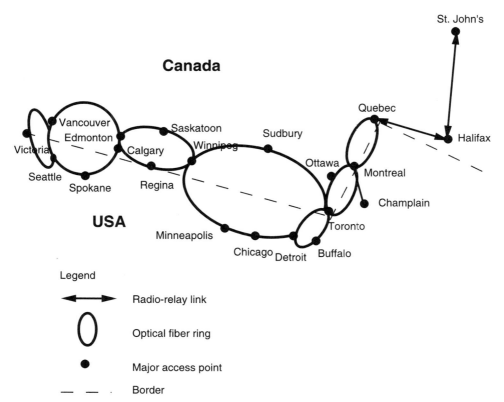

Figure 28.6 AT&T Canada optical fiber chain. (After *Telecommunications*, October 1999, p. C6.)

laying optical fiber cable over long distances. Using its existing infrastructure of 6700 km of rivers and other inland waterways, the French government-owned company Voies Navigables de France VNF (VNF), in cooperation with MCI World-Com and Louis Dryfus Communications (LD Com), laid 60 km of optical fiber cable in the river Seine through the city of Paris and its inner suburbs as the backbone of their city network. Within one year an additional 1000 km of optical fiber cable was laid in the other inland waterways.

Transcontinental optical fiber cable systems were installed in North America and in Australia, over 20 pan-European optical fiber cable networks were installed in Europe,[9] and large optical fiber cable networks are under construction in China and Latin America. Most of these systems operate at a line speed of 2.5 Gbps, and some at 10 Gbps.

The biggest single market for optical fiber systems emerged in China. More than 1 million kilometers of fiber has been installed, and currently, a number of backbone rings are being installed, each up to 5000 km long for the operation of 32 wavelengths each with 10-Gbps signals. In 2000, the Chinese optical fiber market surpassed \$1 billion in value and was distributed as follows: Chinese companies: Huawei Technologies,[10] 32%; Zhongxing Telecom, 12%; Datang Telecom (with Marconi), 6%; FiberHome, 4%; and foreign suppliers: Nortel, 18%; Lucent, 9%; Alcatel, 6%; Siemens, 4%; and others, 9%. In North America some 670,000 route-km of optical fiber cable had been deployed in long-haul networks by the end of the twentieth century. At an average of 52 fibers per cable, this corresponds to 33.8 million kilometers of fibers.

The world's longest intercontinental terrestrial optical fiber link, the Trans-Asia-Europe Optical Fibre Cable System (TAE), in large part following the old Silk Road, began service on January 14, 1999. The idea for this cable was born in 1992 when the Chinese minister of telecommunications, Yang Taifang, met his Asian colleagues in Beijing. This cable connects Germany and China and 18 countries along the route (Figure 28.7). The cable system is 27,000 km long, of which 17,000 km is for the direct cable between Frankfurt and Shanghai and another 10,000 km of cable branched from the direct cable in the 18 countries to serve the national traffic of those countries. The direct cable has one pair that carries 2×155 Mbps between Frankfurt and Shanghai, and another 10 to 16 pairs of fiber to serve the international traffic of the 18 countries.

Optical fiber cable is also used on overhead power lines, where its insensitivity to electromagnetic interference is a special advantage. Composite power line cables were developed for this purpose with an optical fiber core within a concentric ground wire, called an *optical ground wire* (OPGW), or with a concentric phase wire, called

---

[9] The 20 pan-European networks belonged to Atlantic Telecom Ltd.; Carrier 1 International; BT Ignite; Colt Telecom Group plc; Global Crossing Ltd.; Global Telesystems Ltd.; Iaxis Inc.; I-21, Interoute Telecommunications Ltd.; KPNQWest NV; Level 3 Communications; Pangea Ltd.; Storm Telecommunications Ltd.; Viatel Inc.; and others. At least four of the 20 companies fell victim to the telecommunications depression in 2001.

[10] Huawei Technologies, at Shenzen (China's "Silicon Valley") was founded in 1988 as a distributor of PABX equipment. Currently, it has 15,000 employees worldwide and established research centers in the United States (at Dallas and Silicon Valley); Stockholm, Sweden; and Bangalore, India. Optical transmission equipment in China is also manufactured by Zhongxing Telecom, Datang Telecom (allied with Marconi), and FiberHome, a spin-off of the Wuhan Research Institute.

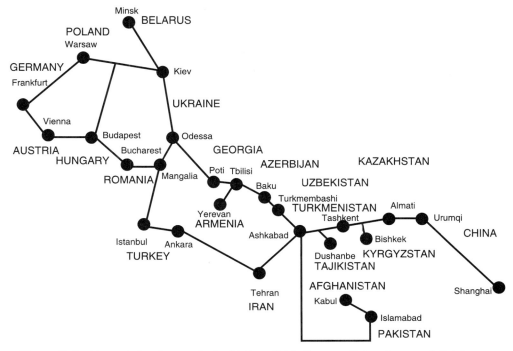

Figure 28.7 Trans-Asia–Europe Optical Fibre Cable System. (After *Post- und Telekommu-nikationsgeschichte*, Vol. 1, 1999, p. 89.)

an *optical phase wire* (OPPW). For exclusive telecommunications services an *all-dielectric fiber optical cable* (ADFOC) was also developed, made entirely of electrically nonconductive material to enable installation in the immediate vicinity of high-voltage lines. This aerial cable required a high tensile strength over a wide temperature range, typically −30 to +70°C for OPGW and ADFOC cable and −30 to +220°C for OPPW cable. Beyond carrying supervisory information for power line operators, these cables were used widely for the implementation of long-distance infrastructures for private telecommunication networks in the 1990s.

Optical fiber cable is also used by railways along rail routes either as ground cable or as self-supporting aerial cable. It is also being used extensively in the automotive and aerospace industries.

28.3 SUBMARINE OPTICAL FIBER CABLE SYSTEMS

The world's first international commercial submarine optical fiber cable went into operation in 1986 between Broadstairs, U.K., and Ostend, Belgium. The cable, 112 km in length, had six single-mode fibers and three regenerators, supplied by STC,[11]

[11] The many manufacturers of the early telegraph cable in the U.K. were consolidated into Submarine Cables Ltd. This in turn was taken over by STC in 1970, in 1991 by Northern Telecom, and in 1994 by Alcatel.

for the operation of three 140-Mbps systems. The first long-distance submarine deepwater cable (at a depth of 2500 m) went into service in 1987 between Marseille on the French mainland and Ajaccio on the isle of Corsica. The link, 392 km long, carried two systems each with a capacity of 280 Mbps operating at 1300 nm on a pair of single-mode fibers. The cable and the nine repeaters were manufactured and installed by Alcatel Submarcom. The U.K. and Ireland were connected by an optical fiber cable in 1989.

At the end of 1995, over 260,000 route-km of submarine cable with a total transmission capacity of 2.5 million 64-kbps circuits connected 70 countries. About half of the transmission capacity was installed in transatlantic cables. A record length of 100,000 km of submarine cable was added in 1996, which provided a further 1.9 million 64-kbps circuits and connected an additional 17 countries. Roughly 70% of this cable was produced by Alcatel and OCC, the remaining 30% by ABB Norsk, GPT, Pirelli, and Siemens.

28.3.1 Transatlantic Optical Fiber Cables

The first transatlantic optical fiber cable, TAT-8, laid in 1988, operated on 2×280 Mbps with a wavelength of 1300 nm on single-mode fibers. The attenuation of the fiber was 0.25 dB/km. The cable outside diameter was 2.5 cm, compared with 5.28 cm for TAT-6. For the first time an underwater branching unit was positioned on the continental shelf at a distance of 350 km from Penmarch in France, 520 km from Widemouth Bay in the U.K., and 6700 km from Tuckerton, New Jersey. The repeater spacing varied from 41 to 67 km. The transmission capacity of TAT-8 with 7680 telephone channels was fully exhausted within one and a half years. TAT-9 was installed in 1991 with double capacity (2×560 Mbps) but was booked up in a few months. A wavelength of 1550 nm was used, which reduced the attenuation of the fiber to 0.21 dB/km and increased the repeater spacing to 120 km. TAT-9 connected the United States and Canada with the U.K., France, and Spain. TAT-10 and TAT-11, laid in 1992 and 1993, respectively, both also had a capacity of 2×560 Mbps. To increase the transmission capacity, digital circuit multiplication equipment (DCME) was used, similar to the way that TASI was used on the coaxial cables TAT-1 to 6. The transmission capacity of 2×560 Mbps, equivalent to 16,128 voice channels, on TAT-9, was increased with DCME to 80,000 voice channels.

TAT-10 connected the United States with the Netherlands and Germany. It was laid in a record-breaking time of 21 days (an average of 270 km/day) by the newest AT&T cable ship, *Global Link*. Three tanks, each with a height of 7.5 m, could store 7600 km of submarine fiber cable. Two lateral propellers at the bow and two propulsion propellers at the stern increased the maneuverability of the ship so that cable laying did not need to stop during bad weather. Over 20 cable ships were used in the Atlantic region for cable laying and maintenance, of which three ships were from Alcatel Submarcom, five ships from AT&T, and 13 ships from C&W. Most of those ships were equipped with high-resolution depth finders to adjust the parameters for laying and to minimize the amount of cable used.

TA-11 connected the United States with the U.K. and France. When TAT-11 was laid, 110,000 telephone channels were in operation on the various transatlantic cables. An additional big step forward was made with the installation of the cables TAT-12 and 13 in 1995. Both cables had two pairs, which each carried a 5-Gbps signal. TAT-12 was laid between the United States and the U.K. and TAT-13 between the United States and France. The two landing points in the United States are interconnected, as are the landing points in the U.K. and France, so that the two cables can be operated as a ring that provides 100% standby: in case of interruption on one of the cables, the traffic is routed automatically via the other cable. The transmission capacity of each cable, and thus of the ring, is 2×5 Gbps, equivalent to 120.000 telephone channels, which doubled the already installed transmission capacity. EDFAs were used for the first time on those cables. They were located at 45-km intervals on the transatlantic section and at 75-km intervals between the U.K. and France.

A consortium of 50 international carriers planned to have a TAT-14 cable in operation by the end of 2000. The cable, with two legs with a total length of 15,000 km and four pairs on each leg, was to have a transmission capacity of $4 \times 16 \times 10$ Gbps = 640 Gbps if used in ring configuration and 1.28 Tbps when used as two parallel cables. The cable should link the United States at Manasquan and Tuckerton, New Jersey, with the U.K. at Bude (Widemouth Bay), France at Saint-Valery-en-Caux, the Netherlands at Katwijk, Denmark at Noerre Nebel, and Germany at Norden. Technical problems and competition from private submarine cable operators delayed implementation.[12] Private companies, taking advantage of the worldwide liberation of telecommunications services, laid transatlantic cables as follows:

- C&W laid PTAT-1 and PTAT-2, each with a capacity of 420 Mbps, in 1989 and 1992, respectively.
- Teleglobe, Canada, laid CANTAT 3, with a capacity of 5 Gbps, in 1993.
- C&W and MCI laid Gemini, consisting of a north and a south cable with a capacity of 40 and 30 Gbps, respectively, in 1998.
- Yellow (a consortium of Global Crossing, Level 3, and Viatel) laid AC-2, 5950 km long and with a capacity of 1.28 Tbps, in 2000.
- The Canadian firm 360networks laid the 360atlantic cable, with a capacity of 1.92 Tbps, in 2000.
- Flag Telecom laid Flag Atlantic-1, with a capacity of 2.4 Tbps in 2001.

Furthermore, C&W is building the 13,000-km transatlantic Apollo cable, with a total capacity of 3.2 Tbps. The cable system will have two transatlantic legs, each leg with four fiber pairs. It is planned to begin service by mid-2002. It will be the first transatlantic cable with 80 wavelengths each with a 10-Gbps signal per fiber pair. Apollo will connect Long Island and New Jersey with Cornwall, U.K., and Brittany in France. The transatlantic cables of the private companies are part of the global cable networks mentioned in Section 28.3.4.

28.3.2 SEA–ME–WE Cable System

The world's longest optical fiber submarine cable system, SEA-ME-WE 2 (Southeast Asia–Middle East–Western Europe second cable), 18,751 km long, was inaugurated

---

[12] TAT-14 began operation on March 21, 2001.

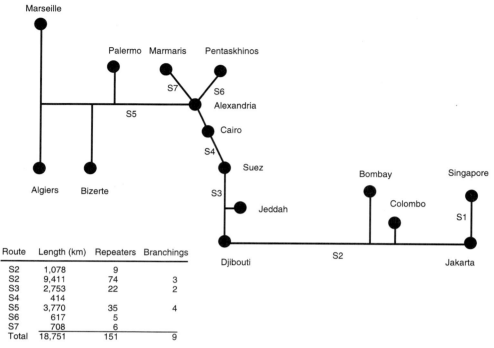

Route	Length (km)	Repeaters	Branchings
S2	1,078	9	
S2	9,411	74	3
S3	2,753	22	2
S4	414		
S5	3,770	35	4
S6	617	5	
S7	708	6	
Total	18,751	151	9

Figure 28.8 Route of SEA–ME–WE 2. (From A. A. Huurdeman, *Guide to Telecommunications Transmission Systems*, Artech House, Norwood, MA, 1997, Figure 4.18; with permission of Artech House Books.)

on October 18, 1994. The cable (Figure 28.8) links 13 countries on three continents: Singapore, Indonesia, Sri Lanka, India, and Saudi Arabia on the Asian continent; Djibouti, Egypt, Tunisia, and Algeria on the African continent; and Turkey, Cyprus, Italy, and France on the European continent. The cable operates with 1.5 μm on two single-mode fiber pairs with a capacity of 2×560 Mbps, the equivalent of fifteen 600-telephone channels. The system includes 160 submerged repeaters and branching units (the coaxial cable SEA–ME–WE 1 operates 820 repeaters!) and serves 52 telecommunication operators.

Five years later, on August 30, 1999, SEA–ME–WE 3 was inaugurated. With a length of 38,000 km, it was also the world's longest submarine cable. The route differs from SEA–ME–WE 2 in that from Jakarta it is extended to Perth in Australia, and from Singapore to Keoje in Japan, with landings in Mersing, Malaysia; Da Nang, Vietnam; Batangas, the Philippines; Macau, Hong Kong, Shantou, and Shanghai, China; Fengshan and Toucheng, Taiwan; and Okinawa, Japan. The cable also has additional landings at Malaysia, Thailand, Myanmar, Pakistan, Oman, and the United Arab Emirates. Instead of terminating in Marseille, it continues to Norden in Germany, with landing points at Sesimbra in Portugal, at Penmarch in France, at Goonhilly in the U.K., and at Ostende in Belgium. The cable, with 39 landing points, connects 33 countries with a total population of almost 2 billion. It is owned by 93 telecommunication operators from 58 countries. The transmission capacity is 8×2.5 Gbps.[13] It is the first submarine cable that offers wavelength routing (of the eight wavelengths) at offshore add/drop units. The companies Alcatel, AT&T Submarine Systems, KDD Submarine Cable Systems, and Pirelli implemented the project.

28.3.3 Caribbean ARCOS Network

At the end of the twentieth century, installation began of a vast optical fiber submarine cable ring in the Caribbean Sea called ARCOS (Americas Region Caribbean Optical-Ring System), owned by a consortium of 28 carriers. The 12-fiber-pair cable, 8600 km long, is to form a ring with 24 landing points in the following 15 countries: United States (at Miami Beach, Florida), Mexico, Belize, Guatemala, Honduras, Nicaragua, Costa Rica, Panama, Colombia, Venezuela, Curaçao, Puerto Rico, Dominican Republic, Turks and Caicos Islands, and the Bahamas. The initial transmission capacity will be 15 Gbps, with an option to upgrade to 80 Gbps. The ring structure largely eliminates the use of submersed repeaters.[14]

28.3.4 Global Submarine Optical Fiber Cable Systems

The worldwide telecommunications liberalization in the last 15 years of the twentieth century opened new opportunities for private companies to lay their own submarine cables and to operate those cables on a global basis. Two groups of global cable operators evolved: companies that installed a network for their own worldwide telecommunications service, and companies, called *carriers' carriers*, installed a network to lease transmission capacity to telecommunications operators. Typical of the first group are the global cable networks of the British company C&W and the Canadian company 360networks. The companies Flag Telecom and Global Crossing, which belong to the second group, implemented networks with very high transmission capacities which they leased to both new and established international telecommunication operators. Thus, transmission capacity became a commodity that could be leased or purchased on demand.

Oxygen Network   *Oxygen Network* was presented by the CTR Group Ltd. in December 1997. A 328,000-km cable, mostly submarine, with a capacity of 1.28 Tbps, was by 2003 to connect 171 countries at 265 landing points at a price of $14.7 billion. High-ranking telecommunications experts joined Oxygen Ltd., including Pekka Terjanne, the former secretary general of the ITU, who became the vice-chairman on March 1, 1999; Neil Tagare was the chairman. The overambitious project was later scaled back to 76 countries and finally abandoned in July 2000 for lack of confidence by the potential investors, who suspected that in the meantime the market had been captured by the four more successful companies noted above.

Global Network   *Global Network* is a global IP network with a transmission capacity of 10 Gbps being implemented on the 460,000 km of submarine optical fiber cable of C&W. The first phase was opened in May 2000 connecting Europe and Asia to its

---

[13] In 2002, the transmission capacity was upgraded to 10 Gbps on seven segments.

[14] ARCOS entered commercial operation in February 2002.

terrestrial backbone network in the United States, which it recently acquired from MCI WorldCom. In Europe C&W bought dark fiber[15] from the new carrier Ipergy and took a 51% stake in Ipergy.

360networks   360networks began in 1987 as the telecommunications division of the Canadian company Ledcor Industries at Leduc, Alberta. It engineered and built communications networks for telephone companies throughout North America. The division became a separate subsidiary in 1998 under the name Worldwide Fiber. The company was named 360networks in March 2000 to reflect its global ambition. The objective of the new company was to implement a global optical fiber terrestrial and submarine cable network 143,000 km long linking more than 100 cities in the Americas, Asia, and Europe by mid-2002. In November 2000, 360networks formed an alliance with Alcatel whereby Alcatel took ownership in 360networks and was to provide the system planning and equipment for network completion. By the end of 2000 in North America, more than 36,000 km of the network was laid, of which 23,000 km was ready for service, including 13 metropolitan networks. The 360atlantic network was almost ready for completion by the first quarter of 2001. The cable runs from Boston, Massachusetts, via Halifax, Nova Scotia and Dublin, Ireland, to Liverpool. The 360americas network connected the United States with Bermuda, Venezuela, and Brazil. About 20,000 km of the planned 28,000 km was laid. Through its strategic partner C2C, a subsidiary of Singapore Telecommunications, 360asia is being developed to connect China, Hong Kong, Japan, the Philippines, Singapore, South Korea, and Taiwan. The network 360pacific, with a length of 22,000 km, will connect 360asia with Canada and the United States.[16]

Flag   Flag (Fiber-optic Link Around the Globe) was founded in 1992 by Nynex Corporation (the owner of New York Telephone and New England Telephone, and now part of Bell Atlantic) along with Marubeni Corp. of Japan, the Dallah–Al Baraka Group of Jeddah, Saudi Arabia, and the Gulf Associates Inc. of New York. It was one of the first privately financed cables in the world. The first cable system, Flag Euro-Asia, was an optical fiber submarine cable connecting 11 countries, running from the U.K., via southern Europe, Egypt, Saudi Arabia, India, and Singapore to Japan. The cable went live on November 22, 1997 with a cable length of 27,000 km and a capacity of 5 Gbps, and carried traffic for over 60 telecommunication operators.

In 1999, Flag Telecom Holdings Ltd. was formed as a holding company based in Bermuda. In the same year the implementation was announced of two cables in a ring configuration called Flag Atlantic-1, with two landing points in the United States on Long Island and European landing points at Cornwall in the U.K. and Brittany in France. Each cable will carry 80 channels with a capacity of 10 Gbps each on three pairs, resulting in an aggregate capacity of 2.4 Tbps. The cables are being laid and jointly owned by the Global Telesystems Group (GTS) of the United States. Commercial service began in June 2001.

In 2000, Flag announced a cable called Flag Pacific-1, which will connect the United States with Japan by early 2002. The cable will be the first transoceanic

---

[15] *Dark fiber* is the name for fiber that is not (yet) used for transmission of a signal.

[16] Unfortunately, it appears that "the arrows" that killed so many pioneers might also kill 360networks: In June 2001 it filed for protection by the Supreme Court of British Columbia.

cable with eight fiber pairs. With a length of 22,000 km, each carrying 64 wavelengths at a capacity of 10 Gbps, thus an aggregate capacity of 5.12 Tbps, it will also be the longest transoceanic DWDM system at such a high capacity. Landing points are planned at Vancouver, Seattle, San Francisco, and Los Angeles on the North American west coast and in Japan at two points east of Tokyo.

Global Crossing   *Global Crossing* was founded in Hamilton, Bermuda, in 1997 with the objective of building a global city-to-city optical fiber cable network. The first part of the network, called Atlantic Crossing-1 (AC-1), connecting the United States with the U.K. and Germany, opened its service on May 26, 1998. AC-1 consisted of two cables, each with four pairs, with a total length of 14,000 km, which formed a self-healing ring to provide an internal restoration possibility in the event of an outage on a pair or on one of the cables. Landing points were at Brookhaven, New York, and in Europe at Whitesands in the U.K., at Beverwijk in the Netherlands, and on the isle of Sylt in Germany. AC-1 began with a capacity of 40 Gbps, which was increased to 80 Gbps within 18 months. Global Crossing expanded enormously in 1999:

- *June:* Global Crossing acquired Global Marine from C&W with 1200 employees, 24 cable ships, and 22 submersible vehicles.[17]
- *July:* Global Crossing merged with U.S. West and acquired Frontier Corporation, both leading long-distance operators in the United States.
- *August:* Global Crossing began South American Crossing (SAC), a 18,000-km ring from the Virgin Islands connecting Rio de Janeiro, São Paolo, and Buenos Aires via the east coast of South America, Caracas to the north, and Bogotá, Lima, and Santiago via the west coast of South America. The Argentinean company Impsat Corporation was bought to provide the land portions of the network, especially a trans-Andean cable from Buenos Aires to Santiago de Chile.
- *September:* Global Crossing acquired 49% of S.B. Submarine Systems Company Ltd. (the stake of C&W and Hong Kong Telecom), with China Telecommunications owning 51%. A joint venture called Asia Global Crossing (AGC) was formed with Microsoft and with Softbank of Japan, with the objective of implementing a 19,500-km network connecting Japan, South Korea, China, the Philippines, Malaysia, and Singapore.
- *October:* Global Crossing opened a Network Operation Center in London for the central technical management of the global network.
- *November:* Global Crossing formed a joint venture with the Hong Kong–based company Hutchison Whampoa Ltd., giving Global Crossing access to networks on Hong Kong and on the mainland of China.
- *December:* Global Crossing opened the first stage of its Pan-European Crossing (PEC), consisting of three rings with a total length of 7200 km and a capacity of 100 Gbps, connecting the U.K., Belgium, The Netherlands, France, Denmark, and Germany.[18]

[17] Global Marine, the oldest submarine company, with a tradition that started with the first transatlantic telegraph cable in 1858, thus lost its British nationality!

[18] Two additional rings were added in 2000, connecting cities in Germany, Switzerland, France, and Italy. A submarine ring called the Irish Ring connected Dublin to the other 23 cities of the PEC.

- *December:* Global Crossing completed its Mid-Atlantic Crossing (MAC), linking New York via Miami with the Virgin Islands. Around this time also the 21,000-km ring Pacific Cable (PC-1) was completed, with landing points in the United States at Seattle and Los Angeles, and in Japan at Tokyo and Osaka.

The expansion continued in 2000. A 10,000-km ring dubbed Pan American Crossing (PAC) was installed linking Mexico, Central America, and the Caribbean. A Mexican ring connected Mexico City, Guadalajara, and Monterrey. Asia Global Crossing went into partial operation with a capacity of 80 Gbps upgradable to 2.56 Tbps. At the end of the twentieth century Global Crossing announced completion of its worldwide IP-based core network connecting 200 cities in 27 countries across the Americas, Asia, and Europe via 160,000 km of cables by the middle of 2001. It had 10,000 employees and access to about 80% of the international traffic.[19]

28.3.5 African Cable Network Africa ONE

The submarine cable project *Africa ONE* was proposed in 1994 by AT&T Submarine Systems Inc. at a meeting of African telecommunications operators during the Africa Telecom exhibition in Cairo. Africa ONE would be a 32,000-km submarine ring with some 30 landing points around the coast of Africa, connected to the noncoastal African countries by satellite, radio relay, and optical fiber cable. The project was to integrate with the Panaftel terrestrial radio-relay and Rascom satellite networks. Connections were also planned with the submarine systems SEA–ME–WE 2, Flag, Sat-2, and Columbus-2. It was assumed that a cable laid in 1998 with two pairs each with a capacity of 10 Gbps would cover international traffic requirements for the next 20 years. Direct telecommunications connections hardly existed in Africa, so that a large portion of inter-African traffic was sent through transit centers outside the continent at an annual cost of about $400,000. It was estimated, therefore, that the total cost for Africa ONE, planned at $1.6 billion, could easily be recovered from the reduction of those transit fees for inter-African traffic. A first project meeting was held at Nairobi in December 1996 with participation of national coordinators from 28 African countries. After a three-year period without noticeable progress, private enterprise rejuvenated the project. Columbia Technologies of New Jersey formed a new company, Africa ONE Ltd., in June 1999. The new company will own, operate, and maintain a cable network with a length of 39,000 km in the form of a self-healing ring around the African coast. The initial capacity of the cable will be 80 Gbps. It will have landing points on the African coast, the Middle East, and southern Europe and connect the African continent with 19 countries and 185 cities around the world. Global Crossing contracted project management and underseas construction, including equipment from Lucent Technologies, with project completion by the end of 2002.

---

[19] On June 21, 2001, Global Crossing reported completion of its worldwide core network "on budget and on time under difficult market conditions." The final connection joined Lima and Peru to the South America Crossing. Alas, on January 28, 2002, Global Crossing filed for Chapter 11 bankruptcy in the fourth-largest insolvency in U.S. history.

28.3.6   Various Submarine Cable Systems

The burgeoning demand for Internet traffic worldwide, requiring massive bandwidth connectivity to the United States but also on the European, Asian, and Latin American routes, caused a veritable boom in submarine cabling toward the end of the twentieth century. Whereas in the mid-1990s some cable factories had to be closed for lack of orders, new factories opened at the end of the century. Some 50,000 km of cable was installed in 1997. The 100,000-km mark for annual deployment was reached in 1998, when the annual production increased to an all-time maximum of about 140,000 km. A summary of major optical fiber submarine cables is given in Table 28.1 (excluding those already mentioned in this chapter).

In the 1980s and early 1990s, practically all the submarine cables were supplied and laid by five large companies which shared the market approximately as follows: the British company STC, 40%; the American company AT&T, 25%; the French company Alcatel Submarcom, 20%; and the Japanese companies NEC and Fujitsu together, 15%. Most submarine cable projects were planned and operated on a consortium basis by the related incumbent operators, such as AT&T, BT, and France Telecom. Generally, the cable suppliers used the cable ships of those operators.

According to KMI Corporation in 1999, the total amount of optical fiber submarine cable in the period between 1988 and 1998 had a value of $21.7 billion and was distributed as shown in Table 28.2. Whereas at the beginning of the twentieth century the world submarine cable telegraphy network was dominated by British private enterprise, which owned almost 70% of the network, at the end of the century the British companies owned less than 25% and the North American companies almost 50% of the optical fiber submarine cable network, as shown in Table 28.3. On the supplier side an even bigger shift took place in the last decade of the twentieth century, when basically only two companies still appear to have complete and fully integrated manufacturing and cable-laying facilities: the French company Alcatel, which now owns 13 cable ships, and the new U.S. company Tyco, which in 1997 acquired the submarine systems business of AT&T.[20] Fifty years of technical progress and fierce competition have resulted in a tremendous decrease in the cost of transmission per equivalent telephone channel (64 kbps) as shown in Table 28.4.

28.3.7   Repeaterless Submarine Cable Systems

The low attenuation of optical fiber made it possible to operate submarine cable over relatively long distances without requiring submerged repeaters around 1990. Repeaterless submarine systems became competitive against radio-relay and underground cable in coastal areas, especially when inland lakes and rivers cause difficulties in terrestrial solutions. Those applications are mainly in shallow or low-depth waters, which facilitate cable laying and allow simpler cable construction. Moreover, the cable for repeaterless systems does not need remote power feeding. Special cable has been developed for repeaterless use, such as the *Minisub* (Figure 28.9), which has been laid at a record depth of 6000 m.

---

[20] Nortel phased out in 1995, but in 2000 it announced a comeback in submarine cable business. NEC made, on March 8, 2001, a "strategic alliance to deliver integrated submarine cable networks" with Global Crossing's Global Marine Systems Ltd. and the Japanese cable manufacturer OCC Corporation.

TABLE 28.1   Major Optical Fiber Submarine Cables

Cable	Year	Capacity (Mbps)	Length (km)	Route (End Points)
TPC-3	1989	280	13,387	Point Arena (U.S.)–Chikura (Japan)
EMOS-1	1990	3 × 280	2,852	Marseille–Tel Aviv
GPT	1989	280	3,738	Guam–the Philippines–Taiwan
HJK	1990	260	4,471	Hong Kong–Japan–Korea
NPC	1991	420	9,400	Seward/Ocean City–Chikura
TPC-4	1993	2 × 565	9,950	Vancouver/Point Arena–Chikura
Tasman-2	1992	2 × 565	2,500	Auckland–Sydney
Pacrim East	1993	2 × 565	13,000	Auckland–Hawaii
Pacrim West	1994	2 × 565	10,000	Sydney–Guam[a]
Americas-1	1994	3 × 565	5,500	Miami–Fortaleza (Brazil)
Columbus 2	1994	—	12,000	Italy/Spain–Mexico/South America
Unisur 1	1994	2 × 565	1,900	Florianapolis–Buenos Aires
Brazilian Domestic	1995	3(24) × 565	3,000	Natal–Florianapolis
Rioja	1994	2 × 2,500	1,800	Alkmaar (The Netherlands)–Santander (Spain)
ECFS	1995	4 × 622	1,730	Eastern Caribbean
Panamericas	1996	2 × 622	11,000	United States–Mexico–Valparaiso

DWDM Cable Systems

Cable	Year	Capacity (Gbps)	Length (km)	Route (End Points)
APCN1	1995	5	12,000	Japan–Indonesia
Jasuraus	1997	5	2,800	Jakarta–Port Hedland (Australia)
FO Gulf	1997	5	1,300	Kuwait–United Arab Emirates
TPC-5	1998	5	22,500	United States–Japan
Southern Cross	1999	40	29,000	Australia/New Zealand–United States
Atlantis 2	1999	20	—	Spain–Argentina
China–United States	2000	80	27,000	China–United States
Japan–United States	2000	640	21,000	Japan–United States
APCN2	2000	2,560	19,000	Japan–Singapore
Maya 1	2000	20	4,524	United States–Colombia

[a] The Pacrim West optical fiber cable was the deepest submarine cable, laid at depths of 8000 to 9000 m.

An experimental repeaterless cable was laid in 1982 between Cagnes-sur-Mer and Juan-les Pins on the French Mediterranean coast. Transmission tests were made at 34 and 140 Mbps, including TV on 34 Mbps. The first commercial repeaterless submarine optical fiber cable was laid in 1988 across the Irish Sea between Dublin and Holyhead. The cable, with a length of 126 km, had six fiber pairs, each carrying a 140-Mbps system operating on 1550 nm. A similar cable with a length of 152 km was laid one year later between Brighton, U.K., and Dieppe, France. A branching unit on the English side routed three systems to a terminal of BT and three systems to the

TABLE 28.2   Optical Fiber Submarine Cable Investment, 1988–1998

Region	Percent
Atlantic	29
Southeast Asia	25
Pacific	19.5
Mediterranean	13.8
Northern Europe	5
Caribbean	4
West Indian Ocean	3.7

new operator Mercury Ltd. In 1991, four cables with lengths of 15, 30, 70, and 185 km, each with a capacity of 140 Mbps, connected Taiwan with surrounding islands. In the same year, a cable 180 km long was laid between Nova Scotia and Newfoundland which carried 560 Mbps on each of its six fiber pairs. A cable with a length of 96 km and a transmission capacity of 560 Mbps was laid between St. Margarets's Bay, U.K., and Dunkirk, France, in 1992. In the same year, 180 km of submarine cable connected Puerto Rica with the Virgin Islands St. Thomas and Tortola. The cable system, named *Taino-Carib* (referring to the Taino Indians, who first inhabited the islands), had six pairs, each carrying a 560-Mbps signal, corresponding to an equivalent capacity of $6 \times 8064 = 48{,}384$ voice channels. DCMS increased the capacity to 225,000 voice channels.

The world's longest repeaterless route, at a length of 309 km, operated at 622 Mbps in 1995. It was laid between the Spanish mainland and the island of Majorca. One year later a transmission capacity of 2.5 Gbps was achieved over 300 km on a cable connecting Jamaica with the Cayman Islands. By that time about 30,000 km of submarine cable was used on repeaterless routes.

Cable for repeaterless systems can be manufactured in standard lengths of 100 km, which significantly reduces the number of joints. A first repeaterless system without an installation joint was laid over a distance of 150 km between Penthashkinos and Yeroskipos along the coast of Cyprus in 1994.

TABLE 28.3   Top Ten Owners of Submarine Cable Systems

Company	Country	Number of Cables
AT&T	United States	100
MCI	United States	89
Teleglobe	Canada	80
British Telecom	U.K.	80
C&W	U.K.	72
Sprint	United States	61
Telefonica	Spain	56
Belgacom	Belgium	51
Singtel	Singapore	50
KDD	Japan	48

Source: Data from a study published by KMI Corporation, 1997.

TABLE 28.4 Summary of Cost per Telephone Channel

Year	Cable System	Number of Channels	Cost per Channel ($)
1956	TAT-1	89	557,000
1957	Hawaii-1	91	378,000
1964	TPC-1	167	406,000
1970	TAT-5	1,440	49,000
1974	Hawaii 2	1,690	41,000
1975	TPC-2	1,690	73,000
1983	TAT-7	8,400	23,000
1988	TPC-3	37,800	16,000
1988	TAT-8	37,800	9,000
1992	TPC-4	75,600	5,500
1993	TAT-10	113,400	2,700
1996	TPC-5	604,000	2,000
1996	TAT-12/13	604,000	1,000
1998	AC-1	2,457,600	<125
1999	China–United States	4,915,200	<200

Source: Data from a study published by Teleography, Inc., 1998.

Repeaterless optical fiber cable systems are also being used to connect offshore oil platforms with onshore facilities. Offshore platforms have used optical fiber cable for communication on the platform since the mid-1980s; however, communication with onshore facilities used to be a domain for radio-relay or satellite, but the increased importance of security against outside tapping and interference may justify the use of submarine cable. In 2000, as an example, a project started connecting 10 oil platforms in the Persian Gulf with onshore facilities in Saudi Arabia and Bahrain via a submarine optical fiber cable in a ring configuration with a length of 300 km and an overall capacity of 2.5 Gbps.

A unique way of laying submarine cables was made available with completion of the 50-km-long Eurotunnel across the English Channel in 1993. A service tunnel runs parallel with the two railway tunnels and provides convenient space for cables. By year-end 2000, French and British telecommunications operators as well as operators of pan-European networks operated 11 optical fiber cables through this service tunnel. The French and British endpoints of the Eurotunnel, Calais, and Folkestone, became important traffic routing stations for various European telecommunication networks.

A combination of EDFAs and Raman amplification further increased the capacities and maximum achievable distances. For this purpose an active pump with an optical source for the Raman circuitry is located at the terminal, and a remote amplifier box (RAB), consisting of the passive part with erbium-doped fiber circuitry, is inserted in the cable. An additional fiber connects the RAB with the pump at the terminal.

In 2000, Alcatel installed a repeaterless submarine cable 250 km long between the U.K. and the Netherlands, with a capacity of 32×10 Gbps per fiber. The same

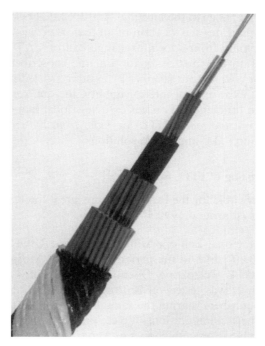

Figure 28.9   Double-armored Minisub cable, 1996. (Courtesy of Siemens.)

company achieved laboratory results with 32×40 Gbps per fiber over 250 km and 32×10 Gbps over 450 km.[21]

28.4 FIBER-IN-THE-LOOP SYSTEMS

The transmission characteristics of optical fiber cable concerning attenuation, bandwidth, and crosstalk are far superior to those of copper cables. Moreover, the raw material, sand, used for fiber production is abundantly available, although extensive purification and processing are required. Those advantages, and the emerging demand for broadband services for business as well as for residential subscribers, destines optical fiber to become the successor to copper cable in the access network. At the end of the 1980s, *fiber in the loop* (FITL) was seen as the near-future technology for the access network. FITL would support universal integrated and interactive broadband communications into the office, workshop, and home. Many solutions were proposed for making fiber economically justifiable to deliver to the desk/terminal of the user: FTTZ, fiber to the zone; FTTR, fiber to the remote unit; FTTC, fiber to the curb; FTTK, fiber to the kerb; FTTP, fiber to the pedestal; FTTB, fiber to the building, basement, or business; FTTF, fiber to the floor; FTTO, fiber to the office; FTTD, fiber to the desk; FTTA, fiber to the apartment; FTTH,

---

[21] In June 2001, at Supercomm in Atlanta, Georgia, Hitachi presented an ultra-long-haul system with a capability of sending 128×10 Gbps unrepeatered over distances up to 8000 km.

fiber to the home; FTTT, fiber to the terminal; and FTTU, fiber to the user. The last letter of the acronym indicates the location of an *optical network termination* (ONT) at the end the fiber coming from a local exchange and/or CATV head end. Continuation from the ONT to the final destination at the subscriber's premises, usually called the *last mile*, for cost reasons is either by existing twisted pair or coaxial cable. If such cable does not exist or is of inferior quality, the quickest and cheapest solution might be by radio [in combination with wireless local loop as described in Section 25.3, called *radio in the loop* (RITL) or *radio from the curb* (RFTC)]. Figure 28.10 shows some major FITL and RITL solutions.

28.4.1   Worldwide Testing of FITL Solutions

The use of optical fiber cable for the local loop drew great interest; worldwide testing of FITL solutions started around 1990.[22]

United States   Most regional Bell operating companies (RBOCs) and independents made tests on over 10,000 lines in the period 1989–1994. Most publicized were the tests performed by GTE Telephone Operations in Cerritos, California. Equipment from four suppliers was tested on 250 residential subscriber lines in point-to-multipoint (several subscribers sharing the same fiber) and point-to-point (individual fiber for each subscriber) configurations. Voice, data, and video services were provided over fiber in combination with copper and coaxial cable. Pacific Bell installed a trial network for 3000 subscribers at Sutter Bay, Sacramento, California, providing telephone, video, and telecommuting services.

Japan   The first major test using optical fiber in the access network began in Japan in 1978 with two-way video service to 168 subscribers.

Australia   Early trials in Centennial Park, Sydney, revealed high cost and power feeding problems. Further trials, including 2-Mbps video, were made on 200 lines in business and residential areas in 1993.

New Zealand   Telecom New Zealand made its first tests in 1989 and connected 100 subscribers in a combined FTTC/FTTH test in 1993.

Singapore   Singapore Telecom deployed FTTB as early as 1987.

Scandinavia   In Denmark, the Copenhagen Telephone Co. started FTTH trials on 25 lines in Ballerup, on the outskirts of Copenhagen, in 1990. In Finland the Helsinki Telephone Co. started a small FTTB trial with six apartment buildings in Helsinki in 1992. In the same year, Norwegian Telecom tested FTTH with 140 subscribers in Oslo. Swedish Telecom made FTTC tests on 30 lines at Örebrö in 1992 and added FTTH tests for 116 subscribers in 1994.

United Kingdom   British Telecom tested FTTC, FTTO, and FTTH on 300 lines at Bishop's Stortford in 1990–1992. The tests included new services, such as videotex

---

[22] Most information on those tests has been taken from a paper called "A light at the End of the Loop," which was published as a supplement to *Communications International*, *TE&M*, and *TelecomAsia* in 1992.

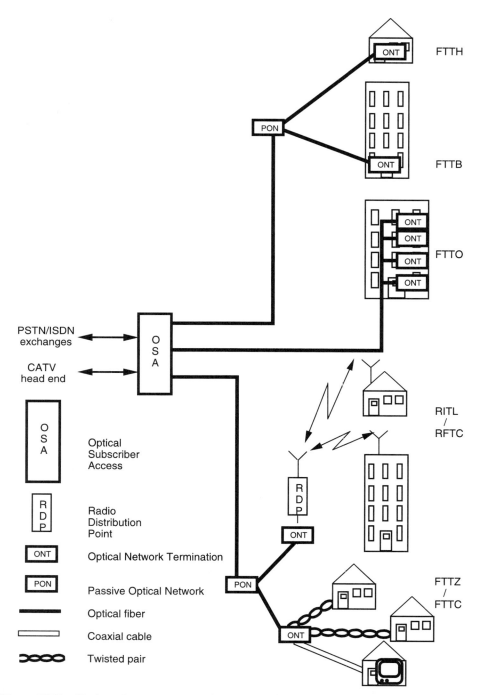

Figure 28.10 Various FITL solutions. (From A. A. Huurdeman, *Guide to Telecommunications Transmission Systems*, Artech House, Norwood, MA, 1997, Figure 4.19; with permission of Artech House Books.)

and video library. The town, some 50 km north of London, was selected for being fairly representative in its socioeconomic mix and its vicinity to British Telecom's Martlesham Heath Laboratories.

France   An optical fiber network operating on 850 nm was installed in the city of Biarritz as early as 1983. The project started with 40 subscribers, who could receive several French and foreign TV programs, videotex, and data retrieval and could use videophone. By 1986 some 1500 of the 5000 households were connected to the network.

Germany   In 1983 the Deutsche Bundespost initiated a FTTH project called BIG-FON (Breitbandiges Integriertes Glasfaser–Fernmelde–Orts–Netz; local broadband integrated optical fiber telecommunication network). BIGFON was carried out in a competition among German manufacturers to find cost-effective ways of transmitting several services to a subscriber over a single optical link.

In 1986, a six-year FTTO/FTTH project called Berkom (Berlin Kommunikation) began in Berlin. This project, also initiated by the Deutsche Bundespost, had the objective of advancing the development of potential services, applications, and systems for the future introduction of broadband-ISDN (B-ISDN) on an optical fiber network. The project was implemented in cooperation with 78 service providers: institutions, universities, local authorities, and manufacturers. The interactive application was tested in telemedicine, telepublishing, cityplanning, computer-integrated manufacturing, broadband information systems, multimedia workstations, and network architectures. Optical fiber lines with capacities between 2 and 140 Mbps connected the 78 participants via three broadband exchanges. The project went synchronously with a research program of the European Community called RACE (research and technology development in advanced communications technologies in Europe).

In 1988, the Deutsche Bundespost made an agreement for the implementation on a competitive basis of FTTC/FTTH pilot projects in seven cities with Raynet Corp. of Menlo Park, New Jersey, and the German telecommunications companies. The pilot projects OPAL 1 to 7 (optical access lines) were implemented in the period between 1989 and 1992 with the objective of evaluating various concepts for a cost-effective implementation of FITL networks.

Italy   The Italian telco SIP started in 1989 with the installation of FTTO in major metropolitan areas. In Rome, FITL was tested in an "optical island" for 100 subscribers between 1990 and 1995.

The Netherlands   The Dutch PTT tested FTTH together with the CATV operator of Amsterdam (KTA) with initially 100, later 250 residential subscribers in an Amsterdam suburb, Sloten, in 1990. WDM was used to separate the broadband and narrowband signals.

Portugal   Telecom Portugal tested FTTH in cooperation with the University of Aveiro in 1995. This test was coordinated with the European RACE II project FIRST (Fiber to the Residential Subscriber Terminal). The objective of FIRST was to find cost-effective fiber-based broadband access solutions that could compete with copper pairs.

Spain Telefonica made FTTC tests in 1990–1991 with equipment supplied by Raynet Corp., with 96 subscribers in the town Tres Cantos, near Madrid, and 76 subscribers at San Cugat des Valles, Barcelona.

Switzerland The Swiss PTT implemented a Baskom project (similar to Berkom) with 30 participants in Basel from 1990 to 1996. Residential FTTC tests, including CATV and ISDN, were made with 80 subscribers each in Sagno and in Les Planchettes in 1992.

28.4.2 Delay of FITL Deployment

The worldwide tests showed that cost parity with conventional copper access under favorable conditions would be possible with FTTC but would be difficult for FTTH in view of the high initial cost of fiber and active optical devices. Moreover, attractive broadband services that would justify a higher cost still had to be developed or could not be implemented due to regulatory obstacles. Large-scale deployment of FITL therefore did not develop after those worldwide tests.[23] Only a few examples are published about significant deployments. In the United States, Ameritech installed a fiber optic network for about 2.5 million subscribers in 1993–1994. Bell-South decided in 1995 to deploy fiber exclusively on greenfield sites and laid some 200,000 FTTC lines in that year. In Germany, as a result of the reunification in 1989, the Deutsche Bundespost had the unique opportunity to replace the over-60-year-old telecommunication network in the previous German Democratic Republic. A new modern telecommunications infrastructure was installed within four years, including FTTC and FTTB, for some 1.2 million subscribers. Figure 28.11 shows optical network termination equipment as used for this project.

In Japan, NTT announced in 1990 an ambitious plan according to which all Japanese households would have broadband access in a national wholly optical network by the year 2015. A smooth evolutionary deployment plan was published whereby FITL deployment starts in the most profitable service to the business area with FTTO. Service to dense residential areas will follow a few years later with FTTC, whereas FTTH will be deployed later, beginning mainly in newly developed residential areas. Gradually, the deployment of FTTC will decrease and migrate to FTTH and FTTO. After estimating that the national implementation of FITL according this plan would cost over $250 billion, NTT clarified that the plan was

---

[23] As an example of this delay, reference is made to *Communications Week* of November 2001, reporting that manifestations of nondeployment of FITL were given by the Brookings Institution and Kushnick's Institute at a conference hosted by the Columbia Institute for Tele-Information (CITI) at Columbia University in New York. It was claimed that Bell Atlantic had announced in 1993 having 8.75 million homes in its eastern seaboard region connected directly to fiber networks by 2000 and that Pacific Telesis announced 5 million homes to be connected by fiber on the other side of the continent. Although those and other announced fiber networks were never built, American telephone customers paid at least $58 billion in additional charges, levied with the approval of state regulatory commissions, to fund fiber network build-out. Actions toward obtaining refunding of these charges were announced at the conference.

Figure 28.11   Optical network termination equipment for FITL operation, 1992. (Courtesy of Alcatel SEL.)

"not a promise but a vision based upon technical possibilities" and that no firm time schedule could be given for implementation. In 2000, NTT once more announced a plan for nationwide implementation of FITL, this time, however, based mainly on FTTC. The objective of the new five-year plan is to make Japan the world's most advanced information technology (IT) country.

In France a positive example was given as to how the cost of FITL implementation could be reduced by use of a micro cabling system (MCS) developed by Siemens for cable laying without tearing up roads. With this system, an optical fiber cable is placed in a pressure-resistant copper tube insulated with a polyethylene jacket that can be laid in a 1- to 2-cm narrow groove with a depth of only 8 to 10 cm below the surface of a road or sidewalk. Once the cable is laid in the groove, it is covered with a small rubber tape and the groove is made watertight with a hot sealing compound. This system was first applied in the town of Castres (100 km east of Toulouse) in 1998. A multimedia city network with a cable length of 38 km was installed within eight weeks. The cable had 192 fibers, of which one fiber each was distributed to the various FTTO/FTTH access points of the local industry, the municipality, and residential subscribers.

The use of urban sewage systems emerged as another way of reducing the cost of fiber deployment. Because the sewers are gravity-driven systems, they are the deepest underground utility infrastructure in any community, providing a secure, well-protected environment for the fiber cables, with access to all buildings. Robotic devices take care of the dirty work, and special junctions ensure a clean cable entrance to buildings. Robots are also used for laying optical fiber cable in existing gas and water pipes.

Cost reduction is also expected from mass-produced plastic optical fiber (POF) for optical indoor networks. The American chemical company DuPont developed the first POF in the 1960s for use with optical sensors. At the end of the twentieth century, progress is being made in the development and manufacture of POFs, which have a minimum, but still significant attenuation of about 50 dB in the range 1300 to 1550 nm. In 1999, Bell Labs, together with the Japanese company Asahi Glass, demonstrated the transmission of signals with a speed of 11 Gbps at a wavelength of 1300 nm over 100 m of perfluorinated graded-index plastic optical fiber (PFGI-POF). In June 2000, Asahi Glass launched a new POF made of a transparent per-fluorinated polymer with a total attenuation below 50 dB over the entire range 1300 to 1550 nm and announced a further reduction to about 10 dB in the near future. Keio University in Japan implemented the world's first local area network operating Gigabit Ethernet across several buildings using POFs in April 2000.

Beyond high cost, another reason for the limited FITL deployment is given by the strong competition that came from the successful development of signal compressing and *digital subscriber line* (DSL) equipment. The copper-wire access networks worldwide have been installed at a cost of billions of dollars. At the beginning of the 1990s it was realized that instead of replacing this valuable copper-wire access network by optical fiber cable, it would cost less to enhance it to data capability by adding the appropriate electronic equipment. Eventually, the copper-wire access network will be replaced by optical fiber cable and wireless radio, but this might happen over a 20- to 50-year time span. The length of this transition time will depend on the cost-effectiveness of the newly developed electronic equipment and on the bandwidth requirements and attractiveness of multimedia services.

The access line between a subscriber and its local exchange is typically 2 km in Japan, between 2 and 3 km in Europe, and 4 to 8 km in North America. Over this limited distance the transmission capacity of the standard twisted copper pairs, orig-

TABLE 28.5   Digital Subscriber Line Equipment

DSL Variant	Acronym	Data Speed Up/ Down (Mbps)	Reach (km)
Asymmetric DSL	ADSL	1.5/8	5
Reduced ADSL	G.lite ADSL	0.512/1.5	3.5
Symmetrical DSL	SDSL	2.3	3
High-speed DSL	HDSL	1.5	3.5
Very high-speed DSL	VDSL	52	0.3
		2/25	≥1.0
ISDN DSL	IDSL	0.144	Any distance from local exchange

TABLE 28.6 Digital Subscriber Lines by Year-End 2000

Country or Region	DSL Subscribers	Country or Region	DSL Subscribers
Korea (Rep.)	4,100,000	Spain	110,000
United States	1,870,000	The Netherlands	100,000
Canada	586,000	Rest of Europe	160,000
Germany	425,000	Latin America	80,000
Taiwan	400,000	Rest of Asia	70,000
Australia	250,000	Total	8,151,000

Source: Market study called "Broadband International 100" undertaken by *Multichannel News International* and *Broadband International* and published as a supplement in those magazines in June 2001.

inally conceived for the transmission of a voice channel with a bandwidth of 4 kHz, can be increased at a relatively modest cost. Signal compressing equipment was therefore developed that reduced the bit rate of a video signal from 140 Mbps to 1.5 or 2 Mbps, whereas DSL equipment was developed, and standardized by the ITU, for transmission speeds on the access line from a few hundred kbps to over 50 Mbps in the versions summarized in Table 28.5. About 8 million DSL lines were installed worldwide at the beginning of the twenty-first century, distributed approximately as given in Table 28.6.

Another competitive solution to FITL, as well as to DSL, comes from the electricity utilities with a new technology called *powerline telecommunications* (PT). The fascinating idea of using electricity outlets in any home, even in any room, for telephony and data came up in the early 1990s. PT is conceived of an outdoor last-mile point-to-multipoint system from a low-voltage transformer to an electricity socket in the home, as well as an indoor system connecting telephone and computer devices through the electrical outlets in a home. Practical application is still delayed by regulatory restrictions, noise generated by switching of electrical equipment, and data-sharing limitations. At the end of the twentieth century, however, solutions to those problems were announced and extensive field trials made in a dozen European countries, the United States, Canada, Singapore, and Hong Kong.[24]

REFERENCES

Books

Huurdeman, Anton A., *Guide to Telecommunications Transmission Systems*, Artech House, Norwood, MA, 1997.

Kanzow, Jürgen, and Helmut Ricke, *Berkom Broadband Communication within the Optical Fiber Network*, Decker's Verlag, Heidelberg, Germany, 1992.

Tenzer, Gerd, *Fiber to the Home*, Decker's Verlag, Heidelberg, Germany, 1991.

---

[24] Commercial powerline telecommunications began July 2001 in Germany in the towns of Essen and Müllheim, where the utility RWE (Rheinisch Westfälische Electrizitätsgesellschaft) offers always-on-Internet. Telephony was to be added in about a year. A backbone network with a capacity of 155 Mbps connects various low-voltage transformers, from which subscribers share a 2-Mbps up-and-down service. At the customer's premises a modem is to be connected to an electrical outlet without any installation.

Articles

Anon., A light at the end of the loop, supplement to *Communications International, TE&M,* and *TelecomAsia,* 1992.

Anon., The path to global standards for optoelectronics, *NTT Review,* Vol. 11, No. 3, May 1999, pp. 10–24.

Anon., The Broadband International 100, *Multichannel News International,* special supplement to June 2001 issue, pp. i–xxiii.

Bigo, S., and W. Idler, Multi-terabit/s transmission over Alcatel TeraLight fiber, *Alcatel Telecommunications Review* fourth quarter 2000, pp. 288–296.

Bigo, S., et al., Road to ultra-high-capacity transmission, *Alcatel Telecommunications Review,* third quarter 2001, pp. 177–180.

Böck, Jürgen, and Lothar Finzel, Get into the groove, *Telcom Report* (Siemens), Vol. 20, No. 2, 1997, pp. 42–45.

Carnevale, Mary Lu, Nynex Corp. plans to build longest fiber-optic link, *Wall Street Journal,* May 14, 1992.

Devaney, John, WorldCom lays down the terabit challenge, *FibreSystems Europe,* March 2001, p. 31.

Felicio, Daniel, Regionale Netze in 6000 Meter Meerestiefe, *Telcom Report* (Siemens), Vol. 19, No. 5–6, 1996, pp. 36–39.

Freeman, Tami, Oil companies connect with subsea fibre cable, *FibreSystems Europe,* October 2001, pp. 63–64.

Gautheron, Olivier, Submarine systems sail towards the terabit horizon, *FibreSystems Europe,* February 2001, pp. 54–60.

Gautheron, Olivier, Submarine optical networks at the threshold of Tbit/s per fiber capability, *Alcatel Telecommunications Review,* third quarter 2001, pp. 171–179.

Golden, Paul, CityNet's fibre roll-out eases last-mile logjam, *FibreSystems Europe,* September 2001, pp. 43–46.

Keller, John J., AT&T to open undersea fiber-optic line, *Wall Street Journal,* August 14, 1992.

Kenward, Michael, Plastic fibre homes in on low-cost networks, *FibreSystems Europe,* January 2001, pp. 35–38.

Kobayashi, Ikutaro, The beginning of terabit/s communication technologies, *NTT Review,* Vol. 11, No. 4, July 1999, pp. 4–38.

McClelland, Stephen, In at the deep end, *Telecommunications International,* Vol. 35, April 2001, pp. 18–24.

Soja, Thomas, Crosscurrents and opportunities: the undersea fibre-optics industry, *Telecommunications International,* Vol. 30, No. 1, January 1996, pp. 53–58, 76.

Thornton, Geoffrey, The 10 megatrends of the submarine cable community, *Telecommunications International,* Vol. 35, June 2001, pp. S1–S12.

Walker, Matt, Huawei makes a move into the telecoms elite, *FibreSystems Europe,* February 2001, pp. 67–70.

Weinzierler, Winfried, A caravan of bits on the silk road, *Telcom Report* (Siemens), Vol. 21, No. 2, 1998, pp. 18–19.

Internet

www.alcatel.com, Information superhighways: Alcatel cables under the Seine—a world technology first.

www.sff.net/people/Jeff.Hecht/chron.html, A fiber-optic chronology.

29

ELECTRONIC SWITCHING

29.1 CONTINUATION OF DEPLOYMENT OF THE PREWAR SWITCHING SYSTEMS

Most telephone networks in Europe and Southeast Asia were heavily damaged during World War II. The reconstruction started basically with improved versions of the prewar crossbar and rotary switching systems.

29.1.1 Crossbar Switching

An improved version of crossbar switching came in the 1960s when the bars were replaced by reed relays. A *reed relay* (developed by Bell Labs) is a small, glass-encapsulated, electromechanical switching device.[1] The contacts latch, so no holding current is required, and they are forced to release upon terminating the communication. Crossbar switching systems were developed and manufactured by most manufacturers of switching equipment, as summarized in Table 29.1. A typical crossbar-switching field is shown in Figure 29.1.

29.1.2 Siemens Rotary Switch

Over a 60-year period, exchanges with the HDW (lift-turning switch) switch of Siemens (Section 16.1.4) were manufactured for 3.9 million subscriber lines in Germany and 2.5 million subscriber lines outside Germany. The production of this two-motion

---

[1] A common switching control selects the reed relay to be closed in response to the number dialed. Pulses sent through a coil wound around the relay capsule change the polarity of plates of magnetic material alongside the glass capsules. The contacts open or close in response to the direction of magnetization of the plates, which is controlled by the polarity of the pulses.

The Worldwide History of Telecommunications, By Anton A. Huurdeman
ISBN 0-471-20505-2   Copyright © 2003 John Wiley & Sons, Inc.

TABLE 29.1   Crossbar Switching Systems

Country	Manufacturer	System	Introduction	Millions of Lines Worldwide (Approx.)
United States	AT&T/Western	No. 1	New York, 1938	6
	Electric	No. 4	Philadelphia, 1943	11
		No. 5	Media, PA, 1948	28
	Kellog	1040/No. 7XB	1950	<0.01
Sweden	L.M. Ericsson	ARF 50	Helsinki, 1950	
		ARM10	Rotterdam, 1923	17
	Licensed to:			
	North Electric (U.S.)	NX 1/2	1956	2
	SFTE/CIT (France)	CP 400	Beauvais, 1956	9
Belgium	BTM	8A	Skien, Norway, 1954	<0.01
France	CGCT/CIT	Pentaconta	Melun, 1955	18
Germany	Mix & Genest	KS/53	1955	1
Norway	STK	8B	1958	0.350
Japan	NEC et al.[a]	C 1–8	Musashifuchu, 1958	60
Eastern Europe	NIITS et al.[b]	ATS-K (M)	Berlin, 1962	9
U.K.	Autelco	5005	Sydney, 1963	2

[a] Fujitsu Ltd., Hitachi Ltd., Nippon Electric Co. Ltd., and OKI Electric Industry Ltd.

[b] The development was carried out by Tesla in Czechoslovakia and BHG in Hungary in cooperation with research institutes in the USSR (NIITS, Research Institute for Urban and Rural Telephone), in the German Democratic Republic (ZLF, Central Laboratory for Telecommunications), and in Czechoslovakia (VUT).

HDW switch ceased in 1968 in favor of a rotary uniselector motor switch called an *EMD switch.* The development of the rotary uniselector motor switch had begun in the early 1930s. A first version was used in the construction of automatic private long-distance networks for the German and Italian railways. World War II and postwar reconstruction, however, caused a 15-year interruption in motor switch development in Germany. In Switzerland, however, the implementation of automatic trunk switching with motor switches, planned in 1938, could, with some delay, still be realized with equipment manufactured by Albiswerk in Zurich. The first exchange was brought into service in Lausanne in 1943, followed by Biel in 1944 and Bern in 1945. The same Swiss motor switch was then used in the Netherlands to equip local exchanges in Amsterdam, one exchange already having a capacity of 40,000 lines.

In the early 1950s, Siemens resumed development of the motor switch and in 1954 introduced the noble metal rotary switch [in German, Edelmetal-Motor Drehwähler (EMD switch)]. The contacts for the speech circuits in this new rotary uniselector motor switch are made of a palladium–silver noble-metal alloy. Moreover, the functions of the contact bank and the multiplying between contact banks are combined,

Figure 29.1   Crossbar switching field. (Courtesy of Alcatel SEL.)

the banks consisting of bronze strips arranged in the form of a helix around a vertical insulating supporting wafer. This does away with the need for soldering. EDM switches are most commonly used as 100-outlet selectors, in keeping with the decimal system. The first public EMD exchange was brought into service in April 1954 in Munich. In the same year, other EMD exchanges were installed in Italy, Luxembourg, Finland (in Salo and Kajaani), and in Venezuela at the Nuevo Granada exchange in Caracas. Figure 29.2 shows the switch, without the contact banks, in two versions: left with four and right with eight parallel contacts. The EMD system was officially approved by the German Administration (then the Bundespost) in 1955 for general use within its network up to the year 2020, thus for a system lifetime of 65 years, almost as long as achieved with Strowger's original step-by-step switch. EMD was not only a success in Germany; within 30 years almost 60 million lines were served by EMD exchanges, about 35 million in Germany and 25 million outside Germany. EMD equipment was produced under license in Argentina, Finland, Greece, Indonesia, Iran, Italy, Pakistan, Republic of Korea, South Africa, and Switzerland.

29.1.3   End of the Strowger Switch

Strowger's step-by-step two-motion switch with various improvements remained the state-of-the-art until crossbar switching in the early 1930s increased capacity, speed, and reliability, and reduced power consumption and noise of automatic telephone exchanges. In 1978, when 99% of the telephone network in the United States was fully automated and electronic digital switching systems emerged, Autelco had supplied Strowger telephone exchanges to AT&T for approximately 10 million lines and exported from their Chicago–North Lake plant exchanges for another 5 million

Figure 29.2 Noble metal rotary switch, 1955. (Courtesy of Museum für Kommunikation, Frankfurt, Germany.)

lines. Even then, supply of step-by-step equipment continued, so that in the United States until 1984 the number of digital exchange lines installed was lower than the number of step-by-step analog exchange lines. Worldwide, the Strowger system has been used in more than 70 countries. In 1987, GTE decided to concentrate on operating and services and to give up its worldwide interest in manufacturing. Siemens bought most of the telecommunications manufacturing facilities, but compatibility between the GTE and the Siemens digital switching systems proved to be difficult to achieve. GTE therefore kept their switching manufacturing plant at Phoenix, Arizona, for the supply of extensions and spare parts. Thus, the almost 100-years tradition of uninterrupted switching equipment manufacturing by Strowger and its successors ended smoothly.

29.2   IMPLEMENTATION OF AUTOMATIC TELEPHONE SWITCHING

The implementation of automatic telephone switching (see Section 16.2) continued with Phase II, national automatic interexchange operation, using electromechanical crossbar and rotary switching equipment. Phase III, international automatic telephone switching, started in 1970, with electronic switching equipment.

29.2.1   National Automatic Switching

The implementation of nationwide automatic switching that had begun in Switzerland before World War II was taken up again in the 1950s and was completed in most industrialized countries around 1980. Table 29.2 summarizes the countries with more than 1 million main lines that had completed nationwide automation by January 1, 1978.

TABLE 29.2   Chronology of Worldwide Completion of Nationwide Automatic Switching

Country	Date of Completion	Automatic Lines, January 1, 1978	Percent Completed, January 1, 1978
Switzerland	December 3, 1959	2,599,000	100
The Netherlands	May 22, 1962	3,933,000	100
Federal Republic of Germany	April 29, 1966	15,748,000	100
Belgium	October 8, 1970	2,050,000	100
Italy	October 1970	>10,000,000	100
Austria	1972	1,747,000	100
Sweden	June 15, 1972	2,599,000	100
U.K.	1977	15,184,000	100
Australia	—	3,795,000	97.1
Canada	—	8,919,000	99.9
France	—	9,913,000	99.9
Greece	—	1,888,000	99.2
United States	1978	>160,000,000	99.9 since 1968
Japan	March 1979	34,959,000	99.7
Mexico	—	2,030,000	3.2
Yugoslavia	—	1,054,000	97.5

29.2.2   International Automatic Switching

For both technical and political reasons, the evolution from local and national public automatic telephone switching to international direct dialing (IDD) took many years. Semiautomatic international switching came first, in which an operator at a center mediated between the calling and the called subscriber in the two countries involved. The various switching systems could not operate with each other, so that special international transit switches had to be developed that could handle both the specific characteristics of the various national switching systems and the different national signaling procedures. To enable worldwide introduction of IDD, the CCITT successively defined a common international signaling system (Table 29.3), a world division into nine switching regions (Figure 29.3), a system of two-digit country codes,[2] and a three-level international network architecture consisting of national, continental, and intercontinental networks.

IDD started between Switzerland and Germany in 1955, when the two neighboring towns of Basel and Lörrach were interconnected. IDD service between Belgium and Germany, and Brussels–Paris, in 1956. With the first transatlantic telephone cable (TAT-1) in service in 1956 and *Intelsat-1* in service in 1965, intercontinental semiautomatic switching started on March 30, 1963 between New York and London. Full automatic intercontinental switching started in March 1970 between New York and London. Intercontinental operation using signaling system No. 6 began between New York, Sydney, and Tokyo on July 17, 1978.

---

[2] As an exception a common one-digit country code, the "1," was given to the United States, Canada, and several Caribbean islands. All other countries have as first digits the number of their region (e.g., 54 for Argentina, 55 for Brazil, 61 for Australia, 91 for India, and 965 for Kuwait).

TABLE 29.3   CCITT Signaling Systems

System	Year	Specifics
No. 1	1934	One-frequency, for manual operation
No. 2	1938	Two-frequency, for semiautomatic operation
No. 3	1949	One-frequency, for automatic operation
No. 4	1953	Multifrequency code (MFC), for automatic operation
No. 5	1964	2-out-of-6 MFC, with confirmation, for intercontinental operation via submarine cable and satellites
R2	1968	In 1958, defined by CEPT as European MFC, after confirmation by CCITT as "regional" (R) system in 1968 worldwide used, except in North America
No. 6	1970	2400-bit/s signaling on a separate common channel for analog switching systems
No. 7	1980	64-kbps signaling on a separate common channel for digital switching systems

29.3   ELECTRONIC SWITCHING SYSTEMS

Progress in switching up to World War II was made exclusively by electromechanical inventions. Whereas electronic vacuum tubes were used widely for transmission systems beginning in the early 1920s, research on the application of electronics for switching did not begin until the 1930s. The electromechanical switching systems, especially those with common control techniques as introduced with crossbar switching, required numerous relays. Both the electromechanical switching elements and the relays, however, operate relatively slowly, are voluminous, noisy, and require regular maintenance. Replacing relays by electronic valves could increase the operational speed of a switch, however, at the cost of tremendous heat dissipation by the thousands of tubes required and would require even more maintenance. Therefore, a program started in the 1930s in Bell Labs to search for a replacement for the vacuum tube, which resulted in the invention of the transistor in 1947 (Section 24.1) and opened the way to electronic switching.

29.3.1   Evolution toward Electronic Switching

Electronic implementation of a switching function was introduced in the United States with a *common control* (CC) for crossbar switching. One controller, called a *marker*, established calls through idle coordinate switch paths, thus considerably reducing, if not eliminating, the individual switch controls. The advent of integrated circuitry and microcomputers spurred the application of electronics to switching so that the CC for electromechanical switching could be extended to *stored program control* (SPC) electronic switching. The basic understanding of stored program control is explained briefly in Technology Box 29.1.

The first call through a SPC system could be placed in 1958 at Bell Labs, followed by commercial service in Morris, Illinois, in June 1960. While SPC and electronics brought improvements in the engineering of switching networks, they also brought in new telephone services, such as abbreviated dialing, call forwarding, and call waiting.

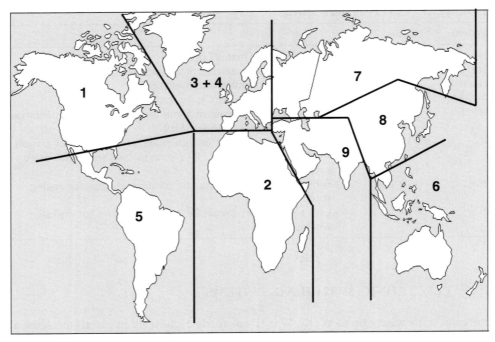

Remark: Regional borderlines are approximate only; e.g., the border between region 7 with regions 8 and 9, and 3 + 4 follows the political borders between the CIS states and the adjacent countries.

Figure 29.3   Nine switching regions defined by the ITU. (From A. A. Huurdeman, *Guide to Telecommunications Transmission Systems*, Artech House, Norwood, MA, 1997, Figure 1.9; with permission of Artech House Books.)

To distinguish the services available up to that time from the enhancements that SPC made possible, the phrase *plain old telephone service* (POTS) became a euphemism for electromechanically switched networks.

Until late 1970, speech and associated signaling, which was transmission of dialing only, were kept together from subscriber to subscriber throughout the switching network, either in-band or out-band, in the same channel. This led to fraudulent use of the network. Various techniques were devised to "fool" switching systems down the line as to the actual progress status and routing of a call, which resulted in illegal "free calls." The fraud was often perpetrated from a public coinbox telephone using circuitry in a small, self-contained box commonly called a *black box*, *blue box*, or *red box*, depending on the differing circuitries that were employed.

With the advent of SPC switching it became possible to separate the speech and signaling paths and combine the signaling for groups of calls in a separate *common channel-signaling* (CCS) *network* within the public switching network. This CCS can now combine the signaling information relating to a group of telephone and/or data circuits with additional network supervisory information. Both the signaling and supervisory information are not in a sequence of tones but are coded digitally as messages. Due to this coding and the higher available bandwidth on the separate channels, much more detailed information about the call, the routing, and the type of

---

TECHNOLOGY BOX 29.1

Stored Program Control

Basically, switching equipment must interpret and react to signals incoming from subscriber lines, trunks, and other inputs. The logic that accomplishes these functions, as well as others required to process information held within the system, such as in a database, is known as the *control*. Databases include information for translating directory numbers and parameters that indicate the constituent elements of a particular exchange. A *program* is understood to mean the call processing that takes place by sequences for which the orders are read from a memory. To *store* the control program of an exchange, a general-purpose processor or microprocessor is generally used. Stored program controlling is then achieved by a set of instructions or orders that may be performed by the processor and directed to control the peripherals, including random access memories. The orders are stored in those usually erasable programmable read-only memories (EPROMs).

Stored program control (SPC) is a method of control that places most of the control logic in a permanent memory as a stored program. The common control thus becomes a general-purpose unit that is programmed to provide the features, options, and traffic patterns for each particular exchange.

---

service can be processed across the network. Other advantages of CCS are the faster signaling and the simultaneous application of signaling during conversation.

The latest version of CCS is CCITT signaling system No. 7 (SS 7). It operates in digital networks on 64-kbps channels and supports the integrated services digital network (ISDN), Intelligent Network (IN), Internet Protocol (IP), and cellular radio networks. SS 7, introduced in 1980, is adopted worldwide for setting up, managing, and clearing calls.

Whereas electromechanical switching systems had been developed at a cost of some $100,000, the development cost of electronic systems incorporating SPC went up at an annual rate of roughly 20%. AT&T had expected to develop their first electronically controlled SPC exchange No. 1 ESS within a five-year period at a budget of $25 million; in fact, it took seven years and absorbed more than $100 million. The development cost of the digital switching systems introduced 20 years later surpassed $1 billion. To keep development cost at an affordable level, a better international dissemination of specialist's knowledge became vital. Fortunately, before World War II, switching manufacturers utilized mutual patent licensing, enabling worldwide spreading of new technologies. Now, to obtain better mutual exchange of information with a cross-fertilization of new ideas, resulting in less work duplication and quicker development progress, the institution of international switching symposia was created.

AT&T held an international symposium at Whippany and Murray Hill from March 4 to 7, 1957—at the time still mainly for its patent licensees—presenting the world's first electronically controlled exchange No. 1 ESS with SPC and electronic speech paths. In 1963 a second symposium was organized by AT&T at Holmdel, New Jersey, reporting on the first production of electronic switching systems for

---

TECHNOLOGY BOX 29.2

Basic Switching Modes

Simultaneous switching of a number of lines through an exchange can be done either in a space- or time-division mode. With *space-division switching*, each telecommunication message, thus each speech path, is continuous through the exchange in such a way that it is spatially separated from all other simultaneous speech paths, once set up, for the entire duration of the call.

With *time-division switching*, all telecommunication messages are separated from one another in time and use a common physically continuous connection known as a *bus* or a *highway*. Time division implies that speech information is sliced into a sequence of time intervals called *time slots*, each corresponding to a *sample* of speech information. In most time-division systems, samples have been the subject of quantization, and in the case of pulse-code modulation (PCM) systems, of digital coding of the quantified values. Time-division switching is practically synonymous with digital switching.

Space-division switching applies to all analog switching systems; however, digital switching can also be in the space-division mode in such a way that a multiplicity of paths, buses, or highways is provided.

---

central office (No. 1 ESS) and PABX (No. 101 ESS) application. In England, the IEE, together with the GPO, organized a general Conference on Electronic Telephone Exchanges in London in November 1960. In 1966, the first *International Switching Symposium* (ISS) took place in Paris. Since then, the ISS, which takes place in various countries at two- to three-year intervals, became the world's premier technical and scientific forum for telecommunications technology. The ISS which took place in Toronto, Canada in 1997 was visited by over 3000 delegates from around the world, discussing the contents of some 140 papers presented around the theme "Global Network Evolution: Convergence or Collision?"

The subjects discussed during the first six ISS symposia give a good view of switching development over the related period:

- *1966, Paris:* presentation of experiments on space and time division, and production of space-division systems (Technology Box 29.2)
- *1969, London:* as above, however, with emphasis on production
- *1972, Cambridge, Massachusetts:* development of digital time-division systems, the change from space- to time-division switching, and the initial appearance of digital technology in the speech path
- *1974, Munich:* new space-division systems in production; modular architecture for SPC systems supporting fast developments in hardware
- *1976, Kyoto:* distributed control

At the end of this period, in 1979, at an ISS in Paris, 2000 participants from 21 countries discussed achievements in electronic switching, presented in 210 papers, and agreed that the time was appropriate for a worldwide change from analog to digital switching.

29.3.2 Preliminary Electronic Switching Systems

With the transistor available in an increasingly reliable and versatile form, various preliminary electronic switching systems appeared around the early 1950s in the United States, the U.K., Sweden, Germany, France, the Netherlands, and Japan.

United States   In the United States in 1949, Bell Labs initiated studies on an electronic central office (ECO) project, resulting in the first call switched through an electronically operated laboratory system in March 1958 in their laboratories in Whippany, New Jersey. The ECO switch used reed relays as the switching cross-point element, a barrier-grid tube as random access memory (RAM), a cathode ray tube with photographic-plate storage as read-only memory (ROM), and the principles of stored program control. A first ECO trial system was then put into general service on June 15, 1960 at the small farming town of Morris, Illinois, near Chicago (and near La Porte, where in 1892 the first Strowger exchange began operation). The exchange, which contained 12,000 transistors and had taken 750 person-years of development effort, served 30 subscriber lines and remained in operation until January 24, 1962. The ECO included advanced services such as call transfer, add-on, conference, abbreviated dialing, code calling of extension telephones, and series completion of calls to nonconsecutive line numbers.

In parallel with ECO, Bell Labs developed an electronic private automatic branch exchange (EPABX), which included time-division switching. Independent of Bell Labs, the Stromberg–Carlson Co. also developed time-division pulse amplitude modulation (PAM) switching systems for mobile use, for community dial offices, and for PBXs.

Sweden   L.M. Ericsson was the first company to present an electronic trial exchange, the EMAX (Electronic Multiplex Automatic Exchange), in 1954. EMAX was a laboratory mockup made up of diodes and cold-cathode tubes and was the first system to apply time-division switching. The digital information was processed in a pulse amplitude modulation (PAM) mode. Remarkably, the speech signal was digitized with a sampling rate of 8 kHz and a corresponding time slot of 125 μs, the standard PCM values that were to be adopted by both AT&T for its T transmission system and by the CCITT in its PCM recommendations.

The Swedish Administration Televerket also developed and manufactured its own switching equipment in its industrial division, known as TELI. In 1956, an agreement was reached between Televerket and L.M. Ericsson to set up a *Joint Electronic Council* (JEC) to implement shared development work on electronic switching. In 1968 two electronically controlled space-division exchanges incorporating SPC were installed as follows:

- The A210 of TELI, installed at Storangen, a purely experimental exchange using a crossbar-switching network with magnetic latching.
- The AKE 11 of L.M. Ericsson, installed at Tumba, south of Stockholm, using a new type of crossbar switch, a miniselector known as a *code switch*, serving 4300

subscriber lines and 640 transit circuits. The AKE 11 was the first SPC exchange operated outside the United States.

United Kingdom   In the U.K., research into electronic switching began very soon after World War II at Dollis Hill, the research station of the GPO, and in the laboratories of British switching equipment manufacturers. Research at Dollis Hill[3] concentrated from the beginning on time-division electronic switching. In 1956, the GPO and five British switching equipment manufacturers founded the *Joint Electronic Research Committee* (JERC).[4] JERC developed a time-division switch using pulse amplitude modulation (PAM).[5] A laboratory mockup was installed at Dollis Hill in November 1959. This was followed by a 600-line exchange which was installed in the North London suburb of Highgate Wood and inaugurated in December 1962 as the world's first public time-division switch, precisely 50 years after the first British automatic telephone exchange had been installed at Epsom.

The system had been designed in 1956 at a time when sufficiently reliable transistors were still unavailable, so that in addition to the 26.000 transistors and 150,000 diodes, the use of some 3000 electronic tubes resulted in difficult heat dissipation problems. The time-division mode also still created more problems than anticipated.

France   The beginning of electronic switching studies in France can be dated to 1957, with the establishment by CNET of an *Electronic Machine Research Department* (RME). CNET (Centre National d'Etudes des Télécommunications) was founded in 1944, inspired by Bell Labs and by GPO's laboratories at Dollis Hill in the U.K., with the objective of becoming a national telecommunication research body in close cooperation with the telecommunication operation services. CNET was organized under the *Directorate Général de Télécommunications* (DGT) of the Ministry of PTT. The first research assignments were concerned primarily with problems resulting from the severe war damage to the national telecommunication network, system restoration, and network expansion. Research focused essentially on transmission, with studies on radio-relay links and multiplexing, particularly digital multiplexing, but with very little on switching apart from some research on electromechanical solutions for rural automatic exchanges. As a result of a policy of decentralization initiated by President de Gaulle, part of the CNET staff was moved from headquarters at Issyles-Moulineux (a Paris suburb) to Lannion[6] in Brittany. To ensure quick implementation and industrialization of the ideas of the CNET, an industrial company, the *Société Lannionaise d'Electronique* (SLE), was founded in Lannion by CIT (Compagnie Industrielle des Téléphones; in 1970, CIT-Alcatel).

---

[3] Benefiting from the experience made on the wartime Colossus computer used to decipher the Enigma code of the German military messages.

[4] The five companies were Associated Electrical Industries Ltd. (AEI), at the time Siemens Edison Swan Ltd.; Automatic Telephone & Electric Co. Ltd. (ATE); Ericsson Telephones Ltd.; General Electric Co. Ltd. (GEC); and Standard Telephone and Cables Ltd. (STC).

[5] Speech was sampled at a 10-kHz frequency and according to the PAM principle, its amplitude was equal to that of the speech current at the instant of sampling. Interspersed pulses corresponding to 100 speech channels, with each pulse occupying a time slot of 100 µs, were fed along transmission channels known as *highways*.

[6] Lannion is located near Pleumeur-Bodou, where in 1962 the first Earth station on the European continent was put into service.

In 1958, Socotel (Société mixte pour le développement de la technique de la Commutation dans le domaine des Télécommunications) was founded along the lines of Sotelec, founded earlier in the transmission field. The members were the administration and its main suppliers of switching equipment. Socotel divided the research work so that the RME[7] worked on full electronic exchanges and the research departments of the manufacturers on semielectronic switching systems. RME produced two entirely different electronic switching prototypes, known as Aristotle and Socrate. Aristotle (Appareillage Réalisant Intégralement et Systématiquement Toute Opération de Téléphonie Électronique) was a model to be used in setting up a high-capacity system organized around one central processor and a number of peripheral secondary processors under the exclusive use of electronic components. Its central processor, known as *Ramses*, was an original achievement of RME. Aristotle was developed in 1963 and put into operation first by CNET at Issy-les-Moulineux and than transferred to Lannion, where it handled the internal telephone service of the CNET sites up to 1969. Aristotle was also used in 1978 to test a new common channel signaling system that was adopted by CCITT as signaling system No. 6 (and slightly modified by AT&T for their common channel interoffice signaling).

The Socrate (Système Original de Commutation Rapide Automatique à Traitement Électronique) project was carried out in close cooperation with the French manufacturers. Socrate was a semielectronic exchange using the crossbar-switching network of the CP-400 system. Ferrite cores and magnetic drums were used for the memories. The control system used a new method known as *load sharing* between duplicated processors. The Socrate experimental exchange was in service at CNET-Lannion from early 1965 to 1972.

Germany   In the late 1950s, research on electronic switching in Germany was done primarily at Siemens AG and to a much lesser degree at SEL and Telefonbau und Normalzeit (TN, now Marconi). Siemens developed a time-division electronic switching system called *EVA* (Elektronische Vermittlungs Anlage; electronic switching equipment). A 1000-line prototype entered into service in early 1962 at Munich in a private network of Siemens. EVA was also used to test pushbutton dialing and 2-out-of-5 voice-frequency-code signaling between subscriber and exchange. The EVA installation operated satisfactorily for a long time, but the time was not yet ripe for cost-competitive production of time-division switching equipment.

In the 10-year period 1958–1968, Siemens developed five different prototype models using space-division switching:

- In 1958, a mockup of the KAMA exchange (studied from 1952) using ESK-relay and hot/cold electronic tubes.
- In 1959, ESM I, a PABX with decimal trunking using direct-wired logic, reed relays, and semiconductors.
- In 1962, ESM II, operational equipment installed at the Munich Farbergraben exchange with decimal trunking using reed relays and semiconductors.

---

[7] The RME department was placed under the direction of Louis-Joseph Libois, who became the Directeur Général des Télécommunications in 1971 and is the author of books on the history of telecommunications: *Genése et croissance des télécommunications* (Masson, 1983) and *Les Télécommunications technologies, réseaux, services* (Eyrolles, 1994).

- In 1965, ESM III, first register-controlled (nondecimal trunking) exchange using ESK relay and wired semiconductor logic. This exchange, with a capacity of 10,000 lines, was installed in Rome.
- In 1968, System IV, the first (mockup) model applying SPC using multi-processors, integrated circuits, steel-enclosed reed relays in ESM technology, and ferrite-core memory with magnetic tape backup.

The ESM II exchange, inaugurated on November 9, 1962, was the world's first electronic-type exchange opened for public service (about one month before the world's first electronic time-division exchange was inaugurated at Highgate Wood, U.K.). The exchange, operating 500 lines initially, was soon extended to 3000 lines.

Those preliminary electronic exchanges constituted a transition from electro-mechanical to electronic switching. The classical EMD motor switch was replaced by a coordinate field consisting of ESK (Edelmetall-Schnell-Kontakt; high-speed relay with noble-metal contacts) relays, or by an ESM (Elektronisches System mit Magnetfeldkoppler; electronically controlled magnetic field switching matrix with dry reed contacts) matrix. The ESK relay had been developed in the Siemens laboratories in Munich in the mid-1950s. It was said to mark the "apogee" of electromechanical switching research in the quest for faster and more reliable relays. The operating time was on the record order of 2 ms, the same order of speed as later obtained by reed relays. The number of ESK relays used in Siemens equipment was estimated to be over 50 million in 1968. Initially, the ESK relay was used primarily for PABXs sold throughout the world; by 1982 it was estimated that more than 6 million extension lines were operated by such PABXs.

Encouraged by the success of the ESK relays in the PABXs, Siemens developed the ESK 10,000 E system for public switching. The electronically controlled exchange was used for small rural applications as well as for very large exchanges with up to 100,000 subscriber lines and 50,000 trunk lines. The German PTT used it widely from 1966 for their international trunk exchanges. The greatest installation at the time, with 24,000 incoming and outgoing trunk lines, was in Frankfurt.

The first ESK 10,000 E public local exchange was installed in Absdorf in Vienna, Austria, in 1966, followed by Hong Kong, Austria again, then Denmark, Finland, Indonesia, Italy (the transit exchanges in Florence, Milan International, and Turin), Nigeria, Switzerland (locally produced by the Albiswerk), South Africa, and the United States. All ESK public exchanges in 20 countries served over 6 million lines by 1982.

The ESM matrix, also a Siemens invention, provided grouped magnetic control of a number of reed relays accommodated in a common unit. Magnetic control of the reed relays was affected by means of coils having their control wires (separately for the horizontal X and vertical Y axes) and their holding wire arranged around the rectangular coordinates of the matrix to select the point of contact. This ESM matrix became the nucleus of the first commercial electronic switching system from Siemens; the EWS (Elektronisch gesteuertes Wähl System; electronic switching system).

On July 12, 1963, in Stuttgart, SEL introduced its first electronically controlled space-division exchange. The Herkon-Electronic telephone switching system, called *HE 60*, served 2000 lines and was accompanied by the introduction of pushbutton telephone sets.[8] The signaling used a 2-out-of-6 voice-frequency code. The exchange was designed as an indirectly controlled system with registers and reed-relay matrices. The Herkon exchange provided a clear departure from the step-by-step procedure still adhered to by the Siemens ESM II. In 1964, two further exchanges of this type were installed at the international transit exchanges in Stuttgart and Nuremberg. The Austrian Administration took another trunk exchange into operation at Vienna in 1966.

Telefonbau und Normalzeit developed a space-division switch with indirect control using magnetic core registers and reed-relay matrices. A 1000-line exchange was installed at Frankfurt in 1965.

The Netherlands   In 1956, Philips invented a new type of transistor component, the *PN-PN component*, consisting of an arrangement of two junction transistors in a hook connection providing a bistable switch function. The PN-PN element could be used as a switching cross-point. *Philips Telecommunication Industrie BV* (PTI) started experimental research on exchanges with electronic cross-points. An experimental exchange using the PN-PN element was field-tested as an ETS3 system in Utrecht, The Netherlands, and in Aarhus, Denmark, from 1967 to 1972. To cope with the fact that these cross-points could handle signals only at a low current level, special low-current telephone sets had to be used. The results obtained with PN-PN components were not up to expectations. The experience gained by PTI with the PN-PN components, however, was used successfully for a telex switching system, DS 714. The PN-PN elements were also used in Japan by Hitachi for the HITEX-1, in France for Aristotle and by LTC for the French Navy, and in Sweden by Televerket and L.M. Ericsson for a system called *TEST1*.

Japan   Research on electronic switching started in 1954 at the NTT and KDD (Kokosai Denshin Denwa Corp.) Laboratories as well as at the University of Tokyo and the development departments of Fujitsu, Hitachi, NEC, and OKI. Initially, research concentrated on wired logic electronic systems using a new type of component, the *parametron*, in the control unit. The parametron was an ingenious device invented at the University of Tokyo in 1954. Made up only of resistors, capacitors, and inductance coils, it offered two bistable states, which made it suitable for computer logic functions.[9] Five electronic switching systems using the parametron were tried successively. The first, alpha, in 1954, was very short-lived. The second, beta, for small exchanges, and the third, gamma, for large exchanges were experiments in space division with crossbars. The fourth, omega, also used space division but with semiconductors as cross-points. The fifth and last of the systems, tau, on which research began in 1959, was an experiment in time-division switching based on the concept of four-wire PAM technology, which had been researched at the Tokyo University.

The Japanese telecommunications manufacturers then brought several wired-logic electronic-control switching systems into service:

---

[8] *Herkon* was the trade name for SEL-designed version of reed relays: "hermetically sealed kontakt relays."

[9] A first computer, the M-1 (Musashino I), using parametrons was completed in 1957 in the Electrical Communication Laboratory of NTT.

- Space-division crossbar systems, in 1960, by NEC types NS-1B, NS-2A, and NS-3A, and by Hitachi types HITEX-3SD
- Time-division switches based on four-wire PAM, in 1961 by Fujitsu type FEAX-302A and by Hitachi type HITEX-3TD and in 1964 by OKI with type ACT-1108

In 1960, NTT realized the advantages of stored program control instead of wired logic and replaced the parametron by superior transistors.

29.3.3   Commercial Electronic Switching Systems

With the above-mentioned preliminary electronic switching systems in experimental operation, substantial experience was gained, and commercial electronic switching systems could be brought into operation from the late 1960s, as summarized in Table 29.4.

TABLE 29.4   Commercial Electronic Switching Systems

Country	Manufacturer	System	Introduction	Millions of Lines Worldwide (Approx.)
United States	AT&T and Western Electric	No. 1 ESS	Succasunna, NJ, 1965	6
		No. 2 ESS	1970	5
		No. 3 ESS	1976	
	GTE	EAX No. 1	St. Petersburg, FL, 1976	3
		EAX No. 2	Mahomet, IL, 1977	
U.K.	AEI, ETL, GEC	TXE-2	Ambergate, 1966	2
		TXE-4	London, 1969	4
	STC	TXE-4A	Belgrave, 1980	8
Canada	Northern Electric	SP-1	Aylmer, 1971	3
France	LMT	Péricles	Paris, Clamart, 1970	—
		Metaconta	Roissy, Ch. de Gaulle, 1972	—
Sweden	L.M. Ericsson	AKE 13	Rotterdam, 1971	—
		AXE 10	Sondertalje, 1977	—
Japan	NEC	D-10	Tokyo, Ginza, 1972	10
		XE1	Tokyo, 1977	—
		ND 10/20	Connecticut (U.S.), 1977	—
		NXE 10/20	Singapore, 1980	—
	Fujitsu	FETEX 100	Singapore, 1978	—
Belgium	ITT	Metaconta	Wavre, Belgium, 1973	7.5 (1975)
The Netherlands	Philips	PRX	Utrecht, 1973	3
Germany	SEL, Siemens, TN	EWS	Munich, 1976	1

Figure 29.4   Electronic switch, type EWS, 1970. (Courtesy of Alcatel SEL.)

As a typical example, an electronic switch type EWS installed in 1970 in the German town Schwäbisch-Hall is shown in Figure 29.4.

29.4   DIGITAL SWITCHING SYSTEMS

Alec H. Reeves at the Laboratoire Central des Télécommunications (LCT) in Paris invented the digitalization of analog signals in 1938 (Section 20.8). The first ideas behind pulse-type switching were published by L. Espenschied of Bell Labs and resulted in U.S. patent 2,379,221, covering switching of pulsed voice signals, for which Espenschied had applied in October 1942.

The basic idea of digital switching was conceived by Maurice Deloraine (director of LCT from 1927 to 1940), reportedly during an overnight rail journey from New York to Chicago in 1945. Deloraine envisaged that it should be possible to steer the extremely short pulses of a digital signal individually through separate electronic gates to different channels. He discussed his idea with P. R. Adams and D. H. Ransom, with whom he worked during World War II at the ITT Company *Federal Telephone and Radio* (FTR). On November 14, 1945, they applied together for a patent under the title "Pulse Delay Communication System," application number

628,613. U.S. patents 2,584,987 and 2,492,344 were granted as late as February 12, 1952. French patent 930,641 had been granted on August 18, 1947. The patents covered "switching by pulse displacement," a principle later defined as *time-slot interchange*. Deloraine also presented his ideas at a convention of the Institute of Radio Engineers in New York in March 1947 and in March 1949 as a doctoral thesis at the Sorbonne in Paris.

Thomas. H. Flowers of GPO's Research Laboratories at Dollis Hill also developed early ideas on digital switching. Flowers started as early as 1935 with research on the use of electronics for telephone exchanges. During World War II he participated in the development of the electronic cryptoanalytic machine "Colossus" (Section 22.2). After World War II, Flowers elaborated ideas on switching of pulse amplitude modulation signals and applied in October 1948 for patents in the U.K. and the United States (U.S. patent 2,666,809).

Between 1947 and 1960, under the direction of Deloraine, then technical director of IT&T, further research on digital switching was made at IT&T laboratories in Belgium, France, and the U.K., resulting in further patents on time-division switching. The most important patent was filed on October 21, 1958 in France (resulting in French patent 1,212,984), defining the principle of time-slot displacement of PCM speech pulses for time-division switching.[10]

A first working model of a digital switch was made at the Bell Laboratories, where H. Earle Vaughan started a project in 1956 known as ESSEX (experimental solid-state exchange). Vaughan made a *time-separated integrated communication switching model* making use of the initial progress made by Bell Labs on digital transmission. A completely electronic time-division switching laboratory model of ESSEX made in 1958 demonstrated the technical feasibility of time-division switching. It showed that switching of digitized telephone channels could be performed without conversion to the analog speech band at the exchange. Although economical industrial production of the required complex technology was not yet possible, the results obtained with the ESSEX were widely published and served for many years as an important reference for the development of digital switching models.

In 1951, LCT made the first mock-up model of a time-division multiplex highway connecting subscriber lines by electronic gates handling amplitude-modulated pulses (PAMs).[11] The 100-line model provided multiplexed pulse amplitude paths through two stages using an amplified bus technique that divided the two directions of transmission. One stage acted as line finder and the other as a line connector. There were 16 channels—now called time slots—that represented the multiplex positions in which pulses could be placed. This model was the basis for two systems: the military system RITA and the public system A. System A was the first of three pulse-type electronic switches developed by ITT derived from two basic patents of Deloraine. The military system RITA (Réseau Intégré de Transmission Automatique) comprised nodal switches to handle the traffic of 24-channel PCM systems. In 1962, with a mockup model, the world's first demonstration was given of combining digital transmission with digital switching by time displacement of coded pulses with the help of memories. The memories stored speech parts for a variable time until they

[10] U.S. patent 3,049,593 was also filed for this principle on October 19, 1959.

[11] PAM does not require an analog-to-digital codec for each line but allows the modulation of a small number of speech samples only and has poor crosstalk characteristics.

could travel across a switching cross-field and leave it in an outgoing channel at a time different from the arrival time of that speech part. Some prototypes of RITA were developed in 1968, leading to a standardization of the system for French military use in 1971. The U.S. Army adopted the same system in 1985.

Commercial digital switching started around 1970, as the progress in digital transmission with an increasing number of digital trunks—and the promising prospects of an integrated digital switching/transmission network—justified the tremendous effort to develop digital switching systems. Switching of a digitized speech signal from the time slot of the digital frame of one PCM system to a different time slot of the digital frame of another PCM system required two new switching elements:

- A device for time-slot interfacing, called the T-stage, that effects the desired changes of time-slot allocation, including the delay required to cover the time difference between the two time-slot locations
- A device with the function of a matrix, called the *S stage*, to effect the space-division switching (for, typically, 32 to 1024 speech channels in parallel) of speech-information samples from subscribers calling, coming from specific time slots, to the corresponding time slots of the subscribers called (and vice versa)

A digital switch usually applies these stages in a T–S–T sequence, although occasionally the S–T–S sequence is used for smaller exchanges. Figure 29.5 illustrates the digital switching in T–S–T configuration. The drawing indicates how subscriber X allocated to time slot 28 of PCM system I is connected with subscriber "b," who is allocated to time slot 2 of PCM system II. A further explanation of the principle of digital switching is given in Technology Box 29.3.

Digital switching was first applied between transit exchanges that were interconnected by PCM multiplex links in the late 1960s. The introduction of digital switching on local exchanges in the 1970s required special *subscriber-line-interface circuits* (SLICs). For almost a century, with analog switching, telephone sets were connected to the exchange by two wires drawing 15 to 250 mA from a 60- to 90-V central battery at the exchange. For digital switching a new set of interface functions was required between the analog telephone and the digital switch. These were called BORSCHT functions (by J. E. Iwerson of Bell Labs) because they define in a new way: B, battery feed, O, overvoltage; R, ringing; S, supervision; C, codec for analog-to-digital coding and decoding; H, hybrid to split the two-wire analog speech circuit into two separate two-wire circuits for sending and receiving the coded digital signals; T, testing. The BORSCHT requirements are different in various countries, and technological progress changed the potential solutions so that it took over 10 years to come to a satisfactorily solution for worldwide applicable subscriber line interface circuits. In fact, the development of those line interface circuits absorbed a significant part of the total digital switching system development, and these circuits still determine the cost of local digital switches. On the other hand, transmission systems became cheaper, as PCM equipment was no longer required on the transmission lines between digital exchanges, due to the use of codecs in the subscriber-line-interface circuits.

The first-generation digital switching systems manufactured in the period from 1970 to 1985 were mainly with *centralized* SPC for switching of digital voice, image, and data lines, with all control elements centrally located. The second generation,

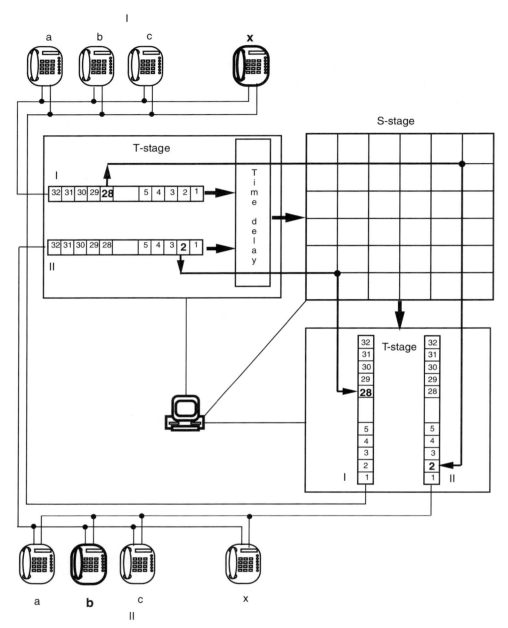

Figure 29.5   Principle of a digital exchange with T–S–T stages.

starting around 1985, uses mainly *distributed* SPC switching, with call processing distributed toward the line modules, thus reducing the initial investment and adding control with the increased capacity. The term *distribution* also refers to remote switching in a remote switching unit (RSU) located nearer large groups of subscribers. Whereas large SPC switches may serve some 100,000 to 200,000 lines, an

TECHNOLOGY BOX 29.3

Basic Principles of Digital Switching

The switching network performs switching between time-multiplexed buses. To allow connection between different time slots in different buses, switching in both time and space are necessary.

The *time-switching stage* (the *T stage*) consists of an incoming speech memory, where PCM words will be delayed at an arbitrary number of time slots (less than one frame). The writing of information on incoming time slots into the cells of the incoming speech memory is normally sequential, and each cell receives an 8-bit word of the PCM incoming channel. The reading of the incoming speech memory will be controlled by a control memory associated with the *T stage*. This control memory orders the reading of a specific cell in the incoming speech memory after a *time-slot interchange delay*. The effective delay is obviously the time difference between writing into the speech memory and reading out of the memory. According to the arrangement of the T stage, the transfer of the 8-bit word of a PCM channel from the incoming memory can take place either to an outgoing speech memory or to an outgoing PCM bus. In the first case, the transfer is generally performed not serially but in parallel through eight wires connecting the incoming and outgoing speech memories.

A *space-switching stage* (the *S stage*) consists of a cross-point matrix, $n \times n$, where the individual cross-points consist of electronic gates. To each cross-point column is assigned a cell of the control memory associated with the " stage, which has as many words, F, as there are time slots. Typical figures for F are from 32 up to 1024. During each individual time slot the cross-point matrix works as a normal, space-divided matrix with full availability between incoming and outgoing buses, the cross-points being controlled by certain cells in the control memory. Just as a time slot shifts to another, the control memory is advanced one step, and during the new time slot a completely different set of cross-points is activated. This goes in cycles of F steps. This time-divided behavior increases the utilization of cross-points on the order of $F = 32$ to 1024 times compared to normal space-divided switching.

Comment: For better understanding, it should be observed that the time-switching stage, T, does not work in a time-divided mode: In the incoming and outgoing speech memories, the same cells are used exclusively for a certain call during the entire connection; the space-switching stage, S, on the other hand, works completely in a time-division mode. The action in an S stage is sometimes called, especially in U.S. terminology, *time-multiplexing switching* (TMS). In TMS, a number of frame-synchronized digital channels are connected as required to pass information in the same time slot from one channel to another.

Source: Adapted from a paper presented by M. Hoshi, Professor of the University of Tokyo, at the ITU Switching Seminar, Singapore, 1978, and published in *Telecommunication Journal*, April 1979, on the occasion of the International Switching Symposium, Paris.

RSU can accommodate some tens to as many as 50,000 lines. The RSU may be located up to about 200 km from the central "host" switch. Usually, part of the control remains in the host exchange. A *stand-alone RSU* can take over from the host the major switching control for its lines in case of emergency.

Whereas prior to introducing public electronic switching, experimental exchanges were installed first, the new digital exchange systems with very complex software configurations were developed for direct public application. Table 29.5 summarizes the major systems manufactured in the last two decades of the twentieth century. During that period, L.M. Ericsson installed 120 million lines AXE 10 in 125 countries, of which 45 million in fixed wire-line exchanges and 75 million in cellular radio mobile switching centers. Siemens installed 200 million lines EWSD in 105 countries, of which 30 million were in China. Alcatel installed 250 million digital subscriber lines in over 100 countries with the systems E10 (originally from the French company CIT), S12 (originally from European ITT companies), and Alcatel 1000 (introduced in 1991 as the integration of System 12 and System E10). Figure 29.6 shows an early version of the S12 system.

At the beginning of 1990 about 50% of all main telephone lines worldwide were connected to digital exchanges. New Zealand was the first country to achieve 100% digitization at that time. By year-end 2000, all industrialized countries had achieved 100% digitization, and the total worldwide digitization of telephone switching was over 95%.

29.5   DATA SWITCHING

The advent of computers capable of communicating with each other led to the introduction of a new type of network: the *public-switched data network* (PSDN). The first computers were, and many small computers today still are, connected to the PSTN (public-switched telephone network) by means of a modem for analog-to-digital conversion of data. For large computers not operating at kbps but at Mbps, the ordinary telephone line and the PSTN[12] were much too slow, so that data switching networks had to be created.

In the early 1940s, an initial form of data switching evolved from the telex, coupling paper-tape perforators and readers so that messages could be stored while awaiting the availability of transmission lines between the corresponding stations. Thus was born the concept of one-way data transmission and storing of data at the switching center. This technique allowed delaying delivery of messages on a first-in, first-out basis until transmission facilities became available. Fewer transmission circuits are thus required, as the circuits can be used to their maximum capacity. One of the first data communications services between distant computers was offered in Japan by NTT in 1968 for money order communication between the national Gunma bank and its regional banks. Subsequently, dedicated data communication lines were made available for business operations.

In 1964, store-and-forward switching assumed a new dimension. To ensure secrecy of military communications, rather than receive-store-and-retransmit each entire message without interruption, messages were *packetized* and combined,

---

[12] Currently, in many countries the PSTN is being upgraded with xDSL equipment (Section 28.4.2), which also enables Mbps transmission.

TABLE 29.5 Digital Switching Systems

Country	Manufacturer	System	Introduction	Millions of Lines Worldwide (Approx.)
France	CNET + SLE	Platon	Perros–Guirec, 1969	
	CIT	E.10A	Lannion, 1972	3
		E.10B	North Yemen, 1980	32
	LMT	MT20/25	Paris–Berny, 1982	6
United States	AT&T	No. 4 ESS	Kansas City, MO, 1976	—
		No. 5 ESS	Seneca, IL, 1982	30 (1990)
	APT	5 ESS-PRX	Lasliki, Saudi Arabia, 1985	—
	Stromberg–Carlson	DCO	Richmond Hills, GA, 1977	3
	TRW–Vidar	IMA/IT24	Ridgecrest, CA, 1976	0.3
ITT	North Electric	DSS	Emlenton, PA, 1978	1
	GTE–AEL	EAX No. 3	Rice Lake, PA, 1978	—
		GTD–No. 5	Banning, CA, 1982	10 (1990)
		GTD5C	Mons, Belgium, 1983	—
	GTE–Italcom	GTD-3 EAX-I	Genoa, Italy, 1982	—
	NTI	DMS 10	Union, KY, 1980	3 (1990)
		DMS 100/ 200/250	Ottawa, Canada 1979	26 (1990)
Canada	Northern Electric	DMS10	Disney World, Fort White, FL, 1977	3 (1990)
		DMS100–300	Ottawa, 1979	26 (1990)
Sweden	L.M. Ericsson	AXE 10	Finland/Italy/Saudi Arabia, 1978	120
Germany	Siemens	EWSD	South Africa, 1980	200
U.K.	Plessey	X	London, Baynard House, 1980	—
Japan	NEC, Fujitsu, OKI and	D60	Tokyo, Othemachi, 1982	—
	Hitachi	D 70	Nagoya, Daido, 1983	13 (1986)
	NEC	NEAX-61	Manteca, CA, 1979	15 (1988)
	Fujitsu	FETEX-150	Singapore, 1981	—
	Hitachi	HDX-10	Sri Lanka, 1981	—
Finland	Nokia	DX 200[a]	Finland, 1982	—
Brazil	Telebras	Tropico	1983	6 (1990)
India	ITI	ILT	Kerala, 1984	—
		RAX	Kittur, Kamataka, 1986	—

(Continued)

TABLE 29.5 *(Continued)*

Country	Manufacturer	System	Introduction	Millions of Lines Worldwide (Approx.)
		MAX/TAX	Ulsoo, Bangalore, 1990	—
Italy	Italtel	UT10–60	1984	5 (1990)
	Telettra	AFDT1	Turin, 1976	—
South Korea	Daewoo, Goldstar, Otelco, and Samsung	TDX-1	Daejon, 1984	4 (1990)
China	SBTMC	System 1240	Hefei, Anhui, 1986	—
ITT		System 12	Brecht, Belgium, 1981	25
Alcatel		Alcatel 1000	1991	190

[a] Derived from a technology transfer agreement with CIT-Alcatel of France on their E.10B system.

Figure 29.6   System 12 switch installed at Stuttgart, 1982. (Courtesy of Alcatel SEL.)

TECHNOLOGY BOX 29.4

Aloha Protocol

The Aloha protocol was developed by the University of Hawaii (hence the name) for the purpose of connecting various terminals to a common computer using radio-relay or satellite links. The Aloha protocol was first applied on the *Alohanet*, a network of the University of Hawaii that connected computers sited on four different islands by radio-relay links. With the Aloha protocol, any user can send packets of data without checking that the line is free. If the line is free, the user will obtain an acknowledgment that the information has been sent. Should such an acknowledgment not be received within a specified time, collision of these data with other data has occurred, and a second attempt has to be made.

resulting in continuous transmission of parts of several messages. The message subdivisions were made uniform in size; an address was placed at the start of each subdivision, which became known as a *packet*. At the source of information, messages are divided in several equally long packages that are transmitted independently across the network and reassembled at the point of destination. In the switching equipment the packages can be buffered and recombined with other groups of packages to obtain optimal use of the transmission links between switches.

One of the first packet-switched networks introduced in 1971 was Arpanet, now popular worldwide as the Internet (Section 34.4). Arpanet used a common communication architecture between Unix computers. Other computer companies also developed protocols and communication architecture for their computers. In 1971, IBM developed their *System Network Architecture* (SNA), which is still widely used for communication between IBM and IBM-compatible computers. DEC developed the less well known DEC–Digital Network Architecture used in DECnet.

In 1974, Robert Metcalfe at Xerox developed Ethernet, a low-cost wideband way of sending packets of data between office machines, printers, and computers connected via coaxial or optical fiber cable in *local area networks* (LANs). The name *Ethernet*, now a registered trademark of Xerox Corporation, refers to the initial application in a radio data network based on the Aloha random-access protocol developed in 1970 (Technology Box 29.4).

At the end of the twentieth century, Ethernet was the dominant LAN technology, accounting for more than 80% of all LAN connections and comprising an installed base of well over 70 million users. To cope with the rapidly growing amount of data transmission, caused by the unprecedented growth of Internet traffic, the traditional capacity of 10 Mbps was extended to 100 Mbps in the early 1990s and called *Fast Ethernet*. By the middle of the 1990s the capacity was increased to 1 Gbps, called *Gigabit Ethernet*. At the end of the twentieth century, 10-Gbps Ethernet, called *10-GigE* or *Optical Ethernet*, was introduced for transmission on optical fiber.

In 1975, IBM developed a token passing[13] concept. This uses a special control

---

[13] Inspired by an old railway collision prevention procedure on single-track sections, whereby the engine driver personally had to take a token before entering the track.

packet that circulates around the network from node to node. The token thus avoids the collision inherent in Ethernet by requiring each node to defer transmission until it receives a "clear-to-send" message: the token. Each node monitors the network constantly to detect any packet that is either addressed to it or which (without an address) is the token. The token is passed along to the next node when the token is received by a node that has nothing to send. If the token is accepted, it is passed on after the node has completed transmitting the data. The token must be surrendered within a specific time so that no node can monopolize the network resources. The token-passing concept is termed the *token ring method*, although bus operation is also possible. The IBM token ring method supports speeds up to 16 Mbps operating on twisted pairs.[14]

Rather than staying with numerous proprietary data-handling protocols, CCITT in 1976 defined a generally applicable protocol for packet-switching data interface called *X.25*. This protocol has been extended to a family of protocols from X.1 to X.34 for handling data communication at speeds of 4.8 to 64 kbps. In 1984, CCITT standardized X.400 as the standard for electronic message handling, followed in 1988 by X.500 as the standard for global directory service. In the same year, *Frame Relay* was standardized as a protocol mainly for LAN-to-LAN data traffic up to 2 Mbps. Frame relay provides an alternative to leased lines. It offers a guaranteed bandwidth plus the benefit of *bursting* beyond that rate when additional bandwidth is available from other users. This bandwidth sharing translates into cost savings of at least 30% over leased lines. While frame relay was conceived for data transmission, real-time voice over frame relay appeared in the mid-1990s. ISDN (Section 29.6), currently evolving globally for combined voice, image, and data transmission and switching, might make the foregoing protocols and separate data switching networks gradually superfluous.

A recent add-on to data switching comes from mobile data. International mobility supported by satellite and cellular radio transmission requires immediate global accessibility of data. At the end of 1998, some 3 million customers were each transmitting an average of 1.1 Mbps per month. Data switching is being implemented on digital cellular radio systems and is available on the LEO satellite system Globalstar, whereas other LEO systems are under development exclusively for switched satellite data transmission. The mobile switching centers of cellular networks can be used to provide direct access to existing data networks so that the mobile user can use the same software whether at home, in the office, or traveling.

The latest add-on to data switching comes from the growing interest in telephony over Internet with the *Voice over Internet Protocol* (VoIP). In a VoIP call, two kinds of communication take place in the IP network between the calling and the called parties: a bidirectional media flow that conveys the actual speech and the signaling messages that control the establishment and the characteristics of the voice flow. The *Realtime Transport Protocol* (RTP) is used to transport the voice information. The RTP is a packet-switching protocol that contains voice frames with typical length between 10 and 30 ms. Special voice encoding compresses the digital speech samples, thus saving bandwidth. *Gateways* are used to take care of the interworking functions, translating information between packet-based VoIP and the continuous signal of the

---

[14] The token-passing method has been standardized for speeds up to 4 Mbps in 1985 under IEEE 802.4 (since 1989 ISO 8802-4) for bus access and in IEEE 802.5 (ISO 8802-5) for ring access.

PSTN. VoIP appeared first around 1995 and became widely applied as the quality of service on Internet improved and the waiting periods decreased. As an example, at the end of the 1990s AT&T introduced VoIP for its complete corporate network, supporting some 120,000 users via 10 gateways in the United States and another 28 outside the United States. To improve the service, special IP private branch exchanges (IP-PBXs) are being installed to replace the conventional digital switches and new types of IP voice terminals to replace today's telephone sets.

29.6 INTEGRATED SERVICES DIGITAL NETWORK

Switching and transmission networks presently undergo worldwide a transition from analog to digital. Installed analog equipment might still be used for another 20 to 30 years; production of switching and transmission equipment, however, has changed almost fully from analog to digital. One obstacle for a fully integrated digital telecommunication network is now being removed: The telephone itself, which has been analog for over 100 years, is now becoming digital—the solution being ISDN (integrated services digital network).

With ISDN the access network between a local switch and the subscriber's telecommunication equipment also becomes digitalized. The main purpose of ISDN, however, is not the digitalization of the telephone and the access network, but combining the various existing public-switched networks for telephony, data, telex, and video into a single integrated network.

The concept of ISDN was first studied in Japan and published by Y. Kitahara of NTT in *Japan Telecom Review*, No. 2 of 1982, under the title "Information Network System: Infrastructure for an Advanced Information Society." A model of this information network system (INS) was put into service in Tokyo–Mitaka in 1984. CCITT drafted the first ISDN recommendations in the same year, and with NTT bringing in their experience with the INS network, a full set of ISDN recommendations was agreed upon in 1986. The recommendations defined ISDN in two versions: narrowband ISDN (N-ISDN) and broadband ISDN (B-ISDN).[15] For N-ISDN, two types of subscriber access are recommended: *basic rate access*, comprising two channels for voice and/or data, and one channel for signaling and optionally low-speed packet data, with a total capacity of $2 \times 64 + 16 = 144$ kbps; and a *primary rate access*, comprising 30 channels for voice and data and two channels for signaling, synchronization, timing, and control, with a total capacity of $32 \times 64 = 2048$ kbps, which is equivalent to the 2-Mbps primary level for 32 PCM channels. For B-ISDN, two types of subscriber access, at 155 and 622 Mbps, are recommended. The new ISDN specifications were implemented first by NTT, with commercial operation on its Tokyo–Nagoya–Osaka network in 1988.[16]

Introduction of ISDN is possible without replacing the existing access network, which worldwide consists primarily of telephone copper cable. For N-ISDN the

---

[15] The distinction between narrowband, wideband, and broadband signals is made in terms of bit rates. *Narrowband* is any signal up to 64 kbps, signals between 64 kbps and 2 Mbps are commonly called *wideband*, and signals above 2 Mbps are called *broadband*.

[16] Two years later, on November 19, 1990, international ISDN traffic began between Germany and France.

TABLE 29.6   ISDN Penetration

Country	ISDN Lines	Percent of Total Main Telephone Lines
France	3,601,000	10.6
Germany	13,776,000	28.4
Italy	3,050,000	11.5
Japan	15,214,000	21.5
The Netherlands	2,280,000	23.8
Switzerland	1,420,000	28.4
U.K.	2,596,000	7.7
United States	11,422,000	6.2

existing copper-line access network requires only special ISDN line terminations (LT) on the exchange side and network terminations (NT) on the subscriber side. For B-ISDN, an optical fiber access cable is normally used, but broadband radio-relay or coaxial cable systems are also possible. The *basic rate* is the preferred solution for SOHO (small office and home office) applications. It enhances the existing two-wire telephone access for simultaneous use of two telephones and one fax machine or computer. The *primary rate* is the adequate digital upgrading for offices with a digital PABX.

In the United States, in 1986, the first No. 5 ESS digital switch with ISDN capability was placed in service at Oakbrook, Illinois. The worldwide deployment of ISDN went much slower than originally expected. By year-end 1998, only 14 million of the 850 million main telephone lines operated as ISDN. The emerging xDSL (Section 28.4.2) transmission on existing copper-wire subscriber access lines with transmission capacities between 144 kbps and 52 Mbps is a particularly serious competitor of ISDN. Table 29.6 shows the equivalent number of basic-rate ISDN connections and the corresponding percentage of total installed main telephone lines at year-end 2000 for the countries with over 1 million ISDN lines.

29.7   BROADBAND SWITCHING

The introduction in the late 1980s of various new services required switching equipment operating at higher speeds, thus higher bit rates. Typical new services that require broadband switching are desktop publishing, medical imaging, video (library) retrieval, color facsimile, CAD/CAM (computer-aided development and manufacturing), multimedia service (voice, text, graphics, and moving pictures), video conferencing, high-fidelity music, and high-definition TV.

Digital networks used to operate at 64 kbps with switching (and transmission) based on the *synchronous transfer mode* (STM). In this mode a fixed number of bits is available periodically to each connection within a network. That implies that the capacity of the connection is constant even if the information flow is bursty, as is usual with computer communication. With the increase in data transmission, STM, therefore led to a waste of capacity; moreover, in STM (which was developed mainly for transmission), the switching functions were difficult to handle at different bit rates

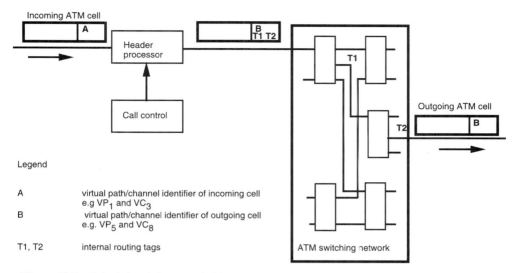

Figure 29.7 Principle of ATM switching. (From A. A. Huurdeman, *Guide to Telecommunications Transmission Systems*, Artech House, Norwood, MA, 1997, Figure 1.16; with permission of Artech House Books.)

(155, 622 Mbps). In the interest of more flexible broadband switching, therefore, a new *asynchronous transfer mode* (ATM) was developed in the late 1980s that can handle traffic relating to services that require widely differing bit rates. In ATM, basically the information is put in fixed-length *cells*. Each cell consists of a 5-byte header field and a 48-byte information (also called *payload*) field (Figure 29.7). The header field contains the data for routing and control of the payload through the telecommunication network. The cells of a particular message are switched and transported independently through the broadband network and at the point of destination reconstituted in their original synchronous form. At Telecom 95, NEC was the first company presenting ATM public switching exchanges, at speeds ranging from 10 to 160 Gbps.

29.8 PRIVATE SWITCHING

After World War II, private switching evolved from manual switching to automatic switching. The private branch exchange, the PBX, became a private automatic branch exchange, the PABX. PABXs are similar to public exchanges except that they include only a few operational and network management functions. In fact, the first PABXs used to be scaled-down public switches. On the other hand, PABXs gradually obtained many functions and features that were not available in the public switch, such as office automation facilities, data-processing applications, restricted dialing into the public network, automatic route selection on the public network, radio connectivity (cordless service), and generation of billing and traffic reports.

Large corporate users with office facilities in buildings at several distant locations interconnected the various PABXs located at the various sites for the operation of an intercompany private network called a local area network (LAN). Interconnection between the PABXs of a LAN is normally via leased lines from the public network. With increasing deregulation and privatization of the public operators, the use of proprietary links between the PABXs also became possible.

The convergence of telephony and computer technology in the 1990s thoroughly changed the features of the PABX. Data switching expanded rapidly in parallel with telephone switching. The intelligence and processor power of PABXs therefore moved from telephone operation to data handling.

In private switching, as in public switching, after more than 100 years of useful applications, Strowger's rotary dialing has been replaced almost worldwide by touch-tone dialing. This far more comfortable way of calling was replaced in the 1990s in many private switching systems by even more comfortable natural voice signaling. *Interactive voice response* (IVR) technology using touch-one detection was introduced to replace personal repetitive routine services such as answering directory and travel schedule inquiries. This touch-tone detection, however, limited to a 12-item vocabulary (0–9, *, and #), was only a forerunner for real *automatic speech recognition* (ASR) devices for signaling, which came on the market at the end of the twentieth century. ASR allows incoming callers phoning an organization to speak the name of the person to whom they wish to speak. The system transfers the call to the appropriate extension and announces the caller when the call is answered. If the call is not answered, the caller may leave a voice message or ask to be transferred to someone else or to the operator. At least for a transition period, an operator is required to obtain a seamless transfer of calls in the case of nonstandard queries.

REFERENCES

Books

Anon., *Alcatel Public Telephone Systems Worldwide*, references as of January 1, 1992, Alcatel Network Systems, Paris. [Booklet]

Anon., *50 Jahre für die schweizerische Nachrichtentechnik, 1935–1985 STR*, Standard Telephon und Radio, Zurich, 1985.

Anon., *100 Jahre SEL von Mix & Genest und C. Lorenz zur Standard Elektrik Lorenz AG*, draft, August 1978.

Chapius, Robert J., *100 Years of Telephone Switching (1878–1978)*, Part 1: *Manual and Electromechanical Switching (1878–1960s)*, North-Holland, New York, 1982.

Chapius, Robert J., and Amos E. Joel, Jr., *Electronics, Computers and Telephone Switching: A Book of Technological History*, Part 2: *1960–1985* of *100 Years of Telephone Switching (1878–1978)*, North-Holland, New York, 1990.

Huurdeman, Anton A., *Guide to Telecommunications Transmission Systems*, Artech House, Norwood, MA, 1997.

Minoli, Daniel, *Telecommunications Technology Handbook*, Artech House, Norwood, MA, 1991.

Libois, Louis-Joseph, *Genése et croissance des télécommunications*, Masson, Paris, 1983.

Oslin, George P., *The Story of Telecommunications*, Mercer University Press, Macon, GA, 1992.

Siegmund, Gerd, *Grundlagen der Vermittlungstechnik*, Decker's Verlag, Heidelberg, Germany, 1993.

van Hemert, D., and J. Kuin, *Automatische Telefonie*, Uitgegeven door de Vereeniging van hoger personeel der PTT, Corps Technische Ambtenaren, the Netherlands, 1933/1953.

Articles

Anon., New signals on the horizon in German telecommunications, *Discovery* (Nokia), Vol. 44, second quarter 1997, pp. 6–11.

Anon., *Alcatel: Daten und Fakten*, 1998.

Green, P. E., Jr., Computer communications: milestones and prophecies, *IEEE Communications*, Vol. 22, No. 5, May 1984, pp. 49–63.

McGarvey, Brendan, Where PBX meets ATM, *Telecommunications*, Vol. 29, No. 9, September 1995.

Peckham, Jeremy, IVR and the art of conversation, *Telecommunications*, Vol. 29, No. 9, September 1995, pp. S153–S159.

Roscitt, Rick, AT&T commits to a next-gen network, *Telecommunications*, Americas Edition, Vol. 33, No. 8, August 1999, pp. 64–65.

Weber, Jürgen, EWSD innovations: the driving force, *Telcom Report* (Siemens), Vol. 20, No. 1, 1997, pp. 12–15.

30

TELEX

30.1 CONTINUATION OF TELEPRINTER DEPLOYMENT

After World War II the teleprinter service outside the United States and the U.K. was interrupted totally and restarted slowly. In Germany, teleprinter service started again in 1947 with 25 subscribers. After the currency reform in 1948, the number of subscribers increased to 3000. It reached 15,000 subscribers in 1955. In Japan, subscriber teleprinter service was inaugurated with experimental service between Tokyo and Osaka on October 25, 1956 with 65 subscribers in Tokyo and 63 in Osaka. The Australian Post Office (APO) introduced a teleprinter network called the *teleprinter reperforator exchange switching system* (TRESS) in 1959. Creed teleprinters were generally used on this network. Morse telegraphy had already been replaced by teleprinter service on the main routes before World War II. In the U.K. the GPO in 1947 enhanced the public teleprinter network from manual to automatic switching as an important incentive to the creation of a worldwide automatic telex network.

30.2 TELEX SERVICE

Teleprinter service was enhanced to telex service in the United States in 1962. By that time Western Union had integrated its manually switched timed wire teletype network into its nationwide telephone network using modems to convert the telegraph code signals to voice-frequency tones. The name *Telex* (standing for "telegraphy exchange") was registered as a trademark of Western Union in the same year.[1] Since

---

[1] In 1966, Western Union bought the automatic teleprinter service of AT&T, called TWX, and AT&T bought Western Union's telex service in 1990.

The Worldwide History of Telecommunications, By Anton A. Huurdeman
ISBN 0-471-20505-2   Copyright © 2003 John Wiley & Sons, Inc.

Figure 30.1   Telex in Calcutta, 1960. (© Siemens-Archiv, Munich.)

then, telex has come to refer to the service of a worldwide switched network of machines that provide a documentary record (called a *telex*) of the communication. A telex used to be cheaper than a telephone call and could be received by unattended telex machines. A telex document was legally binding in most countries. A unique feature of the telex service was that address information was supplied by the teleprinter keyboard rather than by use of a dialing disk. A telex subscriber merely typed out the telex number for the subscriber called, awaited an acknowledgment from the machine called, and then transmitted its message even if the telex machine called was not staffed. The communication could also operate as a conversation, each subscriber being able to respond with an answer message during the call.

Very soon, telex became the most common text transmission service used by operators in local telegraph offices and switching centers and by the press, travel agencies, airlines, government agencies, embassies, and business enterprises in general. Figure 30.1 shows telex operation at a travel office in Calcutta, India.

Intercontinental telex service started in 1970 between North America and Europe. By that time, electronic parts gradually replaced the electromechanical parts of telex machines, and microprocessors introduced intelligent facilities, including word processing. For further convenience, electronic devices such as magnetic memory and video display were added to telex machines.

With the advent of computers, a new coding scheme was developed called the *American Standard Code for Information Interchange* (ASCII). ASCII employs seven coded pulses and thus is able to provide 128 combinations and to transmit messages at speeds up to 150 words per minute, compared to a maximum of 75 words for machines using the Baudot code.

As produced in the 1980s, a telex machine typically could be connected with external word processors, computers, or electronic typewriters. Electronic memories could store standard text, addresses, and out- and incoming messages. A *visual display unit* facilitated text editing and reading of messages, and a *floppy disk unit* allowed important messages to be stored permanently on removable magnetic diskettes. Telex terminals appeared with Arabic, Chinese, Cyrillic, Hebrew, Hindi, Japanese, and other characters. In addition to impact printing with rotary wheels (daisywheels) or type balls, both interchangeable for different languages, nonimpact printing appeared in which a character is formed out of a selection of dots from a 5×5 up to a 9×9 array, resulting in less operation noise and lower maintenance costs. These advanced facilities substantially simplified use of the telex. Within one decade the number of telex subscribers increased from about 800,000 to 1.8 million in 233 telex networks operating in 206 countries. About 65% of the networks applied automatic telex switching and covered 99% of the international telex traffic.

30.3   TELETEX

To meet the requirements of a higher transmission speed for data transmission between word processors and computers, a new service called *teletex* was introduced in the 1980s. Teletex is a form of office-to-office communication for electronic exchange of documents. Teletex applies the full typewriter character set of 309 characters and numbers. The messages are transmitted page by page (contrary to the endless paper used for telex) at a speed of 2400 bps. The standards for teletex were set by CCITT around 1980. A conversion facility was introduced to enable communication between telex and teletex terminals on a memory-to-memory basis. Teletex is a network-independent system, so that it can be used on public-switched telephone networks, on packet-switched data networks, and on circuit-switched data networks. Teletex was no success, however. It was introduced in 1982 in Germany, where even at its peak in 1988 it had only 19,000 subscribers. Worldwide, only 30,000 subscribers were served in that year, mainly in France (4606), South Africa (2475), Austria (1576), and a few hundred subscribers each in Turkey, Norway, Denmark, and Switzerland.

30.4   TERMINATION OF TELEX SERVICES

The highly competitive services of telefax, and in the 1990s even more of e-mail via the Internet, had replaced both telex and teletex almost entirely at the end of the twentieth century. The total number of telex subscribers worldwide, 1,728,000 at its peak in 1988, went rapidly down to less than 100,000 in 1999. Telex services were terminated in the United States in 1992 and in most industrial countries before the end of the century. In countries without adequate data transmission facilities, how-

TABLE 30.1   Decline of Telex Operation within One Decade

Country	Number of Telex Subscribers by Year End:				
	1988	1990	1993	1997	1998
Total worldwide	1,728,000	1,350,000	530,000	295,000	190,000
Australia	25,000	12,100	5,430	—	—
Austria	25,030	18,460	10,000	2,500	—
Belgium	25,000	17,720	9,360	3,224	—
Brazil	121,170	142,860	114,225	17,000	10,000
Canada	21,600	9,000	—	—	—
China	10,100	10,700	15,885	12,000	—
Denmark	11,690	8,600	4,870	2,150	1,670
France	147,285	133,250	—	—	—
Germany	175,640	134,370	55,400	9,000	6,900
Hong Kong SAR	26,760	21,500	15,050	4,640	3,720
India	41,350	46,830	—	25,200	—
Italy	72,770	59,155	—	—	—
Japan	43,000	—	—	—	—
Mexico	17,560	13,680	—	—	—
The Netherlands	33,100	22,000	11,500	—	—
Poland	33,545	38,410	35,460	11,700	8,700
Portugal	27,650	26,640	9,065	1,890	1,440
Russia	—	—	—	40,400	26,500
Saudi Arabia	13,765	10,925	8,290	5,280	4,745
Singapore	16,520	—	—	2,745	2,140
South Africa	22,605	13,914	6,770	—	—
Spain	41,185	30,970	18,520	—	—
Sweden	18,320	14,270	9,200	—	—
Switzerland	38,250	27,145	13,480	—	—
Taiwan-China	17,850	12,180	—	—	—
Turkey	22,225	21,105	17,205	13,185	11,160
U.K.	115,000	—	—	18,000	—
United States	81,055	62,335	—	—	—
Venezuela	13,400	12,675	10,685	8,745	—

Sources: Data from ITU and Siemens.

ever, telex is still being used as a slow but reliable instant written communication. In fact, the last international telex exchange was put into operation as late as 1996, in Minsk (Belarus). Table 30.1 shows the reduction in the number of telex subscribers worldwide in countries that used to have over 10,000 subscribers.

REFERENCES

Books

Gööck, Roland, *Die großen Erfindungen Nachrichtentechnik Elektronik*, Sigloch Edition, Künzelsau, Germany, 1988.

Kobayashi, Koji, *Computers and Communications*, MIT Press, Cambridge, MA, 1986.

Oslin, George P., *The Story of Telecommunications*, Mercer University Press, Macon, GA, 1992.

Articles

International Telecom Statistics (Siemens), for the years 1992, 1993, and 1995.

Internet

To HTML from *The Early History of Data Networks*, by Gerard J. Holzmann and Björn Pehrson.

http://Japan.park.org/Japan/NTT/Museum, History of telegraph and telephone.

31

TELEFAX

31.1 TECHNOLOGICAL DEVELOPMENT OF TELEFAX

Despite various improvements over a 100-year period, tele-facsimile remained a specialized service mainly for the transmission of press photographs, for meteorological charts for weather forecast services, and for ships at sea. Western Union started the first public facsimile service in the United States before World War II (see Section 18.5). A major improvement came in 1948, when engineer Garvice Ridings of Western Union developed a small, easy-to-operate low-cost desktop facsimile machine named *Desk-Fax*. Within a few years, 40,000 Desk-Fax machines were in operation; twice as many as the number of telex machines at that time in the United States. Western Union also introduced a high-speed facsimile service between New York and Washington on March 13, 1951. Ninety pages could be transmitted per hour without requiring processing before or after reception.

Rapid progress in electronic engineering after Word War II with the availability of transistors and especially of light-emitting diodes (LEDs) and lasers in the 1960s enabled the introduction of a new generation of facsimile machines which dramatically improved the quality and reduced the cost of facsimile service. In the mid-1960s, the Xerox Rand Corp. introduced a compact desktop *Telecopier* machine. For transmission of the signal over the telephone line, the telephone transmitter was to be stacked onto the Telecopier. The Telecopier needed about 6 minutes for the transmission of a DIN A4-size page.

In Japan, research on facsimile telegraphy using electrical instead of photographical reproduction began around 1942. Facsimile telegraph service with photographical reproduction was introduced on December 22, 1946, between Tokyo and Osaka. Phone–fax service with electrical reproduction machines started in Japan in

The Worldwide History of Telecommunications, By Anton A. Huurdeman
ISBN 0-471-20505-2 Copyright © 2003 John Wiley & Sons, Inc.

1972, and a facsimile communications network called *F-Net* started its services in September 1981. F-Net stored data (documents) and provided a wide range of services, such as broadcast communications and facsimile boxes. Fax became most popular in Japan, as neither the hiragana script nor the kanji characters used in Japanese writing lend themselves well to the keyboards of telex and computers.

Over the years, different manufacturers adopted operational procedures that allowed their machines to communicate with one another. In contrast with the standardized international telex network, however, there was no worldwide standard for facsimile. In 1968, CCITT issued its first telefax standard (Recommendation T.2), later named the standard for group 1 fax machines. This standard specified fax operation at a transmission speed of one A4 page in 6 minutes and a resolution of 3.85 lines/mm. Interworking between models of different makes, however, was not yet covered sufficiently by this standard. After two amendments in 1972 and 1976, group 2 fax standard T.3 followed in 1976, which reduced the transmission time of one A4 page to 3 minutes at the same resolution, due to an improved modulation method (vestigial-sideband amplitude modulation with phase modulation). Telefax machines meeting the recommendations of group 2 were compatible internationally. Thanks to this standard, some 500,000 telefax machines were in operation worldwide by the time the CCITT issued recommendation T.4 for group 3 fax machines in 1980.

Whereas the group 1 and 2 standards were for analog transmission, the group 3 standard specified a digital transmission at data rates between 2400 and 9600 (later increased to 14,400 and 28,800) bps. The transmission time of an A4 page was typically reduced to 40 seconds at a data rate of 9600 bps. The operation of a group 3 fax machine is explained briefly in Technology Box 31.1.

TECHNOLOGY BOX 31.1

Group 3 Fax Machine Operation

At the transmitter end of a modern group 3 fax machine, the image of a page to be transmitted is focused on a charge-coupled device (CCD), a solid-state scanner that has 1728 photo sensors in a single row. The photo sensors measure the brightness of spots, or pixels, in a line 0.01 inch high across the width of the page. After a line is scanned, the scanner is advanced by one line. The output of the scanner is supplied to an analog-to-digital converter where the spot intensity is converted to a single bit of information.

Two lines are stored in a buffer memory, and a source compression algorithm then reduces the two lines of image information to a fraction of the original number of bits required to represent the image. The source compression applies the modified Huffman code (MHC) or, for higher transmission speeds, the modified Read code (MRC). The digital representation of the image is then transmitted over the PSTN using a voice band modem.

At the receiver, another modem receives the signal, which then undergoes source decompression for reconstitution of the original image information. The image received is then printed by a thermal printer on special temperature-sensitive recording paper or on plain paper via a xerographic process.

TABLE 31.1 Worldwide Fax Penetration, 1997

Country	Number of Fax Machines	Country	Number of Fax Machines
Australia	900,000	Italy	1,800,000
Brazil	500,000	Japan	16,000,000
Canada	1,000,000	Malaysia	150,000
China	2,000,000	Mexico	285,000
Colombia	173,000	The Netherlands	600,000
Czech Republic	103,000	Norway	220,000
Finland	198,000	Pakistan	206,000
France	2,800,000	South Africa	150,000
Germany	5,600,000	Thailand	150,000
Hong Kong SAR	346,000	Turkey	108,000
Hungary	120,000	United States	21,000,000
India	150,000	Rest of world	818,000
Indonesia	185,000	Total worldwide	55,562,000

With the issue of the group 3 standard, mass production of easily operable, good-quality, low-cost fax machines was begun, mainly in Japan, Korea, and Taiwan. Within five years the number of fax machines increased to 120,000 in Europe, 550,000 in the United States, and 850,000 in Japan. Fax operation via the telephone network became very easy. By dialing the number of the fax subscriber desired, a connection is made and a handshake procedure starts in which the two fax machines exchange signals to establish common features such as modem speed, source code, and printing solution. If in agreement, scanning of the message starts and the coded information creates a copy on the receive side. Upon receipt of a signal that no more pages follow, the receiving machine confirms receipt of the message and disconnects the calling machine from the line.

A further standard, group 4, was introduced in the early 1990s for a new generation of fax machines that can operate on ISDN, at speeds up to 64 kbps at a resolution of 7.7 or 11.5 lines/mm. At that speed a single A4 page can typically be transmitted in 3 seconds.

31.2 WORLDWIDE TELEFAX PENETRATION

Whereas the number of telex machines worldwide in operation went down from 1,728,000 at its peak in 1988 to below 100,000 in 1999, the number of telefax machines surpassed 20 million by 1990, went up to 55 million in 1997, and reached the 100 million mark at the end of the twentieth century. This, however, might be the peak, as in the meantime, telefax faces heavy competition from e-mail and in the future possibly even more from the emerging office printers using the *Internet Printing Protocol* (IPP). This protocol, defined in 1999 by the *Printer Working Group*,[1] enables transmission of files via the Internet to high-resolution printers. Similar to

---

[1] The Printer Working Group is an industry body created in 1997 comprising all major printer and printer server vendors, including Hewlett-Packard, Lexmark, Microsoft, Novell, and Xerox.

fax 30 years ago, however, a common international standard for IPP still had to be agreed upon at the end of the twentieth century. Table 31.1 shows the number of fax machines in operation at the end of 1997 worldwide in countries with over 100,000 fax machines according to estimates of the ITU.

REFERENCES

Books

Gööck, Roland, *Die großen Erfindungen Nachrichtentechnik Elektronik*, Sigloch Edition, Künzelsau, Germany, 1988.

Kobayashi, Koji, *Computers and Communications*, MIT Press, Cambridge, MA. 1986.

Oslin, George P., *The Story of Telecommunications*, Mercer University Press, Macon, GA, 1992.

Articles

Gengler, Barbara, The death of the fax? *Communications International*, September 1999, pp. 74–76.

Stevens, Ian, Text communications: the matter of fax, *Developing World Communications*, Grosvenor Press International, London, 1986, pp. 382–383.

Wenger, P.-A., The future also has a past: the telefax, a young 150-year old service, *Telecommunication Journal*, Vol. 56, No. 12, 1989, pp. 777–782.

Internet

http://Japan.park.org/Japan/NTT/Museum, History of telegraph and telephone.

www.britannica.com, (3), Modern facsimile.

32

CELLULAR RADIO

32.1 EVOLUTION OF CELLULAR RADIO

Bell Labs conceived the concept of cellular radio in the early 1940s. The prevailing technology, however, could not cope with the complexities involved, and World War II imposed other priorities. After the war, commercial public mobile radio operating in the 150-MHz band began in 1946 in St. Louis, Missouri, United States (Section 17.3). A system operating in the 450-MHz band followed in 1956. There were almost 1.5 million mobile users in 1964, when AT&T introduced an *Improved Mobile Telephone System* (IMTS), system MJ operating at 150 MHz and system MK operating at 450 MHz via 25- to 30-kHz FM channels. Various mobile radio systems followed, mainly in North America and Europe. Those systems operated on a limited number of channels in the 40-, 80-, and 160-MHz bands. Operation was in a manual mode; an operator could establish a call between two mobile subscribers or between a mobile subscriber and a subscriber on the PSTN.

The mobile terminals equipped with electronic vacuum tubes were large, heavy, had a high power consumption, needed shock-protected mountings, and were expensive. Transistorizing in the 1970s brought substantial improvements, but the limited number of radio channels constrained the network capacities. Improvement in the equipment and services and the opening of the 450- and 800-MHz bands for public mobile radio in the 1980s enlarged the number of subscribers to about 600,000 worldwide by 1985 but caused even more network congestion.

An approach toward a better use of the available radio spectrum, a trunking technique, was introduced in the late 1970s. Trunking provides organized sharing of a small number of RF channels by a large number of professional mobile radio users. These users were mainly taxi drivers, transport enterprises, emergency services, and

The Worldwide History of Telecommunications, By Anton A. Huurdeman
ISBN 0-471-20505-2   Copyright © 2003 John Wiley & Sons, Inc.

government organizations.[1] The subscribers share access to a mobile radio network without being aware of other users. Tens of thousands of subscribers shared trunking networks in places such as London, Los Angeles, Rotterdam, Sydney, Tel Aviv, and Toronto in the early 1980s.[2]

Trunking improved the use of the radio spectrum, but it did not solve the problem of congestion on the public mobile radio networks. The creation of an altogether new concept became a necessity. About 150,000 subscribers used mobile radio in 1978, when the FCC invited the U.S. telecommunications industry to bring forth proposals for a more effective land/mobile telephone system. AT&T responded with a proposal, originating from the above-mentioned Bell Labs' idea in the 1940s, called the cellular *Advanced Mobile Phone System* (AMPS), whereby a set of frequencies could be reused within a system of cells. Each cell would have its own transmitter/receiver station that would be connected to the PSTN. Mobile telephone users traversing the cells would change automatically to the frequency of the cell entered, thus constantly keeping in touch with, or in reach of, the PSTN. AMPS was demonstrated successfully in Chicago in 1978, but matters of competence of the relevant authorities and interference with frequency bands used by the military and for TV delayed the introduction of cellular radio in the United States until 1983.

In the meantime, the concept of cellular radio communication was discussed in a paper called "Fundamental Problems of Nationwide Mobile Radio Telephone System," written by K. Araki and published in NTT's *Electrical Communications Laboratories Technical Journal*, Vol. 16, No. 5, in 1967. Japan, a very densely populated country with plenty of public pay telephone booths, never had a public mobile radiotelephone system apart from public radio paging, introduced in 1968. Research into a land mobile telephone system suitable for Japan was initiated in 1953 by the Electrical Communications Laboratories of NTT and resulted in 1967 in the above-mentioned publication in which a nationwide cellular radio system operating in the 800-MHz band was proposed. In the same year, NTT began development of the proposed cellular system. Trials with laboratory equipment were made in the Tokyo metropolitan area in 1971. A pilot system could then be tested successfully, in 1975. After further system improvements, the world's first cellular radio service was begun in Japan on December 3, 1979 with a cellular radio system called MCS (mobile control station). *Thus the cellular radio era started in 1979 in Japan*.

Cellular radio, even in its first analog version, is the most effective version of mobile radio. It provides radio coverage of large geographical areas by seamless overlapping cells in a honeycomb configuration. Cellular radio networks provide more versatile services to mobile users as obtained at fixed locations from the PSTN. Figure 32.1 shows the basic configuration of a cellular radio network and Technology Box 32.1 gives a brief explanation. Cellular radio became the fastest-growing public telephone service. In the industrial world, it offers the highly mobile user con-

---

[1] Trunked radio networks are complementary to cellular networks and offer corporate users with specific services such as group calling, priority calling, direct mobile-to-mobile communication without infrastructure, limitation of access to PSTN or geographical coverage, fast call setup, conference calls, call transfer, and status reporting. Trunked radio networks provide coverage mainly along roads, highways, waterways, and inside industrial and transport areas.

[2] Trunking has survived despite cellular radio. A digital trunking system named *Tetra* (terrestrial trunked radio), specified by ETSI, was being implemented almost worldwide at the end of the twentieth century.

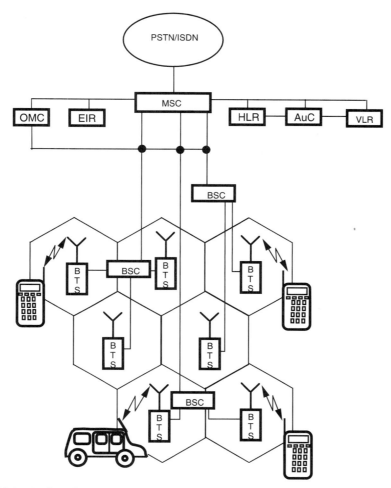

Figure 32.1 Basic cellular radio network. (From A. A. Huurdeman, *Guide to Telecommunications Transmission Systems*, Artech House, Norwood, MA, 1997, Figure 6.5; with permission of Artech House Books.)

tinuous accessibility, whether on foot or in a vehicle. In the developing world, it offers a quick, low-cost substitute for poor-quality or nonexisting fixed telephone networks, thus solving the problem of long waiting times for network access.

32.2 ANALOG CELLULAR RADIO

In addition to AMPS and MCS, were cellular radio systems such as the Scandinavian *Nordic Mobile Telephone* (NMT), the *Total Access Communication System* (TACS) in the U.K., *C 450* in Germany, RadioCom 2000 in France, and Radio Telephone Mobile System (RTMS) in Italy. All those cellular radio systems operated in an analog mode and were not compatible with each other. The mobile terminals

TECHNOLOGY BOX 32.1

Basic Cellular Radio Network

Cellular radio networks provide radio coverage to *mobile stations* (MSs) in large geographical areas by seamless overlapping cells. Each cell has its own radio base station, usually called a *base transceiver station* (BTS). A number of base stations are connected with a *common base station controller* (BSC). All BSCs of a coverage area are connected with a *mobile-service switching center* (MSC), which performs switching between the mobile subscribers and constitutes the interface to the PSTN/ISDN. Mobile subscribers traveling from cell to cell are switched over automatically to the base station of the next cell by a procedure called *handover*, so that uninterrupted conversation is granted and no calls will be missed. The mobile subscriber reports its location automatically to the MSC via the BTS of the cell in which he or she has arrived. The MSC registers this in location registers so that calls to network subscribers can be routed to the cell where the subscriber called has reported its presence.

   The locations of subscribers belonging to the coverage area of the MSC are permanently registered in a *Home Location Register* (HLR). Subscribers arriving from another coverage area are temporarily registered in a *Visitor Location Register* (VLR). A special *Equipment Identity Register* (EIR) checks the status of the MS's identity number. Each MS has its *International Mobile Station Identity* (IMEI) number, which enables the EIR to identify an unauthorized (stolen, cloned, or other type of fraud) MS. An *Authentication Center* (AuC) checks the authentication of a subscriber attempting to use the network. The corresponding information is stored in a *subscriber identity module* (SIM) card which the subscriber has to insert in a MS before it can be used. Instead of a SIM card, some cellular systems use a *user identification module* (UIM) or *subscriber identity security* (SIS) card. An operation and maintenance center (OMC) controls the system elements, provides security management, and collects billing and accounting information.

Source: Adapted from A. A. Huurdeman, *Guide to Telecommunications Transmission Systems*, Artech House, Norwood, MA, 1997; with permission of Artech House Books.

were conceived for installation in vehicles, with the possibility of removing them from vehicles for temporary transportable use. For exclusive portable use, a simplified version of cellular radio was developed called *cordless telephone* (CT), with small cells, mainly in pedestrian areas. Table 32.1 shows the distribution of cellular radio at the end of 1997. Beginning that year, analog cellular penetration stagnated and gradually decreased, due to the superior quality of digital systems.

32.2.1   Analog Cellular Radio in Japan

The Japanese analog cellular radio systems operated in the 800-MHz band. The first system, MCS L1, had 600 radio channels with a channel separation of 25 kHz. The

TABLE 32.1   Analog Cellular Radio by Year-End 1997

System	Number of Subscribers	System	Number of Subscribers
AMPS	69,612,000	NMT-900	2,627,000
TACS	16,107,000	Others	696,000
NMT-450	1,854,000	Total	91,408,000

Source: Data from *Mobile Communications International*, October 1999, p. 84.

mobile units still had a volume of 6600 cm^3 and a weight of 7 kg. Transportable units introduced in 1985 weighed 3 kg, had a volume of 2300 cm^3, and 5 W of output power. The first portable units came in April 1987, with a weight of 750 g, a volume of 500 cm^3, and 1 W of output power. A second version, MCS L2, was introduced in 1988 with a channel spacing of 12.5 kHz and 6.25 kHz with interleaved channels, resulting in 2400 radio channels.

Initial growth was slow until nationwide roaming was introduced in 1984 and cellular radio was liberalized in 1985. There were 100,000 subscribers when two new operators, Daini Denden Inc. (DDI) and Nippon Indou Tsushin Corp. (IDO), joined the cellular market in 1988. IDO also used the MCS L2 system, whereas DDI introduced J-TACS (Japan TACS), based on the British TACS system (Section 32.2.4). In 1989, DDI introduced the Handy-Phone, based on Motorola's TAC portable. With a volume of 221 cm^3 and a weight of 303 g, it marked the beginning of handset miniaturization. In 1991, DDI and IDO introduced N-TACS (narrowband TACS) from Motorola with 1200 channels and a bandwidth 12.5 kHz.

32.2.2   Analog Cellular Radio in Scandinavia

The Swedish engineer Mäkitalo conceived the *Nordic Mobile Telephone* (NMT) system when working for the Swedish Telecommunications Administration, Televerket.[3] The system, operating in the 450-MHz band, has 180 radio channels with a spacing of 25 kHz or 225 channels with a spacing of 12.5 kHz. A second system, NMT 900, operating in the 900-MHz band, was introduced in 1986, with 1000 channels at 25-kHz spacing or 1999 interleaved channels when using 12.5-kHz spacing.

The NMT system was put into operation by the telecommunications administrations of Denmark, Finland, Norway, and Sweden in 1981–1982 and soon found applications beyond the Scandinavian countries: first in The Netherlands, Saudi Arabia, and Spain in 1982; Malaysia, Oman, Tunisia, and Turkey in 1985; and successively worldwide.

32.2.3   Analog Cellular Radio in North America

The American AMPS operates in the 800-MHz band with 832 radio channels at 30-kHz spacing. Narrow-AMPS (N-AMPS), with 2580 radio channels at 10-kHz spac-

---

[3] Mäkitalo, holding some 20 patents, was also one of the pioneers of the GSM system.

ing, followed in 1991. Commercial operation began in the United States in 1983. The AMPS system was used all over the Americas, starting in Canada in 1985, one year later in Mexico and Colombia and outside the Americas in Korea and Australia in the next years, followed by China, Hong Kong, Indonesia, Laos, Malaysia, New Zealand, Pakistan, the Philippines, Singapore, Taiwan, Thailand, Vietnam, and all Latin American countries.

32.2.4   Analog Cellular Radio in Western Europe

The British *Total Access Communication System* (TACS) was introduced in the U.K. in 1985. The U.K. government, as a pacemaker in telecommunications liberalization, had invited potential operators, excluding the fixed network operators British Telecom and Mercury, to apply for a 25-year license for the operation of a cellular radio network. The successful tenderers were Vodafone (then a subsidiary of Racal Millcom Ltd., since 1988 independent Vodafone Group plc), and Cellnet (a BT/Securicor partnership). They agreed on a modified version of the U.S. AMPS developed by Motorola, which was then named TACS. The system operated 1000 radio channels with a spacing of 25 kHz in the 900-MHz band. An extended TACS (called E-TACS) with 1640 radio channels at 25-kHz spacing, followed in 1988. Outside the U.K., TACS was introduced in Ireland and Hong Kong in 1985, and one year later in Bahrain and Kuwait, soon followed by Austria, China, Italy, Kuwait, Pakistan, Spain, and many other countries.

In Germany the cellular radio C 450 system was introduced in 1985. The nomination C indicated that it was the third national public mobile network, after the noncellular networks A, introduced in 1958, and B, introduced in 1972. The A and B networks operated in the 160-MHz band. The A-Net, with terminals weighing 16 kg, had 10,500 users in 1972 (served by 600 operators), when the B-Net was introduced. The B-Net reached a maximum of 26,911 subscribers in 1986. The cellular system C 450 operated 222 radio channels with a channel spacing of 20 kHz in the 450-MHz band. It was the first system that applied a subscriber identity card, the forerunner of the SIM (see Technology Box 32.1). The C 450 system was also introduced in South Africa in 1986.

In France in 1985 a system was introduced called RadioCom 2000, which also operated in the 450-MHz band using 170 radio channels at a spacing of 20 kHz.

In Italy, in addition to TACS, a 450-MHz band system called RTMS (Radio Telephone Mobile System) was used.

32.3   DIGITAL CELLULAR RADIO

In the mid-1980s, when analog systems were introduced, the first market forecasts predicted 500,000 subscribers by 1990. The actual demand, however, far superseded this expectation, and there were 10 million analog cellular radio subscribers in 1990 and two years later, almost 15 million subscribers in 30 countries. By that time, cellular networks in Chicago, Los Angeles, and New York each had over 500,000 subscribers and London over 750,000, which again caused substantial congestion. Once more, a new approach was required: a change from analog to digital operation. Digital operation not only improves transmission quality but facilitates data transmission and speech encryption and allows different multiple-access modes, which substantially increase the capacity of a cellular network. Three multiple-access methods are used, as illustrated in Figure 32.2 and explained briefly in Technology Box 32.2.

Five different second-generation (2G; the analog cellular radio systems were then 1G systems) digital cellular radio systems were developed; the *global system for mobile* communication (GSM) in Europe, two *digital-AMPS* systems (D-AMPS TDMA and CDMA) in the United States, and two *personal digital cellular* systems (PDC 800 and PDC 1500) in Japan. Furthermore, the analog CT systems were succeeded by digital *personal communications networks* (PCNs), in which person-to-person communication was provided using a single national—and eventually international—personal telephone number, independent of home or office.

Introduction of these digital cellular radio systems in the early 1990s caused an enormous boost in telecommunications, especially after prepaid service replaced subscription contracts. Prepaid service was first offered by the German operator D1 (later T-Mobil) in 1995 to increase Christmas sales. One year later, the Italian operator Telecom Italia Mobile (TIM) offered rechargeable prepaid cards with the slogan "no activation fee, no subscription, no monthly bill," which made cellular radio worldwide a mass consumer product. Handsets bundled with prepaid cards were sold in the supermarket like washing detergent. By year-end 2000, 61% of all European cellular users and almost 30% of all cellular radio users worldwide used a prepaid card.

Another impetus to cellular radio was given by the introduction of nonvoice *small message services* (SMSs), which offer text messages showing a maximum of 160 characters on the display of the handset at a marginal cost and with a minuscule impact on network capacity. SMS was introduced on GSM networks in Europe in the mid-1990s, soon followed in Asia and since the end of 2000 in the United States on non-GSM networks. SMSs in the Chinese language became possible in the year 2000 with the introduction of handsets on which eight basic strokes of the Chinese characters are allocated to the 1 to 0 keys of the keypad in an eZiText[4] mode. At the end of the century, SMS messages were being sent worldwide at a volume of 20 billion per month and still rising.[5] In 1996 the analog NMT system was enhanced to include SMS.

Cellular radio became the fastest-growing sector of telecommunications. Deregulation and privatization were first applied on cellular radio operation so that competition became a dominant factor for growth. Most countries got at least two competing operators. Optimistic market observers, including the author of this book, dared to forecast in 1990 around 36 million subscribers by 1995 and a maximum of 100 million by the beginning of the new century for cellular radio, including PCN. By 1995, instead of the 36 million forecasted, there were 80.3 million analog and digital cellular subscribers worldwide, of which there were 36 million in North America, 22.5 million in Europe, and 21.8 million in the Asia-Pacific region. The

---

[4] eZiText was developed by the Canadian Zi-Corporation and endorsed by the Chinese government. When writing a character, the user presses the keys that correspond to the necessary strokes. After a couple of strokes, the system "intuitively" creates the most likely character intended.

[5] According to information from the GSM Association, SMS reached a landmark 1 billion messages per day in October 2001.

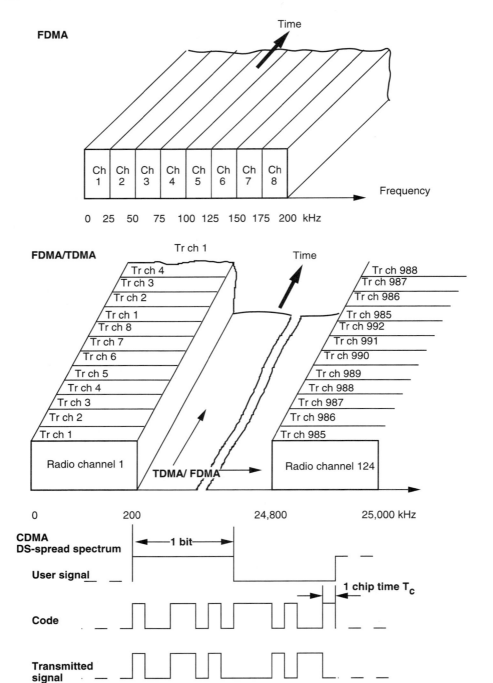

Figure 32.2   Cellular multiple access modes. (From A. A. Huurdeman, *Guide to Telecommunications Transmission Systems*, Artech House, Norwood, MA, 1997, Figure 6.9; with permission of Artech House Books.)

TECHNOLOGY BOX 32.2

Cellular Multiple Access Modes

With analog cellular radio, a separate radio channel is required for each call. Thus, for simultaneous calls for eight subscribers within a cell, eight radio transceivers are required to operate in parallel in a *frequency-division multiple access* (FDMA) mode. With digital engineering, however, those eight calls can be handled in eight different time slots by one transceiver, and eight such digital transceivers can thus handle 64 simultaneous calls in a *time-division multiple access* (TDMA) mode. With digital operation a combination of FDMA and TDMA is also possible: for example, with eight parallel transceivers as used by the GSM system, thus increasing the number of simultaneous calls in a cell to 64. In the GSM system, a total of 124 RF channels each of eight full-rate 13-kbps channels or 248 RF channels each of eight half-rate 6-kbps channels can be used within a 200-kHz band, thus supporting 992 full-rate or 1984 half-rate channels per network.

Another way of increasing the capacity of a digital cellular system was derived from a military spread-spectrum (SS) technology. With this technology the signal is spread over the available transmission spectrum to make it less vulnerable to deliberate jamming or other interference. This spreading is made by *time hopping* (TH-SS), *frequency hopping* (FH-SS), or in *direct sequence* (DS-SS). For identification, each call is given an individual code, and this method is referred to as *code-division multiple access* (CDMA). Due to this spreading, a higher path loss is permitted, which allows operation of larger cells than with TDMA systems. The capacity of a CDMA system is not fixed but determined by the signal-to-interference ratio. This means that with the increase in simultaneous calls, the background noise increases and thus the transmission quality degrades.

Source: Adapted from A. A. Huurdeman, *Guide to Telecommunications Transmission Systems*, Artech House, Norwood, MA, 1997; with permission of Artech House Books.

distribution was AMPS, 51%; TACS, 17%; GSM, 14%; NMT, 6%; PDC, 4%; D-AMPS-TDMA, 2%; PCN, 2%; and others, 4%.[6] The tremendous development of cellular radio in the last decade of the twentieth century is documented in Table 32.2, whereas Table 32.3 shows the distribution to the major cellular systems of the worldwide 490 million cellular subscribers by year-end 1999. One year later there were 740 million subscribers: 39.4% in Europe, 34% in the Asia-Pacific region, 16% in North America, 8.5% in Latin America, and 2.1% in Africa.[7]

[6] Market figures as published in *Mobile Europe*, May 1996. This unexpectedly rapid deployment inspired the world's major mobile radio manufacturers in 1995 to predict for the year 2000 a total number of analog and digital cellular radio users, including PCN as follows: Motorola forecast more than 200 million subscribers, AT&T more than 300 million, L.M. Ericsson 350 to 450 million, and Nokia perhaps 400 million. In fact, there were 740 million.
[7] Data from ITU Telecommunication Indicators Database on the Internet.

TABLE 32.2 Worldwide Cellular Radio Subscribers

Year	Millions of Subscribers	Year	Millions of Subscribers
1990	11	1996	144
1991	16	1997	215
1992	23	1998	307
1993	34	1999	490
1994	55	2000	740
1995	91		

Source: Data from ITU Telecommunication Indicators on the Internet.

Cellular growth surpassed growth in the fixed telephone networks. In 1993, Cambodia was the first country where the number of mobile subscribers exceeded the number of fixed lines (albeit at a mobile teledensity of 0.81 and a fixed-line teledensity of 0.25), followed by Finland in 1998 (with a mobile teledensity of 67 and a fixed teledensity of 55). Austria, Côte d'Ivoire, Hong Kong SAR, Israel, Italy, Korea (Republic), Paraguay, Portugal, Uganda, and Venezuela followed in 1999, and a further 23 countries in 2000. In that year the end users bought a total of 413 million cellular handsets,[8] of which 30.6% were made by Nokia, 14.6% by Motorola, 10% by Ericsson, 6.5% by Siemens, 5% by Samsung, and 33.2% by others.

The 10 operators with the highest number of subscribers at the end of the twentieth century are summarized in Table 32.4. The largest market was the United States with 110 million subscribers, followed by China with 85.3 million; Japan, 66.8; Germany, 48.1; Italy, 42.2; and the U.K., 40 million cellular radio subscribers.

32.3.1 Global System for Mobile Communication

In 1981, a joint Franco–German study was initiated to develop a binational cellular radio system, with the desired side effect of counterbalancing the Scandinavian NTM system. Fortunately, the CEPT (Committee of European Post and Telecommunications) recognized the challenge of a pan-European cellular system and used the study as an incentive to extend cooperation from Franco–German to pan-European.

TABLE 32.3 Cellular Radio Systems by Year-End 1999

System	Percent	System	Percent
GSM	53	TACS	2.4
AMPS	15.5	NMT 450	0.4
CDMA	10.9	NMT 900	0.2
PDC	10.2	Total	100
D-AMPS/TDMA	7.4		

Source: Data from *Communications International*, February 2000.

[8] Figures published by *Garda Dataquest* in March 2002.

TABLE 32.4   Major Cellular Operators by Year-End 2000

Operator	Country	Number of Subscribers
NTT DoCoMo	Japan	36,030,000
Verizon	United States	27,505,000
TIM	Italy	21,600,000
Cingular	United States	19,681,000
Mannesmann	Germany	19,245,000
T Mobil	Germany	19,141,000
AT&T	United States	15,716,000
Omnitel	Italy	14,920,000
France Telecom	France	13,941,000
Total		253,039,000[a]

Source: Data from ITU Telecommunication Indicators on the Internet.

[a] Represents 35% of total cellular subscribers.

CEPT decided to set up a working group under the French language name *Groupe Spéciale Mobile*, hence the abbreviation (GSM), which after worldwide penetration of the system was changed to *Global System for Mobile Communication*. This group was given the task of defining and working out the specifications for a cellular system that could be introduced across Europe in the early 1990s. In line with the ongoing evolution from analog to digital in switching and transmission, it became obvious that this new system should be a digital system.

In 1986, the European Union issued a directive to all members states to clear the frequency band of 890 to 915 MHz for transmission from mobile to base station and 935 to 960 MHz for the reverse direction for the operation of 124 radio channels for the future pan-European cellular radio network. On September 7, 1987, a *memorandum of understanding* (MoU) was signed in Copenhagen by 13 European countries[9] committing them to implement a common digital cellular radio system on their networks, including roaming with the networks of the other MoU-signature operators, by July 1, 1991. The completion date was later advanced by one year. In 1989 the responsibility for the GSM specifications passed from CEPT to the newly created *European Telecommunications Standards Institute* (ETSI), located at Sophia Antipolis in the south of France.

By 1990, the GSM working group, embracing hundreds of engineers, could complete the specifications.[10] A race began for the first networks to open commercial operation by July 1, 1992. The Finnish operator OY Radiolinja AB was the winner starting in December 1991. A further seven operators met the July 1, 1992 date.[11]

[9] The original signatories were Belgium, Denmark, Finland, France, Germany, Ireland, Italy, The Netherlands, Norway, Portugal, Spain, Sweden, and the U.K. Additional signatories were Cellnet and Racal-Vodafone of the U.K.

[10] The GSM specifications consisted of some 5000 pages (mainly software related) in 140 documents filling a row more than 10 m long.

[11] The seven operators meeting the target date were the Danish operators Sonofon and Tele Denmark Mobil, the Finnish operator Telecom Finland, the French operators France Telecom and *Société Française de Radiotéléphonie* (SFR), and the German operators DeTeMobil (later called T-Mobil and operating the D1 network) and Mannesmann Mobilfunk (operating the D2 network; since 2000, Vodafone).

TABLE 32.5   Worldwide GSM Deployment

Country or Region	1995		2000	
	Number of Subscribers	Percent of Total	Number of Subscribers	Percent of Total
Europe	6,948,000	78	283,590,000	62.4
Middle East	243,500	2.7	6,972,000	1.5
North Africa	16,500	0.2	5,273,000	1.2
South Africa	425,000	4.8	7,833,000	1.7
Rest of Africa	6,850	0.1	3,015,000	0.6
Asia-Pacific	1,266,100	14.2	137,066,000	30.1
United States	—	—	8,290,000	1.8
Canada	—	—	923,000	0.2
Rest of Americas	—	—	2,062,000	0.5
Total worldwide	8,906,000	100	455,022,000	100

Source: Data from information published in *GSM World Focus*, 1996 and 2001.

Within one year all other original MoU-signature operators had begun commercial GSM operation, and surprisingly, also operators outside Europe in Australia,[12] New Zealand, and Hong Kong. Full national coverage and international roaming has been reached for at least 90% of the population in countries with GSM networks.

In 1991, ETSI extended the GSM specification and added 374 radio channels in the frequency band 1710 to 1880 MHz for a *Digital Cellular System 1800* (DCS 1800), to be used for the personal communications networks (PCNs; see Section 32.4). The British company Mercury One-2-One launched the first network with DCS 1800 equipment on September 7, 1993. In Asia the first network based on DCS 1800, supplied by Nokia, was launched in Thailand on September 28, 1994, under the name *Worldphone 1800*. It started with an initial capacity of 60,000 subscribers served in 140 cells in Bangkok and its suburbs. In 1997, the various DCS 1800 systems using GSM technology were renamed GSM 1800 to differentiate them from networks applying CDMA or another technology in the 1800-MHz band.

GSM planning in the 1980s had been based on 10 million subscribers all over Europe, North Africa, and part of the Middle East by the end of the century. In fact, this number was almost reached at the end of 1995, with few subscribers in North Africa (only 16,500 in Morocco) but an unexpected large number in South Africa and even more in the Asia-Pacific region.[13] Table 32.5 summarizes worldwide GSM deployment by the end of 1995 and 2000.

One of the key advantages of GSM is the use of sophisticated encryption derived from military technology known as A5, which at least until early 1999[14] was never cracked. Export of GSM equipment with the A5 algorithm was restricted temporar-

[12] The Australian operator Telstra became the first non-European GSM-MoU member in 1992.

[13] The chairman of the GSM MoU Association, Mike Short, predicted to a disbelieving audience that the number of GSM subscribers would increase to a fantastic 100 million by the year 2000. He was too pessimistic; that figure was reached in July 1998, there were 500 million subscribers by May 2001 and 750 million in 184 countries by September 2002.

[14] In May 1999, David Wagner and his team at the Weizmann Institute in Israel claimed to have reconstructed the A5 algorithm.

ily by COCOM regulations[15] and a simplified A5/2 algorithm had to be used, which, however, was still very secure against call interception.

GSM has a transmission speed of only 9.6 kbps, which is suitable for voice, SMS, and low-speed data services. From mid-1990, systems with more advanced data transmission were developed as *2.5G* (between the second and third generations) solutions:

1. *High-speed circuit-switched data* (HSCSD), by combining channels offering full-ISDN 64 kbps in circuit-switched mode
2. *General packet radio service* (GPRS), by combining channels and a new coding offering up to 115 kbps in packet-switched mode
3. *Enhanced data rates for GSM evolution* (EDGE), for speeds up to 384 kbps in wide-area applications and up to 554 kbps in local areas (EDGE is also compatible with the U.S. D-AMPS-TDMA system)
4. *GSM/EDGE radio access network* (GERAN), as Phase 2 EDGE, offering data rates up to 1920 kbps

The next addition to GSM was a facility for Internet access on GSM terminals called the *Wireless Access Protocol* (WAP). This protocol was developed by cooperation between Ericsson, Motorola, Nokia, and UP in 1999. To enable the presentation of Internet on the small screen of a mobile handset, a special *wireless markup language* (WML) was developed as a counterpart to the Internet HTML (hypertext markup language). WAP is not only for GSM but is compatible with all digital cellular standards. Figure 32.3 shows a handset with a WAP Internet browser as the state of the art in 1999. A further facility is location-based service using triangulation on signals either from multiple base stations or from a GPS (see Chapter 27, footnote 32).

GSM not only went far beyond its originally planned geographical borders, it also succeeded in entering into other domains, such as conventional *Private Mobile Radio* (PMR), railways, airplanes, and maritime communications. Conventional PMR services include features such as group call services, dispatch capabilities, fleet management, and flexible private group communication. In 1998, L.M. Ericsson launched the GSMPro solution, which added those and other PMR facilities to the worldwide GSM network.

Railway operating systems in Europe still differ from country to country, with different power supply, signaling, and incompatible analog communication systems. Now, at least a pan-European rail communication system is under development based on a GSM-R(ail) standard which was accepted by 32 countries in a MoU in 1997. The majority of the rail operators agreed to implement GSM-R by 2003. The standards for GSM-R are being elaborated by ETSI. The radio spectrum will be in the 900-MHz band, slightly above the GSM band.

The use of GSM on airplanes was launched on *Malaysian Airlines* (MAS) and *Singapore Airlines* (SIA) in 1999. This service was offered by Skyphone Mobil Con-

---

[15] COCOM (*Coordination Committee for East–West Trade Policy*) was founded by 14 NATO countries and Australia and Japan at Paris in 1949 to control military relevant technology transfer. After the Cold War, the policy was terminated on April 1, 1994.

Figure 32.3   State-of-the-art handset with WAP Internet browser, 1999. (Courtesy of Siemens Press Photo.)

nect, a subsidiary of BT Skyphone System, in a combination of GSM and the Inmarsat satellite network. The radio-frequency part of the GSM handset is switched off so as not to interfere with an airplane's electronics, but facilities such as SMS, Internet, and central billing can be used. In 2000, the same company started trials for use of GSM on cruise liners.

By the end of the twentieth century there were almost half a million GSM subscribers (500,000,000 in May 2001) in 350 networks in 170 countries. While GSM was a European invention and almost 98% of the West European population is covered by GSM, 40% of all GSM customers were based outside Europe, and the largest single-country customer base, about 83 million, was in China.

32.3.2   D-AMPS System

The development of digital cellular radio in the United States started much later than in Europe, for the obvious reason that for over 90% of AMPS subscribers in the United States, transborder roaming was no issue, and once it became an issue it could easily be implemented between the United States and Canada and Mexico in the common analog AMPS system. International roaming on the analog AMPS has been established on the North American cellular networks and also with AMPS operators in Hong Kong and Australia. In the late 1980s, however, concern arose that capacity problems would also arise with analog AMPS system and that a digital system should be developed.

In March 1988, the *U.S. Telephone Suppliers Association* (USTSA) and the information technology group of the *Electronic Industries Association* (EIA) formed the *Telecommunications Industry Association* (TIA), which then created subcommittee TR-45, given the task of producing a digital cellular standard. A vigorous debate started as to whether to base the new system on TDMA or on CDMA technology.[16] TIA made a balloted selection by its members in favor of TDMA in 1990. In the same year the standard IS-54 was issued, specifying the transition from AMPS to digital AMPS (D-AMPS), increasing the capacity by a factor of 3, but still using an analog control channel (to prevent a further three-year delay on implementation). To facilitate the transition it was decided to apply TDMA and to use the same frequencies as were used for AMPS. Moreover, dual-mode operation was mandated on the old analog and the new digital networks. The first commercial D-AMPS network began operation in 1993. A second digital cellular standard, named IS-95 or cdmaOne, was developed by Qualcomm in 1991, based on the CDMA access mode. In 1994, with IS-136, a full digital specification for D-AMPS-TDMA became available, including enhanced TDMA, which accommodates up to 10 voice/data channels per RF channel. Motorola developed a noncompatible digital standard called iDEN (integrated digital enhanced network), providing voice and text.

In addition to the frequencies in the 800-MHz band, the band from 1850 to 1950 MHz was allocated using the same or other technologies for the newly planned *Personal Communications Services* (PCS; Section 32.4), called PCS 1900. Surprisingly, the first PCS 1900 network did not use the American technologies IS-95 or CDMA but the European GSM adapted to U.S. conditions. This GSM network began operation in Washington, DC, in November 1995. Quickly, half a dozen operators implemented PCS 1900 networks based on GSM in the United States and Canada. In 1997, those networks were named GSM 1900 to differentiate them from PCS networks using CDMA or IS-95. By 1995, *Cellular Digital Packet Data* (CDPD) data transmission at a speed of 19.2 kbps was introduced on AMPS networks.[17]

At the end of the twentieth century, one analog and four noncompatible digital systems were in operation, including the PCS networks, distributed approximately as summarized in Table 32.6. To enable communication in the networks with different technologies, dual-band, triband, and dual-band dual-mode handsets were introduced.[18] More complicated interstandard handsets, enabling roaming between networks with GSM, TDMA, CDMA, and iDen technologies, were announced for introduction by late 2001 at the *GSM Global Roaming Forum* in Miami in 2000.

Outside the United States, D-AMPS-TDMA was adopted widely in countries that had already installed analog AMPS systems, especially in Latin America. Ironically,

[16] CDMA technology for commercial application was invented by Irwin Jacobs in 1988, three years after he founded Qualcomm in San Diego.

[17] CPDP is a wireless packet data standard developed by a U.S. consortium consisting of Airtouch (formerly PacTel), Ameritech, Bell Atlantic, GTE, McCaw Cellular, Nynex, and Southwestern Bell.

[18] Dual-band handsets for 900- and 1900-MHz bands; triband handsets for 900, 1800, and 1900 MHz; dual-band dual-mode handsets to support traffic across 800- and 1900-MHz bands on analog AMPS and digital TDMA networks.

TABLE 32.6   Cellular Radio Subscribers in the United States by Year-End 2000

Millions of Subscribers	System	Operator
29	CDMA	Verizon Wireless Inc.
15.8	TDMA	Cingular Wireless Inc.
15.8	TDMA	AT&T
10.8	CDMA	Sprint PCS Corp.
6.7	iDEN	Nextel Communications Inc.
4.4	GSM	VoiceStream Wireless Corp.
3.2	GSM	Cingular Wireless Inc.
22.9	AMPS	Various
1.5	GSM	Powertel and others
110		

Source: Data from *IandC World* (Siemens), No. 3, September 2001, combined with the total number of 110 million as published by ITU in Telecommunication Indicators Update, April 2001.

the world's first network with the American CDMA system was put into operation not in the United States, but in Hong Kong, where the private operator Hutchison launched a CDMA network on September 28, 1995. With dual-mode handsets supplied by Qualcomm, subscribers could freely use analog AMPS and digital CDMA, which became progressively available with 120 cells in Hong Kong. In January 1996, CDMA was launched in Japan and Korea. In Korea it became the national standard for operation in the 1700-MHz band, supported by early technology transfer agreements from Qualcomm with the Korean companies Samsung and Hyundai. Remarkably, in October 2000, the state-owned Chinese manufacturer Zhongxing Telecom Equipment Co. (ZTE) developed the world's first handsets with interstandard SIM cards for global roaming in CDMA and GSM networks.

32.3.3   Personal Digital Cellular System

The Telecommunications Council of the Japanese *Ministry of Post and Telecommunications* established a *Digital Mobile Telephone System Committee*, which issued a report in June 1990 on the technical conditions for a future digital cellular radio system. Detailed standards were elaborated by the *Research & Development Center for Radio Systems* (RCR) and issued in April 1991. To meet the expected high penetration, a *Personal Digital Cellular* (PDC) system was specified based on TDMA in two frequency bands: PDC 800 operating in the 800-MHz band, and PDC 1500, operating in the 1.5-GHz band. Moreover, an 11.2-kbps full-rate codec was specified for three traffic channels per RF channel, and a 5.6-kbps half-rate codec for six traffic channels per RF channel.

Beginning in 1992, digital cellular licenses were issued for PCD 800 to the existing operators NTT, DDI, and IDO, and for PCD 1500 to the new operators Tu-Ka Cellular Group and Digital Phone Group. By year-end 1995, over 5 million subscribers used cellular radio, almost 40% of them digital and slightly over 50% still operated by NTT.

On February 22, 1999, NTT DoCoMo, the first operator worldwide, introduced a narrowband Internet access service for mobile cellular users called *i-mode*. It uses a compact HTML (cHTML) as a wireless markup language. It became a tremendous success; the number of subscribers reached 5 million within one year and 20 million within two years. By year-end 2000 there were 17.2 million i-mode subscribers. By that time there were two competing companies, which also had launched mobile Internet services in 1999: KDDI launched EZWeb, which then had 5.7 million subscribers, and J-Phone launched J-Sky, with 5 million subscribers. By then, mobile Internet access had surpassed the approximate 15 million home users of Internet. In view of this success, NTT DoCoMo and AOL agreed on joint development of Internet services in Japan.

At the end of the twentieth century, Japan had 58 million digital cellular subscribers distributed as follows: NTT DoCoMo, 34.2 million; KDDI, 14.3 million; and J-Phone, 9.5 million. In addition, there were 5.8 million PHS subscribers (see Section 32.4.2).

32.4 PERSONAL COMMUNICATIONS NETWORK

Personal communications network (PCN) is the name for telecommunication networks in which person-to-person communication is possible under a single national —and eventually, worldwide—personal telephone number independent of home or office. In a PCN, in order to offer *personal communications service* (PCS, a generic term prevailing in North America for digital microcellular services), the telephone subscriber should always be near a switched-on terminal. This terminal may change during the day and season: using the residential phone at home, the office phone at work, the cellular phone while commuting, and a handset while away from those terminals, always accessible under the same personal telephone number. Numerous microcells (diameter 100 to 1000 m and RF power below 1 W), and picocells (diameter 10 to 100 m and RF power typically 100 mW) are required in a PCN to ensure full coverage, not only for traffic roads but also for office buildings, hotels, shopping centers, sporting grounds, and residential areas.

Whereas analog and digital cellular networks originally were optimized for quick-moving vehicle-mounted terminals, the emphasis in a PCN is on low-cost lightweight handsets in a slow-moving densely populated environment. Basically, a digital cellular radio network can evolve into a PCN by enhancing the capacity (e.g., by using 1.8- or 1.9-GHz bands, which provide three to four times more RF channels than do 800- and 900-MHz bands) and by adding micro- and picocells.[19]

The British system CT2, introduced in 1988, was one of the first systems where digital low-cost lightweight handsets were used in an early version of personal communications services. The service did not, however, include handover between cells. Genuine cordless personal communications systems were developed in the early 1990s as follows:

---

[19] Strictly speaking, the above-mentioned cellular radio networks, operating in the 1.8-GHz band with GSM equipment, in the 1.5-GHz band with PDC, and in the 1.9-GHz band with D-AMPS equipment, belong in the PCN category. In this section, however, only the cordless versions are covered. Anyway, in the late 1990s the differentiation between cellular radio, PCN/PCS, and cordless disappeared and it became common to refer to all three versions as cell phones or simply as mobile radio.

TABLE 32.7   **Summary of Personal Communications Systems**

System	Year of Introduction	Frequency Band (MHz)	Access Mode	Channels RF	Channels Traffic
CT2	1988	864–868	FDMA/TDD	40	40
CT3	1990	862–864	TDMA/TDD	32	32
PHS	1995	Indoor: 1895–1906.1 Outdoor: 1895–1918.1	TDMA/TDD	77	308
DECT	1996	1881.792–1897.344	TDMA/TDD	10	120
PACS	1997	Uplink: 1850–1910 Downlink: 1930–1990	TDMA/FDD	200	1600

- The *Digital European Cordless Telecommunications* (DECT) System
- The Japanese *Personal Handyphone System* (PHS)
- The North American *Personal Access Communications System* (PACS)

All three systems include handover and roaming and use a 32-kbps speech channel ADPCM coded in line with ITU-T Recommendation G.721, supporting voice, data (up to 64 kbps at a later stage), and multimedia services. Dynamic channel selection is used, whereby the handset or the base station selects the best available channel at calling, thus obviating the need for specific frequency plans and frequency coordination. The RF output power of the base stations ranges between 10 and 100 mW. DECT is conceived for walking speed, whereas both PHS and PACS can handle downtown motorcar speeds. Table 32.7 summarizes the major data of those personal communications systems.

32.4.1   CT1–CT3 Systems

In the early 1980s, an analog cordless telephone system was used in North America, the U.K., and a few other European countries which operated eight RF channels below 50 MHz. In 1985, CEPT specified another analog system, dubbed *CT1*, and the previous system was then named *CT0*. CT1 was used in 11 European countries.[20] CT1 had 40 channels at a spacing of 25 kHz operating in the 900-MHz band. A second CT1 version, called *CT1+*, was introduced with 80 channels.

CT2 was the first system developed as one-way portable digital radio access to suitably placed base stations in pedestrian areas over very short distances, typically up to about 200 m. In an enhanced version, incoming calls could be received once the subscriber had logged in on the local base station. Services with CT2 handsets were offered under various names, such as Telepoint, Callpoint, Phonepoint, Fonepoint, and Zonepoint in the U.K., BiBop and Pointel in France, and Greenpoint in the Netherlands. The CT2 system was not very successful in Europe, apart from BiBop, which reached almost 100,000 subscribers; in all the other European countries together, there were fewer than 20,000 subscribers. It was successful in Southeast

---

[20] CT1 was used in Austria, Belgium, Denmark, Finland, Germany, Italy, Luxembourg, The Netherlands, Norway, Sweden, and Switzerland.

Asia, especially in Hong Kong, were it reached a peak of 180,000 subscribers in 1995, and in Singapore, where Callzone got 80,000 subscribers. About 60,000 CT2s were in operation in various towns in China. It was also used in Australia, Malaysia, Taiwan, and Thailand. A modified version called *CT2 Plus*, later changed to *CT2 Canada*, was introduced in Canada in 1992, operating 100 channels around 940 MHz, but also found little use. *CT3*, developed around 1987 by Ericsson, was used primarily in Brazil, Canada, Germany, and the United States. The unexpectedly high penetration of the more versatile digital cellular radio systems caused a complete interruption of the CT services by 1999.

32.4.2   Japanese Personal Handyphone System

NTT developed the *personal handyphone system* (PHS), which was put into operation on July 1, 1995 by NTT and DDI in the Tokyo and Hokkaido areas. PHS is not a separate network like DCS 1800 but connects areas around the home, office, and outdoors to a PSTN/ISDN, thus extending the services of the fixed-line network as shown in Figure 32.4.

When introduced in 1995, PHS set new standards with its very compact handyphone with a 95-g weight and 98-cm^3 volume, offering five hours of communication time and two weeks of standby operation. Three years later, NTT introduced a PHS wristwatch phone with a volume of 30 cm^3 and a weight of 40 g, offering 60 minutes of call time and 100 hours of standby, also providing speech recognition for voice dialing. The cell stations are very compact, too, enabling installation on traffic signal poles, public telephone boxes, and other public gathering places. PHS, with a maximum hand-off speed[21] of about 40 km/h, was planned for pedestrians and motorcar speed in town, not for highways or trains. It can be used for voice, e-mail, fax, digital picture transmission, and mobile video-on-demand.

PHS was introduced with a monthly rental of one-third of the prevailing cost for cellular and one-fifth of cellular call charges. Over 80,000 subscribers acquired a PHS handset in the first month. A peak of 7.07 million was reached in September 1997 but then declined to 5.7 million in 1998. By that time, NTT DoCoMo took over the loss–making PHS unit of NTT and added various facilities, such as dual-mode PDC/PHS handsets, 64-kbps data transmission, a hand-off speed up to 100 km/h, GPS-supported location services, SMS handwritten on the screen with a stylus, a built-in camera for image transmission, and i-mode Internet service.[22]

32.4.3   Digital European Cordless Telecommunications

The U.K. submitted the specifications of the CT2 system to ETSI for approval as a European standard. Simultaneously, L.M. Ericsson proposed its more advanced

---

[21] *Hand-off speed* is the traveling speed of the terminal at which switching over from one cell to the other does not function correctly anymore.

[22] As a further improvement in cellular radio, NTT DoCoMo announced in early April 2002 their plan to introduce within the next five years a lip-reading telephone that could eliminate the annoyance of loud cell phone conversations. A prototype has been developed, in which contact sensors near the mouthpiece in the phone detect tiny electrical signals sent by muscles around the mouth and a speech synthesizer converts the signals into spoken words or text for a message or e-mail.

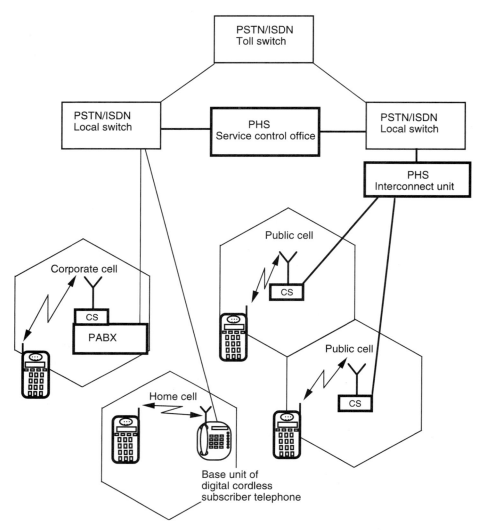

Figure 32.4 PHS network. (From A. A. Huurdeman, *Guide to Telecommunications Transmission Systems*, Artech House, Norwood, MA, 1997, Figure 6.13; with permission of Artech House Books.)

system, called CT3 or DCT-900 (digital cordless telephone), which was able to handle large traffic densities. ETSI accepted CT3 but changed the frequency band from 800 MHz to 1800 MHz and made it the base for the future *Digital European Cordless Telecommunications* (DECT) system. In 1991, ETSI presented two solutions for a pan-European PCN operation in the 1.8-GHz band: DECT (Digital European Cordless Telecommunications) and DCS 1800 (Digital Cellular System).

DECT is basically for cordless on-site roaming with wireless PABX operation and for cordless access via microcells to the PSTN/ISDN, including wireless local loop operation. DCS 1800 (see Section 32.3.1) is an enhancement to the original GSM

Figure 32.5 DECT prototype handset for videophone transmission, 1998. (Courtesy of Siemens Press Photo.)

specification for full cellular operation of handheld or pocket terminals in the 1.8-GHz band.

DECT evolved to a worldwide standard for high-density cordless systems and was renamed the *Digitally Enhanced Cordless Telecommunications System*. A *GSM Interworking Protocol* enables interworking between GSM and DECT. Dual-mode DECT/GSM handsets came on the market in 1996. With about 20 million terminals in use in 1998, DECT was the most successful of all cordless systems. Figure 32.5 shows a prototype DECT handset for videophone transmission.

DECT also found worldwide use in wireless local loop systems (see Section 25.3), especially in urban and suburban areas with high subscriber densities, such as in China, Indonesia, and Cambodia.

32.4.4 Personal Access Communications System

The American *Personal Access Communications System* (PACS) is a microcellular system developed by Hughes Network Systems that combines the characteristics of PHS and the *Wireless Access Communications System* (WACS) recommendations of Bellcore.

TABLE 32.8   International Proposals for IMT-2000

Proposal	Description	Source
CDMA I	Multiband synchronous DS-CDMA	Korea, TTA
CDMA II	Asynchronous direct sequence CDMA	Korea, TTA
cdma2000	W-CDMA (IS-95)	United States, TIA TR45.5
UWC-136	Universal wireless communications	United States, TIA TR45.3
WIMS/W-CDMA	Wireless multimedia and messaging services W-CDMA	United States, TIA TR 46.1
DECT	Digital enhanced cordless tele-communications	ETSI
NA W-CDMA	North American W-CDMA	United States, T1P1-ATIS
TD/SCDMA	Time-Division Synchronous CDMA	China, CATT
UTRAUMTS	Terrestrial Radio Access	ETSI, SMG2
W-CDMA	Wideband CDMA	Japan, ARIB

Source: Data from *ITU News*, September 1998 p. 53.

32.5   INTERNATIONAL MOBILE TELECOMMUNICATION SYSTEM

In 1996, the ITU presented a concept for a third-generation (3G) *International Mobile Telecommunication System* dubbed *IMT-2000* (see Section 27.4.5), as an "always-on" service offering "anywhere, anytime" personal high-speed wireless access to the global telecommunications infrastructure via low-cost multimedia handsets at an affordable service cost. ITU was confronted with one of its most difficult tasks: to define a single standard for mobile multimedia communication across frontiers and technologies. The problems were of less technical than commercial and political significance, which required due consideration for national as well as continental ambitions. Members of the ITU were requested to submit respective proposals. By the June 30, 1998, deadline, 15 proposals were received, 10 for terrestrial and five for satellite operation. The proposal submitted by ETSI had been the result of a difficult half-year process of selecting among the five proposals. On January 28, 1998, ETSI reached a consensus which, as a novelty, was obtained with the active support of representatives of Japanese manufacturers, of the operator NTT DoCoMo, and of the Japanese *Association of Radio Industries and Business* (ARIB). The solution was the *Universal Mobile Telecommunications System* (UMTS), based on wideband CDMA (W-CDMA) and compatible with GSM networks with two access modes: one for wide-area coverage and high mobility, and a second mode for local area (indoor) coverage and low mobility.

Table 32.8 summarizes the 10 proposals which the ITU received for terrestrial operation. Selection of a consensus system was aggravated because some of the proposing parties, in particular Qualcomm and L.M. Ericsson, had serious disputes regarding international property rights (IPRs) concerning the CDMA proposals. The ITU set a deadline of December 31, 1998 to solve the patent issues. The deadline was missed, but the concerned parties established a *3G Partnership Programme* (3GPP)[23]

---

[23] The 3GPP is made up of *China Wireless Telecommunications Standard Group* (CWTS), the Japanese *Association of Radio Industries and Business* (ARIB), the Japanese *Telecommunications Technology Committee* (TTC), the Korean *Telecommunications Technology Association* (TTA), the Standards Committee T1 of the U.S. *Telecommunications Industry Association* (TIA), and ETSI.

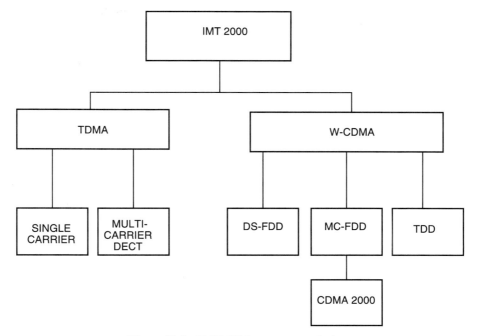

Figure 32.6   IMT 2000 consensus solution.

with the aim of harmonizing the various CDMA proposals. Simultaneously, the North American parties formed a 3GPP2 partnership with the objective of finding a common solution for their cdma2000 and UWC-136 proposals. Finally, mutual interest prevailed. A consensus decision could be taken at a task group meeting of ITU-R in Fortaleza, Brazil, March 8–19, 1999. It was decided to base IMT-2000 on a single W-CDMA standard, albeit with three optional access modes and two optional TDMA modes.[24] Figure 32.6 shows the complex consensus solution. The transmission speed was defined as a range between 384 kbps and 2 Mbps and above.[25]

As a result of the IMT-2000 consensus, UMTS will offer broadband voice and data in a frequency-division duplex (FDD) mode for large areas with a transmission speed up to 384 kbps, and in a time-division duplex (TDD) mode in areas with heavy traffic with speeds up to 2 Mbps. Figure 32.7 shows a prototype UMTS handset.

In 1992, the frequency bands 1885 to 2025 and 2110 to 2200 MHz had been allocated to IMT-2000. At that time, voice services were considered the main source of

---

[24] The three W-CDMA modes are DS-FDD, basically ETSI's UTRA (UMTS terrestrial radio access), MC-FDD (multicarrier frequency-division duplex, FDD indicating go and return signals on separate frequencies), TIA's cdma2000, and TDD (time-division duplex, indicating go and return signals in separate time slots on the same RF channel), a combined TDMA/CDMA solution harmonized between ETSI and China. The TDMA modes are the frequency-time mode of DECT, and the single-carrier mode proposed by TIA with TR 45.3.

[25] As a surprise settlement of the two-year IPR battle, the two major opponents not only agreed after the foregoing decision to declare all IPR issues solved but even agreed that L.M. Ericsson would buy from Qualcomm the cdmaOne infrastructure business and R&D unit.

Figure 32.7   UMTS prototype handset, 2000. (Courtesy of Siemens Press Photo.) See insert for a color representation of this figure.

traffic; in the meantime, however, it became obvious that more bandwidth would be required for rapidly emerging multimedia services, such as Internet, intranet, e-mail, e-commerce, and video. At the World Radiocommunication Conference (WRC-2000) in Istanbul, Turkey, from May 8 to June 2, 2000, therefore, three new frequency bands were added: 806 to 960, 1710 to 1885, and 2500 to 2690 MHz. As a next action, ITU-R started to work out basic principles to obtain unhindered global circulation of the terminals without having to pay customs duties or needing national approvals and individual licenses.

Operators in North America, Europe, and Southeast Asia installed trial networks with 3G-prototype equipment. The world's first commercial networks using cdma2000 version 1x were launched in Korea in October 2000.[26] Licensing for the operation of IMT-2000 networks from 2001 started in 1999. Basically, two different approaches were used: by auction to the highest bidders and by "beauty contest" to bidders with the best qualification.[27] The first licenses for the UMTS system were awarded in a beauty contest in March 1999 to four operators in Finland for 20-year

---

[26] Version 1x provides a transmission speed of 144 kbps. The next version, 1xEV (EV for "evolution"), with a transmission speed up to 2.4 Mbps, recognized as meeting IMT-2000, was announced for the year 2002.

[27] The government of New Zealand introduced in 2000 the world's first auction for 3G on the Internet, using an entirely Internet-based iBid Spectrum Auction Service for issuing 3G licenses.

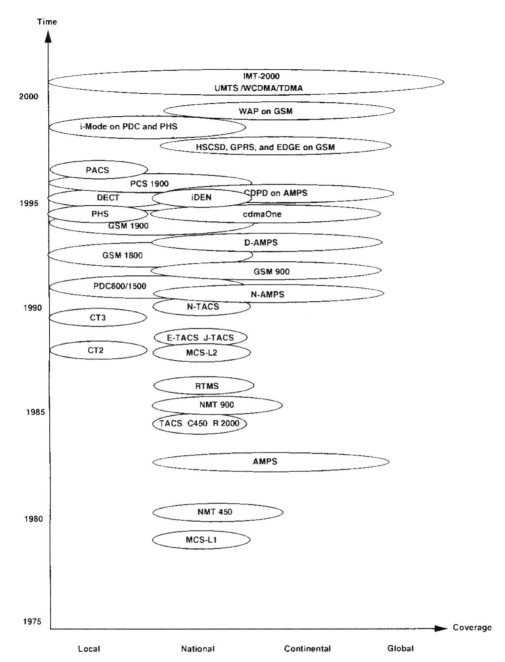

Figure 32.8 Cellular radio development. (From A. A. Huurdeman, *Guide to Telecommunications Transmission Systems*, Artech House, Norwood, MA, 1997, Figure 6.15; with permission of Artech House Books.)

operation of 3G networks. In the following year, excessive fees were paid at a total value of $105 billion for 44 UMTS licenses in 10 auctions held in Europe. The experience with the unexpected high penetration of the 2G systems and the prevailing optimism about Internet development and future e-commerce obviously encouraged the operators to expect high returns on future 3G services. Those overpaid licenses, which still have to follow investments at even higher amounts, were one of the reasons for the dramatic collapses in share prices of telecommunications operators from the second half of 2000. Considered positively, this might force operators to accelerate implementation of the new networks even more in order to achieve a satisfactory return on investments.[28]

A summary of the development of cellular radio systems over the last quarter of the twentieth century is given in Figure 32.8.

REFERENCES

Books

Balston, D. M., and R. C. V. Macario, editors, *Cellular Radio Systems*, Artech House, Norwood, MA, 1993.

Hadden, Alan David, *Personal Communications Networks: Practical Implementation*, Artech House, Norwood, MA, 1995.

Huurdeman, Anton A., *Guide to Telecommunications Transmission Systems*, Artech House, Norwood, MA, 1997.

Articles

Anon., The shrinking world: European cellular survey, *Mobile Europe*, May 1996, pp. 37–40.

Anon., US–Canada GSM alliance announced, *GSMQ, The GSM Industry Journal*, October 1997, pp. 7–8.

Anon., IMT-2000 standardization within the ITU, *ITU News*, August 1998, pp. 52–54.

Anon., World telecommunication development report, 1999: mobile cellular, *ITU News*, August 1999, pp. 35–39.

Anon., Cracking the algorithm, *Mobile Europe*, February 2000, p. 14.

Anon., The world's mobile users—by technology, *Communications International*, February 2000, p. 78.

Anon., Getting them talking: interstandard roaming between GSM and other standards, *GSMQ, The GSM Industry Journal*, March 2001, pp. 4–6.

Anon., Östen Mäkitalo, "father" of wireless networks, *On—The New World of Communication* (Ericsson), April 2001, p. 7.

Awoyagi, Tadashi, PHS provides innovation in Japan, *Mobile Communications International*, October 1999, pp. 96–97.

Borman, Bill, The needs for mobile communications, *Developing World Communications*, Grosvenor Press International, London, 1986, pp. 44–547.

---

[28] NTT DocoMo began commercial UMTS service in October 2001; Manx Telecom began with life testing of UMTS on December 5, on the Isle of Man, U.K.; and Monaco Telecom began on December 12 in the Principality of Monaco.

Cadet, Max-Henry, et al., Global circulation of IMT-200 terminals, *ITU News*, April 2000, pp. 11–13.

Channing, Ian, Making history, *GSMQ, The GSM Industry Journal*, October 1997, pp. 10–20.

Channing, Ian, UMTS agreement clears the air, *GSMQ, The GSM Industry Journal*, March 1998, pp. 5–7.

Daniels, Guy, A brief history of 3G, *Mobile Communications International*, October 1999, pp. 106–107.

Daniels, Guy, The single 3G standard, *Mobile Communications International*, December–January 1999–2000, p. 4.

Fletcher, Peter, The office goes cordless, *The Mobile Revolution*, supplement to *Communications International*, 1993, pp. 24–27.

Foster, Keith, Character sketch: a new system is bringing Chinese to a keypad near you, *On—The New World of Communication* (Ericsson), May 2000, p. 18.

Hibberd, Mike, The end of an era, *Mobile Communications International*, December–January 1998–1999, p. 80.

Hibberd, Mike, Bowing out gracefully, *Mobile Communications International*, October 1999, pp. 83–84.

Kandiyoor, Suresh, and Peter van den Berg, DECT'S potential for public network operators, *Mobile Communications International*, December 1995, pp. 87–96.

Kaneshige, Thomas, The answer to a prayer (PACS), *Communications International*, December 1995, pp. 24–25.

Kelly, Tim, et al., Asia-Pacific mobile markets from handyphone to i-mode, *ITU News*, November 2000, pp. 24–30.

Langer, Johan, and Gwenn Larsson, CDMA2000: a world view, *Ericsson Review*, March 2001, pp. 150–158.

Leong Joe, Orderly revolution, *Mobile Asia–Pacific*, June–July 1995, pp. 11–16.

Strunz, Günter, Überall erreichbar sein: die Entwicklung des öffentlichen Mobilfunks, *Archiv für deutsche Postgeschichte*, Vol. 1, 1989, pp. 85–94.

Suanders, Jake, GSM: the road ahead, *GSMQ, The GSM Industry Journal*, March 2000, pp. 8–14.

Wong, Peter, It's useful, and it comes in a packet, *Mobile Asia–Pacific*, December–January 1994–1995, pp. 20–22.

Young, W. R., Advanced mobile phone service: introduction, background, and objectives, *Bell System Technical Journal*, Vol. 48, January 1979, pp. 1–14.

Internet

www.totaltelecom.com.

Ingelbrecht, Nick, NTT chief sounds death knell for loss-making PHS, March 2, 1998.

DoCoMo launches i-mode-enabled PHS phone, May 1, 2000.

AOL and DoCoMo to team for wireless Internet in Japan, August 1, 2000.

CDMA gets global roaming boost, October 16, 2000.

Japan's mobile phone users rise 20% in 2000, January 11, 2001.

Japan Telecom delays 3G launch, March 6, 2001.

GSM subscribers numbers reach half-billion mark, May 11, 2001.

33

TELEPHONY AND DEREGULATION

33.1 TELECOMMUNICATIONS DEREGULATION AND LIBERALIZATION

By 1950, there were 75 million telephones worldwide, of which almost 60% were in the United States. Reconstruction of the telecommunications infrastructure, which had been heavily damaged during World War II, had been completed. Within six years about 33 million additional telephone lines were installed. Table 33.1 shows the approximate distribution of the 108 million telephones in use on January 1, 1957, for countries with more than 500,000 telephones. At that time the following 10 cities with a population of more than 100,000 had a teledensity (number per 100 population) above 50%: Washington, DC, 65.3%; Los Angeles, 64.9%; San Francisco, 58.2%; Basel, 55.9%; Stockholm, 54.8%; Bern, 54.1%; Geneva, 53.5%; Hartford, Connecticut, 53.2%; Pasadena, California, 52.2%; and Wilmington, Delaware, 51.8%. By January 1, 1976, one hundred years after the invention of the telephone by Graham A. Bell, the total number of telephones reached 380 million (Table 33.2). More than half of the world's population was still living in countries with a teledensity below 1 when Sir Donald Maitland wrote his much publicized report, "The Missing Link," in January 1985.[1] In that report, the term *telecommunications gap* was coined and the objective postulated "to bring all mankind within easy reach of a telephone by the early part of the 21st century."

Bridging the telecommunications gap and offering universal access then became a major objective of the ITU and other organizations. Simultaneously, and indepen-

---

[1] Maitland was chairman of an Independent Commission for Worldwide Telecommunications Development formed by the Plenipotentiary Conference of the ITU in Nairobi, Kenya, in October 1982 with the objective of recommending actions for the development of telecommunications in developing countries.

The Worldwide History of Telecommunications, By Anton A. Huurdeman
ISBN 0-471-20505-2   Copyright © 2003 John Wiley & Sons, Inc.

TABLE 33.1   Worldwide Telephone Distribution, January 1, 1957

Continent and Country	Thousands of Telephones	Distribution (% of worldwide total)
The Americas		
United States	60,200	55.9
Canada	4,500	4.2
Mexico	500	0.5
Central America	100	0.1
Caribbean	300	0.3
South America	2,600	2.4
Total Americas	68,200	63.4
Asia		
Japan	3,500	3.3
Rest of Asia	1,550	1.4
Total Asia	5,050	4.7
Oceania		
Australia	1,760	1.6
New Zealand	570	0.5
Rest of Oceania	50	<0.1
Total Oceania	2,380	2.1
Africa		
North Africa	500	0.5
South Africa	765	0.7
Rest of Africa	200	0.2
Total Africa	1,465	1.4
Europe		
Austria	540	0.5
Belgium	930	0.9
Denmark	925	0.9
France	3,315	3.1
Germany (Dem. Rep.)	1,070	1.0
Germany (Fed. Rep.)	4,325	4.0
Italy	2,610	2.4
The Netherlands	1,230	1.1
Norway	615	0.6
Spain	1,200	1.1
Sweden	2,310	2.1
Switzerland	1,295	1.2
U.K.	7,220	6.7
USSR	1,000	0.9
Rest of Europe	2,000	1.9
Total Europe	29,585	27.5
Total worldwide	107,680	100

Source: E. I. Green, Telephone, *Bell System Technical Journal*, March 1958; Robert J. Chapius, *100 Years of Telephone Switching, 1878–1978*, Part 1, North-Holland, New York, 1982. Both sources are based on *Telephone Statistics of the World*, which was then published annually by AT&T.

TABLE 33.2   Worldwide Telephone Distribution, January 1, 1976

Continent or Region	Thousands of Telephones	Teledensity (%)	Percent Distribution	
			1957	1976
Africa	4,616	1.1	1.4	1.2
The Americas	175,465	30.9	63.4	46.2
Asia	59,559	2.6	4.7	15.7
Europe	132,201	19.5	27.5	34.9
Oceania	7,683	34.9	2.1	2.0
Total worldwide	379,524	9.6	100	100

Source: Data from *Siemens International Telephone Statistics, 1977.* The numbers of telephones refer to "telephones of all kinds" and thus include main telephone stations, extensions, and for several countries parallel stations as well.

dently, a worldwide wave of deregulation, privatization, and liberalization of telecommunications networks emerged. Four milestones highlight this process:

1. *In the United States:* divestiture of the monolithic AT&T into the regional Bell operating companies (RBOCs) in 1982
2. *In the U.K.:* change from monopoly to duopoly in 1982 and complete liberalization of telecommunications in 1991
3. *In Japan:* privatization of NTT in 1985
4. *In the European Union:* the end of the voice service monopoly on January 1, 1998

Deregulation, in fact, was re-regulation, as it replaced established monopoly practice by regulated competition. Deregulation and liberalization stand for loosening the ties between the incumbent operator and the government, and opening the telecommunications market by permitting new companies to offer telecommunications services in competition with the incumbent. The activities of the deregulated incumbent were no longer restricted to the national market but were also opened to the international telecommunications and financial markets. Free competition is not governed by social responsibility and provides a challenge to universal access. Along with the opening of the telecommunications market, therefore, came the requirement to establish new regulating agencies to safeguard quality of service, universal access, and fair competition. Regulatory agencies were established in 104 countries at the end of 2000.

Privatization usually took place by converting the incumbent into a public company. The government then sold part or all of its shares of the incumbent. In many cases those shares were bought by other, usually international telecommunication companies, which thus gained access to larger markets. As a positive effect for the developing countries, the new owners, with their experience and capital, could significantly improve the national network and its performance. The most impressive privatization occurred in Brazil, where in 1998 the government sold its rights in Telebras for a record value of $19 billion.

The competition was first applied to the emerging mobile cellular radio services in the late 1980s. Competition in international telephony came in 1990, first in Japan, New Zealand, the U.K., and the United States. This was followed in 1995 by Australia, Canada, Chile, Colombia, Denmark, Finland, Korea (Republic), Malaysia, the Philippines, and Sweden. An additional 24 countries followed in 1998. By the end of the century, competition was introduced in approximately 120 countries for the operation of cellular radio, in 70 countries for international communications, in 65 countries for local communications, and in about 60 countries for national long-distance communications.

Deregulation, privatization, and liberalization led to the emergence of many new telecommunication operators. In the United States, as in the early days of telephony, over 700 operators share the market, which used to belong to AT&T. Some of those new operators invest heavily in installing their own networks, including local access to the premises of subscribers. Other new operators, however, rely on the existing telecommunications infrastructure by leasing lines or buying transmission capacity on existing and newly planned transmission networks. Various companies that possessed regional, national, or even international infrastructures, such as transport companies, utilities for water, gas, and electricity, as well as motor and waterway companies, started to use their infrastructure for the operation of new telecommunications networks. In conjunction with the creation of these new operators, a worldwide reshuffling of the telecommunications market occurred, with over 500 mergers and acquisitions, with the purpose of enlarging geographical coverage and service diversification. More than one-third of these mergers and acquisitions were led by U.S. corporations and a little less than one-third by U.K.-based enterprises. In addition to these mergers, various international alliances were made between national operators. The biggest alliances were WorldCom, an alliance of AT&T with major operators in North America, Europe, and Asia-Pacific; Global One, an alliance of Deutsche Telekom, Sprint, and France Telecom (which was the sole owner at the end of 2000); and Concert, formed by BT and AT&T.[2] An indication of this reshuffling can be obtained from Table 33.3, which compares the ranking based on the minutes of international traffic in 1997 and in 2000 for the 10 top international telecommunications operators.

The heavy competition that entered into telecommunications services resulted in price reductions beyond 70%, even up to 100%, as some Internet service providers offered Internet surfing time free of charge. As a typical landmark, the average price for a 1-minute call from a Western European country to the United States fell from over $2 in 1990 to less than $1 in 1997 and to $0.50 in 2000. In Switzerland, for example, the price of $0.07 per minute for international calls was almost down to the price of $0.04 for local calls or national long-distance calls. The international traffic rapidly increased from 10 billion minutes at the beginning of the liberalization in 1982 to 100 billion in 2000. This increase went together with a shift of international traffic originating from Europe to North America and Asia (Table 33.4).

As a result of the deregulation and liberalization, a new type of telecommunications operator appeared on the scene with *indirect access* to their subscribers. In addition to the established system of separate codes for each country, each exchange, and each subscriber, a *carrier access code* was therefore introduced around 1995.

---

[2] The Concert alliance was dissolved in October 2001.

TABLE 33.3   Top 10 International Operators

Operator	Traffic (millions of minutes)		Ranking	
	2000	1997	2000	1997
MCI WorldCom	12,400	7,300	1	2
AT&T	9,680	10,330	2	1
BT	4,569	2,710	3	6
Deutsche Telekom	4,525	4,819	4	3
France Telecom	4,390	3,100	5	4
Sprint	3,920	2,760	6	5
C&W Communications	3,490	970	7	15
Telecom Italia	2,710	2,210	8	7
China Telecom	2,050	1,630	9	>10
Swisscom	2,050	1,960	10	8

Source: For 2000, Industry outlook, *Telecommunications Americas*, January 2002, p. 11; for 1997, Direction of traffic, *ITU News*, August 1999, pp. 14–20.

Instead of being connected automatically with the incumbent national operator, telephone users in a country with deregulated telecommunications services can now select the carrier of their choice by dialing the respective carrier access code before dialing the intended destination telephone number. Alternatively, in some countries, such as in the United States and in Germany, carrier preselection is possible. In that case telephone customers agree with the incumbent operator and a new operator of their choice that all calls are routed automatically from the local exchange of the incumbent operator to the network of the new operator. In that case the customer does not need to use the carrier access code. The incumbent operator continues to charge the customer for the access network and usually still handles the local calls, while the new operator charges the customer for national and international calls.

As a result of the substantially reduced prices, the introduction of new services, and experienced foreign operators building modern reliable telecommunication networks in developing countries, the total number of telephone subscribers increased globally from about 450 million at the beginning of the liberalization to 987 million

TABLE 33.4   Global Share of International Traffic

Continent or Region	Percent of International Traffic	
	1983	1997
Europe	68	43
North America	16.8	32.5
Asia Pacific	9.7	18.5
Latin America and Caribbean	3.6	4.1
Africa	1.9	1.9

Source: Direction of traffic, *ITU News*, August 1999, pp. 14–20, and *Telecommunication Indicators Update*, July–September 2000.

at the end of the century. Remarkably, about 20% of this increase came from China, where the teledensity was raised from below 1 to over 11.

To obtain universal access, the ITU postulated in 1998 the goal that by the year 2010, a teledensity of 10 and service coverage of 50% of the households should be achieved for developing countries. For low-income countries the target was set at a teledensity of only 5% and service for 20% of the households. Since Maitland's "missing link" report, much progress has been made, but Kofi Annan, Secretary-General of the United Nations, observed at Telecom 99 in Geneva on October 9, 1999: "A quarter of all countries have not yet achieved even a basic level of access to telecommunications (at a teledensity of 1), and half the world's people have never made or received a telephone call." Yoshio Utsumi, Secretary-General of the ITU, also confirmed in the *ITU News* of December 2001 that an information gap still exists because 55% of telephones are used by the 15% of the world population that live in high-income countries[3]; 14% are used by the 10% of the population living in upper-middle-income countries; 11% by the 15% of the population living in lower-middle-income countries, and only 20% of the telephones are used by the 60% of the population living in low-income countries. In fact, the missing link has been replaced by a *digital divide* with a teledensity of around 50 in developed countries and around 1.5 in developing countries. A summary of the distribution of telephones, together with cellular radio and Internet, in relation with the population and the teledensity achieved at the end of the twentieth century is given for 205 countries in Appendix B.

33.2 TELEPHONY AND DEREGULATION IN THE AMERICAS

The telephone deployment in the Americas is determined primarily by the United States and to a lesser degree by Brazil and Canada. An impression of the growth is given in Table 33.5.

33.2.1 Telephony and Deregulation in the United States

The second half of the twentieth century began very positively for the American Telephone & Telegraph Company (AT&T; also called the Bell System; see Figure 33.1). AT&T could increase dividends above $9 for the first time since 1921, take the first transatlantic telephone cable into operation in 1956, and place in orbit the first commercial communications satellite, *Telstar I*, in 1962. In 1976, after 100 years of telephony, the United States had 149 million telephones, of which 85% were operated by the Bell System, 10% by GTE, and the rest by 1700 independents, the smallest of them with 19 subscribers. Efforts to break the near-monopoly of AT&T have been made since the beginning of the twentieth century. As early as 1907, Theodor N. Vail, then president of AT&T could prevent privatization by promoting the idea that AT&T functioned as a legally sanctioned regulated monopoly. He stated: "The telephone, by the nature of its technology, would operate most efficiently as a monopoly providing universal service. Government regulation, provided it is independent, intelligent, considerate, thorough and just, is an appropriate and acceptable substi-

---

[3] The following 23 countries are rated as high-income countries: Australia, Canada, the 15 countries of the European Union, Iceland, Japan, New Zealand, Norway, Switzerland, and the United States.

TABLE 33.5   Telephone Growth in the Americas

Country or Region	Millions of Telephones					Percent of Telephones, 2000
	1957	1976	1990	1997	2000	
Canada	4.5	13.2	15.3	17.5	20.8	7.2
United States	60.2	149	136.9	167.4	192.5	66.2
Mexico	0.5	2.9	5.4	9.3	12.3	4.2
Central America	0.1	0.4	1.0	2.2	3.2	1.1
Caribbean	0.3	1.3	2.9	4.0	4.6	1.6
Argentina	1.1	2.5	3.5	6.7	7.9	2.7
Brazil	0.8	3.4	8.8	17.0	30.9	10.6
Chile	0.2	0.5	0.7	2.7	3.4	1.2
Colombia	0.2	1.3	2.3	5.4	7.2	2.5
Venezuela	0.1	0.6	1.5	2.8	2.6	0.9
Rest of South America	0.2	1.4	1.7	4.0	4.8	1.7
Total Americas	68.2	176	180	239	290.2	100

Source: Archiv für deutsche Postgeschichte, Vol. 1, 1977; ITU, *Yearbook of Statistics, 1990–1999*, and *Telecommunication Indicators*, 2000; Siemens, *International Telecom Statistics, 1977*.

tute for the competitive marketplace." The government accepted this principle in 1913 in an agreement called the *Kingsbury Commitment*. A new approach toward breaking AT&T's near-monopoly was made in 1934, when the U.S. government passed the Communications Act, which created an independent regulatory agency, the Federal Communications Commission (FCC). The FCC, the world's first regulatory agency, was charged with "regulating interstate and foreign commerce in communication by wire and radio so as to make available, so far as possible, to all the people of the U.S., a rapid, efficient nationwide and worldwide wire and radio communication service with adequate facilities at reasonable charges." The Great Depression and World War II interrupted the privatization efforts. In 1958, Tom Carter made the first successful step toward deregulation of AT&T. Carter had developed a two-way radio, which was used for communication with oilfield trucks. He made a phone-patching device, the *Carterfone* (also called *Hushaphone*), which enabled communication between the trucks and the public network. AT&T objected to the use of the Carterfone, but finally, in 1968, FCC permitted the "non-carriers to provide terminal equipment to be attached to any of the telephone networks."

A further important step to break the AT&T near-monopoly came one year later, when a new company, *Microwave Communications Incorporated* (MCI), founded by William McGowan (1928–1992), obtained FCC approval to install and operate their own radio-relay link between St. Louis and Chicago. The radio-relay terminals in both cities were connected with a telephone exchange such that subscribers of the Bell System in those cities could bypass the Bell long-distance system by dialing a special MCI number and use the MCI line at a cheaper rate. The FCC was generous enough to consider this to be a private line to which, according to prevailing legislation, AT&T could not object. Encouraged by this success, MCI bought some 100,000 km of lines in the United States from Northern Telecom and also obtained FCC coverage for those *private lines* in 1971. Moreover, FCC opened up the entire private-line market to *specialized common carriers*.

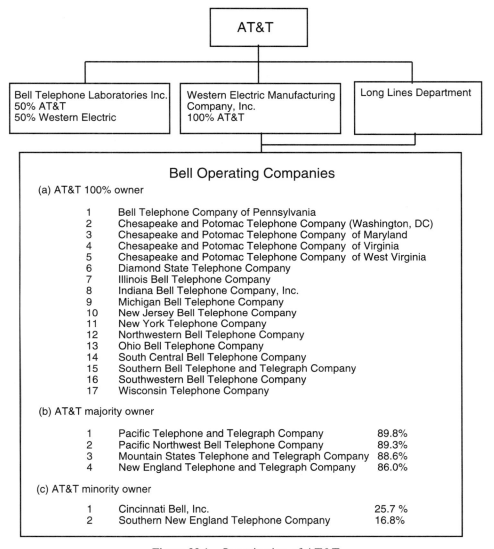

Figure 33.1   Organization of AT&T.

AT&T strongly objected to this interpretation of private lines and refused to grant MCI low bulk long-distance rates. MCI filed an antitrust suit against AT&T in March 1974. Even worse for AT&T, the U.S. Justice Department filed another suit against AT&T in November of the same year, charging monopolization and conspiracy to monopolize the supply of both telecommunications service and equipment. Thus the largest antitrust suit in the history of the United States began. An almost 10-year legal battle followed, which nearly bankrupted MCI, but in 1981, AT&T was charged to pay MCI compensation of about $1 billion. Finally, in 1982, the U.S. Department of Justice made an agreement in which AT&T accepted to give up its monopoly in the interest of global competition and to divest its operating companies

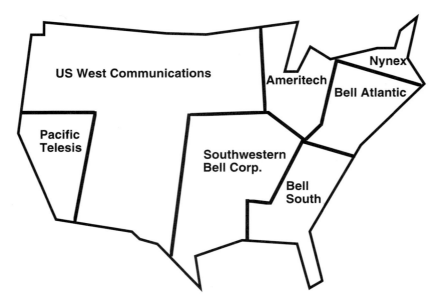

Figure 33.2   Geographical location of the seven Baby Bells.

into seven regional *Baby Bell companies*. The Baby Bells were a redistribution of the RBOCs listed in Figure 33.1 as follows: Ameritech, a7, a8, a9, a13, and a17; Bell Atlantic, a1–a5 and a10; Bell South, a14 and a15; Nynex, a11 and b4; Pacific Telesis, a6 and b1; Southwestern Bell, a16; US West, a12, b2, and b3. The geographical location of the seven Baby Bells is shown in Figure 33.2. The Baby Bells provided telephone services to about 80% of the 120 million subscribers; the remaining 20% were served by some 1500 independent companies scattered throughout the United States and covering 51% of the territory. They differed greatly in size from small family concerns serving little more than 100 subscribers in rural areas to giant operators such as Continental Tel, Centel, Mid-Continental Tel, Rochester (New York) Telephone Corp., United Tel, and the largest, GTE,[4] which alone served almost 18 million subscribers.

The AT&T divestiture was the most radical change in U.S. corporate history. The famous Bell logo, 636,000 of AT&T's 1,009,000 employees, and $116 billion of the $150 billion in assets were to transfer to the new Baby Bells. AT&T had to get used to worldwide competition; over 400 companies entered the long-distance market. Prices for long-distance services went down by 60% and AT&T's market share dropped from 90% to around 50% over the next 10 years. By then AT&T regained market presence: In 1991 it merged with the computer company NCR (National Cash Register), in 1994 with the cellular operator McCaw, and in 1999 with a leading provider of local telephone service to business customers, TCG, and with the leading provider of global data networking services, IBM Global.

---

[4] GTE evolved from the Richland Center Telephone Company, founded in 1918, and was named General Telephone and Electronics (GTE), in the 1950s. After the merger of GTE and Bell Atlantic in 1998, the new company was named Verizon.

The Telecommunications Act of 1996 allowed further opening of the telecommunications market and permitted long-distance operation and equipment manufacture to the Baby Bells, while AT&T was allowed to enter into local operations. Adapting to the situation, AT&T made a voluntary breakup into three new companies on December 31, 1996.

- A new AT&T, providing communication services, including that part of Bell Labs that supported the communications services business
- Lucent Technologies, consisting of the previous Western Electric and major part of Bell Labs
- NCR Corporation

Within 20 years AT&T transformed from a long-distance to an "any distance" company, maintaining a place in the world top 10 telecommunications companies. Despite all efforts, however, AT&T had to cut the dividend (by 83%) in the year 2000 for its 4 million shareholders for the first time in the company's over-100-year history.

MCI, an initiator and major beneficiary of the AT&T divestiture, was acquired by WorldCom in 1998, and as shown in Table 33.3, the new company, MCI World-Com,[5] overtook AT&T in international traffic.

GTE merged with Bell Atlantic into a new company named Verizon[6] in March 2000 and thus became the largest local telephone company, operating about 63 million telephone lines in 31 U.S. states.

At the end of the century the local telephone market was still dominated by the Baby Bells and Verizon, with about 90%, and the number of independents was reduced to about 1000. On the long-distance telephone market the AT&T share was reduced to slightly less than 60%, MCI WorldCom had 20%, Sprint[7] 10%, and the remaining 10% was shared by about 400 independents.

33.2.2   Telephony and Deregulation in Canada

At the Canadian centenary of the telephone in 1974, some 1900 telephone companies served 12.4 million telephones, of which 97% were connected to automatic switches and 75% operated by Bell Canada (prior to 1968 called Bell Telephone Company of Canada).

Deregulation brought a dramatic change in the telecommunications landscape of Canada. So far, each telecommunications company operated on an exclusive basis in its geographic region. This convenient situation changed suddenly when the Canadian regulatory authority, the *Canadian Radio, Television, and Telecommunica-*

---

[5] MCI WorldCom restructured in 2001, creating two groups: MCI Group for residential long-distance operations and dial-up Internet, and WorldCom Group for corporate telephony and international businesses. With this change MCI Group ranked again behind AT&T in revenue.

[6] The name *Verizon* was selected from over 8500 possibilities. Apparently, it is a blend of *horizon* and the Latin word *veritas* (truth).

[7] Sprint originates from the Brown Telephone Company founded at Abilene, Kansas, in 1899. It was named United Telephone and Electric in 1911, United Utilities in 1939, United Telecommunications in 1972, and Sprint Corporation in 1992.

tions Commission (CRTC, established in 1976), opened the market for long-distance operation in 1992. Immediately, new operators appeared backed by U.S. operators such as Telus, backed by GTE and Bell Atlantic, operating in British Columbia and Alberta; and Sprint Canada, backed by the U.S. Sprint Corporation. Adapting to the new situation, the major telecommunications operators[8] formed an alliance named *Stentor* in 1992. Each company still operated on an exclusive basis in its geographic region, and Stentor was to provide national operational support to its members.

Further changes appeared in 1994 when CRTC deregulated local operations, and in 1998, long-distance and international operations. Virtually restoring the situation prevailing before 1932, CRTC allowed traffic routing through the United States for overseas traffic as well as for traffic between Canadian locations. Taking advantage of the new rules, AT&T Canada[9] bought the optical fiber network operator Metro-Net Communications, which operated fiber networks in 10 Canadian cities. As a consequence of the new situation, the Stentor alliance broke up in 1998 and competition began among the erstwhile members, resulting in the creation of new companies by Bell Canada:

- *Bell Nexxia*, offering national broadband service via a coast-to-coast optical fiber network across southern Canada and northern United States similar to AT&T Canada's network
- *Aliant*, an alliance of four of the Maritime operating companies with 41% Bell ownership
- *Intrigna*, a joint venture of Bell Canada, Maritime Telegraph and Telephone, and Bell Nexxia, operating in British Columbia and Alberta

Furthermore, Bell Canada, which already owned Telesat Canada, increased its share of international traffic by acquiring the international long-distance operator Teleglobe Inc. in 2000. In the meantime, Telus, Canada's second-largest telecommunications company, activated a national optical fiber network linking Vancouver to Quebec and extending into the United States (see Figure 28.6).

At the end of the twentieth century, over 98% of all households had telephone service, of which almost 95% were connected by cable; they enjoy one of the world's lowest tariffs, including free local calls.

33.2.3   Telephony and Deregulation in Mexico

The national operator, Telmex, became a public company in 1972, when the Mexican government obtained 51% of the shares. Besides Telmex, a few other smaller telephone companies existed, such as the Teléfonica Nacional S.A. The number of telephones reached 3 million by August 20, 1976, a teledensity of 4.8.

Telephony in Mexico experienced a substantial upheaval in the last decade of the century. With only 5.4 million telephones at the beginning of that decade, 70% of

---

[8] Stentor members were Bell Canada, British Columbian Tel, Island Tel, Manitoba Telecom Services, Maritime Telegraphy & Telephony, New Brunswick Tel, NewTel Communications, NorthwestTel, QuebecTel, Saktel, and Telus.

[9] AT&T Canada belongs 22% to AT&T Corp. of the United States after AT&T Corp. sold 10% of the shares to British Telecom in 1999.

small towns still had no telephone service at all. Waiting times for a telephone line was typically 1.6 years outside Mexico City and six years in the capital. To improve the situation, Telmex was privatized in 1990; 51% of the voting stock was sold to a consortium of the American Southwest Bell, French Cables and Radio, and Mexican Grupo Carso.[10] Telmex obtained a six-year extension of its basic service monopoly provided that it met a set of stringent network improvements, including an annual 12% network extension and the provision of at least one public telephone for each of 9000 villages of 500 people or more by 1994. At the same time, Telecomm (Telecomunicaciones de Mexico) was created to operate Mexico's satellite, telegraph, telex, and packet-switched networks.

Telmex partially exceeded the mandates and increased the teledensity from 6 in 1990 to 9.2 in 1994. To achieve this quick increase in teledensity, various rural public fixed wireless systems were installed in the interior regions of Mexico, serving 8.5 million people.

The federal regulatory authority, Comisión Federal de Telecomunicaciones (Cofetel), established in 1996 based on the Telecommunications Act of 1995, authorized competition on local and long-distance services in 1997. A few new long-distance operators were established, backed primarily by U.S. companies, the two biggest being Alestra, 49% owned by AT&T Corp., and Avantel, backed by MCI–WorldCom. A four–year dispute on interconnecting rates and other points followed between Telmex and the Mexican government on one side and Alestra, Avantel, and the U.S. government on the other side. The dispute culminated in a complaint filed with the *World Trade Organization* (WTO) until an agreement was reached in December 2000. By that time Telmex had lost about 25% of its long-distance traffic to its new competitors, but it had also extended its network to another 100,000 locations in Mexico and achieved 98.6% coverage of the population with a teledensity of 12.5 and a fully digital network. About 400,000 pay phones were in operation, of which over 200,000 were multimedia phones offering e-mail, Internet, and fax services.

In an attempt to access the 50 million Mexicans living in the United States, Telmex extended its activities to the southwestern United States, in cooperation with Sprint, and started operation in Tucson, Arizona, in 1999. Furthermore, Telmex acquired an operating contract with Guatemalan's main long-distance operator, Telgua.

33.2.4   Telephony and Deregulation in Central America

Central America had about 50,000 telephones in 1950. Civil wars, earthquakes, hurricanes, and a strong population increase slowed telephone penetration until the mid-1990s. The waiting list in El Salvador for 600,000 lines and that in Honduras for 400,000 lines were larger than the number of subscribers served and caused waiting times of up to 16 years. Privatization, begun at the end of the 1990s, improved the situation substantially.

In Guatemala, the operator Telgua in 1997 came under control of Luca S.A., a Guatemalan–Honduran investment group with 51% shares, which subsequently were resold almost completely to Telmex of Mexico.

---

[10] The Grupo Carso belongs to Latin America's richest billionaire, Carlos "Slim" Helu.

In Belize, the complete telephone network was destroyed by hurricanes in 1931 and 1961. C&W installed and operated a new network beginning in 1963, but this network was nationalized in 1971 and was eventually operated by *Belize Telecommunications Limited* (BTL). BTL was privatized in 1989, when British Telecom bought a 25% share. This was the first telecommunications privatization in Central America and, after Chile, the second in Latin America.

In Nicaragua, efforts made since 1996 to privatize the national operator *Empresa Nicaraguense de Telecomunicaciones* (Enitel) were unsuccessful.[11]

Honduras had a few private companies that served only 7400 telephones in 1955. The companies were nationalized and eventually became the semigovernmental *Empresa Hondureña de Telecomunicaciones* (Hondutel). Privatization of Hondutel was attempted in 2000 to meet the requirements of the World Bank. The sole offer received was so low that privatization has been postponed.

In El Salvador, the private companies were nationalized and eventually became the governmental *Administración Nacional de Telecomunicaciones* (Antel), which was privatized in 1998 when it came under the control of France Telecom with 51% ownership.

In Costa Rica, the current operator, *Instituto Costarricense de Electricidad* (ICE), was founded in 1962. At that time Costa Rica had fewer than 20,000 manually switched telephone lines. ICE started an effective modernization and extension program, adding almost 100,000 automatically switched lines within 15 years, thus increasing the teledensity from 1.4 to 5.6 and creating the most advanced telecommunications infrastructure in Latin America. The Autoridad Reguladora de Servicios Públicos was established as a regulator in 1996, but ICE was still wholly government owned at the end of the century.

In Panama, the major private company, *Compañia Panamena de Teléfonos* (CPT), served about 70,000 telephones in 1969 when it was nationalized. Beginning in 1973, CPT was gradually integrated with the other private companies in the *Instituto Nacional de Telecomunicaciones* (Intel). Intel was the first telephone operator in Central America to be privatized, in 1997.

33.2.5 Telephony and Deregulation in the Caribbean

The Caribbean had about 300,000 telephones at the beginning of the 1950s, of which 150,000 were in Cuba and 60,000 in Puerto Rico. C&W long had a dominance on international but also national telephone traffic in many Caribbean Islands. Currently, privatization and liberalization are being considered by some of the Caribbean states, but it appears that the choice will be a balance between negotiating better conditions from C&W or accepting future dominance by U.S. operators.

33.2.6 Telephony and Deregulation in Brazil

The *Emprêsa Brasileira de Telecomunicações* (Embratel) was founded in 1962. At that time the country had about 1200 telephone companies which served over 1 million telephones. Embratel interconnected all state capitals and other major cities. Between

---

[11] The privatization objective is to achieve 40% foreign ownership, with employees allowed to buy 10% and receive a donation of 1%. The government would reserve the right to sell the remaining 49% within three years of the initial sale.

TABLE 33.6   Brazilian Fixed-Line Operators

Region	Baby Bras		Mirror Companies	
	Company	Main Owners	Company	Main Owners
North, east	Tele Norte Leste	Inepar, Macal	Canbrá	Bell Canada, Qualcomm, Velocom
Central, south	Tele Centro Sul	Telecom Italia	Global Village Telecom	Bell Canada, Qualcomm, Velocom
State of São Paulo	Telesp	Telefónica	Megatel	National Grid U.K., Sprint
National and international	Embratel	MCI World-Com	Intelig	France Telecom, British Telecom

Source: Latin America and Mexico: a change in focus, *Telecommunications*, March 2000, pp. 51–54, 64.

1969 and 1973 it also took over international long-distance services as the concessionary periods terminated for the foreign companies. The Ministry of Communications nationalized telephone services in 1972 and founded the *Telecomunicações Brasileiras* (Telebras), which gradually absorbed the about 900 independent telephone companies and consolidated them into 27 state companies, which operated 2 million telephones.

The General Telecommunications Law of July 16, 1997, defined the principles of free competition among all telecommunications service providers. Under this law the *Agência Nacional de Telecomunicações* (Anatel) was established as the regulatory agency on November 5, 1997, with the specific task of implementing privatization, with an obligation for universalization of services. To meet the Brazilian principle of universalization, the future private operators were obliged to ensure that "everyone in society, no matter their location or socioeconomic status, should have access to telecommunications."

The long-distance operator Embratel was the first to be privatized. MCI WorldCom acquired a 52% voting interest in Embratel in July 1998. Telebras, the largest operator in Latin America, with 27 local operators serving 13 million telephone lines, was privatized in the same year at a record price of $19 billion. The 27 local operators were grouped into three large regional *Baby Bras*. Embratel and the Baby Bras each got a competing "mirror" operator (see Table 33.6). Figure 33.3 shows the regions of the Baby Bras and their operating companies. The Baby Bras were also allowed to compete for domestic long-distance traffic in June 1999, Embratel was authorized to offer local and intrastate phone service.[12]

33.2.7   Telephony and Deregulation in Chile

A national long-distance public telecommunications company, the *Empresa Nacional de Telecomunicaciones S.A.* (Entel Chile), was established on December 30, 1964, as an affiliate of CORFO.[13]

---

[12] The mirror competition duopoly was limited until January 1, 2002, when the Brazilian telecommunications market was completely liberalized.

[13] The Corporación de Fomento de la Producción was established as a government agency for national development in 1965.

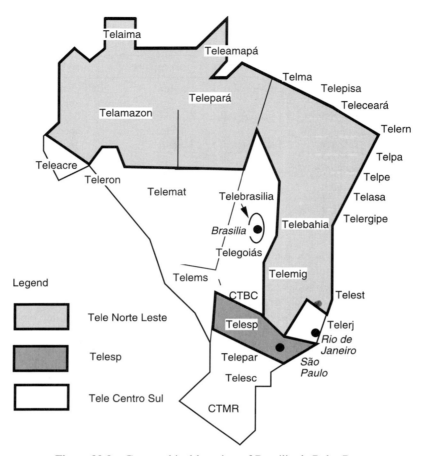

Figure 33.3   Geographical location of Brazilian's Baby Bras.

Deregulation in South America began in 1982 with privatization in Chile of two smaller companies: the *Compañía Nacional de Teléfonos* and the *Compañía de Teléfonos de Coyhaique*. Both Entel and Chiltelco were privatized in 1985. Entel sold 20% of its shares to the Spanish operator Telefónica and 12% to the Chase Manhattan Bank in 1988. In the same year, the Chilean Bond Corporation bought 52% of Chiltelco and sold those shares later to Telefónica of Spain. Shares of ENTEL and Chiltelco were also obtained by private Chilean pension funds. Liberalization began in 1988 when long-distance services were opened to competition. Three competitors, Bell South Chile, VTR, and CTC Mundo, and the national satellite company, Chilesat, reduced Entel long-distance traffic share from almost 99% to merely 35% in 1995.

Competition in local operation came more slowly. In 1991, Chiltelco still provided 92% of the local traffic; four local companies provided the other 8%. By that time the Chilean regulator Subtel (Sub-Secretary of Telecommunications) extended the scope of local operation competition, and in the next four years more telephone lines were installed (800,000) than in Chile's entire previous telephone history.

33.2.8 Telephony and Deregulation in Argentina

The *Empresa Nacional de Telecomunicaciones* (Entel) was founded in 1956 and absorbed Teléfonos del Estado and the other private companies in the next four years. Argentina was the second South American country to introduce telecommunications privatization. In 1990, Entel was divided into Telco Sur (60% sold to Telefónica de España and Citibank), later named Telefónica de Argentina; and Telco Norte (60% sold to France Telecom, Italian STET, J. P. Morgan, and Perez Company), later named Telecom Argentina. The two companies increased the teledensity from 10.7 to 21 in 2000. They lost their monopoly in November 2000, when eight additional companies also obtained licenses for local and long-distance services.

33.2.9 Telephony and Deregulation in Peru

The ITT-owned *Compañía Peruana de Teléfonos* (CPT) was nationalized in 1969. CPT operated mainly in Lima and central Peru, the *Compañía Nacional de Teléfonos del Peru* (CNT) operated in the north, and the *Sociedad Telefónica del Peru, S.A. Arequipa*, later called the *Compañía Telefónica de Sur* (CTS), operated in the south of Peru. The *Empresa Nacional de Telecomunicaciones* (Entel) was established in 1969 and absorbed CNT and CTS.

One of the first regulatory bodies in South America was the *Organismo de Inversión Privada en Telecomunicaciones* (Ospitel), established in Peru in 1991. Privatization began in Peru in 1994, when 35% of Entel and 20% of CPT were sold to a consortium led by the Telefónica of Spain. Both companies then merged into *Telefónica de Perú* (TdP). TdP got exclusive rights until the end of 1998 with the commitment of doubling the number of telephones in that time span, thus increasing the teledensity from 3.6 to 6.7. Both long-distance and local services were liberalized in 1998. By the end of the century, a total of 209 companies obtained telecommunications licenses, of which 28 companies offered long-distance services.

33.2.10 Telephony and Deregulation in Venezuela

CANTV was nationalized in 1953 and privatized in 1991. A consortium headed by the U.S. GTE Corporation bought 40%, and 11% of the shares were sold to employees. CANTV got a nine-year monopoly. The teledensity of 8 in 1991 increased to 12 in 1998 but then stagnated and fell to 10.8 in 2000, when mobile radio teledensity reached 22.

Liberalization of local line services started on November 28, 2000, when Conatel (Comisión Nacional de Telecomunicaciones, Venezuela's regulator) auctioned 15 licenses for three parallel wireless local loop (WLL), networks in five regions. This was probably the world's first nationwide WLL project. The WLL solution was chosen to avoid a costly cable infrastructure and to reduce implementation time.

33.2.11 Telephony and Deregulation in Colombia

The *Empresa Nacional de Telecomunicaciones* (called Telecom) was founded in 1956 to provide long-distance service. Colombia had 1 million telephone lines in 1971 operated by 58 different companies, most of them still existing, which teamed up in

Figure 33.4 Public telephone in the center of Bogotá, 1970. See insert for a color representation of this figure.

the *Asociación Nacional de Empresas de Teléfonos de Columbia* (ANET). Privatization of Telecom planned in 1991 was blocked by a trade union protest action that caused a nine-day telecommunications interruption that isolated the country. Liberalization therefore came before privatization, when in 1998 competition was allowed on long-distance operation. Colombia's largest telephone company, the municipal-owned *Empresa de Telecomunicaciones de Bogota* (ETB), began long-distance operation in the same year. Beginning in 1999, various efforts were made to privatize ETB, but legal and financial problems and violence from drug gangs and guerillas caused postponements beyond the end of the century. Figure 33.4 shows a public telephone of ETB.

33.2.12 Telephony and Deregulation in Ecuador

The public company *Empresa de Radio, Telégrafos y Teléfonos de Ecuador*, later renamed *Empresa Nacional de Telecomunicaciones* (Entel), was founded in 1958. In 1997, Entel was divided into the state-owned Andinatel, operating in the highlands and jungle, and Pacifictel, operating in the coastal area and on the Galápagos Islands. Both companies got a five-year regional exclusivity in fixed-line and other services. To meet the requirements of the International Monetary Fund and after privatization failed twice, the government decided to end the regional exclusivity in

2001 and to appoint international administrators to improve the management of Andinatel and Pacifictel.[14]

33.2.13   Telephony and Deregulation in Bolivia

The *Empresa de Teléfonos de Bolivia* (Entel) was established in 1976 with an exclusive license for national and international long-distance services. Entel also provides local services in areas with no telephone cooperatives. Entel was privatized in 1995. Unlike most other countries, proceeds from the privatization did not go to the state but were reinvested in Entel. Telecom Italia, through STET International, obtained 50% of the shares by investing the counter value in Entel. The national pension funds own the remaining 50%. Entel and the regional cooperative operators obtained a six-year period of exclusivity.

33.2.14   Telephony and Deregulation in Uruguay

In Uruguay the national operator *Administración Nacional de Telecomunicaciones* (Antel) successfully increased the teledensity from 13 in 1990 to the highest South American value, 27.8, in 2000 and kept the cost of services among the lowest in South America. The government issued a law in 1993 that foresaw privatization of Antel. However, in a public referendum, some 75% of the population demonstrated their satisfaction with Antel's service and voted against this law, so that Antel was still not privatized by the end of the century.

33.2.15   Telephony and Deregulation in Paraguay

In 1976, 38,000 telephone subscribers of Paraguay were served by the public company *Administración Nacional de Telecomunicaciones* (ANTELCO). This number of telephones corresponded to South America's lowest teledensity, 1.6. At the end of the century its teledensity of 5 was still the lowest in South America.

33.3   TELEPHONY AND DEREGULATION IN AFRICA

Telephony in Africa came very slowly and was concentrated in the large cities. Even at the end of the twentieth century, two-thirds of the 20 million telephone lines served only those 30% of the African population who lived in urban areas. The majority of the African population never saw a telephone. A rural African typically has to walk 50 km to reach a telephone. According to the ITU in 1998, over 450,000 villages in Africa, with a total population of 400 million people, had no telephone access at all. They still used indigenous means of communication, such as drums, gongs, elephant tusks, bells, and whistles. An excellent book on telecommunications in Africa, *Tam Tam to Internet*, published in 1998 with forewords by Nelson Mandela, and Kofi Annan, states:

---

[14] By then, Etapa, the municipal operator of the southern city Cuenca, may also enter into competition with Andinatel and Pacifictel.

Drums are generally used as instruments for communicating over long distances. The beating of the drum and the resulting resonance can travel as far as 45 km although, for practical purposes, it is more effective to use relay. The distance between relays is usually two or three or ten kilometers. The huge drums convey language as clear to the initiated as vocal speech. The key to understanding the language of talking-drums of Africa lies in the bi-tone nature of most African languages. The drummer is not an artist, he is not inventing, he is not improvising, he is merely a technician. The drummer takes the role of a messenger. He transmits messages (births, deaths, marriages, summons, alerts, instructions, imminent attack, events, etc.) within the village and between villages. If he changes the text, he could be punished by death.

Drums thus still serve the needs of communications for small communities. Since the beginning of the 1990s, this principle is also applied to telephone development in African rural areas. Instead of bringing the telephone directly into living quarters, common telephone facilities are located within x kilometers of inhabited areas (e.g., within 10 km in Malawi) or at walking distance, as approached in Kenya, or in the center of each community of more than 500 people, as in Ghana. For this purpose, transportable telephone kiosks, village pay phones, community public telephones, and multipurpose community centers, generally connected by satellite with public networks, are being installed in various rural areas all over Africa. In Senegal, for example, private companies were allowed to operate *Telecentres privés* (TCPs). More than 5000 were in operation at the end of 1996, accounting for 5% of all telephone lines in the country.

The African Ministers of Communication at Africa *Telecom 98* (a telecommunications exhibition and forum organized by the ITU) solemnly expressed their determination within the scope of a general "African Renaissance" to connect every village, every school, and every clinic in Africa by 2003.

The number of telephones in Africa increased from less than 1.5% of the worldwide number of telephones in the 1950s to only 2.0% at the end of 2000. The growth is achieved mainly in North Africa and South Africa, which together always had more telephones than the rest of Africa, also called Sub-Saharan Africa (Table 33.7). Privatization has taken off slowly in Africa, but with some encouraging results. Countries such as Ghana, Guinea, and Senegal, which were among the first to privatize their telecommunications operator in 1996, one year later achieved the highest growth in telephone lines. At the end of the century, about 20 operators have some degree of private and/or foreign ownership.

TABLE 33.7   Telephone Growth in Africa

Year	Millions of Telephones				Percent of Worldwide Total
	North Africa	South Africa	Sub-Saharan Africa	Total	
1957	0.5	0.8	0.2	1.5	1.4
1976	1.1	2.1	1.4	4.6	1.2
1990	3.2	3.3	2.4	8.9	1.7
1997	7.5	4.7	3.3	15.5	1.9
2000	10.1	4.9	4.6	19.6	2.0

Source: Archiv für deutsche Postgeschichte, Vol. 1, 1977; ITU, *Telecommunications Indicators*, 2000; Siemens, *International Telecom Statistics*, 1993 and 1999.

International telecommunications, even African intrastate communication, is still routed mainly via London, Paris, and Brussels. This will change at the beginning of the twenty-first century when the Africa-ONE submarine cable project (Section 28.3.5) and the Rascom satellite project (see Section 27.4.2) will connect all African capitals and most other major towns.

Despite slow progress, the year 2000 was, according to the ITU, to be remembered as the year in which Sub-Saharan Africa achieved a teledensity of 1. However, this achievement is not due to improvement in fixed-line networks but to the sudden, rapid deployment of mobile telephony. While in the last decade 2.2 million additional fixed-line telephones were installed, the number of mobile telephones in Sub-Saharan Africa increased from practically zero to 3.2 million. As a result, the teledensity was 0.75 for fixed-line telephones plus 0.53 for mobile telephones: 1.28 together.

33.3.1 Telephony and Deregulation in North Africa

In Egypt, the telephone network, with around half a million telephone lines, was in a desolate state in the 1970s.[15] The telephone operator, the *Arab Republic of Egypt Telecommunication Organization* (Arento), in five-year plans beginning in 1981, increased the number of telephone lines to 5.5 million by the end of the century and made out of the formerly highly inefficient network the most advanced telecommunications network in Africa. About every fourth household had a telephone, but the waiting list remained fairly static, at around 1.2 million over the last decade. Arento was renamed *Egypt Telecom* in 1998 and its privatization was initiated in 2000 when 5% of the shares were offered to the company's employee union, 10% to the local market, and another 10% to the international market.

Libya had about 5000 telephones in 1951 when it became an independent monarchy. C&W obtained a concession for the international telephony in 1952, but that service was nationalized 15 years later. The national telephone network was extended to the major towns and oases in 1961 and also connected with Egypt and Tunisia. The network served about 40,000 telephones in 1969 when Libya became a Socialist People's Republic.

Tunisia had 26,000 telephones in 1950. The telephone network is still under government control.

Algeria was a department of France until it became independent in 1962. It appears that the first statistics published on the development of telephony in Algeria refer to the size of the telephone network in 1955, which then had 135,000 telephones. This number increased to about 200,000 in the year of independence. The number of telephones then diminished to 185,000 in 1970 but increased to 250,000 in 1976, and reached 1,600,000 by the end of the century.

Morocco had 120,000 telephones in 1956, the year of independence. The international operator Itissalat Al Maghrib (IAM) was privatized in 1999. Privatization of the national telecommunications operator, Maroc Telecom, in 2000 was not successful.

---

[15] Wealthy Egyptians reportedly took airplanes to Cyprus to make international telephone calls.

33.3.2   Telephony and Deregulation in South Africa

The *South African Post Office* (SAPO), later named *the South African Post and Tele-communications* (SAPT), became a state-owned corporation called Telkom on October 1, 1990. By the middle of the 1990s, South Africa had 10 million telephones, and a teledensity of 10, the highest in Africa.[16] This relatively high telephone penetration was, however, very unequally divided between 64 in white areas, to less than 1 in rural black areas. To obtain a more balanced solution with universal and affordable telecommunications service, the Telecommunications Act 103 of 1996 was issued. Under this act were established the *South African Telecommunications Regulatory Authority* (SATRA)[17] to secure fair competition, the *Universal Service Agency* to recommend ways of achieving universal service, and the *Human Resource Fund* to define and support universal service projects. Telkom obtained a 25-year license to build and operate public-switched telecommunications services. This license, with a five-year exclusivity on fixed lines,[18] included a commitment to install 2.8 million telephone lines (of which 1.7 million were to be in underserved areas), including 120,000 pay phones, and to upgrade 1,242,000 telephone lines by 2002. In addition, Telkom agreed to connect 24,000 priority customers, comprising villages, educational and medical establishments, and libraries.

Telkom was privatized in April 1997 when the U.S. Bell South Corp. obtained 18% and Telekom Malaysia 12%—combined in a consortium called Thintana Communications—and a further 10% of the shares were obtained locally. By that time more than 85% of white households had a telephone but only 29% of urban African and 2% of rural African households. Since 1999, to improve telephone access in the African urban and rural areas, a nationwide network of telecenters is being installed. Those *public information terminals* (PITs), installed in various post offices, offer telephony as well as e-mail and other online services upon insertion of a "smart" telephone card.

At the level of 5.8 million, the number of mobile telephone subscribers surpassed the number of fixed-line telephone subscribers in August 2000. By that time about 42% of South Africans had universal service with a telephone at home, and 80% had universal access with a telephone within 30 minutes walking distance.

33.3.3   Telephony and Deregulation in Sub-Saharan Africa

In Nigeria, the *Nigerian External Telecommunications Ltd.* (NET) took over the international traffic from C&W in 1963. NET opened the first direct inter-African line between Nigeria and Benin on August 7, 1962. NET was integrated into the *Nigerian Post and Telecommunications* (NP&T), which is presently called Nitel (Nigerian Telecommunications).

By December 2000, Nitel's telephone network had 720,000 lines, but only 492,000 lines were connected to subscriber, and only six cities had international direct-dial services.

The Republic of Congo (Brazzaville) had 5046 telephones when it became inde-

[16] Apart from the islands of Mauritius, which had a teledensity of 16, Reunion with 34, and the Seychelles with 19.6.

[17] Satra merged with the Independent Broadcasting Authority (IBA) into the *Independent Communication Authority of South Africa* (Icasa) in 2000.

[18] The exclusivity will be extended to six years if Telkom meets 90% of the targets within four years and commits itself to installing 3 million lines in the six-year period.

pendent in 1960. The *Office National des Postes et Télécommunications* was in charge of the national telecommunications services from January 1, 1965. The *Compagnie Française de Câbles Sous-Marins et de Radio* provided international telecommunications until 1972, when it was nationalized and became the *Office de Télécommunications Internationales* (Intelco).

The Democratic Republic of Congo had 28,300 telephones in 1960, when it became independent and temporarily was named the Democratic Republic of Zaire. The *Office Congolais des Postes et Télécommunications* (OCPT, established in 1968, later called the Office Nacional des Postes et Télécommunication) took over the service from Bell Congo in 1972. By that time the number of telephones was reduced to 22,000. The number of telephones increased slowly to 36,000 in 1996, but a civil war then dropped the number of telephones to 20,000 by year-end 2000, which gives Africa's lowest teledensity of 0.04.

Kenya, Tanzania, and Uganda had a common operator, the *East African Post and Telecommunications Administration* (EAPTA), which was dissolved when those countries became independent in 1960.

In Tanzania, the *Tanzania Telecommunications Company Ltd.* (TTCL) was the successor of EAPTA with 15,000 telephones. A consortium of Germany's Detecon GmbH and the Dutch *Mobile System International* (MSI) obtained management control and a 35% stake in TTCL in 2000. The privatization included a commitment to increase the number of telephone lines from 150,000 to 800,000 by the end of 2003, with at least 30% of the new lines in rural areas.

In Uganda, the *Uganda Telecommunications Ltd.* (UTL) was the successor to EAPTA. It operated a network with 15,000 lines and it had 57,000 telephone lines in 1998 when a second operator was introduced. MTN Uganda, a subsidiary of South Africa's MTN (Mobile Telephone Networks Holdings) obtained a license for fixed and mobile telephone operation. In the interest of quick implementation, MTN first concentrated on wireless networks, and within one year the number of mobile telephone subscribers in Uganda increased from 26,000 to 87,000, compared with an increase of 2000 fixed telephone subscribers. Uganda was the first country in Africa in which, in July 1999, the number of mobile subscribers surpassed the number of fixed telephone subscribers (then 59,000). The incumbent UTL was privatized in 2000 when another consortium of Detecon with Egyptian Orascom and Swiss Telcel acquired a 51% stake in UTL and accepted the commitment to install 100,000 telephone lines in the next five years.

In Kenya, the *Kenya Post and Telecommunications Corporation* (KPTC) took over EAPTA's telephone network with 42,000 lines. In 1998, KPTC was prepared for privatization; post and telecommunications were separated and the latter called Kenya Telekom.[19]

33.4   TELEPHONY AND DEREGULATION IN ASIA

Asia's share in the worldwide number of telephones increased from less than 5% in the 1950s to over 35% at the end of 2000. This increase was mainly in Japan and

---

[19] In February 2001, 49% of Telkom Kenya was acquired by a consortium led by Econet Wireless Holdings of Zimbabwe with South Africa's Eskom Enterprises and the Dutch KPN (Koninklijke PTT Nederland NV).

TABLE 33.8 Telephone Growth in Asia

Year	Millions of Telephones				Percent of Worldwide Total
	China	Japan	Rest of Asia	Total	
1957	<0.3	3.5	>2.2	5.1	4.7
1976	<0.3	43.2	>16.1	59.6	15.7
1990	6.9	55.3	43.9	106.1	20.4
1997	69.3	63.3	100	232.6	29.5
2000	144.8	74.3	130.2	349.3	35.4

Source: Archiv für deutsche Postgeschichte, Vol. 1, 1977; ITU, *Telecommunications Indicators*, 2000; Siemens, *International Telecom Statistics*, 1977, 1992, and 1998.

China, which together always had more telephones than the rest of Asia (Table 33.8). Deregulation in Asia was confined primarily to the Asia-Pacific region, which became a very competitive region and contains about one-third of the world's total telecommunications market. The *Asia Pacific Economic Cooperation* (APEC) Forum made the year 2020 the target date for full liberalization of telecommunications services throughout the region. Competition in international telecommunications services was implemented by at least 10 countries, which by March 1999 had the following number of operators with international services: Hong Kong, 60; Japan, 44; Australia, 28; South Korea, 24; New Zealand, 17; the Philippines, 12; Malaysia, 5; Thailand, 2; Indonesia, 2, and Brunei, 2.[20] Universal service obligations are required from new operators at least in Australia, Cambodia, China, Indonesia, Iran, Malaysia, Pakistan, the Philippines, South Korea, and Sri Lanka.

33.4.1 Telephony and Deregulation in India

India had slightly over 100,000 telephones at independence in 1947. During the following 30 years, about 1.73 million telephones were installed and an annual average of 200,000 persons queued on the waiting list. By 1990 the network had increased to 4.6 million telephones, the waiting list to 1,9 million, and the teledensity to 0.57.

The Department of Post and Telecommunications was separated in 1985 into the telecommunications operator, the *Department of Telecommunications* (DoT), and a postal service department. The following year, DoT was divided into two government-owned corporations: *Videsh Sanchar Nigam Ltd.* (VSNL) for long-distance operations and *Mahanagar Telephone Nigam Ltd.* (MTNL) for local operations. The government defined liberalization principles in the National Telecom Policy of 1994. Unlike most other countries, which initially privatized long-distance telephony and then introduced competition in the local market, the Indian government acted the other way around. VSLN's exclusive license for long-distance operations was extended to the year 2000 for domestic services and to 2004 for international services. A duopoly was proposed for local operation over a 15-year period. The stringent license conditions, which did not apply to the incumbent MTNL, kept private interest low, so that at the end of the century only six states of 22 states have a regional operator competing with MTNL. These new operators are:

---

[20] Figures are from *Communications International*, June 1999, "Indicators," p. 61.

- Bharti Telenet Ltd.[21] in the state of Madhya Pradesh
- Tata Teleservices in the state of Andhra Pradesh
- Hughes Ispat Ltd. in the state of Maharashtra
- Reliance Telecom Ltd. in the state of Gujarat
- Essar Commvision in the state of Punjab
- Shyam Telelink, India in the state of Rajastan

In Delhi and Mumbai, telephones are now available on demand, but rural areas still have to cope with waiting times between six and 15 years. To accelerate the telecommunications development, the *Telecommunications Regulatory Authority of India* (TRAI), established in 1997, was reconstituted with more power in 2000, and better conditions were defined for the new operators. Domestic long-distance services were opened up to the private sector on India's fifty-third independence day, August 15, 2000. Foreign companies may hold up to a 49% share of Indian operators. Domestic long-distance charges in India are among the highest in the world. Licenses are for 20 years and include 15% revenue sharing with the government.

To improve the quality and reliability of the telephone network, by January 2000 all electromechanical exchanges (many over 50 years old) were replaced by mainly Indian-made electronic switching equipment. Special attention has been given to introducing telephone service in small villages. In 1990, almost none of the 600,000 villages had a telephone. At the beginning of 2000, over 360,000 villages had at least one telephone. It is planned that all other villages will be connected by 2002 and to increase the teledensity in rural areas from 0.4 to 4 by 2010 (compared with 15 for all India).

33.4.2 Telephony and Deregulation in China

China had fewer than 250,000 telephones in the early 1950s, and this low number decreased further during the Cultural Revolution, from 1965 to 1976. Privately owned telephones hardly existed; a public pay phone, if available, had to be used. Later, when private telephones were allowed, the cost was well outside the range of most Chinese salaries. The situation improved with the economic reform program introduced under President Deng Xiaoping in 1978. All telecommunication activities from research via manufacturing to operation were under the responsibility of the *Ministry of Post and Telecommunications* (MPT). About 3 million telephones were in use when, in 1983, local production of telephones and telephone switching equipment was begun in the Shanghai Bell Telephone Equipment Manufacturing Company (SBTEMC) and at Tong Guang-Nortel. These were the first telecommunications joint ventures of MPT with Bell Telephone of Belgium and Nortel of Canada. An impressive expansion of the local telecommunications industry at an average annual growth rate of 51.5% was achieved under the 1991–1995 national five-year plan. The

---

[21] Bharti Telenet belongs to Bharti Enterprises, founded in 1976 by Sunil Bharti Mittal, who graduated from Punjab University and was named telecom man of the year in 1997. Bharti Telenet is the first private telecommunications company in India since independence. Bharti Telesonic Ltd. was established in a joint venture with Singapore Telecommunicatiions Ltd. (SingTel) in 2000 to provide domestic long-distance services. Bharti and SingTel also decided to build and operate a 2500-km submarine cable between India and Singapore.

number of telephones increased from 6.8 million, corresponding to a teledensity of 0.6 in 1990, to 40.7 million and a teledensity of 3.3 in 1995, of which about 75% were residential. An even more impressive expansion followed under the ninth five-year plan, resulting in 144.8 million telephones and a teledensity of 11.8 at the end of 2000.

Deregulation in China started in 1993 with the MPT acting as regulator. At first, China Telecom (CT) was established as an independent national operator under the MPT; non-MPT entities were permitted to operate secondary networks. Successively, five further companies obtained licenses for mobile operation, paging, Internet, and partly also for fixed-line operation: China Unicom, Jitong Corporation, Great Wall Telecom, Spacecom, and China Netcom. China Unicom (China United Telecommunications Corporation) was established in 1994 with the objective of offering competition to CT. In the same year the Ministry of Electronics established Jitong Corp. with the objective of providing data communication services. Great Wall Telecom was established in 1995 by the People's Liberation Army to introduce a national mobile network based on CDMA. SpaceCom (China Space Mobile Telecommunications Co. Ltd.) was established in 1998 to provide national and international satellite communications. China NetCom was founded by MPT with the city of Shanghai, the Chinese Academy of Science, and the state agency for radio film and TV. As a novelty, a small foreign participation was permitted.

By year-end 2000 two new companies had obtained fixed service licenses: China-Sat for satellite operation and the *China Railway Telecom* (CRT) for domestic fixed network and international data and voice operation. CRT connects 500 cities with 120,000 km of fiber optic cable.

At the end of the twentieth century, still over 99% of the revenues on fixed-line operations were obtained by China Telecom. This may soon change. China Unicom, not having a circuit-switched PSTN like China Telecom, developed within a few months a network connecting over 320 cities based on the Internet Protocol (Section 34.4), hailed to be the world's biggest Voice over IP (VoIP) network.[22]

Hong Kong, since June 1, 1997, a *Special Administrative Region* (SAR) of China, by 1990 had the reputation of having the world's most advanced telecommunications network with the cheapest domestic telephony services. The network at that time was operated by the *Hong Kong Telephone Company* (HKTC), which belonged to C&W (58.4%), the general public (21.6%), and the *China International Trust and Investment Corporation* (CITIC, 20%). Subscribers could have unlimited local calls for a flat monthly fee of about $6. In 1993, the *Office of the Telecommunications Authority* (OFTA) was established as a regulator. In 1995, three companies were licensed to compete with HKTC: Wharf Holdings New T&T (Hong Kong) Ltd., Hutchison Communications Ltd., and New World Telephone Ltd. In addition to national service, initially only international callback service was permitted, followed by full international competition from January 1, 2000. In August 2000, HKTC merged with *Pacific Century Cyber Works*, whereby PCCW acquired most of the shares of C&W.

---

[22] VoIP was first used illegally in China in 1998 by the Chen brothers from their computer shop in Fujian. They made VoIP popular in China and could not be fined, as IP telephony was not explicitly prohibited under existing rules and regulations. Thus, the MPT, on March 20, 2000 (then the MII, Ministry of Information industry), issued licenses for commercial operation of VoIP to all three fixed network operators, which was then used primarily by China Unicom.

33.4.3 Telephony and Deregulation in Japan

The responsibility for the entire telecommunication services was taken away from the administration (which from then on retained only a regulatory and rate-control function) in 1952 and placed in the hands of two state-funded, yet privately operated public corporations: Nippon Telegraph & Telephone Public Corporation (NTT), responsible for the domestic service, and Kokusai Denshin Denwa Corporation Ltd. (KDD), responsible for the international services. The autonomy granted to the two telecommunications corporations (and their independence from the postal services and other financial and administrative constraints experienced by most national telecommunication administrations) resulted in a tremendous development of telecommunications in Japan.

Upon its foundation in 1952, NTT initiated a plan to construct a nationwide telephone network, with the objective "to provide telephones promptly when requested," "to provide connections anywhere," and "to eliminate the waiting time for telephone installation." The objectives were met within 25 years, thanks to an aggressive adoption of all domestic and foreign technological achievements, especially with the rapid implementation of transistor technology. The number of telephones increased from 1.7 million in 1950 to 10 million in 1968, and to 36 million by March 14, 1979, when the telephone network was fully automatic. The last manual exchanges replaced by automatic equipment were on the islands of Izu and Okinawa.

Video communication via the PSTN had begun in 1970 on leased circuits, but fewer than 1500 lines were in operation 20 years later. With the introduction of the prepaid telephone card in 1982, NTT initiated a new service that became popular worldwide. Within 10 years, NTT sold over 1 billion cards, the number of telephone calls increased tenfold, and the maintenance cost of the telephone booths—no longer an object of vandalism—was reduced to a minimum.

As the world's third telecommunications operator, NTT was privatized in 1985 as a publicly owned corporation, although still 65% owned by the government. Telephone subscribers no longer needed to rent the standard telephone from NTT but very soon could make their choice from a large selection offered in the stores.

The Ministry of Post and Telecommunications introduced competition with two new types of operators, called *new common carriers* (NCCs): type I NCCs, which use their own telecommunications infrastructure, and type II NCCs, which provide telecommunications services using the infrastructure of type I operators. New operators were, however, allowed only in the long-distance market, which represented, at most, 15% of the total telecommunications market. Two type I operators opened competition on international traffic with KDD, ITJ (International Telecommunications of Japan), and IDC (International Digital Communications). Three type I operators competed on the national long-distance market with NTT: DDI (Daini Denden Inc.), TWJ (TeleWay Japan Corp.; in 1999 TWJ merged with KDD), and JT (Japan Telecom).[23] Numerous type II operators started offering value-added services such as public database access, voice mail, and leased-circuit resale. By 1989, 768 type II NCCs existed.

The long-distance liberalization resulted in a 75% reduction of cost of calls originating in Japan and a substantial growth of international traffic. The local access

---

[23] Japan Telecom uses the optical fiber infrastructure of Japan Railways.

market, although opened to competition in 1996, was still 95% controlled by NTT in 1999. Strong protests from the World Trade Organization, from the U.S. government, and from the major international operators,[24] finally opened this last Japanese telecommunications bastion, on February 1999, when the Japanese telecommunications market, representing about 65% of the total Asian telecommunications market, was opened for competition. In July 2000, NTT reduced interconnection rates by 50% for long-distance access and 20% for local access. Many foreign companies immediately entered the market. IDC, owned by Toyota and Itochu, was the first Japanese telecommunications company to be bought by a foreign company. In a hostile takeover in June 1999, C&W obtained majority control of IDC, then named C&WIDC. This company established a nationwide network covering 80 cities and including metropolitan networks in Tokyo and Osaka, thus reaching 70% of the country's business customers and interconnecting with C&W's global network.

BT and AT&T acquired 30% of Japan Telecom. Moreover, in 1999, BT, in a joint venture with Marubeni, installed and operated broadband wireless networks in Tokyo, Osaka, and Nagoya.

MCI WorldCom, in a joint venture with Mori Building Co., installed and operated its own fiber optic network in Tokyo with direct access to major office buildings, thus offering direct local access and worldwide telecommunications using MCI WorldCom's network.

In 1990, the Deutsche Bundespost Telekom founded the *Deutsche Telekom Japan K.K.*, which established its own network in Japan, offering domestic and international long-distance services.

Flag Telecom of Japan, wholly owned by Bermuda-based Flag Telecom, opened an international telecommunications service connecting Tokyo with Flag Telecom's "link around the world."

In addition to those companies of foreign origin, new Japanese operators also entered the market. An example is *Crosswave Communications*, with Sony and Toyota as large shareholders. In 2000, Crosswave began to build a national backbone with rings in the major cities. Remarkably for the changes going on in Japan, Crosswave used imported material for its network, such as wavelength-division multiplex from the U.S. company Cisco and Sonet from Canadian Nortel. Another example is *PowerNets Japan Communications, Inc.*, a consortium of Japanese utility-based carriers, which linked their networks to provide data communications between Tokyo and Osaka.

NTT reacted to the new competitive situation on July 1, 1999, with a breakup into the following three companies:

1. *NTT East*, providing local (intra-prefecture) communications in he northeast Japan, including Tokyo
2. *NTT West*, providing local communications in southwest Japan
3. *NTT Communications*, providing long-distance communications

---

[24] The Japanese Ministry of Posts & Telecommunications received complaints about excessive connection fees from AT&T, British Telecom. Deutsche Telekom, France Telecom, Global One, and MCI World-Com.

One year later, NTT Communications, although conceived as a long-distance carrier, also decided to enter into the local market and thus to compete against its own two regional sister companies. Simultaneously, the two international operators KDD and Japan Telecom announced that they would start local services. As a novelty, since 1976 NTT has reduced its rates. NTT Communications reduced the out-of-town cost by about 20% in April 2000. In August 2000, NTT East and NTT West reduced the standard 3-minute rate for calls beyond 20 km from 90 to 40 yen but kept the 3-minute local call at the traditional 10 yen.

33.4.4 Telephony and Deregulation in Other Asian Countries

The most progressive deregulation, according to the ITU, was made in Malaysia, where in the *Communications and Multimedia Act* of 1999 the regulation of all communication and information technology functions was given to a single agency. Telekom Malaysia got five major competitors: Celcom, Binariang, Mobikom, Tine Telekom, and DiGi Telecommunications. Each of the operators has to contribute significant amounts to a central fund for universal service obligations, from which Telekom Malaysia is to serve rural and other less profitable areas.

Singapore has one of the most advanced telecommunication networks in the world. The national operator, Singapore Telecommunications Ltd. (SingTel), was partially privatized in 1993. In 1999, the *National Computer Board* and the *Telecommunication Authority of Singapore* (TAS) merged into the new regulator *Infocomm Development Authority* of Singapore (iDA). iDA awarded the company StarHub a license for public basic telecommunications services in a duopoly with SingTel from April 1, 1998 to April 1, 2002. Later this period was shortened, and full competition in the telecommunications market was introduced in April 1, 2000. Both SingTel and StarHub received compensation from the government for the advance ending of their duopoly. Starhub is a joint venture of Singapore Technologies Telemedia, Singapore Power, NTT, and BT. By that time, fixed services were also offered by MobileOne, a joint venture of C&W with Singapore Press Holdings, and Keppel Corporation. A license for full public telecommunications services was also given to the monopoly CaTV operator *Singapore Cable Vision* (SCV), which operates a nationwide broadband TV cable network that passes all households. A total of 58 licenses were issued for telecommunications services by April 1, 2000.

Thailand was probably the world's first country to separate telephone service from post and telegraph services when the *Telephone Organization of Thailand* (TOT) was established in 1954. Within 11 years, TOT increased the number of telephone lines from 10,000 to 55,000. Another world's first event took place in Thailand in January 2000 when TOT opened a nationwide VoIP network. Initially, the network with 100,000 subscribers connected Bangkok with 20 cities all over Thailand. Privatization of TOT is planned for 2002.

The *Korean Telecommunications Authority* (KTO) became a public company in 1982. In the meantime, privatized and named *Korea Telecom*, it lost its monopoly when the existing company Dacom (Data Communications Corporation of Korea) was permitted to offer international telephone services in 1995 and national services in 1997.

Indonesia, which in 1990 had 863,000 telephones, a teledensity of only 0.48, slowly increased its teledensity to 1 in 1993, to 2 in 1996, and to 3 (6.4 million tele-

TABLE 33.9   Telephone Growth in Europe

| Year | Millions of Telephones | | | Percent of Worldwide Total |
	Eurostat	Rest of Europe	Total	
1957	27.6	2.0	29.6	27.5
1976	107.1	25.1	132.2	34.9
1990	159.5	56.9	216.4	41.5
1997	201.1	87.6	288.7	36.6
2000	215.0	104.4	315.2	31.9

Source: Archiv für deutsche Postgeschichte, Vol. 1, 1977; Telecommunication Indicators in the Eurostat Area, 2000; ITU, Telecommunications Indicators, 2000; Siemens, International Telecom Statistics, 1993 and 1999.

phones) in 2000. The incumbent domestic operator PT Telkom got competition in the early 1990s when two international satellite operators were licensed: PT Indosat and Satelindo, Indonesia. To accelerate the network development, a *joint operations scheme* was initiated in 1995 whereby five joint operation partners with local and overseas firms were to build and operate 2 million telephone lines each until 2010. In August 2000 the government announced that PT Indosat would be given a license to operate on the local market beginning in August 2002 and one year later on the long-distance market. At that date, PT Telkom would also be permitted to operate on the international market.

33.5   TELEPHONY AND DEREGULATION IN EUROPE

Europe's share in the worldwide number of telephones increased from 25% (20 million telephones) in 1950 to 40% (315 million telephones) at the end of 2000. This growth was achieved primarily in the Eurostat countries,[25] which together always had more than double the number of telephones in the rest of Europe (Table 33.9).

The roots of telecommunications deregulation in Europe are in the U.K. The *General Post Office* (GPO) had had a monopoly since 1911. In 1968, the GPO became the British Post Office (BPO). In 1979, the conservative Prime Minister, Mrs. Thatcher, came to power with a firm commitment to privatization. British Petroleum (BP) was one of the first to be privatized, followed by C&W in November 1981. In the same year, *British Telecom* was divested from the BPO and became a public company. In 1982, a license was granted to C&W to run a public telecommunications network through its subsidiary, Mercury Communications, in competition with British Telecom. In 1984, British Telecom was privatized and 50.2% of its shares were sold to the public at an attractive price. All the government's remaining shares were sold in 1993. The duopoly of British Telecom and Mercury came to an end on March 5, 1991, when the telecommunications market in the U.K. was fully liberal-

[25] The Eurostat countries are a group of 18 countries for which the Statistical Office of the European Communities collects statistics. The countries are the 15 member countries of the European Union: Austria, Belgium, Denmark, Finland, France, Germany, Greece, Ireland, Italy, Luxembourg, The Netherlands, Portugal, Spain, Sweden, and the U.K.; and the countries of Iceland, Norway, and Switzerland.

ized.[26] From that year British Telecom took its new trading name, BT. In the U.K., now the "classical country of telecommunications liberalization," almost 100 operators compete with BT.[27]

33.5.1   Telephony and Deregulation in the European Union

In 1987, the *Commission of the European Communities* published a paper entitled "Towards a Dynamic European Economy: Green Paper on the Development of the Common Market for Telecommunications Services and Equipment," which outlined the way to gradual liberalization of the European telecommunications market. As foreseen in that paper, the impact of liberalization was evaluated in 1992, and the date of January 1, 1998, was set for full liberalization of telecommunications services in all countries of the European Union. Greece, Ireland, and Portugal got a grace period up to 2003.

Within 18 months, over 1000 licenses were issued to new telecommunications service providers. The incumbent operators acquired significant participation in those new national and foreign service providers. BT acquired 20 to 100% ownership in 10 new operators in 10 European countries; France Telecom in 10 operators in seven countries; Swisscom in seven operators in seven countries; Telia, Sweden in five operators in five countries; Tele Danmark in four operators in five countries; and Telenor, Norway, Telecom Italia, and Deutsche Telekom[28] each acquired ownerships in operators in three countries. Within that time, according to the European Commissioner in charge of liberalization, Mario Monti, the telephone charges went down by an average of 35% and 500,000 jobs had been created in the telecommunications sector. In Scandinavia the prices went down by 50% and in Germany by 70%. The liberalization increased the market volume significantly. One problem, however, was the reluctance of the incumbent operators to open the local loop to the new competitors, which the Commission of the European Communities made mandatory by January 1, 2001. By that date, in Germany less than 2% of the local loop was opened to competition, and in the U.K. after 18 years of competition, still only 20%.[29]

33.5.2   Telephony and Deregulation in Eastern Europe

Telephony in the Eastern European countries up to 1991 was dictated by the interests of the Communist Party. As such, telephony had a low priority, the telephone

---

[26] A special case is Kingston Communications (Hull), which is 100% owned by the council and thus is the only locally owned telecommunications operator in the U.K. The operator traditionally distinguishes itself from BT with cream-colored telephone booths instead of BT-red.

[27] Early in 2002, as a novelty in its history, BT got a non-British subject as its chief executive officer, Ben Verwaayen, who is Dutch.

[28] The German incumbent telecommunications operator was called Deutsche Bundespost up to December 31, 1989, Deutsche Bundespost Telekom beginning in January 1, 1990, and Deutsche Telekom AG beginning January 1, 1995.

[29] The European Competitive Telecommunications Association observed on February 2, 2002, that one year after mandatory local loop unbundling took effect, "less than 0.01 per cent of European incumbent's lines have been unbundled to new entrants" and "3 per cent of the 4.1 million DSL lines in operation are provided by new entrants over unbundled local loops."

TABLE 33.10   Teledensity Increase in Eastern Europe

Country	Teledensity 1991	Teledensity 2000	Country	Teledensity 1991	Teledensity 2000
Albania	0.18	3.91	Hungary	9.44	37.25
Bulgaria	23.05	35.04	Poland	8.64	23.24
Czechoslovakia	14.91	—	Romania	9.46	17.46
Czech Rep.	—	37.79	USSR	11.92	—
Slovak Rep.	—	31.42	Belarus	—	26.88
Yugoslavia	16.04	—	Estonia	—	36.33
Bosnia	—	10.29	Latvia	—	30.31
Croatia	—	36.49	Lithuania	—	32.11
Macedonia	—	25.49	Moldova	—	13.33
Slovenia	—	38.63	Russia	—	21.83
Yugoslavia	—	22.61	Ukraine	—	19.89

Source: For 1991: Political change opens telecommunication markets, *International Telcom Report* (Siemens), Vol. 4, 1993, pp. 19–21; for 2000: ITU, Database, Basic indicators 2000, *Telecommunication Indicators*, January 9, 2002.

networks were badly maintained, and outdated material was generally used for extensions if such extensions were made at all. The prevailing COCOM regulations (Section 32.3.1) prohibited access to modern technology and provided a barrier against Western competition for the national industry, which thus had little incentive for improvements. With the collapse of the communist system in 1991, both national telecommunication operators and telecommunication equipment manufacturers faced a tremendous task to modernize the telecommunications infrastructure and to extend its capacity to meet the increased demand. To obtain access to state-of-the-art equipment, modern management, and capital for financing the huge investments required, many of the national telecommunications operators and manufacturers made joint ventures with Western enterprises.

After decades of neglect, within one decade, most countries established modern digital telecommunication networks, which increased teledensity from about 10 to around 30 (Table 33.10). Deregulation was undertaken in Poland, the Czech Republic, and in Hungary. In Poland the incumbent Telekomunikacja Polska lost its monopoly on local voice services in 1991, when a second operator was licensed in most areas. The long-distance market was opened in 1999 and opening up of international services is announced for January 1, 2003. In the Czech Republic, Cesky Telecom lost its monopoly on fixed-line telephony on January 1, 2001 and the incumbent Màtav in Hungary on January 1, 2002.

After the collapse of the USSR, nearly 100 local monopolies, called *Electrosvyaz*, operated an estimated 25 million telephone lines in inefficient fixed-line networks. In Russia, 88 of those operators were grouped in a new holding company called Svyazinvest. This holding was partially privatized (25%) in 1997. By that time, Rostelecom provided the long-distance operation. Rostelecom, which was 51% owned by Svyazinvest, started a long-term infrastructure development, including a "50 × 50 project" for the implementation of 50 digital switching centers and a national 50,000-km fiber optic cable network. Modest improvements in telephone networks were

TABLE 33.11   Telephone Growth in Oceania

Year	Millions of Telephones				Percent of Worldwide Total
	Australia	New Zealand	Rest of Oceania	Total	
1957	1.8	0.6	<0.01	2.4	2.1
1990	8.0	1.5	0.2	9.7	1.8
1997	9.6	1.8	0.4	11.8	1.5
2000	10.1	1.9	0.4	12.4	1.3

Source: *Archiv für deutsche Postgeschichte*, Vol. 1, 1977; ITU, *Telecommunications Indicators*, 2000; Siemens, *International Telecom Statistics*, 1993 and 1998.

achieved in Russia, Modavia, and the Ukraine; significant improvements were achieved in the Baltic states of Estonia, Latvia, and Lithuania.

33.6   TELEPHONY AND DEREGULATION IN OCEANIA

The number of telephones in Oceania increased by about 10 million in the period from 1950 to the end of 2000 (Table 33.11). Australia and New Zealand took different deregulation approaches. New Zealand deregulated telecommunications in 1986 without establishing a regulator to control deregulation and without obligation for social subsidies. The incumbent *Telecom New Zealand* (TCNZ) was privatized in 1990 and got competition from Clear Communications Ltd., which then belonged to MCI, Bell Canada, TV New Zealand, and Todd Corporation and in the meantime is in full ownership of BT. Clear Communications used the optical fiber network of the railways and the radio-relay network of TV New Zealand. A second competitor, which entered the market a little later, was Saturn (then owned by the Australian TV operator Austar, later by Telstra, and then named TelstraSaturn). Saturn, in addition to its own optical fiber cable networks, uses poles of electricity utilities. In September 1990, TCNZ was sold to the Baby Bells Ameritech and Bell Atlantic. By year-end 2000, after several legal suits, mainly on interconnecting charges, TCNZ still had 99% of the local telephone market, with Clear Communications and TelstraSaturn sharing the rest.[30]

In Australia the Australian Telecommunications Authority (AUSTEL), in 1990 changed to the Australian Communications Authority (ACA), was established as an independent agency under the Telecommunications Act of 1989 with the objective of implementing deregulation under fair competitive conditions and safeguarding social objectives with basic telephone services. In 1991, the monopoly carriers Telecom Australia, for domestic services, and OTC (Overseas Telecommunications Commission), founded in 1946 for international services, merged to become Telstra Corporation. Simultaneously, a second operator, C&W Optus (52.5% owned by C&W), obtained a license for domestic operation. C&W Optus was also permitted to purchase the domestic satellite operator AUSSAT. The duopoly was fixed up to the end

---

[30] Early 2002, TelstraSaturn acquired Clear and then had 11% of the New Zealand's telecommunications market.

of 1997, when full competition started. By mid-1999, the Australian government sold 16.6% of its stock in Telstra, reducing its stake to 51%. At that time Telstra's part of the telecommunications market had been reduced to 75%.

REFERENCES

Books

Chapius, Robert J., *100 Years of Telephone Switching (1878–1978)*, Part 1: *Manual and Electromechanical Switching (1878–1960s)*, North-Holland, New York, 1982.

Chapius, Robert J., and Amos E. Joel, Jr., *Electronics, Computers and Telephone Switching: A Book of Technological History*, Part 2: *1960–1985* of *100 Years of Telephone Switching (1878–1978)*, North-Holland, New York, 1990.

Seeleman, Claus, *Das Post und Fernmeldewesen in China*, Gerlach-Verlag, Munich, 1992.

Solymar, Laszlo, *Getting the Message*, University Press, Oxford, 1999.

Terrefe Ras-Work, *Tam Tam to Internet: Telecoms in Africa*, Mafube Publishing and Marketing, Johannesburg, South Africa, 1998.

Articles

Anon., Global communications, Telecom 91 special issue, *Telecommunications International*, October 1991, pp. 1–246.

Anon., The global communications landscape, Telecom 95 special issue, *Telecommunications International*, October 1995, pp. 1–178.

Anon., The communications cosmos: a country-by-country survey of a global industry, Telecom '99 special issue, *Telecommunications International*, October 1999, pp. 1–191.

Anon., Direction of traffic trading telecom minutes, *ITU News*, August 1999, pp. 14–20.

Anon., Trends in telecommunication reform: convergence and regulation 1999, *ITU News*, August 1999, pp. 39–41.

Anon., Europe: new network operators, *ITU News*, Vol. 1, 2000, pp. 10–11.

Anon., Telecommunications in India, *ITU News*, Vol. 3, 2000, pp. 24–26.

Anon., IP telephony: China country case study, *ITU News*, Vol. 8, 2000, pp. 16–20.

Anon., Brazil country case study, *ITU News*, Vol. 10, 2001, pp. 23–28.

Anon., Africa reaches historic telecom milestone, ITU Telecommunication indicators update, *ITU News*, July–October 2001.

Basse, Gerhard, Die Verbreitung des Fernsprechers in Europe und Nordamerika, *Archiv für Post und Telegraphie*, Vol. 1, 1977, pp. 58–103.

Basse, Gerhard, Die Verbreitung des Fernsprechers in Lateinamerika, Afrika, Asien und Ozeanien, *Archiv für Post und Telegraphie*, Vol. 1, 1978, pp. 24–93.

Budde, Paul, Asian perspective: New Zealand decline and fall, *Communications International*, January 2000, pp. 42–43.

Green, E. I., Telephone, *Bell System Technical Journal*, Vol. 37, March 1958, pp. 289–325.

Kobelt, Christian, Ein Jahrhundert Telefon in der Schweiz, *Archiv für Post und Telegraphie*, Vol. 1, 1977, pp. 104–137.

Lerner, Norman C., Telecom privatization in Latin America: trends in competition and regulation, *Telecommunications International*, February 1998, pp. 59–67.

Lerner, Norman, Latin America and Mexico: a change in focus, *Telecommunications International*, March 2000, pp. 51–54, 64.

Ministry of Communications and Information Technology (Singapore)/IDA, Singapore towards full market competition in telecommunications, *ITU News*, Vol. 3, 2000, p. 31.

Ministry of Post and Telecommunications, China, Telecommunications rocketing in China, *ITU News*, Vol. 2, 1997, pp. 11–12.

Newlands, Mike, Advantage China, *Communications International*, January 2002, p. 33.

Shetty, Vineeta, Are you being served? Universal service, *Communications International*, March 2000, pp. 21–26.

Steffens, Marec Béla, and Lisa Sundrum, Political change opens telecommunication markets, *Telcom Report* (Siemens), Vol. 4, 1993, pp. 19–21.

Telecommunications Industry Division, Department of Communications and the Arts, Australia, Liberalization of the telecommunications sector: Australia's experience, *ITU News*, Vol. 1, 1998, pp. 21–26.

Utsumi, Yoshio, Facing our future head-on: end of year message, *ITU News*, Vol. 10, 2001, pp. 2–3, 7–8.

Wheelwright, Geoff, Canadian shield, *Communications International*, September 1997, p. 17.

Wittering, Stewart, Privatising Central America, *Communications International*, February 1994, pp. 46–50.

Wittering, Stewart, Antel's vote of confidence, *Communications International*, September 1996, pp. 25–26.

Internet

www.europe.eu.int/comm/eurostat/Public/datashop, Market structure in network industries, fixed telecommunications, January 11, 2001.

www.itu.int/itu-D/ict/statistics/index/html, ITU telecommunication indicators.

www.totaltelecom.com.

Brooks, Andrew, Billion-dollar alliance of four to take on Bell Canada, March 22, 1999.

Bourne, Emily, U.S. marriage sees its future on the "Verizon," April 4, 2000.

Hurst, Richard, South Africa sets up new media regulatory body, June 9, 2000.

U.S. ready to approve mammoth merger, June 15, 2000.

Reuters staff, Indonesia brings forward end of telephone monopoly, August 3, 2000.

Ortiz, Fiona, Mexico's Avantel demands end to Telmex concession, August 3, 2000.

Melnbardis, Robert, Bell Canada, BCE mirrors AT&T with four-way restructure, October 26, 2000.

Reuters staff, China Netcom launches network and confirms financing, November 16, 2000.

Taaffe, Ouida, New Chinese operator on track, December 4, 2000.

34

MULTIMEDIA

34.1 EVOLUTION LEADING TO MULTIMEDIA

Multimedia results from the convergence of information technology and telecommunications. Basically, multimedia is the technology of processing voice, audio, text, still and live images, and data in an integrated format for presentation at a common terminal. Multimedia networks give access to databanks, bank accounts, shopping catalogs, educational services, medical assistance, and various other services.

In the 1980s, two early forms of multimedia services emerged, teletext and videotex, with electronic still-image transmission services which provided access to computer-supported databases. In *teletext*, still images were transmitted in several scan lines of a television signal and stored in a memory for subsequent display on a television screen. Teletext is still used in television systems.

In *videotex*, still images are transmitted digitally over the public-switched telephone network. At the receiver, the digital signal is recovered from a modem and stored in a local memory, again for subsequent display on a television screen. The first videotex system was the French Télétel system, which after a decade of development was introduced in 1982. It became very popular. In a unique approach the French administration provided the terminals, named *Minitel*, free of charge to the subscribers. Initially, the service offered an electronic telephone directory and a limited number of information pages. It expanded quickly, and creating a Minitel service became a priority for French businesses. It reached 3.3 million subscribers within seven years and 6.3 million subscribers and over 17,000 servers in 1992.

Minitel use peaked in 1998, and despite fierce competition from the Internet, 26% of French households were still connected with Minitel and only 17% to the Internet at the end of the century. The Minitel services can be divided into four broad cate-

The Worldwide History of Telecommunications, By Anton A. Huurdeman
ISBN 0-471-20505-2   Copyright © 2003 John Wiley & Sons, Inc.

gories. Thirty-six percent of the services are practical: timetables, e-commerce, general information, and directories. Entertainment, including chat, dating services known as *Minitel rose*, astrology, and games, accounts for 21%; financial services amount to 22%; and 21% are professional database services.

Even if Minitel, with only 1 million videotex subscribers outside France, was not successful internationally, it obtained worldwide fame. U.S. Vice-President Al Gore, a long-time promoter of global information flow, wrote his thesis on the Minitel system. By the end of the century, videotex, with the exception of Minitel, was integrated into Internet.

In the 1990s, the telecommunication networks went through an evolution whereby in a convergence of *computers and communications* and in interaction with the construction of a *global information infrastructure*, a global network emerged called the *Internet*, which hopefully will unite the world population in a *global village* in which *multimedia* services can give everybody the right answer to her or his question in a matter of seconds at an affordable cost.

34.2   COMPUTERS AND COMMUNICATIONS

The concept of *computers and communications* (C&C) was first presented in 1977 by Koji Kobayashi, then chief executive officer and president of the board of NEC, at Intelcom 77 (International Telecommunication Exhibition) held in Atlanta, Georgia. Kobayashi envisaged that computers would become an integral part of communications networks. He predicted that computers would merge to processors distributed in total communications systems with digitized switching and transmission. He gave lectures in Atlanta and at other international conferences: in 1978 in San Francisco, 1981 in Washington, 1982 in Helsinki, and 1983 at the 4th World Telecommunication Forum held in Geneva. The lectures were compiled in an interesting book published in 1985 by the Massachusetts Institute of Technology (MIT) under the title *A Vision of C&C Computers and Communications*. In that book, Kobayashi also describes the basic principles of information generation, storage, and transfer for the media: voice/acoustics, numerical data, text, still images such as graphics, scanned images and natural pictures, and moving images with animation and natural moving pictures. Most of his vision has been realized in current multimedia systems.

34.3   GLOBAL INFORMATION INFRASTRUCTURE

The concept of a *global information infrastructure* (GII)[1] was presented by Al Gore (then U.S. Vice-President) at the World Telecommunication Development Conference of the ITU held in Buenos Aires, Argentina, March 21–23, 1994. In his proposal,[2] Gore set forth five principles that he believed to be essential for successful construction of the GII:

---

[1] Al Gore presented the idea of an *information superhighway* in 1985 on the occasion of the thirtieth anniversary of the signing of the Interstate Highway Act. In 1991 he wrote: "A nationwide network of information superhighways is needed to move the vast quantities of data that are creating a kind of information gridlock."

[2] Published in *ITU News*, September 1998, "Al Gore's Five Challenges to the Telecom World," pp. 3–7.

1. To encourage private interest
2. To promote competition
3. To create a flexible regulatory framework that can keep pace with rapid technological and market changes
4. To provide open access to the network for all information providers
5. To ensure universal service

Four years later, on October 12, 1998, speaking at the Plenipotentary Conference of the ITU in Minneapolis, Gore reported on the enormous progress made on these five principles in constructing the GII:

1. *Private interest.* Over $600 billion of private capital has been invested in telecommunications, and over 48 telecom operators were privatized.
2. *Competition.* Whereas in 1994, only seven countries had competitive markets for the basic voice service, in 1998 as many as 47 countries either had already full competition or were committed to it. In the United States alone, competition in telecommunications created over 50,000 new jobs.
3. *Regulatory framework.* Additional independent regulatory agencies, promoting competition and investment while protecting the interest of the customers, have been established as follows: 18 in the Americas, 17 in Africa, and 11 in the Asia-Pacific region.
4. *Open access to the telecommunication network.* Open access enabling every user of the GII to reach thousands of different sources of information from various countries, in every language, was in daily practice in 1998 for over 100 million Internet users.
5. *Universal service.* More than 200 million telephone lines could be added worldwide, thus, over the last four years more telephone lines were added than were installed in the first 100 years of telephony.

It is in the United States, U.K., France, Germany, and Japan that plans for broadband information highways are most advanced and a dozen experimental information highways have been launched.

In the United States an *Information Infrastructure Task Force* (IITF) coordinates the technological and legislative activities to construct a *National Information Infrastructure* (NII). In Europe, the European Union worked out guidelines and priorities for a *European Broadband Infrastructure* to interconnect with all the European telecommunications and CATV networks and to provide file transfer, e-mail, video, and other multimedia services to all European Union citizens. In Japan, some 20 companies, including NTT, NEC, Hitachi, and Fujitsu, established a multimedia observatory for the elaboration of a national information network. Taiwan and Hong Kong also implemented national multimedia networks. In China, a nationwide optical fiber network 200,000 km long was installed connecting the Chinese provinces with each other and with the GII.

By the end of the century, much progress has been made on implementation of the GII as described in Chapters 27 and 28. However, the telecommunication infrastructure in developing countries needs substantial improvement to prevent a *digital divide*.

34.4   INTERNET[3]

The origin of the Internet can be traced back to October 4, 1957, when *Sputnik* shocked U.S. scientists, engineers, and politicians. President Dwight David Eisenhower appointed James A. Killian (then president of MIT) as presidential assistant of science, and one year later, the U.S. Department of Defense established its *Advanced Research Projects Agency* (ARPA). The origin of the new technology applied in the Internet of *packet transmission and switching* goes back to July 1961, when Leonhard Kleinrock of MIT wrote the first paper on packet-switching theory, entitled "Information Flow in Large Communication Nets." He also published those ideas extended with network design criteria in his book *Communication Nets* in 1964.

The concept of the Internet goes back to August 1962, when Joseph Carl Robnett Licklider and Welden Clark of MIT wrote a paper about a computer network entitled "On-Line Man–Computer Communication." Licklider also wrote memos in which he proposed a *galactic network* with interconnected computers which would quickly provide every user with access to data and programs from any site. In October, ARPA began a computer program and appointed Licklider as first director of its newly founded *Information Processing Techniques Office* (IPTO). Licklider left ARPA in September 1964. A historic meeting then took place at Homestead, Florida, in November 1964, where Licklider could convince his ex-colleague at MIT, Lawrence C. Roberts, of the merits of a reliable computer network based on packet switching. In February 1965, Ivan Sutherland, successor to Licklider at ARPA, placed a contract with Roberts for the establishment of such an experimental computer network. Together with Thomas Marill, Roberts connected a TX-2 computer at MIT Lincoln Laboratories successfully via an acoustic coupler and a telephone line with a Q-32 computer at Santa Monica, California, in October 1965. This was the first time that two computers talked to each other by interchange of packets of information. The results were published one year later in a paper entitled "Toward a Cooperative Network of Time-Shared Computers."

Roberts left MIT and joined ARPA as the chief scientist of IPTO in December 1966 with the task of implementing a reliable computer network. In October 1967, at a symposium of the *Association for Computing Machinery*, he presented a proposal for a packet-switched computer network named *Arpanet* in a paper entitled "Multiple Computer Networks and Intercomputer Communication." At that symposium, papers on packet switching for computer networks were also presented by the British inventor of packet switching, Donald Watts Davies, and Roger Scantlebury of the *National Physical Laboratory* (NPL) of the U.K., and by Paul Baran and others of the American Rand Corporation. Davies, in his paper, introduced the concept of a message *packet*. Baran had already written a paper on secure packetized voice in 1964 entitled "On Distributed Communication Networks," in which he made a proposal for a network configuration that would survive nuclear attacks.[4]

[3] The history of the Internet can be found on the Internet in at least 20 versions with partially confusing contradictory information. In an effort to summarize the history of the Internet as accurately as possible, this section is based primarily on the information presented in the Internet by Internet co-founder Lawrence G. Roberts, the Computer Museum History Center in California, and the IEEE.

[4] From this paper it is frequently claimed that the Internet was created by the military to withstand nuclear war. In his *Internet Chronology*, Roberts clarifies that the Arpanet and Internet stem from MIT work of Licklider, Kleinrock, and Roberts and had no relation to Baran's work.

ARPA released a *request for quotations* (RFQ) in August 1968 for the main packet-switching design element of Arpanet called the *Interface Message Processor* (IMP). The company, Bolt, Beranek & Newman (BBN) in Cambridge, Massachusetts, with a team under Frank Heart,[5] won the RFQ and supplied the first four IMPs in 1969. Arpanet was established with those IMPs in the same year. Leonhard Kleinrock[6] at the *University of California–Los Angeles* (UCLA) made the first node in September with a Sigma 7 computer used as the *Network Measurement Center* of Arpanet. A second node followed in October with a SDS 940 computer at the *Stanford Research Institute* (SRI), which was used by Dough Engelbart's[7] group while working on a project called *Augmentation of Human Intellect*. With the completion of those two nodes, the world's first long-distance computer host-to-host[8] communication was made on October 25, 1969. A third node was connected in November at the *University of California–Santa Barbara* (UCSB), where computer graphics were developed, and a fourth node in December at the University of Utah. The transmission speed was 50 kbps. Arpanet spanned the United States from coast to coast from March 1970 when a fifth node was connected at BBN. In December, Steve Crocker at UCLA completed an initial Arpanet host-to-host protocol, called the *Network Control Protocol* (NCP). Ray Tomlinson at BBN sent the first e-mail, with a program called Readmail in March 1972. For addressing e-mails he introduced the @[9] symbol between user name and host name (being a secure symbol not used in any of those names). In the same year the first public demonstration of the Arpanet was given at the *International Conference for Computer Communication* held at Washington in October. It was a great success and proved the feasibility of packet switching.[10]

ARPA changed its name in 1973 to DARPA, the "D" standing for Defense (the Department of Defense). The University of Hawaii installed a packet-switched radio net, named *Packet Radio Net*, which connected seven computers on four islands using the Aloha mode of access (see Technology Box 29.4). Around the same time a satellite network named Satnet was established. It connected Arpanet with University College, London, via an Earth station at Etam, West Virginia, the Intelsat satellite 4A, and Earth station Goonhilly Downs, Cornwall, U.K. Another connection was with the Norwegian Royal Radar Establishment "Norsar" via the Earth station

[5] The team under Frank Heart that developed the IMP included Bob Barker, Bernie Cosell, Will Crowther, Bob Kahn, Severo Ornstein, and Dave Walden.

[6] The team that produced the software to enable communication between their computer and the IMP consisted of Vint Cerf, Steve Crocker, and Jon Postel.

[7] Dough Engelbarth was also the inventor of the computer mouse.

[8] The term *host* is used for a computer that is connected permanently to the Internet. Internet users, by contrast, create a temporary Internet connection when they log on.

[9] The @ symbol is used in the English language as "commercial a" and as such was already included on the keyboards of the first American typewriters. The origin of this symbol dates at least back to the year 1555, when it was used in Spain as a measure unit for a weight of about 11.5 kg called "arroba." Other contemporary names for @ are "aape staart" (monkey tail) in Dutch, "Affenklammer" (monkey clip) in German, "kanelbolle" (cinnamon pastry) in Norwegian, "kukac" (worm) in Ungarian, "Miukumauku" (cat's tail) in Finnish, "petit escargot" (little snail) in French, "shtrudel" (a "bended" pastry) in Israeli, and "sobachka" (little dog) in Russian.

[10] A highlight of the ICCC was a conversation between Eliza, Joseph Weizenbaum's artificially intelligent psychiatrist located at MIT, and Parry, a paranoid computer developed by Kenneth Colby at SRI.

at Tanum, Sweden. In that year, Robert E. Kahn came from BBN to DARPA, where he developed a protocol for communication between Arpanet, Packet Radio Net, and Satnet in cooperation with Vinton G. Cerf, who was then a professor at UCLA. Cerf and Kahn named communication among the three networks *internetting*, from which the word *Internet* was derived. In September they presented a first paper on the new protocol, called *Transmission Control Protocol* (TCP), at a meeting of the newly formed *International Networking Group* at the University of Sussex, U.K. TCP was developed further into a combination of two protocols named TCP/IP. One was a new TCP controlling transmission and error-correcting functions, and the other was an *Internet Protocol* (IP), controlling the addressing, routing, and forwarding of individual packets.

Lawrence Roberts left DARPA in 1974 and joined BBN, where he directed the world's first public packet-switching company, called Telenet. Licklider agreed to return temporarily from MIT to replace Roberts. Vinton G. Cerf also joined DARPA in 1976 and demonstrated one year later, together with Robert Kahn, the use of TCP for communication between Arpanet, Packet Radio Net, and Satnet. Messages were sent by radio from a van in California, across the United States on Arpanet, then via Satnet to University College, London, back via satellite to Virginia, and through Arpanet to the *University of Southern California's Information Sciences Institute* (USC/ISI).

In 1978, when over 100 hosts were connected to Arpanet, Vint Cerf initiated the foundation of an *Internet Configuration Control Board* chaired by Dave Clark of MIT and an *International Cooperation Board* chaired by Peter Kirstein of University College, London.

In 1980, the *National Science Foundation* (NSF) of the United States established a *Computer Science Research Network* called CSNET, which within six years connected the computer science sites of almost all universities and academic institutions in the United States. The network included an interconnection based on TCP/IP with Arpanet, Telenet, and an e-mail service called PhoneNet.

In January 1983, DARPA finalized the transition from NCP to TCP/IP protocol throughout Arpanet and subsequently separated the Arpanet into a public network, then formally called the Internet, and a new military network called Milnet. In the meantime, various other computer networks were established which all gradually incorporated TCP/IP, and as a loose collection of networks, evolved into the Internet. To facilitate communication among the almost 1000 hosts within the Internet, a *Domain Name System* (DNS) was developed in 1983 by Jon Postel and Paul Mockapetris of USC/ISI and Craig Partridge of BBN. Seven generic *top-level domain* names (TLDs) were allocated as a symbol consisting of a point plus three letters as follows: open or unrestricted: .com, for commercial organizations; .net, for networks; and .org, for organizations; and restricted: .edu, for education; .gov, for government; .int, for international organizations; and .mil, for military. The standard e-mail addressing introduced by Ray Tomlinson in 1972 extended with this domain thus became *name@host.domain*.

In 1985, the NSF added five supercomputing centers to its CSNET, in the meantime named NSFNET, which were accessible by the other computers at a transmission speed of 56 kbps.

The Internet, so far, apart from satellite connections with University College, London and Norsar, was limited to the United States. In 1988, when over 10,000 hosts were connected, the Internet was opened to Canada, Denmark, Finland, France, Iceland, Norway, and Sweden. The domain name system was then extended by a point plus two-letter symbol indicating the country[11]: for instance, ".ca" for Canada and ".se" for Sweden. The next year the following nine countries got access to he Internet: Australia, Germany, Israel, Italy, Japan, Mexico, The Netherlands, New Zealand, and the U.K. One year later, the Internet was joined by Argentina, Austria, Belgium, Brazil, Chile, Greece, India, Ireland, South Korea, Spain, and Switzerland. With this international opening, and also because the NSF opened Internet to commercial use, the number of hosts increased to 300,000 in 1990. In that year, Arpanet was formally closed and its service was taken over by the NSFNET. The transmission speed on the NSFNET was increased to 45 Mbps in 1993, and one year later to 155 Mbps.

Around 1990, a revolutionary contribution toward the worldwide penetration of the Internet came from Switzerland. A British computer scientist, Tim Berners-Lee, working at the European Laboratory for Particle Physics in Geneva, Switzerland, dubbed CERN (from its French name, Conseil Europeen pour la Recherche Nucleaire), developed a concept for a user-friendly navigation method providing access to information wherever it is stored within the Internet. He started in 1989 with the development of a *hypertext markup language* (HTML). In the following two years he developed in cooperation with Robert Caillau the *HyperText Transfer Protocol* (HTTP) and a *uniform resource locator* (URL). Furthermore, he completed the concept with a browser developed by a British student, Nicola Pellow, then a technical student at CERN; and a server developed by Bernd Pollermann of the Computing and Network Division of CERN. Much simplified, using HTML, words, pictures, and sound can be combined on a computer screen using hypertext,[12] whereby individual pieces of text within a page can be given a URL address leading to related text. The term *hypertext* probably was coined by Theodor Holm Nelson in 1963 and used first in his book *Literary Machine* in 1965. Upon activation of that address with a mouse, a browser follows hyperlinks using the HTTP until it has found the related page in a "web" of further pages at a server which may be located at the same or at different hosts around the world. The name World Wide Web (WWW), referring to the web of distributed information, was coined by Tim Berners-Lee in 1990.

In 1992, Marc Andreesen and a group of student programmers developed an improved browser called *Mosaic* on a summer program at the University of Illinois at Urbana–Champaign organized by the *National Center for Supercomputing Applications*. Commercialization of Mosaic soon followed. Jim Clark used it to establish the browser company Netscape Communications Corp. in 1994, followed by the Microsoft Explorer, integrated in Windows 95 in 1995. With this browser facility,

---

[11] The two-letter code was taken from ISO 3166, a standard of the International Organization for Standardization. At the beginning of the twenty-first century there were 243 country code top-level domains (ccTLDs).

[12] Hypertext became the subject of a patent lawsuit in February 2002. BT claims that the version of hypertext for which it obtained a U.S. patent in 1989 was developed by BT in the 1970s and thus should not have been used without license by the oldest U.S. online access service provider, Prodigy, in 1984 nor subsequently by millions of other companies and persons. A serious argument against the BT claim might be the existence of a video clip in which Douglas Engelbart in 1968 at the SRI demonstrates how by clicking on certain words in a computer program a new page of text appears. This could be the first example of hypertext linking.

Internet found worldwide use on millions of scientific, commercial, and residential PCs and Apple Macintosh computers.

Introduction of the WWW and its browser facilities was a most valuable contribution toward construction of the GII. To reach 50 million subscribers took the telephone close to 75 years, the radio 38 years, TV 13 years, and the Internet less than four years from the introduction of the WWW in 1992. In the same time, Internet evolved from a scientific to a mainly commercial network, but what exactly is the Internet? The *Federal Networking Council* (FNC) of the Internet saw the necessity to answer this question. On October 24, 1995, in a resolution, it defined the term *Internet* (rather heavily) as follows: Internet refers to the global information system that:

1. Is logically linked together by a globally unique address space based upon the Internet Protocol or its subsequent extensions/follow-ons;
2. Is able to support communications using the Transmission Control Protocol/Internet Protocol (TCP/IP) suite or its subsequent extensions/follow-ons; and/or other IP-compatible protocols; and
3. Provides, uses or makes accessible, either publicly or privately, high level services layered on the communications and related infrastructure described herein.

In the same year, the National Science Foundation announced that it would no longer allow direct connection to the NSFNET. It contracted private companies to provide that access against payment. Since that time, thousands of *Internet service providers* (ISPs) worldwide provide access to the Internet on a commercial basis.

Vinton G. Cerf, often called the "father of the Internet," and Robert E. Kahn were awarded the U.S. National Medal of Technology in December 1997 by President Bill Clinton for having founded and developed the Internet.[13]

Around the same time, digital watermarking was introduced to protect copyrights of digital images and to reduce digital pirating. The digital watermark is visible only with special software. It operates two-way: It provides additional information to the user and informs the owner where their images are copied and reappear in the Internet.

To keep pace with this tremendous development and the increasing internationalization of the Internet, the U.S. government decided in 1998 to take administration of the domain name system and some other tasks away from the *National Telecommunications and Information Administration* and bring it under the responsibility of a new independent nonprofit organization, the *Internet Corporation for Assigned Names and Numbers* (ICANN). In October 2000, ICANN had five of its nine directors elected in a worldwide online election which was open to anyone who cared to register for the election. This was the first worldwide democratic election, although as a demonstration of the shortcomings of the Internet: Only 158,000 people registered, only 76,000 of them managed to proceed to the voting, and only 34,035 of them succeeded in casting their vote. One month later, the new board of ICANN selected seven additional top-level domain names for introduction in 2001: .biz, for business

---

[13] One year earlier, Sigrid Cerf, who was profoundly deaf, got a cochlear implant and could for the first time after 30 years of marriage talk naturally to her husband, Vint Cerf, who is hearing-impaired. The Internet played a vital role in selecting the successful bioelectronics solution.

purposes; .info, for unrestricted; .name, for personal names; .aero, for the aviation community; .coop, for cooperatives; .museum, for museums; and .pro, for professionals.

In the same month, ICANN, in coordination with *the Internet Engineering Task Force* (IETF), extended internationalization of the Internet. It opened the domain name system, which was confined to the English language,[14] to Arabic, Chinese, Japanese, Korean, Portuguese, Spanish, and the Scandinavian languages.[15] Multilingual domain names were first developed at the National University of Singapore. In July 1998, the Asia-Pacific Networking Group started a project called *Internationalized Multilingual Multiscript DNS* (iDNS). By year-end 1999, the results were promoted and commercialized under *www.idns.net*. The first Internet portal in the Spanish language, T1msn, was opened in March 2000 with the cooperation of Microsoft and the Mexican telecommunications operator Telmex. In the same month, the Brazilian media company Globo opened a portal in the Portuguese language under *www.globo.com.br*. One month later, domain name registration began in the Japanese language through the Japan registry service under *www.jprs.jp*. In the same month, *Emirates Internet and Multimedia* of the United Arab Emirates and the U.S. Compac Computer Corporation launched the first Internet search machine in the Arabic language called Arabvista, on *www.arabvista.com*. The service includes Arabic-to-English translation, allowing Arabic-speaking users to search English content on the Internet using their native language, and it offers a virtual Arabic keyboard for non-Arabic systems.[16]

A major milestone was reached in the year 2000 when the ITU and the IETF, after one year of intensive work, agreed on a single standard, H.248/Megaco, for a gateway between circuit-switched networks and IP-based networks. With devices based on this standard it will become possible to make voice calls via the Internet (dubbed VoIP, voice over Internet Protocol) from a conventional telephone. VoIP existed since the mid-1990s, but with nonstandard devices, of which a common version is required for the two participants of a VoIP call. VoIP found limited application because of this limitation, the poor quality, and above all the significant price reductions of telephone calling on the PSTN due to increased competition upon deregulation of telephone services. As further support for VoIP, the IETF released the Stream Control Transmission Protocol (SCTP), which facilitates the transmission of signaling system 7 signals (SS7; see Section 29.3.1) via the Internet.

When the domain name system was introduced in 1983, a universal addressing system was developed to allow any type of user to identify itself on the Internet. For this purpose, a special protocol, named IPv4, was developed using a 32-bit code that offered 4.2 billion addresses. It is expected that there will be a lack of addresses around 2005 due to the introduction of multimedia services on the 2.5G and 3G

[14] DMS was confined to the characters defined by ASCII (American Standard Code for Information Interchange). In an ASCII file, each alphabetic, numeric, or special character is represented with a 7-bit binary number (a string of seven 0s or 1s).

[15] Currently, ICANN is struggling for survival. ICANN's president, Stuart Lynn, concluded in March 2002 that ICANN has failed to meet its objectives and that the most productive course for ICANN's role in Internet affairs would be to discontinue.

[16] Deployment of multilingual domain names since 2001 is promoted by the Arabic Internet Names Consortium (AINC), the Chinese Domain Name Consortium (CDNC), the International Forum for IT in Tamil (INFITT), and the Japanese Domain Name Association (JDNA).

TABLE 34.1   Internet Development, 1990–2000

Year	Number of Countries	Millions of Hosts	Millions of Users	Year	Number of Countries	Millions of Hosts	Millions of Users
1990	22	0.4	—	1996	165	22	54
1991	34	0.7	—	1997	191	30	90
1992	43	1.3	—	1998	200	43	149
1993	60	2.7	—	1999	211	72	231
1994	81	5.8	—	2000	214	104	315
1995	121	14	34				

Source: ITU Telecommunication Indicators Update, January–February–March 2001.

mobile radio systems, the application of Internet for control of "smart houses," and other advanced services. Due to initial generous allocations, IPv4 cannot be used to its full capacity, and a new protocol, IPv6, is being developed by the IETF. IPv6 will use a 128-bit code which can offer 340×10^{36} (340 billion billion billion billion[17]) Internet addresses! The new protocol will also improve security, quality of service for real-time data, and facilitate roaming on mobile networks.[18]

Internet started in the brains of scientists at the MIT as an instrument for scientific research and information exchange. It became the world's biggest commercial service network. Now a group of U.S. universities has begun a project named *Internet 2* (I2) with the objective of developing a second-generation Internet. The project is coordinated by the *University Cooperation for Advanced Internet Development* (UCAID), which brings together over 150 universities working with partners in industry and government. One of the priority items will be the deployment of IPv6.

Aiming even further, Vint Cerf, in cooperation with the Jet Propulsion Laboratory at Pasadena, California, is working on the design of an *Interplanetary Internet* (II). With the Interplanetary Internet, each planet will get its own addressing system. One of the challenges is the 40-minute round trip of a radio signal between Earth and Mars. Table 34.1 shows the enormous development of the Internet.

34.5   GLOBAL VILLAGE

The idea of the *global village* was first conceived by the Canadian communications theorist Marshall McLuhan (1911–1980) in 1960. Looking at the emerging spread of electronic telecommunications and mass media such as television, McLuhan wrote that these technologies are transforming the world's nations into a global village in which distance and isolation are eliminated. In the global village, McLuhan envisaged that all information will become available for everyone and that local events will acquire worldwide significance.

Despite the tremendous development of the Internet, which connects practically all countries of the world, at the end of the twentieth century the world was far from being a global village. The major reason was, and still is, that a lack of infrastructure

---

[17] The exact number is 340,232,366,920,938,463,463,374,607,431,768,211,456.

[18] NTT communications started commercial Ipv6 service on April 1, 2002.

TABLE 34.2   Ten Countries with the Highest Internet Penetration, Year-End 2000

Country	Thousands of Users	Percent Penetration	Country	Thousands of Users	Percent Penetration
Iceland	168	60	Finland	1,927	37
Norway	2,200	49	Denmark	1,950	37
Sweden	4,048	46	United States	95,354	35
Canada	12,700	41	Australia	6,600	34
Korea (Rep.)	19,040	40	Singapore	1,200	30

Source: ITU Telecommunication Indicators Update, January 9, 2002.

and a lack of education cause a *digital divide* between countries that are rich in information and countries that are poor in information. In a study released in May 2000, the UN *Economic and Social Council* observed: "There are more Internet hosts in New York than on continental Africa, more hosts in Finland than in Latin American and the Caribbean, and, notwithstanding the remarkable progress in the application of information and communication technology in India, many of its villages still lack a working telephone." The ITU, in the *Telecommunications Indicators Update* of January 2001, state that in the developed countries almost a third of the people are online compared with less than 2% in the developing countries. Table 34.2 shows the 10 countries with the highest Internet penetration. An impression of the digital divide can be obtained from Table 34.3, which indicates the very unequal worldwide distribution of the 366,611,200 Internet users and the 106,710,508 Internet hosts by year-end 2000.

A further indication of the orientation of the Internet toward the United States is given in Table 34.4, which for the 10 top cities shows the bandwidth capacity of the links that connect those cities on interregional links with each other and with other, smaller hub cities. Although Europe and Asia have major Internet hubs, most Internet traffic between Asia and Europe still passes through the United States.

34.6   MULTIMEDIA SERVICES

Subscribers in a multimedia network can access various services on demand, which are offered by different content providers via a multimedia service provider. Figure 34.1 shows the basic configuration of a typical multimedia network. The first multimedia services on the Internet started in the United States in the late 1980s with teleworking, soon followed by telemedicine, telebanking, telebooking, teleshopping, teleconsulting, telelearning, tele-entertaining, and others.

The driving force for teleworking from a business point of view was cost saving. From the residential point of view, it was the reduction of traveling time and a better balance between business and private life. In the United States there were some 8 million teleworkers by 1995 and 35 million by year-end 2000, when Europe had about 9 million teleworkers. The highest rates of teleworking, about 15% of the workforce, was in the northern countries: Finland, The Netherlands, and Sweden, whereas in Italy and France this rate was only about 4%.

TABLE 34.3   Worldwide Distribution of Internet Hosts and Users

Country, Region, or Continent	Population (millions)	Percent Penetration		Percent of World Distribution		
		Users	Hosts	Population	Users	Hosts
Africa						
South Africa	43.69	5.5	4.3	0.7	0.7	0.2
Rest of Africa	748.94	0.3	<0.01	12.2	0.6	<0.1
Total Africa	792.63	0.6	0.3	13.0	1.3	0.2
The Americas						
Canada	30.75	41.3	7.7	0.5	3.5	2.2
United States	275.13	34.7	29.3	4.5	26.0	75.5
Rest of the Americas	519.26	3.4	0.4	8.5	5.2	1.9
Total Americas	825.14	15.4	10.3	13.5	34.7	79.6
Asia						
China	1295.33	1.7	<0.01	21.3	6.1	<0.01
Hong Kong SAR	6.73	38.6	3.4	0.1	0.7	0.2
India	1012.40	0.5	<0.01	16.6	1.4	<0.01
Israel	6.27	20.3	2.9	0.1	0.4	0.2
Japan	126.92	37.1	3.7	2.1	12.8	4.3
Korea (Rep.)	47.3	40.3	0.8	0.8	5.2	0.4
Malaysia	23.27	15.9	0.3	0.4	1.0	0.1
Singapore	4.02	29.9	4.4	<0.1	0.3	0.2
Taiwan–China	22.28	28.1	4.9	0.4	1.7	1.0
United Arab Emirates	2.61	28.2	1.3	<0.1	0.2	<0.1
Rest of Asia	1105.74	0.8	0.02	18.1	2.4	0.2
Total Asia	3648.85	3.3	0.2	59.9	32.2	6.6
Europe						
Eurostat countries	389.44	24.2	2.3	6.4	25.7	10.5
Russia	146.93	2.1	0.2	2.4	0.8	0.3
Rest of Europe	261.89	4.5	0.4	4.3	3.2	0.9
Total Europe	798.26	13.7	1.6	13.1	29.7	11.7
Oceania						
Australia	19.16	34.4	8.4	0.3	1.8	1.5
New Zealand	3.83	21.7	9.9	0.1	0.2	0.3
Rest of Oceania	7.47	2.7	0.2	0.1	0.1	0.1
Total Oceania	30.46	25.1	6.5	0.5	2.1	1.9
Total worldwide	6095.34	6.1	1.8	100	100	100

Source: ITU Telecommunication Indicators: Basic Indicators, and *Information Technology, Update*, January 2, 2002.

Teleworking produced two new office types: virtual offices and nonterritorial offices. A *virtual office* is any individual place of teleworking: at home, at the beach, at a business lounge of an airport, and others. A *nonterritorial office* is a corporate office where teleworkers can have occasional social contact with their superiors, colleagues, and customers. A nonterritorial office has a few workplaces which are available on a random time-sharing basis and have ample contact zones, such as

TABLE 34.4   Top Ten Internet Hub Cities

City	Internet Bandwidth[a] (Gbps)	City	Internet Bandwidth[a] (Gbps)
New York	150	Tokyo	17
London	86	Washington	13
Amsterdam	25	Miami	12
Paris	23	Los Angeles	11
San Francisco	21	Copenhagen	10

Source: Data from TeleGeography Inc. as reported in *www.totaltele.com*, October 30, 2001.

[a] The bandwidth capacities indicated apply for mid-2001 and refer to international routes only; domestic routes are not included.

lounges, coffee shops, and rooms for conferences, discussions, and general corporate identity cultivation.

Teleworking also has a positive effect on the environment. AT&T reported that the increased amount of teleworking of their employees in 2000 has avoided 170 million kilometers of car travel, with a subsequent reduction of 50,000 tons of carbon dioxide emission. BT reported that in the same year its staff saved over 240 million kilometers of traveling by means of audio and video conferencing, thus contributing to a saving of 1 million tons of carbon dioxide emission.

Telelearning makes the Internet into a virtual classroom with a very effective sharing of resources. Countries such as Canada, France, Germany, Italy, and the United States have already connected most schools to the Internet. In Germany in 2000, 34,000 schools were connected with the Internet in a free-of-charge program called T@School. In the United States all telecommunication operators contribute to a $4 billion central fund which is used to provide free Internet access for schools and libraries. South Africa launched SchoolNet in 1997, with the objective to train over 2000 teachers in 1035 schools all over the country. The Catholic University in Chile launched an Enlaces program, providing a wireless connection with two computers at schools in remote indigenous areas. MIT launched the University of the World in the mid-1990s as an interactive Master of Business Administration (MBA) learning program for a number of universities, initially in Singapore and in China, and later for other Asian countries. Real-time, two-way, full-motion videoconferencing is transmitted via satellite with large screen projection in the classrooms. Thus, Asian students can have question-and-answer sessions with teachers at MIT without anybody having to undertake intercontinental traveling. In the U.K., there are reported to be more and more MBA recipients who obtained their degree without setting a foot in a classroom; thanks to distance learning they can complete degrees while keeping career and family life in balance.

Telemedicine with teleconsulting, remote medical diagnosis, and even surgery assistance is improving health care in developing countries. As a pilot project in Mozambique, the ITU, through its Telecommunication Development Bureau, established a telemedicine link via radio-relay and satellite between the central hospitals in the capital Maputo and in Beira, the second-largest city at a distance of 1000 km from Maputo. Doctors in Beira can now send x-rays and other medical records

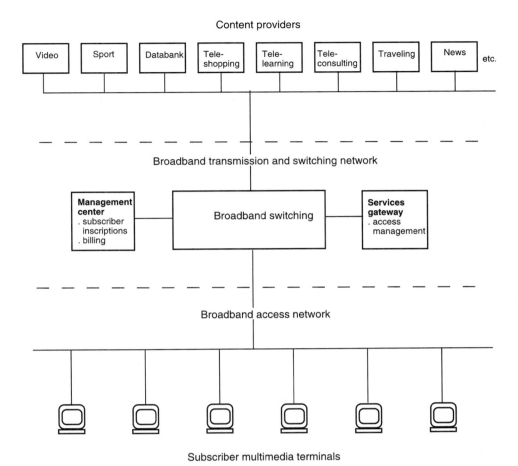

Figure 34.1 Typical multimedia network.

for immediate advice to experts in Maputo. Similar telemedicine projects are initiated in other African countries. As another example, in Georgia, the Research Institute of Radiology and Interventional Diagnosis at Tbilisi, since 1998, sends medical files, including x-rays, via Internet to the Center of Imaging Diagnostic in Lausanne, Switzerland, which than gives treatment recommendations.

To differentiate Internet services from conventional distance services by mail and telephone, the prefix *tele-* has gradually been replaced by *e-* (electronic). Tele-shopping, used successfully to sell consumer products such as books, for instance, introduced worldwide by *www.amazon.com*, evolved to e-commerce with *business-to-consumer* (B2C) and *business-to-business* (B2B) transactions made legally binding by an e-signature.[19] On July 4, 2000, President Bill Clinton gave e-signatures a status

---

[19] An *electronic signature*, also called a *digital signature*, is a way to authenticate the source of an electronic message by using an encryption system, *encryption* being a way of encoding a message to hide its content from unauthorized persons.

similar to that of handwritten signatures, when he approved the Electronic Signatures in Global and National Commerce Act. Further support for more secure e-commerce came one month later when the *International e-commerce Security Standard* (IES 2000), developed by the *Customers Service Institute of Australia*, received approval from the *International Standards Accreditation Board*. As a measure to prevent a further digital divide, the ITU in cooperation with the Geneva-based World Trade Centre (WTC) and WISeKey S.A. (International Secure Electronic Transaction Organization) started an Electronic Commerce for Developing Countries (EC-DC) project in 2000. Under this project, an e-commerce infrastructure for developing countries is being established whereby participating countries can share secure electronic payment and e-signature facilities in a network covering over 100 countries.

Another variation of e-service, namely *e-government*, also called *e-governance*, is being implemented widely, notably in two of the world's smallest countries; Singapore and Qatar. The government of Singapore began in 1997 a multimedia pilot project called Singapore One in which 5000 households and offices got access to 50 multimedia applications, including government services. In 2000 followed a three-year e-government action, combined with implementation of a *national information infrastructure* covering the country, enabling high-speed online execution of the majority of public services. The Arabian Gulf state of Qatar launched an e-government plan on its twenty-ninth anniversary of independence on September 3, 2000. Under this plan, 10 public institutions and some ministries began to offer online public services such as issuance and renewing of passports (which can then be collected at the Post Office) and residence permits, and collection of fees. Qatar, which did not get its first telephone service with 150 lines until 1953, got the first Earth station in the region in 1976, the first cable television network in the Middle East in 1993, the first GSM network in the Middle East in 1994, and was one of the earliest in the region to get Internet in 1996. E-government services are also being introduced in the neighboring United Arab Emirates in the state of Dubai, where fines for traffic offences are claimed by SMS, and divorces can be settled by a mouse click. In October 2000, the city of Dubai opened the world's first e-commerce free-trade zone, called *Dubai Internet City* (DIC). DIC was developed within 15 months on a 25-km^2 desert area just outside Dubai City. Currently, it provides always-on Internet access via optical fiber to 6000 businesspeople. The national operator Etisalat established a regional Internet *network access point* (NAP) in Dubai. This NAP has a direct 155-Mbps optical fiber connection with New York using the Flag undersea cable system. Adjacent to DIC, a new residential town called Dubai Marina is being developed with 100 apartment buildings up to 39 stories each and 4000 houses to attract an additional 100,000 residents. Fiber-to-the-apartment provides the new residents with high-speed Internet access with advanced services such as smart washing machines and refrigerators and e-health service for consulting the doctor using videoconferencing. Dubai Marina is announced to become the world's first fully online city, offering e-government, e-commerce, e-health care, and e-entertainment in all homes, offices, and 3G handsets. A Virtual Internet Academy will cover e-learning for some 5000 students.

An even more impressive example of an ambitious e-government project is the *Multimedia Super Corridor* (MSC) in Malaysia. The Malaysian government initiated the MSC project in 1996 within the scope of its Vision 2020 objective of turning Malaysia into a developed country by 2020. The MSC covers an area of 15 × 50 km between the world's highest Petronas twin towers (452 m) in Kuala Lumpur and a new international airport. Half a dozen new towns will include the city of Putrajaya, which will accommodate Malaysia's e-government, and the city of Cyberjaya, which will be residential and the location of multimedia industries, R&D centers, a multimedia university, and smart schools. A fiber optic ring with a transmission speed of 2.5 Gbps has been installed, and a further 10-Gbps Hypernet Internet backbone was under construction at the end of the century. An area that used to be a mixture of jungle and tin mines and in the twentieth century was cultivated with rubber and palm trees is being developed into a high-technology park that is 60% environmentally friendly green. MSC has the elements to become an example of the global village at its best: with immediately accessible worldwide information, the finest Southeast Asian architecture, and an international multicultural population in a cultivated tropical landscape.

REFERENCES

Books

Kobayashi, Koji, *Computers and Communications*, MIT Press, Cambridge, MA, 1986.

Tatipamula, Mallikarjun, et al., *Multimedia Communications Networks*, Artech House, Norwood, MA, 1998.

Articles

Anon., From telephony to multimedia in the 21st century, *ITU News*, Vol. 7, 1997, pp. 41–42.

Anon., Taking telemedicine to the people: focus on Mozambique, *ITU News*, Vol. 2, 1998, pp. 18–20.

Anon., Al Gore's five challenges to the telecom world, *ITU News*, Vol. 9, 1998, pp. 3–7.

Anon., E-commerce for developing countries, *ITU News*, Vol. 8, 2000, pp. 28–29.

Anon., Interview with Vinton G. Cerf, long and winding road: the evolution of the Internet, *Nokia Link*, fourth quarter 2000, pp. 24–29.

Anon., Internet and health: is there a doctor? *ITU News*, Vol. 4, 2001, pp. 4–7.

Anon., Internet and education: virtual classrooms for everybody? *ITU News*, Vol. 4, 2001, pp. 12–15.

Anon., Joint ITU/WIPO symposium creating a wider understanding of the complex issues surrounding the implementation of multilingual domain names, *ITU News*, Vol. 1, 2002, pp. 9–13.

Carr, J. Scott, Watermarking the digital age, *Telecommunications International*, December 1997, p. 62.

Cederquist, Lars, IP upgrade, *On—The New World of Communication* (Ericsson), February 2001, p. 49.

Warwick, Martyn, Last of the big spenders, *Communications International*, July 1997, pp. 50–52.

Wold, Robert N., Europe's pursuit of Internet and broadband revenues, *Via Satellite*, September 2001, pp. 36–41.

Internet

www.computerhistory.org, History of the Internet, presented by the Computer Museum History Center at California.

www.ieee.org/organizations/history-enter/board/messages/28.html, A brief history of the Internet.

www.itu.int/itu-D/ict/statistics/index/html, ITU telecommunication indicators.

www.totaltelecom.com.

Riseborourgh, Steve, Lunatics invade the asylum of ICANN, October 12, 2000.

Osborn, Alan, ICANN takes a mantle of multilingual DNS, November 7, 2000.

Osborn, Alan, ICANN chooses seven new domain names for the net, November 17, 2000.

EIM, Compac launch Arabic search engine, May 2, 2000.

I-DNS to launch Japanese domain name service, May 11, 2000.

ITU and IETF agree on H.248 gateway standard, August 7, 2000.

World e-commerce standard approved, August 22, 2000.

Singapore to spend S$1.5 billion on e-government plan, June 6, 2000.

Qatar gets e-government, September 4, 2000.

Dubai Internet City e-commerce zone full, September 18, 2000.

New York is the Internet capital: TeleGeography, October 30, 2001.

Warner, Bernhard, and Eric Auchard, Reuters, BT goes to court over hypertext patent, February 27, 2002.

www.ziplink.net/lroberts/internetchronology.html, Internet chronology by Lawrence G. Roberts, March 22, 1997; updated October 24, 1999.

35

INTERNATIONAL COOPERATION

After World War II, the United States took the initiative to encourage the same spirit of international cooperation that had prevailed at the first convention of the International Telegraph Union in 1865. It proposed that the government of the USSR should invite the other four of the Big Five victorious powers—China, France, the U.K., and the United States—for a conference in Moscow. The conference took place in 1946: where important proposals were made to bring the ITU into close association with the newly established *United Nations* (UN) and to form an elected permanent administrative council.

The first Plenipotentiary Conference of the ITU was held at Atlantic City, New Jersey, in 1947. The 600 delegates came from 76 countries and had to decide on the proposals made in Moscow. A permanent administrative council was agreed upon, composed of 18 members of the union "elected by the Plenipotentiary Conference with due regard to the need for equitable representation of all parts of the world." The language question arose again and was solved in such a way that now, in alphabetical order, the official languages of the ITU are Chinese, English, French, Russian, and Spanish. English, French, and Spanish are the working languages, and "in case of dispute, the French language shall be authentic." Concerning the relationship with the UN, it was agreed that to safeguard the "technical and universal" tasks independent of political influences, the ITU became the "specialized telecommunications agency" of the UN. Instead of Berne, Geneva was chosen as the new seat for the secretariat of the ITU, mainly because French-speaking Geneva was a subsidiary headquarters of the UN.

During World War II, the use of radio frequencies had been extended widely in an uncoordinated way for military radiotelephony, radio-relay systems, radar, radio navigation, and for aeronautical communication. At the Atlantic City conference, a

The Worldwide History of Telecommunications, By Anton A. Huurdeman
ISBN 0-471-20505-2   Copyright © 2003 John Wiley & Sons, Inc.

new *International Frequency Registration Board* (IFRB) was created. The IFRB was given the difficult task of securing efficient and orderly use of the frequency spectrum such as to avoid interference between the various radio services, especially between the radio services of the different countries. The frequency spectrum was extended to 10.5 GHz. An *International Frequency List* was introduced with the classifications: notification and registration. Under *notification* the date of receipt of the notice that a new station used a particular frequency for a specific purpose was entered. Under *registration* a frequency assignment was given with the "right to international protection from harmful interference" for stations complying with specific requirements.

Since World War II, radio transmission covered an astonishingly wide range of applications. In 1959, therefore, at an administrative radio conference in Geneva, the frequency allocations of the radio spectrum were extended to the range from 10 kHz to 40 GHz. To better adopt the frequency allocation to regional service requirements and regionally varying conditions for radio propagation, the world was divided into three regions: *region 1*: Europe Africa, and the USSR; *region 2*: the Americas; and *region 3*: Asia, Australia, and the Pacific.

A major task of the ITU has always been to elaborate recommendations for the standardization of telecommunications services, operation, performance, and maintenance of equipment, systems, networks, and services. Adherence to the recommendations guarantees interconnectivity and interoperability on a global scale of networks, systems, and equipment from various operators and manufacturers. To work out the standards to ensure that, for example, the quality of an international very long distance call on the order of 20,000 to 30,000 km is the same as for a national call over a few kilometers was quite a challenge. From the Atlantic City conference onward therefore, substantial efforts have been made to improve the efficiency of the standardization work. In 1957, in view of the basic similarity of many of the technical problems, the two International consultative committees for telegraphy (CCIT) and for long-distance telephony (CCIF) merged into a new *International Telegraph and Telephone Consultative Committee*, officially the *Comité Consultatif International Télégraphique et Téléphonique* (CCITT). Experts from all over the world who work together in the various study groups of CCITT, and similarly in study groups of CCIR, published the results of their studies as their *Recommendations*, with detailed justifications and explanations at regular four-year intervals— after being agreed upon at a CCITT or CCIR *plenary assembly*—in a set of books. The 10 to 60 books of a particular four-year study period of the CCITT all had a common cover color that was specific for that period. The first were red books, issued following the II Plenary Assembly at New Delhi in 1960. These were followed by blue books after the III Plenary Assembly in Geneva at 1964; white books in 1968 after the IV Plenary Assembly in Mar del Plata, Argentina; green books in 1972 at Geneva; orange books in 1976 at Geneva; yellow books in 1980 in Geneva, and red books again after the VIII Plenary Assembly at Malaga–Torremolinos, Spain, in 1984. It was customary, therefore, for telecommunications engineers worldwide to refer to the recommendations of the red book if they referred to the series of 10 volumes divided into 35 fascicles (parts of a book), containing all the recommendations valid since 1984. The blue book published in 1988 was the last complete set of such recommendations, consisting of 20 volumes divided into 61 fascicles. The standardization work became so immense, and the technical progress so rapid, that publication of the recommendations at four-year intervals no longer met requirements. Due

TABLE 35.1   ITU Reorganization of the International Standardization, 1992

Old	New
CCITT	ITU-TS (Telecommunication Standardization)
CCIR	ITU-R (Radio Communication)
CCITT/R Recom.	ITU-T Recommendation
CCITT Secretariat	TSB (Telecommunication Standardization Bureau)
CCITT Study Groups	ITU-TS Study Groups TSAG (Telecommunication Standardization Advisory Group)
CCITT Plenary Assembly	WTSC (World Telecommunication Standardization Conference)
International Frequency Registration Board (IFRB)	Radio Regulation Board

to the rapid development of services using radio waves, the CCIR decided very early to set up an *interim meeting* between the plenary assemblies. The results of those meetings were published in *interim reports* of the study groups. Documentation of the CCIR meetings was always light green in color. In those books, in addition to the recommendations, *questions* and related *study programs* were formulated for the forthcoming study period. Intermediate results were given in *reports*. Furthermore, *resolutions, opinions, and decisions* expressed answers to peripheral matters.

The ITU came under great pressure from about 80 specialized forums, interest groups, and organizations active in telecommunications standardization in the 1990s. These included the ATM Forum, the Frame Relay Forum, the GSM Association, the Internet Engineering Task Force, the UMTS Forum, and many others. Those structures served regional interests or specific technologies and applications. Most significantly, the rapidly emerging Internet was based on global standards that had been developed mainly outside the ITU. In 1992, the ITU improved its efficiency by reorganizing the standardization work, as shown in Table 35.1. Individual recommendations are now published immediately upon completion. The average elaboration time of a recommendation, which before 1988 was four years, was reduced progressively, to two years between 1988 and 1993, one and a half years between 1993 and 1996, and nine months beginning in 1997, with the target of reaching a time frame of six months or even less. Simultaneously, the number of recommendations increased substantially, with a peak of 362 in 1999.

In 1971, the ITU began telecommunications exhibitions in Geneva at four-year intervals as showcases of the state of the art and future trends. Those Telecom exhibitions became the most influential gathering of telecommunications professionals worldwide. Operators and manufacturers from all over the world present their latest achievements and planned activities and exchange their ideas. In parallel with the exhibition, a telecommunications forum is organized where representatives from the ITU, operators, and manufacturers and consultants report and provide the opportunity to discuss the latest achievements. In 1971, at the first exhibition, Telecom 71, some 28,000 visitors could inform themselves about commercial electronic switching systems, all-solid-state transmission equipment, geostationary satellite systems, and experimental optical fiber transmission at the stands of 250 exhibitors. At Telecom 95, some 1066 exhibitors, focusing on digital mobile radio, mobile satellite services, and Internet attracted 154,671 visitors. At the last exhibition of the twentieth century, Telecom 99, more than 175,000 persons from 161 countries had the choice of meeting 1123 exhibitors from 47 countries. Telecom 99 focused on Internet, e-commerce, mobile radio data transmission, broadband fixed wireless access, the disappearance of the traditional distinction between wire-line and wireless applications, and the combined impact on society of mobility and the Internet. The surplus income derived from those exhibitions goes into the *ITU Telecom Surplus Fund*. This Fund is used primarily to support countries, such as Afghanistan, that are in special need to rehabilitate their networks.

The ITU, in the *ITU News No. 6/99*, stated: "As a result of privatization and liberalization, telecommunications is no longer just a national and international public service. It is fast becoming a globally competitive business." Consequently, "the ITU is to enlarge the ITU membership and ensure the active participation of the new players who are driving the changes in telecommunications. This particularly includes companies from convergent industries, representatives of user communities, financial institutions and regulators." In fact, by year-end 2000, the ITU had 407 members as follows: 162 operating agencies, 204 scientific and industrial organizations, 38 regional and international organizations, and three other national entities.

Looking back on 135 years of shaping the world of telecommunications, the ITU proudly claimed: "Without the ITU there would be no global telephone network. No Internet. No e-commerce. No geostationary satellite network. No fax standard. No modem standards. No TV and radio broadcasts free of interference. Today we're busy designing the technologies that will drive the information society of tomorrow."

APPENDIX A

CHRONOLOGY OF THE MAJOR EVENTS IN THE TWO CENTURIES OF TELECOMMUNICATIONS

Year	Telecommunications Event
1794	*The era of telecommunications begins* with a message on the Paris–Lille optical telegraph line on August 15.
	Abraham Niclas Edelcrantz constructs the first optical telegraph line outside France in Stockholm, Sweden.
	First books on telecommunications published in Germany.
1795	P. Basilius Sinner builds private optical telegraph system.
1796	Optical telegraph lines are installed in England.
1799	Alessandro Volta stores galvanic electricity.
1809	Von Soemmering constructs an electrochemical telegraph.
1820	Hans Christian Oerstedt discovers the electromagnetic field.
	André-Marie Ampère discovers electrodynamics.
	Johann S. Ch. Schweigger develops the galvanometer.
	Johann Christian Poggendorf develops the multiplicator.
1825	William Sturgeon constructs the first electromagnet.
1826	Georg Simon Ohm defines the basic electrical law $U = IR$.
1830	William Ritchie demonstrates electrical telegraphy.
1831	Michael Faraday discovers electromagnetic induction.
1832	Schilling von Cannstadt develops a five-needle telegraph.
1833	Carl Friedrich Gauss and Wilhelm Weber transmit information via a galvanic chain in Göttingen.
1835	Samuel F. B. Morse demonstrates the writing telegraph.
1836	Carl August von Steinheil develops a writing telegraph and a telegraphic code.

The Worldwide History of Telecommunications, By Anton A. Huurdeman
ISBN 0-471-20505-2   Copyright © 2003 John Wiley & Sons, Inc.

Year	Telecommunications Event
1837	*The era of electrical telegraphy begins.*
	June 12, William Fothergill Cooke and Charles Wheatstone apply for a patent for an electrical five-needle telegraph.
	September 4, Morse demonstrates his electrical writing telegraph at the University of the City of New York.
1839	Electrical telegraph line with five-needle telegraph operates between London–Paddington and West Drayton.
1844	Morse telegraph operates on Baltimore–Washington line.
1847	Louis-François Breguet constructs a step-by-step telegraph.
	Werner Siemens constructs a gutta-percha press for seamless insulation of copper wires.
	Frederick Collier Bakewell makes the world's first successful image transmission in London.
1851	U.K.–France submarine telegraph cable is laid.
	Heinrich Daniel Rühmkorff invents the induction coil.
1852	E. P. Smith coins the word *telegram*.
1854	Charles Bourseul proposes electrical speech transmission.
1855	David Hughes develops a plain-language writing telegraph.
1858	First transatlantic telegraph cable operates three weeks.
	William Thomson develops a mirror galvanometer.
1860	Pony express operates between Kansas City and Sacramento.
1861	U.S. transcontinental telegraphy line is inaugurated.
	Philipp Reis demonstrates his telephon.
1865	International Telegraph Union is founded in Paris.
	Pantelegraph image transmission begins on Paris–Lyon line.
1866	First successful transatlantic telegraph cable is laid.
1867	Charles Wheatstone presents his cryptograph.
1869	Public telegraph service is inaugurated in Japan.
	Indo-European telegraph line begins operation.
1871	Great Northern Telegraph Line begins operation.
1872	Australian overland telegraph line begins operation.
	Bernhard Meyer develops the first telegraph multiplexer.
	Mahlon Loomis receives a patent for an aerial telegraph.
1874	Emile Baudot develops a five-unit telegraph code.
1876	*The era of telephony begins.*
	February 14, Alexander Graham Bell submits a patent application for "improvement in telegraphy," which results in the world's first telephone patent.
	March 10, Bell calls "Mr. Watson, come here, I want you."
1877	First permanent telephone line is installed between Boston and Somerville, Massachusetts.
	Edison develops the carbon telephone transmitter.
	Siemens & Halske start telephone production in Germany.
1878	Manual telephone switchboard operates in New Haven, Connecticut.
	Emma M. Nutt starts as the first woman telephone operator.
	Bell Telephone Company opens telephone service in London.
	Louis-François Breguet starts telephone production in Paris.

Year	Telecommunications Event
1881	International telephone operation starts between Detroit in the United States and Windsor, Ontario, in Canada.
	Bell patents the metallic circuit for telephone lines.
1886	Strowger Company develops the telephone-dialing disk.
1888	Heinrich Herz proves the existence of electromagnetic waves.
	Elisha Gray develops the Telautograph facsimile machine.
1889	Ernest Mercadier proposes voice-frequency telegraphy.
	First coinbox telephone is installed at Hartford, Connecticut.
1891	Edouard E. D. Branly invents the coherer.
1892	Strowger telephone exchange operates in La Porte, Indiana.
1895	Alexander St. Popoff detects thunderstorm lightning.
1896	Marconi obtains patent for wireless telegraphy.
	One million telephones are installed.
1897	*The era of radio transmission begins.*
	Official trial of Marconi's radio is made over the Bristol Channel, U.K.
	Ferdinand Braun invents the cathode-ray tube.
1898	The AEG in Germany starts production of radio equipment developed by Adolf K. H. Slaby.
	Karl F. Braun introduces the inductive antenna coupling.
1899	Michael O. Pupin proposes coils to extend line transmission.
	First radio emergency call is made from a ship to the U.K.
1901	Marconi transmits the Morse signal "S" across the Atlantic Ocean.
	Reginald A. Fessenden receives patent for radiotelephony.
1902	Emil Karup introduces transversal inductance cable.
	Oliver Heaviside and Arthur Edwin Kennelly predict the existence of an ionized layer in the upper atmosphere.
1903	Charles Krum obtains the first teleprinter patents.
1904	Edouard Estaunié creates the word *telecommunications*.
	John Ambrose Fleming invents the diode.
	Arthur Korn develops the telautograph for telephotography.
	Reginald A. Fessenden achieves radiotelephony over 40 km.
1905	Tokyo–Sasebo (1583 km) record phone call is made.
	Reginald A. Fessenden invents the superheterodyne circuit.
1906	*The electronic era begins* when Lee de Forest and Robert von Lieben, independently, invent the triode.
	International Radiotelegraph Union is founded.
	E. Belin develops the Belinograph for telephotography.
1910	Ten million telephones exist worldwide.
1912	Alexander Meissner develops the electronic HF generator.
	First rotary exchange is installed in Landskrona, Sweden.
	Western Union starts teletypewriter service.
1915	New York–San Francisco telephone line opens using telephone amplifiers.
	John G. Roberts and John N. Reynolds obtain the first patent for crossbar switching.
	U.S.–France transatlantic radio conversation is made.
	AT&T engineers develop permalloy cable.

Year	Telecommunications Event
1917	The last optical telegraph line closes on Curaçao.
1918	Four-channel multiplex is used on Baltimore–Pittsburgh line.
	U.K.–Australia radio-telegraphy is made over 17,700 km.
1920	Heinrich G. Barkhausen proposes velocity–modulated vacuum tubes.
1923	Automatic long-distance telephony begins in Germany.
	Albert Scherbius starts production of Enigma.
1924	A. Karolus patents picture transmission using Kerr cells.
1925	CCIT (Consultative Committee International on Telegraphy) and CCIF (Consultative Committee International on Long Distance Telephony) are founded.
	AT&T introduces telephoto service.
1926	Canada–U.K. shortwave radiotelephone service begins.
	Transcontinental radio picture service begins.
	The first crossbar switch is installed at Sundsvall, Sweden.
1927	U.K.–U.S. intercontinental radiotelephony begins.
	CCIR (Consultative Committee International on Radio) is founded.
1928	Hidetsugu Yagi and Shintaro Uda present the Yagi antenna.
1931	AT&T inaugurates the Teletypewriter Exchange Service.
1932	The International Telecommunication Union is founded.
1933	National automatic teleprinter network opens in Germany.
	Edwin Howard Armstrong invents frequency modulation.
1934	Radio-relay transmission era starts with a 54-km commercial link over the English Channel.
	Coaxial cable is installed on a 11.5-km line in Berlin.
1935	Around-the-world radiotelephone conversation takes place.
	Western Union introduces Teledeltos facsimile service.
1936	Bristol–Plymouth, U.K., long-distance cable carries 12 multiplexed telephone channels.
1937	Alec H. Reeves invents pulse–code modulation.
1939	A six-channel carrier-telephone cable operates over 3000 km between Japan and China.
1943	Submarine coaxial cable using submerged telephone repeaters is laid between Anglesey, Wales, and the Isle of Man.
1945	Arthur C. Clarke conceives the idea of placing satellites in geostationary orbit as extraterrestrial relays.
1947	International Frequency Registration Board is founded.
	New York–Boston 480-channel radio-relay route is established.
	Transistor is invented at the Bell Telephone Laboratories.
1950	Key West–Havana submarine coaxial cable is laid.
	James Van Allen proposes an Earth satellite for the IGY.
1955	CCIT and CCIF combine into CCITT.
1956	TAT-1, transatlantic coaxial submarine cable, is laid.
1957	*Sputnik 1* sends its bleep-bleep from space.
1958	*Explorer 1* in space confirms the existence of (Van Allen) radiation belts.
	IDD starts from Germany to Switzerland and Belgium.
1959	Theodore Maiman invents the laser.

Year	Telecommunications Event
1960	Electronic telephone exchange is installed in Morris, Illinois.
1962	*Satellite transmission era begins* with TV transmission between the United States and Europe via *Telstar 1*.
1964	SYNCOM satellite is launched into geostationary orbit.
1965	*Intelsat 1* opens commercial satellite transmission.
1966	K. C. Kao and G. A. Hockham propose optical transmission through pure glass fibers.
1969	Digital radio-relay route begins operation in Japan.
	Arpanet (later Internet) begins operation.
1970	Digital switching begins operation in Lannion, France.
	U.K.–U.S. intercontinental direct dialing begins.
1973	Optical fiber transmission begins on 24-km Frankfurt–Oberursel link, in Germany.
1977	P-MP rural radio begins operation in Canada.
1979	Cellular radio begins operation in Japan.
1980	Bell Telephone Laboratories develops Sonet.
1987	First long-distance submarine optical fiber cable is laid connecting the island of Corsica with the French mainland.
1988	CCITT adopts SDH (Synchronous Transport Modules).
	TAT-8, first transatlantic optical fiber cable, is laid.
	N-ISDN operation begins on Tokyo–Nagoya–Osaka network.
1990	Tim Berners-Lee conceives the World Wide Web.
1991	Digital cellular radio (GSM) begins operation in Finland.
1998	Low-Earth-orbit satellite system Iridium begins operation.
1999	Free-space optics systems connects the museums of the Smithsonian Institution in Washington, DC.
2000	Transatlantic cable 360Atlantic, with a capacity of 1.92 Tbps, begins operation.
	Worldwide there exist 987 million fixed-line telephones, 740 million mobile telephones, and 367 million Internet users.

APPENDIX B

WORLDWIDE STATISTICS OF POPULATION, INTERNET USERS, CELLULAR PHONES, AND MAIN TELEPHONES

Country	Inha-bitants (millions)	Internet Users		Cellular Phones		Main Telephones	
		Thousands	%	Thousands	%	Thousands	%
Africa							
Algeria	30.88	50.0	0.17	86.0	0.28	1,761.3	5.70
Angola	13.13	30.0	0.23	25.8	0.20	69.7	0.53
Benin	6.10	15.0	0.25	55.5	0.91	51.6	0.85
Botswana	1.62	15.0	0.93	200.0	12.33	150.3	9.27
Burkina Faso	11.94	10.0	0.08	25.2	0.21	53.2	0.45
Burundi	6.70	3.0	0.05	16.3	0.24	20.0	0.30
Cameroon	15.09	40.0	0.27	148.0	0.98	94.6[a]	0.64
Cape Verde	0.43	8.0	1.84	19.7	4.54	54.9	12.62
Central African Republic	3.62	1.5	0.04	5.0	0.14	9.5	0.26
Chad	7.65	3.0	0.04	5.5	0.07	9.7[a]	0.13
Comores	0.69	1.5	0.22	—	—	7.0	1.00
Congo (Rep.)	2.94	0.5[a]	0.02	70.0	2.38	22.0	0.75
Congo (Dem. Rep.)	51.65	1.4	0.22	15.0	0.03	9.7	1.52
Côte d'Ivoire	14.79	40.0	0.27	450.0	3.04	263.7	1.78
Djibouti	0.64	1.4	0.22	0.2	0.04	9.7	1.52
Egypt	63.48	450.0	0.71	1,359.9	2.14	5,483.6	8.64
Equatorial Guinea	0.45	0.5[a]	0.11	0.3[b]	0.07	6.1	1.35

The Worldwide History of Telecommunications, By Anton A. Huurdeman
ISBN 0-471-20505-2   Copyright © 2003 John Wiley & Sons, Inc.

Country	Inha-bitants (millions)	Internet Users Thousands	%	Cellular Phones Thousands	%	Main Telephones Thousands	%
Eritrea	3.83	5.0	0.13	—	—	30.6	0.80
Ethiopia	63.49	10.0	0.02	17.8	0.03	231.9	0.37
Gabon	1.23	15.0	1.22	120.0	9.79	39.0	3.18
Gambia	1.30	4.0	0.31	5.6	0.43	33.3	2.56
Ghana	20.21	30.0	0.15	130.0	0.64	237.2	1.17
Guinea	7.90	8.0	0.10	42.1	0.53	62.4	0.79
Guinea-Bissau	1.20	3.0	0.25	—	—	11.1	0.93
Kenya	30.67	200.0	0.65	127.4	0.42	321.5	1.05
Lesotho	2.15	4.0	0.19	21.6	1.00	22.2	1.03
Liberia	3.15	0.5	0.02	—	—	6.7	0.21
Libya	5.61	10.0	0.18	40.0	0.71	605.0	10.79
Madagascar	15.94	30.0	0.19	63.1	0.40	55.0	0.34
Malawi	10.34	15.0	0.15	49.0	0.47	45.0	0.44
Mali	11.23	18.8	0.17	10.4	0.09	39.2	0.35
Mauritania	2.65	5.0	0.19	7.1	0.27	10.0	0.72
Mauritius	1.19	87.0	7.29	180.0	15.08	280.9	23.53
Mayotte	0.14	—	—	—	—	9.7[a]	7.27
Morocco	28.35	200.0	0.71	2,342.0	8.26	1,425.0	5.03
Mozambique	19.68	30.0	0.15	51.1	0.26	85.7	0.44
Namibia	1.67	30.0	1.71	82.0	4.67	110.2	6.27
Niger	10.73	5.0	0.05	2.1	0.02	20.0	0.19
Nigeria	113.86	200.0	0.18	30.0	0.03	492.0	0.43
Réunion	0.70	130.0	18.60	276.1	39.5	268.5[a]	38.86
Rwanda	7.73	5.0	0.06	—	—	17.6	0.23
São Tomé & Príncipe	0.15	6.5	4.37	—	—	4.6	3.10
Senegal	9.52	40.0	0.42	250.3	2.63	205.9	2.16
Seychelles	0.08	6.0	7.40	26.0	32.00	19.0	23.45
Sierra Leone	4.85	5.0	0.10	11.9	0.25	19.0	0.39
Somalia	9.94	0.2[a]	0.01	—	—	15.0	0.15
South Africa	43.69	2,400.0	5.49	8,308.0	19.02	4,961.7	11.36
Sudan	31.10	30.0	0.10	23.0	0.07	32.2	3.19
Swaziland	1.01	10.0	0.99	33.0	3.27	32.2	3.19
Tanzania	35.12	115.0	0.33	180.2	0.51	173.6	0.49
Togo	4.63	100.0	2.16	50.0	1.08	42.8	0.92
Tunisia	9.59	100.0	1.04	55.3[a]	0.58	850.4[a]	8.99
Uganda	22.21	40.0	0.18	188.6	0.85	61.7	0.28
Zambia	10.42	20.0	0.19	98.9	0.95	83.3	0.80
Zimbabwe	13.48	50.0	0.37	309.0	2.29	249.9	1.85
Total Africa	792.63	4,637.9	0.59	15,652.9	1.98	19,660.8	2.48
The Americas							
Antigua and Barbuda	0.08	5.0	6.52	22.0	28.69	38.3	49.95
Argentina	37.03	2,500.0	6.75	6,050.0	16.34	7,894.2	21.23
Aruba	0.10	4.0[a]	4.07	15.0	14.63	38.1	37.16
Bahamas	0.30	13.1	4.32	31.5	10.36	114.3	37.59

Country	Inhabitants (millions)	Internet Users Thousands	%	Cellular Phones Thousands	%	Main Telephones Thousands	%
Barbados	0.27	10.0	3.74	30.0[a]	11.14	117.0	43.74
Belize	0.24	15.0	6.25	16.8	7.0	35.8	14.88
Bermuda	0.06	25.0[a]	39.01	12.5[b]	19.64	56.1	86.98
Bolivia	8.33	120.0	1.44	579.8	6.96	504.2	6.05
Brazil	170.12	5,000.0	2.94	23,188.2	13.63	30,926.3	8.18
Canada	30.75	12,700.0	41.30	8,751.3	28.64	20,802.9	67.65
Chile	15.31	22,537.3	16.58	3,401.5	22.22	3,387.5	22.13
Colombia	42.32	878.0	2.04	2,256.8	5.33	7,158.6	16.92
Costa Rica	4.02	250.0	6.21	209.1	5.2	1,003.4	24.94
Cuba	11.20	60.0	0.54	6.5	0.06	488.6	4.36
Dominica	0.08	2.0[a]	2.61	1.2	1.56	22.7	29.43
Dominican Republic	8.55	55.0	0.64	705.4	8.25	894.2	10.45
Ecuador	12.65	180.0	1.42	482.2	3.81	1,265.2	10.0
El Salvador	6.28	50.0	0.80	743.6	11.85	625.8	9.97
French Guiana	0.18	2.0[a]	1.15	39.8	22.01	49.2[a]	28.26
Grenada	0.09	4.11	4.36	4.3	4.55	31.4	33.2
Guadeloupe	0.46	8.0	1.75	169.8	37.25	204.9	44.93
Guatemala	11.39	80.0	0.70	696.6	6.12	649.8	5.71
Guyana	0.86	4.0	0.47	39.8	4.63	68.4	7.94
Haiti	8.14	6.0[a]	0.07	25.0[a]	0.31	72.5	0.89
Honduras	6.49	40.0	0.62	155.3	2.39	298.7	4.61
Jamaica	2.58	80.0	3.11	367.0	14.24	511.7	19.86
Martinique	0.40	5.0[a]	1.27	162.1	41.03	171.6	43.44
Mexico	98.88	2,712.4	2.74	14,077.9	14.24	12,331.7	12.47
Netherlands Antilles	0.22	2.0[a]	0.93	16.0[b]	7.52	80.0	37.16
Nicaragua	5.07	50.0	0.99	90.3	1.78	158.6	3.12
Panama	2.84	90.0	3.17	410.4	14.46	429.1	15.12
Paraguay	5.50	40.0	0.73	820.8	14.93	268.1[a]	5.0
Peru	25.66	2,500.0	9.74	1,222.6	4.76	1,635.9	6.37
Puerto Rico	3.91	400.0	10.22	926.4	23.67	1,299.3	33.19
St. Kitts and Nevis	0.04	2.0[a]	5.16	1.2	3.12	21.9	56.88
St. Lucia	0.16	3.0[a]	1.95	2.5	1.60	48.9	31.35
St. Vincent	0.11	3.5	3.09	2.4	2.08	24.9	21.96
Suriname	0.43	11.7	2.70	41.0	9.46	75.3	17.35
Trinidad & Tobago	1.29	100.0	7.73	133.2	10.29	299.1	23.11
United States	275.13	95,354.0	34.66	109,478.0	39.79	192,518.8	69.97
Uruguay	3.34	370.0	11.09	440.2	13.19	929.1	27.84
Venezuela	24.17	950.0	3.93	5,526.0	21.75	2,605.6	10.78
Virgin Islands	0.12	12.0[a]	10.03	35.0	28.9	69.0	56.97
Total Americas	825.14	127,234.1	15.42	181,117.3	21.95	290,226.5	35.18

Country	Inha-bitants (millions)	Internet Users		Cellular Phones		Main Telephones	
		Thousands	%	Thousands	%	Thousands	%
Asia							
Afghanistan	22.72	—	—	—	—	29.0	0.13
Armenia	3.52	50.0	1.42	17.4	0.49	533.4	15.15
Azerbaijan	7.73	12.0	0.16	430.0	5.56	801.2	10.36
Bahrain	0.68	40.0	5.84	205.7	30.05	171.0	24.97
Bangladesh	137.44	100.0	0.07	205.0	0.15	491.3	0.36
Bhutan	0.68	1.5	0.22	—	—	13.3	1.97
Brunei Darus-salam	0.33	30.0	9.14	95.0	28.94	80.5	24.52
Cambodia	13.10	6.0	0.05	130.5	1.00	30.9	0.24
China	1,295.33	22,500.0	1.74	85,260.0	6.58	144,829.0	11.18
D.P.R. Korea	24.04	—	—	—	—	1,100.0	4.458
Georgia	5.47	23.0	0.42	185.5	3.39	757.5	13.86
Hong Kong SAR	6.73	2,601.3	38.64	5,447.3	80.92	3,925.8	58.32
India	1,012.40	5,000.0	0.49	3,577.1	0.35	32,436.1	3.20
Indonesia	212.09	2,000.0	0.94	3,669.3	1.73	6,662.6	3.14
Iran (I.R.)	63.66	250.0	0.39	962.6	1.51	9,486.3	14.9
Iraq	22.95	—	—	—	—	675.0	2.94
Israel	6.27	1,270.0	20.25	4,400.0	58.1	3,021.0	48.18
Japan	126.92	47,080.0	37.10	66,784.4	52.62	74,343.6	58.58
Jordan	6.67	127.3	1.91	388.9	5.83	620	9.29
Kazakhstan	16.22	100.0	0.62	197.3	1.22	1,834.2	11.31
Korea (Rep.)	47.3	19,040.0	40.25	26,816.4	56.69	21,031.7	46.37
Kuwait	1.91	150.0	7.84	476.0	24.86	467.1	24.40
Kyrgyzstan	4.88	51.6	1.06	9.0	0.18	376.1	7.71
Lao P.D.R.	5.43	6.0	0.11	12.7	0.23	40.9	0.75
Lebanon	3.50	300.0	8.58	743.0	21.25	681.5	19.49
Macao SAR	0.44	60.0	13.70	132.0	30.14	176.8	40.38
Malaysia	23.27	3,700.0	15.90	4,960.8	21.32	4,634.3	19.92
Maldives	0.27	6.0	2.23	7.6	2.84	24.4	9.08
Mongolia	2.37	30.0	1.26	107.5	4.53	132.2	5.57
Myanmar	47.75	7.0	0.02	13.4	0.03	266.2	0.56
Nepal	23.04	50.0	0.23	10.2	0.04	266.9	1.16
Oman	2.54	90.0	3.55	164.3	6.48	225.4	8.88
Pakistan	141.26	133.9	0.10	349.5	0.25	3,053.5	2.16
Palestine	3.19	35.0	1.10	285.0	8.93	270.0	8.46
The Philippines	76.50	2,000.0	2.61	6,454.4	8.44	3,061.4	4.0
Qatar	0.60	30.0	5.01	120.9	20.19	160.2	26.77
Saudi Arabia	21.61	200.0	0.93	1,375.9	6.37	2,964.7	13.72
Singapore	4.02	1,200.0	29.87	2,747.4	68.38	1,946.5	48.45
Sri Lanka	18.92	121.5	0.64	430.2	2.27	767.4	4.06
Syria	16.19	30.0	0.19	30.0	0.19	1,675.2	10.35
Taiwan China	22.28	6,260.0	28.10	17,873.8	80.24	12,642.2	56.75
Tajikistan	6.13	3.0	0.05	1.2	0.02	218.5	3.57
Thailand	60.61	2,300.0	3.80	3,056.0	5.04	5,591.1	9.23
Turkmenistan	4.46	6.0	0.14	9.5	0.21	358.9[a]	8.19

Country	Inha-bitants (millions)	Internet Users		Cellular Phones		Main Telephones	
		Thousands	%	Thousands	%	Thousands	%
United Arab Emirates	2.61	735.0	28.20	1,428.1	54.80	1,020.1	39.14
Uzbekistan	24.65	120.0	0.49	53.1	0.22	1,655.0	6.71
Vietnam	79.83	200.0	0.25	788.6	0.99	2,542.7	3.19
Yemen	18.35	15.0	0.08	32.0	0.17	346.7	1.89
Total Asia	3,648.85	118,071.1	32.99	240,444.6	6.59	349,339.4	9.57
Europe							
Albania	3.91	3.5	0.09	29.8	0.76	152.7	3.91
Andorra	0.08	7.0	8.97	23.5	30.18	34.2	43.87
Austria	8.21	2,100.0	25.58	6,252.8	76.15	3,832.9	46.68
Belarus	10.24	180.0	1.76	49.4	0.48	2,751.9	26.88
Belgium	10.16	2,326.3	22.89	5,335.9	52.51	5,060.9	49.81
Bosnia	3.97	20.0	0.50	117.9	2.97	408.8	10.29
Bulgaria	8.22	430.0	5.23	738.0	8.97	2,881.8	35.04
Croatia	4.47	250.0	5.59	1,033.0	23.09	1,633.6[a]	36.49
Cyprus	0.68	120.0	17.65	218.3	32.11	440.1	64.72
Czech Republic	10.24	1,000.0	9.76	4,346.0	42.42	3,871.7	37.79
Denmark	5.33	1,950.0	36.59	3,363.6	63.11	3,835.0	71.95
Estonia	1.44	391.6	27.21	557.0	38.70	522.8	36.33
Faroe Islands	0.04	3.0[a]	6.73	17.0	37.71	25.0	55.45
Finland	5.18	1,927.0	37.23	3,728.6	72.04	2,847.9	55.02
France	58.89	8,500.0	14.33	29,052.4	49.33	34,114.0	7.93
Germany	82.26	24,000.0	29.18	48,202.0	58.60	50,220.0	61.05
Greece	10.65	1,000.0	9.39	5,932.0	55.73	5,659.3	53.16
Greenland	0.06	1,000.0	9.39	16.0	28.53	26.2	46.75
Guernsey	0.06	17.8	31.86	21.9	34.90	53.1	84.65
Hungary	10.20	20.0	31.90	3,076.3	30.17	3,798.3	37.25
Iceland	0.28	168.0	59.79	160.5	57.11	197.0	70.10
Ireland	3.79	784.0	20.70	2,490.0	65.75	1,590.0	41.99
Italy	57.30	13,200.0	23.04	42,246.0	73.73	27,153.0	47.39
Jersey	0.09	8.0	9.22	44.7	51.56	73.0	84.11
Latvia	2.42	150.0	6.19	401.3	16.55	734.7	30.31
Lithuania	3.70	225.0	6.08	524.0	14.17	1,187.7	32.11
Luxembourg	0.44	100.0	22.66	380.0	86.11	331.0	75.00
Malta	0.39	100.0	13.03	114.4	29.24	204.2	52.17
Moldova	4.38	52.6	120.0	139.0	3.17	583.8	13.33
The Netherlands	15.99	3,900.0	24.39	10,710.0	66.99	9,879.0	61.79
Norway	4.49	2,200.0	49.05	3,367.8	75.09	2,386.4	53.21
Poland	38.77	2,800.0	7.22	6,747.0	17.40	10,945.6	28.24
Portugal	10.02	2,500.0	24.94	6,665.0	66.49	4,313.6	43.03
Romania	22.33	800.0	3.58	2,499.0	11.19	3,899.2	17.46
Russia	146.93	3,100.0	2.11	3,263.2	2.22	32,070.0	21.83
Slovak Republic	5.41	650.0	12.03	1,109.9	20.53	1,698.0	31.42
Slovenia	1.99	300.0	15.11	1,215.6	61.21	767.3	38.63

Country	Inhabitants (millions)	Internet Users		Cellular Phones		Main Telephones	
		Thousands	%	Thousands	%	Thousands	%
Spain	40.60	5,387.8	13.27	24,736.0	60.93	17,101.7	42.12
Sweden	8.88	4,048.0	45.58	6,369.0	71.72	6,056.8	68.20
Switzerland	7.20	2,134.0	29.62	4,638.5	64.39	5,235.0	72.67
TFYR Macedonia	2.02	50.0	2.47	115.7	5.72	516.0	25.49
Turkey	65.70	2,000.0	3.04	16,133.4	24.56	18,395.2	28.00
Ukraine	50.46	300.0	0.59	818.5	1.62	10,074.0[a]	19.89
U.K.	59.77	18,000.0	30.12	43,452.0	72.70	35,177.0	58.86
Yugoslavia	10.64	400.0	3.76	1,303.6	12.25	2,406.2	22.61
Total Europe	798.26	109,034.6	13.66	291,755.4	36.55	315,145.1	39.47
Oceania							
Australia	19.16	6,600.0	34.45	8,562.0	44.69	10,050.0	52.46
Fiji	0.81	12.0	1.47	55.0	6.76	86.4	10.62
French Polynesia	0.23	15.0	6.43	21.9[a]	9.49	51.5	22.07
Guam	0.17	5.0[a]	3.04	27.2	16.19	80.3	47.80
Kiribati	0.08	1.0[a]	1.22	0.4	0.48	3.4	4.03
Marshall Islands	0.07	0.5[a]	0.80	0.4	0.66	4.0	5.87
Micronesia	0.12	4.0	3.38	—	—	9.6	8.15
New Caledonia	0.22	24.0	11.15	49.9	23.20	51.0	23.69
New Zealand	3.83	830.0	21.67	2,158.0	56.33	1,915.0	49.99
Northern Marianas	0.05	—	—	2.9[a]	5.59	26.8	50.55
Papua New Guinea	4.81	135.0	2.81	7.1[a]	0.15	64.8	1.35
Samoa	0.18	1.0	0.56	3.0[a]	1.69	8.5	4.73
Solomon Islands	0.44	2.0	0.46	1.2	0.26	7.7	1.76
Tonga	0.10	1.0[a]	1.02	0.1[a]	0.14	9.7	9.86
Vanuatu	0.20	3.0[a]	1.61	0.4	0.19	6.6	3.37
Total Oceania	30.46	7,633.5	25.12	10,889.5	35.88	12,375.4	40.62
Total worldwide	6,095.34	366,611.2	6.09	739,859.7	12.14	986,747.2	16.19

Source: With permission of the ITU elaborated from the *ITU Indicators Updates*, updated on January 9, 2002, for the end of the year 2000 on the basic indicators, the Internet indicators, and the cellular subscribers.

[a] Data for 1999.

[b] Data for 1998.

APPENDIX C

GLOSSARY

ABTC	American Bell Telephone Company
ACA	Australian Communications Authority
AEG	Allgemeine Elektricitäts-Gesellschaft
AEI	Associated Electrical Industries Ltd.
AIS	American Interplanetary Society
AM	Amplitude-modulated
AMPS	Advanced Mobile Phone System
APEC	Asia Pacific Economic Cooperation
APO	Australian Post Office
ARCOS	Americas Region Caribbean Optical-Ring System
Arento	Arab Republic of Egypt Telecommunication Organization
ARIB	Association of Radio Industries and Business (Japan)
ARPA	Advanced Research Projects Agency
ARS	American Rocket Society
ASCII	American Standard Code for Information Interchange
ASR	Automatic speech recognition
ATE	Automatic Telephone & Electric Co. Ltd.
ATM	Asynchronous transfer mode
AWG	American Wire Gauge
B2B	Business-to-business

Remark: Acronyms used together with the complete term and not repeated on subsequent pages are not included.

The Worldwide History of Telecommunications, By Anton A. Huurdeman
ISBN 0-471-20505-2   Copyright © 2003 John Wiley & Sons, Inc.

BBN	Bolt, Beranek & Newman
BCE	Bell Canada Enterprises Inc.
BERKOM	Berlin Kommunikation
BIGFON	Breitbandiges Integriertes Glasfaser-Fernmelde-Orts-Netz
BIS	British Interplanetary Society
B-ISDN	Broadband-ISDN
BPO	British Post Office
BTC	Bell Telephone Company
BTCA	Bell Telephone Company of Canada
BTM	Bell Telephone Manufacturing Company
C&C	Computers and Communications
C&W	Cable and Wireless
CC	Common control
CCIF	Comité Consultatif International de Communications Téléphoniques à Grande Distance
CCIR	Comité Consultatif International Technique des Communications Radioélectriques
CCIT	Comité Consultatif International de Télégraphy
CCITT	Comité Consultatif International Télégraphique et Téléphonique
CDMA	Code-division multiple access
CEPT	Committee of European Post and Telecommunications
CERN	Conseil Europeen pour la Recherche Nucleaire
CIT	Compagnie Industrielle des Téléphones
CITA	Canadian Independent Telephone Association
CNES	Centre Nacional d'Etudes Spatiales
CNET	Centre National d'Etudes des Télécommunications
COCOM	Coordination Committee (for East–West Trade Policy)
COMSAT	Communication Satellite Corporation
CRTC	Canadian Radio, Television, and Telecommunications Commission
CSF	Compagnie Générale de Transmission Sans Fil
CSF	Compagnie Sans Fil
CT	China Telecom
CT	Cordless telephone
CWHKT	Cable and Wireless Hong Kong Telecom
D-AMPS	Digital AMPS System
DARPA	Defense Advanced Research Projects Agency
DCME	Digital circuit multiplication equipment
DCS	Digital Cellular System
DDI	Daini Denden Inc. (Japan)
DDM	Direct digital modulation

DEBEG	Deutsche Betriebsgesellschaft für drahtlose Telegraphie
DECT	Initially, Digital European Cordless Telecommunications; later, Digitally Enhanced Cordless Telecommunications
DIC	Dubai Internet City
DNS	Domain Name System
DPLMRS	Domestic Public Land Mobile Radio Service (USA)
DSI	Digital speech interpolation
DSL	Digital subscriber line
DS-SS	Direct sequence spread-spectrum
DTH	Direct-to-home (satellite broadcasting)
DWDM	Dense wavelength division multiplexing
EAPTA	East African Post and Telecommunications Administration
ECL	Electrical Communications Laboratories (of NTT)
ECO	Electronic central office
EDFA	Erbium-doped fiber amplifier
EDGE	Enhanced data rates for GSM evolution
EIA	Electronic Industries Association
EMAX	Electronic multiplex automatic exchange
EMBRATEL	Emprêsa Brasileira de Telecomunicações
EMD	Edelmetal-Motor Drehwähler (noble-metal motor switch)
EPABX	Electronic private automatic branch exchange
EPIRB	Emergency position indicating radio beacon
ESA	European Space Agency
ESK	Edelmetall-Schnell-Kontakt (noble-metal contact high-speed relay)
ESM	Elektronisches System mit Magnetfeldkoppler (electronically controlled magnetic field switching matrix)
ESSEX	Experimental solid-state exchange
ETB	Empresa de Telecomunicaciones de Bogota
ETSI	European Telecommunications Standards Institute
EWS	Elektronisches Wähl System (electronic switching system)
FCC	Federal Communications Commission
FDD	Frequency-division duplex
FDMA	Frequency-division multiple access
FIRST	Fiber to the residential subscriber terminal
FITL	Fiber in the loop
FLAG	Fiber-optic link around the globe
FM	Frequency modulation
FNC	Federal Networking Council (Internet)
FPLMTS	Future Public Land Mobile Telecommunications System
FSO	Free space optics
FTR	Federal Telephone and Radio (U.S.)

FTTH	Fiber to the home
GE	General Electric Company (U.S.)
GEC	General Electric of Coventry (U.K.)
GEO	Geostationary earth orbit
GERAN	GSM/EDGE radio access network
GII	Global information infrastructure
GIRD	Gruppa Isutschenija Reaktiwnogo Dwishenija
GMDSS	Global maritime distress and safety system
GMPCS	Global mobile personal communication by satellite
GPO	General Post Office (U.K.)
GPRS	General packet radio service
GPS	Global positioning system
GSM	Global system for mobile communication
GTO	Geostationary transfer orbit
GVD	Group velocity dispersion
HDW	Heb-Dreh-Wähler (lift-turning switch)
HEO	Highly elliptical orbit
HF	High frequency
HKTC	Hong Kong Telephone Company
HLR	Home location register
HSCSD	High-speed circuit-switched data
HTML	Hypertext markup language
HTTP	Hypertext transfer protocol
IBTC	International Bell Telephone Company
ICANN	Internet Corporation for Assigned Names and Numbers
ICE	Instituto Costarricense de Electricidad
ICO	Intermediate circular orbit
IDD	International direct dialing
iDNS	Internationalized Multilingual Multiscript Domain Name System
IDO	Nippon Indou Tsushin Corp.
IEE	Institution of Electrical Engineers (U.K.)
IEEE	Institution of Electrical and Electronics Engineers (U.S.)
IES	International e-commerce Security Standard
IETF	Internet Engineering Task Force
IFRB	International Frequency Registration Board
II	Interplanetary Internet
ILS	International Launch Service
IMEI	International mobile station identity
IMO	International Maritime Organisation
IMP	Interface Message Processor (ARPA)

IMSO	International Mobile Satellite Organisation
IMT	International Mobile Telecommunication System
IMTS	Improved Mobile Telephone System
INMARSAT	International Maritime Satellite Organisation
INTELSAT	International Telecommunications by Satellite
IP	Internet Protocol
IPP	Internet Printing Protocol
IPR	Intellectual property right
IPTO	Information Processing Techniques Office (ARPA)
IRE	Institute of Radio Engineers
ISDN	Integrated services digital network
ISP	Internet Service Provider
ISS	International Switching Symposium
IT&T	International Telephone & Telegraph Company
ITT	International Telephone & Telegraph Company (from 1958)
ITU	International Telecommunication Union
JERC	Joint Electronic Research Committee (Japan)
JTACS	Japan TACS
KDD	Kokusai Denshin Denwa Corporation Ltd. (Japan)
LAN	Local area network
LCT	Laboratoire Central des Télécommunications
LEO	Low earth orbit
LMT	Le Matériel Téléphonique
MA	Multiple access
MCI	Microwave Communications Incorporated
MEO	Medium earth orbit
MIT	Massachusetts Institute of Technology
MSC	Multimedia Super Corridor (Malaysia)
NAMPS	Narrowband-AMPS
NAP	Network Access Point (Internet)
NASA	National Aeronautics and Space Administration
NBTC	National Bell Telephone Company
NBTM	Nederlandse Bell-Telephoon Maatschappij
NCP	Network Control Protocol
NMT	Nordic Mobile Telephone
NPL	National Physical Laboratory (U.K.)
NSF	National Science Foundation
NSFNET	National Science Foundation Network
NTACS	Narrowband TACS
NTCA	National Telephone Cooperative Association (U.S.)
NTSC	National Television Standard Committee

NTT	Nippon Telegraph & Telephone Public Corporation
o/w line	Overhead wire
OFTA	Office of the Telecommunications Authority (China)
OPGW	Optical ground wire
OTC	Overseas Telecommunications Commission
OTI	Odyssey Telecommunications International
PABX	Private automatic branch exchange
PACS	Personal access communications system
PAM	Pulse amplitude modulation
PBX	Private branch exchange
PCCW	Pacific Century Cyber Works
PCM	Pulse code modulation
PCN	Personal Communications Network
PCS	Personal communications services
PDC	Personal digital cellular
PDH	Plesiochronous digital multiplex hierarchy
PHS	Personal Handyphone System
PM	Pulse modulation
P-MP	Point-to-multipoint
PMR	Private mobile radio
POF	Plastic optical fiber
POTS	Plain old telephone service
PSDN	Public-switched data network
PSTN	Public-switched telephone network
PT	Powerline telecommunications
RASCOM	Regional African Satellite Communications Organisation
RBOC	Regional Bell Operating Company
RF	Radio frequency
RITL	Radio in the loop
RTMS	Radio Telephone Mobile System
RTP	Realtime transport protocol
SAPO	South African Post Office
SAPT	South African Post and Telecommunications
SAT	Stockholm Allmänna Telefonaktiebolag
SCTP	Stream Control Transmission Protocol
SEA-ME-WE	South-East Asia, Middle East, Western Europe cable
SEL	Standard Elektrik Lorenz AG (Germany)
SES	Société Européenne des Satellites
SESA	Standard Elèctrica, S.A. (Spain)
SFR	Société Française de Radiotéléphonie
SFR	Societé Française Radioeléctrique

SGT	Sociètè Générale des Telephones (France)
SIM	Subscriber identity module
SingTel	Singapore Telecommunications Ltd.
SIS	Subscriber identity security
SIT	Sociètè Industrielle des Téléphones (France)
SLE	Société Lannionaise d'Electronique (France)
SMS	Small message service
SNA	System network architecture
SNR	Signal to noise ratio
SOHO	Small office and home office
SONET	Synchronous optical network
SPC	Stored program control
SRI	Stanford Research Institute
STC	Standard Telephone and Cables Ltd. (U.K.)
STET	Società Torinese Esercizi Telefonici (Italy)
STL	Standard Telephone Laboratories (U.K.)
STM	Synchronous transport module
TACS	Total Access Communication System
TAE	Trans-Asia-Europe Optical Fibre Cable System
TAS	Telecommunication Authority of Singapore
TAT	Transatlantic telephone cable
TCNZ	Telecom New Zealand
TCP	Transmission control protocol
TCP/IP	Transmission control protocol/Internet Protocol
TDD	Time-division duplex
TDMA	Time-division multiple access
TdP	Telefónica de Perú
TELEBRAS	Telecomunicações Brasileiras SA
TIA	Telecommunications Industry Association (U.S.)
TLD	Top-Level Domain name
TSF	Société Anonyme de Télégraphe sans Fil (France)
TV-RO	TV-receive only
TWJ	TeleWay Japan Corp.
TWT	Traveling-wave tube
TWX	Teletypewriter exchange service
UCLA	University of California, Los Angeles
UCSB	University of California, Santa Barbara
UMTS	Universal Mobile Telecommunications System
URL	Uniform resource locator
USITA	U.S. Independent Telephone Association
USTSA	U.S. Telephone Suppliers Association

UHF	Ultra-high frequency
VfR	Verein für Raumschiffahrt
VHF	Very-high frequency
VoIP	Voice over Internet Protocol
VSAT	Very small aperture terminal
WAC	Women's Auxiliary Corps
WACS	Wireless access communications system
WAP	Wireless Access Protocol
WDM	Wavelength-division multiplexing
WLL	Wireless local loop
WML	Wireless markup language
WTO	World Trade Organization
WU	Western Union Telegraph Company
WWW	World Wide Web

INDEX

PART I   PERSONS

Adams, P. R., 495
Ader, Clement-Agnes, 169
Albert:
    J. Wilh., 154
    Valentin, 54
Alberti, Leon Battista, 351, 352
Aldrin, Edwin, 409
Alexanderson, Ernst F. W., 278
Ampère, André M., 31
Anderson:
    Carl Johan, 175
    James, 136
Andreesen, Marc, 586
Annan, Kofi, 551, 563
Apostolov, S. M., 193
Arago, Dominique F. J., 11
Arago, François, 72
Araki, K., 520
Arco, George W. A. H. von, 213
Armstrong:
    Edwin, 290
    Neil, 409
    Samuel T., 94
Arnold, H. D., 314
Aschkinass, Emil, 204
Autry, Gene, 145

Bain, Alexander, 65, 66, 72, 76, 147, 149
Bakewell, Frederick C., 66, 149
Bardeen, John, 364
Barkhausen, Heinrich Georg, 338
Barré, Major J. I., 409
Bartiger, Bary, 436
Baudot, Maurice-Émile, 107, 301
Bazieres, Etienne, 351
Becquerel, Alexandre E., 150
Belin, Edouard, 295
Bell:
    A. Graham, 12, 65, 88, 155–181, 195, 231, 270,
        314, 324, 328, 445
    Alexander, 159
    Alexander Melville, 159
    Alice (later Alice Todd), 120, 122–124
    Chichester A., 179
    Mabel, 12, 165, 167, 180
    Melville, 159
Bennett, James Gordon, 135
Bentley, Ernest Lungley, 351
Bergsträsser, Johann A. B., 16
Berliner, Emile, 178
Berners-Lee, Tim, 586
Bétancourt, Augustin, 37, 49
Bethencourt y Mollina, Augustin de, 49
Betulander, Gotthilf A., 262, 264
Bill, Buffalo, 99

The Index covers Chapters 1 to 35, excluding footnotes, figures, tables, and technology boxes.

The Worldwide History of Telecommunications, By Anton A. Huurdeman
ISBN 0-471-20505-2   Copyright © 2003 John Wiley & Sons, Inc.

Blake:
  Jr, Francis, 178
  Dr. Clarence J., 160
Blauvelt, 256
Boerner, M., 446
Boettger, Professor, 154
Bois Duddell, William du, 274
Bonelli, Gaetano, 149
Bose, Jagadis Chunder, 204
Bourseul, Ch., 88, 153, 154, 328
Bragg, William, 123
Branly, Edouard, 205, 206
Braun:
  Karl F., 202, 211, 215
  Wernher von, 407–411
Breese:
  Elizabeth Ann, 55
  Judge Samuel, 55
Breguet:
  Abraham-Louis, 13, 20, 22, 37, 169
  Antoine, 22, 168
  Louis, 22
  Louis F. C., 168
  Louis-François, 70, 73
Brett, John, 106, 129–131
Bright, Charles T., 129–132
Brittain, Walter, 364
Brooke, Lieutenant J. M., 131
Brunel, Sir Isambard K., 97
Buckley, Oliver E., 314
Bulkley, Charles S., 63
Butterfield, John J., 63

C. M. (Charles Marshall?), 48
Caillau, Robert, 586
Callender, Romaine, 238, 252
Cambell, George, 318
Cambell-Swinton, A. A., 208
Cannstadt, Baron Pavel Lvovitch Schilling von,
    30, 54, 55, 66
Carlile, Antony, 30
Carnegie, Andrew, 145
Carr, Major G. A., 210
Carty, John. J., 247, 279, 316
Caselli, Giovanni, 149, 150
Catterton, John, 132
Cedergren, Henrik T., 176, 193
Cerf, Vinton G., 585, 587, 589
Chappe:
  d'Auteroche, Jean, 18
  Abraham, 18, 35, 37
  Claude, 11, 14–18, 34, 35, 55
  Ignace, 34, 37
  Ignace Urbain, 18, 23
  Jean, 18
  Pierre-Fr., 18, 34, 37
  Renè, 18, 37
Chow, Martin, 447

Christie, Samuel Hunter, 67
Clark:
  Dave, 585
  Welden, 583
  Arthur C., 407–411, 419, 435
Clavier, Dr. André, 342, 343
Cleator, Philip Ellaby, 408, 409
Clement, Edward E., 251
Colussy, Dan, 436
Connolly, D. and Th., 193
Cooke, William Forthergill, 33, 55, 66–72, 107
Cooper, Peter, 131
Cornell, Ezra, 60–64, 98
Courrejolles, de, 16
Creed, Frederick G., 66, 303, 304
Crippen, Dr. Henry Hawley, 282
Crookes, William, 205
Crossley, Louis John, 168
Curtis, Austin, 280

Damm, Arvid Gerhard, 355
Davies, Donald Watts, 583
Delaunay, Léon, 20
Della Porta, Giovanni B., 351
Deloraine, Dr. M., 342, 347, 495
Deloy, Leon, 280
Depew, Chauncey M., 165
Depillon, 39
d'Erlanger, Baron Emile, 133
Dolbear, Amos E., 176–178
Dollond, John, 16
Drawbaugh, Daniel, 177
Ducretet, Eugène, 212, 213
Dujardin, Dr. M., 73

Edgwort, Richard Lovell, 16
Edison, Thomas Alva, 145, 159, 168, 178, 192,
    205, 225, 277
Einstein, Albert, 450
Ellis, Alexander J., 159
Ellsworth, Annie G., 61
Elmen, G. W., 314
Elwell, C. F., 275, 290
Engelbart, Dough, 584
Erickson, John and Charles, 195
Ericsson, Lars Magnus, 174
Esnault-Pelterie, Robert, 409–410
Espenschied, L., 495
Estaunié, Edouard, 3
Euler, Leonhard, 16
Evans, Cyril, 282
Everest, George, 390

Faller, Ernst, 251
Faraday, Michael, 32, 66, 134, 199
Fardely, William, 70, 74, 76
Faxton, Theodore S., 63

Feddersen, Berend Wilhelm, 200
Ferrier, Alexandre, 39
Fessenden, Reginald A., 276–277
Field:
    Cyrus W., 130–132, 141, 164
    David Dudley, 130
    Mathew, 131
Fischer:
    Henry C., 169
    J. F., 97
Fisher, J. C., 60
FitzGerald, George F., 201, 203
Fleming:
    John A., 225, 269
    Sir Sandford, 309
Flowers, Thomas H., 355, 496
Forbes, William H., 167
Forest, Lee de, 226, 231, 277, 279, 322, 364
Foucault, Léon, 149
Foy, Alphonse, 37, 39, 73
Franklin, Ch. S., 333
Fred, H., 280
Freidenberg, M. F., 193
Froment, Gustave, 103, 150

Gale, Leonhard, 56–60, 141
Galvani, Luigi, 29
Ganswindt, Hermann, 407, 408
Garbasso, Antonio Giorgio, 204
Garnier, 70
Gates, Bill, 440
Gauß, Carl Friedrich, 50
Gavey, John, 210
Gavin, Paul V., 286
Gerke, Friedrich C., 76, 143, 217
Gifford, George, 178
Gisborne, Frederic N., 104, 130
Goddard, Robert H., 408
Goldschmidt, Rudolf, 278
Gore, Al, 581, 582
Gould, Jay George, 135
Gower, Frederic A., 168
Gray, Elisha, 151, 156, 162, 164, 176, 178
Green, Norvin, 178
Griffin, David, 63
Grocco, General Arturo, 409
Grosvenor, Gilbert M., 179
Gröttrup, Helmut, 407, 411
Grzanna, Gustav, 294
Gurlt, Wilhelm, 81

Hagelin, Boris, 355
Hall, E. K., 231
Halske, Johann Georg, 78, 81, 94
Hartfield, John Charles, 351
Haslam, Pony Bob, 99
Hayes, Hammond V., 192
Heart, Frank, 584

Heaviside, Oliver, 269, 280, 319
Hebern, Eduard Hugo, 352
Hell, Rudolf, 296
Helmholtz, H. von, 159, 201
Helu, Carlos Slim, 557
Henry, Joseph, 32, 57, 61, 67, 93, 141, 160, 199
Hertz, Heinrich R., 12, 90, 201–206, 343
Heurtley, 312
Hockham, George A., 446
Hofer, Andreas, 36
Hohmann, Walther, 407, 408
Holst, Gilles, 227
Hooke, Robert, 15, 16, 153
Hoover, Herbert, 232
Horiguchi, Masaharu, 448
House, Royal E., 65, 98
Hubbard:
    Gardiner G., 160–167, 179
    Mabel, 160, 162, 163, 165
Hughes, David E., 65, 103, 155, 168
Hultman, Axel, 261
Humboldt, A. von, 54, 72, 76
Hunnings, Reverend Henry, 168
Huth, Gottfried, 153
Hutin, Maurice, 324

Iwerson, J. E., 497

Jackson, Charles T., 55, 56
Jacob, Frank, 316
Jameson-Davis, Colonel, 208, 210
Johnson:
    Cave, 61
    William, 168
Jones, Professor Viriamu, 210

Kahn, Robert E., 585, 587
Kao, Charles K., 446
Kapron, Felix, 447
Karolus, August, 296
Keck, Donald, 447
Keith, Alexander E., 195, 237
Kelvin, Lord, 96, 303
Kempelin, Baron de, 159
Kendall, Amos, 58, 61, 63
Kennedy, Captain J. N. C., 210
Kennelly, Arthur E., 269, 280
Kerespertz, Count, 37
Kilby, Jack, 365
Killian, James A., 583
Kirchhoff, Gustav, 201
Kirstein, Peter, 585
Kitahara, Y., 505
Kleinrock, Leonhard, 583, 584
Kleinschmidt, Ernst Eduard, 302
Kleist, Ewald Jürgen von, 49
Klingenstjerna, Samuel, 16

Kobayashi, Koji, 581
Koch, Hugo Alexander, 352
Koenig, Dr. Rudolph, 155
Korn, Arthur, 294
Koroljow, Sergeji, 410
Kramer, Dr. A., 70, 76, 79
Krarup, Carl Emil, 321
Krum, Ch. and H., 301

Langmuir, Irving, 226
Lasser, David, 409
Latour, Marius, 278
Leblanc, Maurice, 324
Leonhardt:
   Dr.-Ing. Fritz, 393
   Ferdinand, 70, 76, 78
Leopold, Raymond J., 436
Lesage, Georges-Louis, 49
Lewert, David Friedrich, 70, 81
Ley, Willy, 407, 409
Lhuys, M. Drouyn de, 219
Licklider, Joseph C. R., 583, 585
Lieben, Robert von, 226, 322
Linnaeus, Carolus, 49
Lippershey, Hans, 15
Lockwood, Thomas D., 177
Lodge, Oliver J., 201, 203, 205
Loomis, Mahlon, 200
Lorimer, G. W., E. and J. H., 252
Lowell, Judge, 177
Lundquist, Frank A., 195

Mackay, John William, 135
MacKinnon, F. B., 231
Madden, O. E., 167
Maiman, Theodore, 445
Maitland, Sir Donald, 546, 551
Majorana, Quirino, 275
Mäkitalo, Östen, 523
Mandela Nelson, 563
Marconi, Guglielmo Marchese, 12, 90, 177, 201,
   207–215, 225, 269, 270, 280–285, 289–291,
   375
Marill, Thomas, 583
Marzi, G. B., 193
Maurer, Robert F., 447
Maury, Matthew Fontaine, 131
Maxwell, James Clerk, 200
McCaw, Craig, 438, 440
McDonough Collins, Perry, 100
McGowan:
   Samuel, 116
   William, 552
McLuhan, Marshall, 589
McMinn, Gilbert, 122
McNair, Archibald, 94
McTighe, Thomas J., 193
Meissner, A., 226, 279, 337, 343

Mélito, Miot de, 20
Melot, Henri, 408
Mercadier, Ernest, 324
Metcalfe, Robert, 503
Meyer, Bernhard, 151
Mills, William, 122
Mockapetris, Paul, 585
Mollenauer, Linn, 452
Moncabrier, 39
Monge, 37
Montgomery, Dr., 94
Monti, Mario, 575
Moore, Gordon E., 366
Morse:
   Jedediah, 55
   Samuel, 11, 33, 55–66, 72, 73, 81, 97, 98, 103,
     123, 131, 141, 160, 167
Morton, Sterling, 302
Mostsisky, K. I., 193
Moulton, Fletcher, 208
Muirhead, Alexander, 206
Muncke, George Wilhelm, 54, 66
Murray:
   Donald, 301, 303
   George, 50
Musashino, I, 493
Muschenbroek, Pieter van, 49

Nakazawa, Masataka, 452
Nebel, Rudolf, 407
Nelson, Theodor Holm, 586
Newman, Maxwell H. A., 355
Niaudet, Alfred, 169
Nicholson, William, 30
Nobili, Leopoldo, 32
North, Charles H., 251
Noyce, Robert N., 366
Nutt, Emma M., 191

O'Shaughnessy, Sir William Brooke, 113, 125
Oberth, Herman, 407, 408, 411
Oerstedt, Hans Ch., 31, 54, 199
O'Etzel, Franz August, 76
Ohm, Georg Simon, 32
Ollson, Herman, 261
Orbeliani, Grigola, 109
O'Reilly, Henry, 63, 65, 98
Orton, William, 141, 165, 178
Osanai, Hiroshi, 448
Oslin, George P., 101
Otschenaschek, Ludwig, 409

Partridge, Craig, 585
Pasley, Charles William, 39
Patocchi, Michele, 182
Pearne, Frank, 301
Peary, Robert Edwin, 270
Pedro II, Emperor of Brazil, 164

Pellow, Nicola, 586
Pender, John, 132, 135, 137
Pendray, G. Edward, 409
Penrose, Major C., 210
Peterson, Ken, 436
Phelps, George M., 178
Phillips, J. G., 282
Pierce, John R., 364
Pirquet, Guido, 409, 410
Playfair, Earl St. Andrews of, 352
Poggendorf, Johan Ch., 31, 155
Pollermann, Bernd, 586
Pope, Frank L., 178
Popoff, Alexander St., 206, 207, 212, 357
Postel, Jon, 585
Poulsen, Valdemar, 274, 290
Pouyer-Quertier, M., 134
Preece, William H., 167, 208, 210
Prescott, George B., 169
Proudfoot, John, 138
Pupin, Michael I., 12, 90, 318
Puskás, Ferenc, 193

Ramsey, Murray, 456
Raps, August, 104
Rayleigh, Lord John W. S., 333
Reeves, Alec H., 12, 327, 347, 366, 446, 495
Reis, Philipp, 88, 154–156, 160, 177, 328
Reuter, Julius, 87, 133
Reynolds, J. N., 264
Rheita, A. M. Schyrleus de, 15
Ridings, Garvice, 515
Riedel, Klaus, 407
Righi, Augusto, 204, 207
Ritchie:
    Foster, 152
    William, 32
Rive, Lucine de la, 204
Roberts:
    F. F., 447
    Lawrence C., 583, 585
    Marshall O., 131
Robinson, William, 76
Rogers, Henry, 61, 351
Ronalds, Francis, 49
Ruhmer, Ernst, 277, 324
Rühmkorff, Heinrich Daniel, 199
Russell, William H., 99
Rutherford, Ernst, 204
Rynin, Nikolai A., 408, 410

Saint-Haouen, Vice Admiral de, 39
Salvá y Campillo, Fr., 30, 49
Sanders, Thomas, 160, 165
Sänger, Eugen, 409
Sarasin, Edouard, 204
Scantlebury, Roger, 583
Schaefer, Bela, 210

Scherbius, Albert, 352
Schönborn, Johann Ph. von, 15
Schottky, Walter, 227
Schultz, Peter, 447
Schweigger, Johann S. Ch., 31
Scott, John, 451
Scribner C. E., 189
Scrymser, James A., 138, 139
Selden, Judge Samuel L., 98
Shaver, William, 447
Shockley, William, 364
Sholes, Christopher L., 300
Short, Mike, 530
Shreeve, Colonel, 280
Sibley, Hiram, 98, 99, 100
Siemens:
    Carl, 125, 134
    Georg, 134
    Walter, 125
    Werner, 78, 94, 125, 130, 171
    William, 78, 94, 97, 125, 134
Sinclair, Dave, 193
Slaby, Adolf, 210, 213
Slingo, Sir William, 324
Smith:
    Chauncy, 177
    Francis Ormond J., 58, 61, 351
    Major General W. F., 138
Soemmering, S. Th. von, 30, 54
Speed, John J., 65
Squier, George O., 324
Staino, Ed, 436
Steinheil, Carl A. von, 52, 75
Stephan, H. von, 108, 169–171
Sterns, J. B., 97
Stöhrer, Emil, 70, 76, 81
Stone, John, 318
Storrow, James J., 177
Strowger:
    Almon B., 12, 90, 193–195, 198, 237
    Walter S., 13, 194
Stuart, John McDouall, 119
Sturgeon, William, 32, 55
Susskind, Professor Charles, 207
Sutherland, Ivan, 583
Symonds Bell, Eliza Grace, 159

Tagare, Neil, 463
Tainter, Charles Summer, 179
Tawell, John, 71
Taylor, Moses, 131
Tellegen, Bernardus D. H., 227
Terjanne, Pekka, 463
Tesla, Nicola, 205, 277
Thomson:
    Alice, 124
    Silvanus, 156
    William, 96, 131, 132, 164, 200, 201, 308

Todd:
  Alice, 120, 123
  Charles Heavitree, 119–124
Tomlinson, Ray, 584, 585
Tschertok, Major Boris, 411
Tuchatschewski, Michael, 408
Tue, William Clausen, 351
Turing, Alan M., 355
Tyndall, John, 445

Uda, Shintaro, 281
Utsumi, Yoshio, 551

Vail:
  Alfred L., 13, 57–61, 141, 143, 167
  Judge Stephen, 57, 167
  Theodor N., 167, 179, 551
Valier, Maximilian, 407, 409
Van Allen, Dr. James, 412
Vaschy, 319
Vaughan, H. Earle, 496
Volta, Alessandro, 29, 30

Wade, Jeptha Homer, 100
Walker, Lucretia Pickering, 56
Watson, Thomas A., 13, 160–163, 179, 232
Weber, Wilhelm Eduard, 50
Wendt, Johann Wilhelm, 76
Westinghouse, George, 193
Weyde, Henri van der, 155
Wheatstone, Charles, 33, 55, 66–72, 76, 93, 107,
  132, 153, 159, 352
White, Chandler, 131
Whitehouse, Edward O. W., 132
Wien, Max, 271
Williams, Charles, 160, 165
Wiman, Erastus, 105
Winkler, Johannes, 407
Winterbotham, F. W., 355
Wise. Raleigh G., 298
Wologdin, Valentin P., 279
Wood, Orrin S., 61, 66
Woodbury, Judge, 141

Yagi, Hidetsugu, 281
Young, William, 94

Zander, Friedrich A., 408
Ziolkovsky, Konstantin E., 408

PART II   COMPANIES AND INSTITUTIONS

360networks, Canada, 463

AB Cryptoteknik, 355
ABB Norsk, 460

ABTC, U.S., 179, 238
ACA, Australia, 577
AEG, Germany, 214, 226, 271
African Direct Telegraph Co., 137
Afrisat, 432
Agence France Press, 88
AirFiber, 390
AIS, U.S., 409
Aktiebologat Cryptograph, 355
Alcatel, 181, 421, 431, 441, 455, 458, 460, 463,
  464, 467, 470
  ATFH, 377
  BTM, 179
  SEL, 170, 267, 380
  Standard Electrica, 377
Alestra, 557
Alexander Graham Bell Association for the Deaf,
  179
Aliant, Canada, 556
All America Cables, 138, 291
All-Canadian Government Telegraph Co., 105
Alpha Lyracom, 430
Amalgamated Radio Telegraph Company Ltd.,
  275
American Cable & Radio, 139
American Machine Telephone Co., 252
American Rocket Society, 409
American Telegraph and Cable Co., 135
American Telegraph Co., 103, 104
Ameritech, 475, 554, 577
ANATEL, Brazil, 559
Andinatel SA, Ecuador, 562
ANET, Columbia, 562
Anglo-Irish Magnetic Telegraph Co., 129
Anglo-Mediterranean Telegraph Co., 136
ANTEL:
  El Salvador, 258
  Uruguay, 563
  Paraguay, 563
AOL, 430, 535
APO, Australia, 510
ARABSAT, 428
Arento, Egypt, 565
ARIB, Japan, 540
Aries, United States, 435
ARPA, United States, 583
Asahi Glass, 477
Asia Global Crossing, 465
AsiaSat, 430
Associated Electrical Industries Ltd., 490
Associated Press, 88
ASTC, United States, 178
ASTRA satellite, 429
Astrium, 421
Astrolink, 440
AT&T, 10, 13, 181, 192, 226–231, 237, 238, 247,
  248, 251, 254, 259, 265, 280, 290, 302, 318,
  323–326, 334, 348, 364, 375, 376, 400, 413,

416, 420, 455, 456, 460, 467, 482, 487, 489,
    491, 505, 520, 527, 548, 549, 552, 554, 557,
    572, 592
  Canada, 457
  Submarine Systems, 463, 466
Atlantic Telegraph Co., 131, 270
Atlantic, Lake, and Mississippi Telegraph Co.,
    63
AUSSAT, Australia, 432, 577
AUSTEL, Australia, 577
Austrian–German Telegraph Union, 81, 88, 141,
    143, 217
Autel, 248
Autelco, 237, 238, 241, 242, 482
Automatic Electric Co., 237
Automatic Telephone Manufacturing Co. Ltd.,
    244
Automatic Telephone & Electric Co. Ltd.,
    490
Avantel, 557

Baby Bell, 554, 577
Baby Bras, 559
BASKOM, 475
BBN, Bolt Beranek & Newman, 584, 585
Belize Telecom. Ltd., 558
Bell Atlantic, 554, 555, 556, 577
Bell Canada, 555, 577
Bell Canada Enterprises Inc., 431
Bell Congo, 567
Bell Nexxia, 556
Bell Patent Association, 160
Bell South, 475, 554, 566
Bell South Chile, 560
Bell Systems Laboratories, 555
Bell Telephone Co. Ltd., U.K., 167
Bell Telephone Laboratories, 13, 181, 234, 326,
    331, 333, 347, 364, 367, 375, 398, 410, 413,
    448, 452, 480, 485, 489, 495, 519
Bellcore, 539
Bharti Telenet Ltd, India, 569
Binariang, Malaysia, 573
Blakey and Emmott, 168
Boeing, 421
Bolivarsat, 430
Böttcher und Halske, 78
BPO, 574
Brasilsat, 429, 433
Brazilian Submarine Telegraph Co., 137
British Aerospace, 433
British Australian Telegraph Co., 136
British Eastern Telegraph Co., 309
British Electric Telegraph Co., 106
British Indian Submarine Extension Co., 136
British Interplanetary Society, 409
British North American Electric Telegraph
    Association, 104
British Telecom, 472, 524, 574, 575

British Telegraph Construction and Maintenance
    Co., 309
British–Indian Telegraph Administration, 125
British–Indian Telegraph Co., 126
BT, U.K., 390, 474, 467, 468, 549, 572–575, 577,
    592
BT Skyphone System, 532
BTC, U.S., 165, 194
BTM, Belgium, 179, 248, 259, 266, 569

C&W, 289–291, 377, 404, 460, 461, 463, 465, 558,
    566, 570, 572–574
C&W Optus, 577
C&WIDC, 572
C. Lorenz AG, 275, 278, 295, 306, 343, 345, 349,
    354, 375, 380
Câbles de Lyon, 327
California State Telegraph Co., 98
Canadian Overseas Telecommunications Co., 400
CANTV, Venezuela, 561
Cape of Good Hope Telegraph Co. Ltd., 116
CCIF, 358, 598
CCIR, 359, 369, 435
CCIT, 358, 598
CCITT, 367, 369, 397, 484, 489, 491, 504, 512,
    516, 598
Celcom, Malaysia, 573
Cellnet, U.K., 524
Celsat, 435
Centel, USA, 554
Central and South America Telegraph Co., 138
CEPT, 528, 529, 536
CERN, 586
Cesky Telecom, 576
Chase Manhattan Bank, 560
Chiffriermaschinen-Gesellschaft Heimsoeth und
    Rinke, 353
Chilean Bond Corporation, 560
Chilesat, 560
China Great Wall Industry Corporation, 424
China Netcom, 570
China Submarine Telegraph Co., 136
China Telecom, 570
China Unicom, 570
ChinaSat, 570
Cisco, 572
CIT, France, 490
Citibank, 561
Clear Communications Ltd., New Zealand, 577
CNES, France, 424
CNET, France, 455, 490
CNT, Peru, 561
COCOM, 531, 576
Cofetel, Mexico, 557
Commercial Cable Co., U.S., 135
Commercial Pacific Cable Co., U.S., 309
Commercial Telegraph Co., U.S., 65
Compac Computer Corporation, U.S., 588

Compagnie Française de Câbles Sous-Marins et de Radio, 567
Compagnie Française des Cables Télégraphique, 135
Compagnie Française du Télégraphe de Paris á New York, 134
Compagnie Générale de Télégraphie, 290
Compagnie Générale Radioélectrique, 278, 279
Compagnie Russe des télégraphes et des téléphones sans fil, 279
Companhia Telegrafica Platino-Brasiliera, 139
Compañía de Teléfonos de Coyhaiqu, Chile, 560
Compañía Nacional de Teléfonos, Chile, 560
COMSAT, 427, 433
COMTELCA, Central America, 377
Conatel, Venezuela, 561
Concert, U.K./U.S., 549
Consolidated Telephone Construction and Maintenance Co., U.K., 167
Continental Tel, U.S., 554
Copenhagen Telephone Co., 472
CORFO, Chile, 559
Corning, 447, 456
CPT, Panama, 558
CPT, Peru, 561
Creed & Co., 304
Crosswave Communications, Japan, 572
CRT, China Railway Telecom, 570
CRTC, Canada, 556
Crypto AG, 356
CSF, France, 343
CTC Mundo, Chile, 560
CTS, Peru, 561
Cuban–American Telephone and Telegraph Co., 324
Cushman Telephone Co., 195
Customers Service Institute of Australia, 594
CWHKT, 432
Cyberstar, 441

Dacom, Korea, 573
DARPA, 584, 585
Datang Telecom, China, 458
DDI, Japan, 534, 571
De Forest Radio Telephone Co., 277
DEBEG, Germany, 273
Det store nordiske Telegraf-Selskab, 110
Detecon, Germany, 567
DeTeMobil, Germany, 529
Deutsche Bundespost, 10, 475, 492
Deutsche Bundespost Telekom, 572
Deutsche Edison Gesellschaft, 214
Deutsche Telekom AG, 549, 455, 575
Deutsche Telekom Japan K.K., 572
Deutsch-Niederländische Telegraphengesellschaft AG, 310
DIC, Dubai, 594

DiGi Telecommunications, Malaysia, 573
Digital Phone Group, Japan, 534
Direct United States Telegraph Co., 134
Direct West India Cable Co., 137
DoT, India, 568
DuPont, 477
Dutch PTT, 474

E. Remington & Sons, 301
EAPTA, East Africa, 567
Eastern & South African Telegraph Co., 137
Eastern and Associated Telegraph Companies, 138
Eastern Extension Australasia and China Telegraph Co., 128, 136
Eastern Telegraph Company, 136, 291
ECL, Japan, 448, 493, 520
Edison Telephone Co. of London Ltd., 168
Egypt Telecom, 565
EIA, U.S., 533
Electric Telegraph Co., U.K., 71, 72, 106
Electrosvyaz, Russia, 576
Ellipsat, 437
Ellipso, 435
EMBRATEL, Brasil, 558
English and Irish Magnetic Co., 106, 136
ENITEL, Nicaragua, 558
ENTEL:
   Bolivia, 563
   Chile, 559
   Argentina, 561
   Ecuador, 562
   Peru, 561
Erie and Michigan Telegraph Co., 65
ESA, Europe, 424
Essar Commvision, India, 569
ETB, Columbia, 562
Etisalat, Dubai, 594
ETSI, 529, 530, 531, 537, 540
Eurasiasat, 431
Europe*Star Ltd., 431
Eutelsat, 429, 435, 441
Euteltracs, 435

FACE, Italy, 349, 380
FAISAT, 441
Falmouth Gibraltar and Malta Telegraph Co., 136
FCC, 286, 440, 520, 552
Federal Telegraph Co., 275
Federal Telephone and Radio, 495
Felten & Guilleaume, 226
FiberHome, China, 458
Fiberworks, U.S., 457
Flag Telecom, 10, 461, 463, 464, 572, 594
FNC, U.S., 587
France Cables and Radio, 557
France Telecom, 10, 377, 467, 529, 549, 561, 575

French Atlantic Cable Co., 133
fSONA, Canada, 390
Fujikura Cable Works Ltd., 448
Fujitsu, 455, 467, 493, 582
Funkentelegraphie G.m.b.H, 215

Galvin Manufacturing Corporation, 286
Garuda, Asia Cellular Satellite International Ltd., 438
GC&CC, U.K., 354
GE Americom, 430, 433, 441
General Electric Co., 277, 289
General Electric Co. Ltd., 490
General Electric Signaling Co., 277
German-Atlantic Telegraph Co., 308
Gesellschaft für drahtlose Telegraphie m.b.H., 271
Gesellschaft für drahtlose Telegraphie System Prof. Braun & Siemens & Halske, 215, 271
GIRD, Gruppa Isutschenija Reaktiwnogo Dwishenija, 408
Glass Elliot & Co., 97, 131, 132, 136
Global Crossing, 463, 466
Global One, France, 549
Global Telesystems, U.S., 464
Globalstar, 435, 437, 504
Globo, Brazil, 588
GMDSS, 142, 434
GPO, U.K., 142, 167, 221, 235, 244, 281, 291, 303, 304, 315, 324, 343, 379, 400, 447, 488, 490, 496, 510, 574
GPT, U.K., 460
Great Northern Telegraph China and Japan Extension, 113, 128
Great Northern Telegraph Co., 128, 309, 310
Great Northwestern Telegraph Co., 105
Great Western Railways, 67
GTE, 456, 472, 483, 551, 554, 555, 556, 561
Gutta Percha Co., 132

Halifax and Bermudas Telegraph Co., 137
Harmonix, U.S., 390
Hasler S.A., Switzerland, 262
Helsinki Telephone Co., 472
Henley & Co, U.K., 101
Hispasat, 430
Hitachi, 493, 582
HKTC, Hong Kong, 570
Homages, Germany, 278
HONDUTEL, Honduras, 558
Huawai Technologies, 458
Hughes:
    Aircraft Company, 419
    Ispat Ltd, India, 569
    Network Systems, 539, 440
Hutchison Com. Ltd, 570
Hyundai, 534

I&IC, U.K., 291
IAM, Morocco, 565
IBM, 503, 554
IBTC, U.S., 174, 179
ICANN, 587
ICE, Costa Rica, 558
ICI, 314
ICO, 435, 438
iDA, Singapore, 573
IDO, Japan, 523, 534
IEE, U.K., 185, 456, 488
IETF, 588, 589, 599
IFRB, 431, 598
IITF, 582
Illinois and Mississippi Telegraph Co., 98
Imperial and International Communications Ltd., 127
Imperial Chinese Telegraph Co., 116
Impsat Corporation, Argentina, 465
IMSO, 434
IMT-2000, 435
India Rubber, Gutta Percha and Telegraph Works Co., 167
Indian Rubber Co., 311
Indo-European Telegraph Co., 126
Indostar, Indonesia, 432
Inmarsat, 433, 434, 438, 532
Insat, India, 432
INTEL, Panama, 558
INTELCO, Rep. Of Congo, 567
INTELSAT, 433, 441
Inter-American Radio Office, 359
International Launch Service, U.S., 423
International Networking Group, U.K., 585
International Ocean Telegraph Co., U.K., 138
International Radio Conference, 358, 359
International Standards Accreditation Board, 594
International Telegraph Conference, 359
International Western Electric, 328
INTERSPUTNIK, 426, 433
Intrigna, Canada, 556
Ionica, U.K., 389
Ipergy, U.K., 464
IPTO, 583
IRE, U.S., 496
Iridium, 13, 368, 435, 436, 438
Iridium Satellite LLC, 436
ISky, U.S., 441
IT&T, 139, 181, 248, 291, 302, 304, 324, 328, 342, 347, 348, 375, 381, 496
ITT, 13, 181, 495, 433
ITU, International Telegraph Union, 88, 104, 107, 143, 219, 221, 229, 260, 359, 351, 357
ITU, International Telecommunication Union, 10, 14, 15, 114, 343, 359, 414, 449, 478, 518, 540, 546, 551, 563, 565, 573, 581, 582, 588, 590, 592, 594, 597

Jitong Corporation, China, 570
John Pender & Co., U.K., 136
Johnson and Nephew, U.K., 120
J-Phone, Japan, 535
JPL, U.S., 410, 589
JT, Japan, 571, 573

KDD, Japan, 463, 493, 571, 573, 535
Kenya Telekom, 567
Keystone Telephone Co., USA, 231
Korea Telecom, 573
Koreasat, 432
KPTC, Kenya, 567
KTO, Korea, 573

L.M. Ericsson, 185, 246, 248, 252, 261, 262, 265,
    489, 493, 527, 531, 537, 540
LCT, France, 328, 342, 495
LD Com, France, 458
LMT, France, 343
Lockheed Martin, 421, 423, 428, 430
Lodge-Muirhead Syndicate, 279
Loewe & Co, 241
London District Telegraph Co., 106
Loomis Telegraph Co., U.S., 200
Loral, 437
LTC, France, 493
LTT, France, 327
Lucent, U.S., 390, 458, 466, 555

Magnetic Telegraph Company, U.S., 61
Magneto Marelli, Italy, 380
Malaysian Airlines, 531
Mannesmann Mobilfunk, Germany, 529
Mapple & Brown, U.K., 70
Marconi International Marine Communication
    Co., 281, 357
Marconi Wireless, 273, 279–286, 291, 357
Marconi Wireless and Telegraph Company of
    America, 211, 290
Maroc Telecom, 565
Marseille, Algiers, and Malta Telegraph Co.,
    137
Martlesham Heath Laboratories, 390, 474
Marubeni Corp., 464
Màtav, Hungary, 576
MCI, 433, 456, 461, 552, 555, 577
MCIWorldcom, 458, 464, 555, 557, 572
MCS, U.S., 433
Mediterranean Extension Telegraph Co., 130
Mercury, U.K., 469, 524, 530, 574
MetroNet Communications, 556
Mexican Telegraph Co., 138
Microsoft, 465, 588
Mid-Continental Tel, 554
MIT, 163, 375, 581, 583, 585, 589, 592
Mitchell & Co., 97
Mix & Genest, Germany, 152, 170, 185, 186, 267

Mizushima Research Laboratory, 448
Mobikom, Malaysia, 573
MobileOne, Japan, 573
Montreal Telegraph Co., 66, 104
Morkrum-Kleinschmidt Co., 302, 306
Motorola, 286, 436, 523, 524
MSI, The Netherlands, 567
MTN, South Africa, 567
MTNL, India, 568

Nahuelsat S.A, Argentina, 433
NASA, 410, 419, 431
National Electric Signaling Co., U.S., 277
National Telephone Co., U.K., 315
National Telephone Exchange Association of the
    United States, 247
NBTC, U.S., 167
NBTM, The Netherlands, 191
NCR, U.S., 554, 555
NEC, 181, 327, 347, 349, 365, 377, 380, 381, 388,
    455, 467, 493, 507, 582
NET, Nigeria, 566
Netsat 28, U.S., 441
Netscape Communications Corp., 586
New England Telephone Co., 167
New World Telephone Ltd., 570
New York and Mississippi Valley Printing
    Telegraph Co., 98
New York Telephone Co., 192
New York, Albany and Buffalo Telegraph Co.,
    66
New York, Newfoundland & London Electric
    Telegraph Co., 131
Newall & Co., U.K., 129, 130
Newfoundland Electric Telegraph Co., 130
Nilesat, Egypt, 432
Nitel, Nigeria, 566
Nokia, 527, 530
Norddeutsche Seekabelwerke, 333
Norsar, Norway, 585
Nortel, 390, 455, 458, 467, 569, 572
North California Telegraph Co., 99
North Electric Company of Galion, 251, 252
Northern Electric, 181
Northern Telecom, 552
Norwegian Telecom, 472
Nott and Barlow, U.K., 70
NP&T, Nigeria, 566
NPL, U.K., 583
NSF, U.S., 585, 587
NTCA, U.S., 233
NTT, 432, 448, 452, 455, 475, 476, 493, 494, 505,
    520, 534, 548, 571–573, 582
NTT DoCoMo, 534, 537
Nynex, 464, 554

Ocean Telegraph Co., U.S., 133
Oceanic Telegraph Co., U.K., 106

OCPT, Dem. Rep. of Congo, 567
Odyssey, U.S., 435
Office National des Postes et Télécommunications, Congo Rep., 567
OFTA, Hong Kong, 570
OKI, 493
OmniTrac, 435
Orascom, Egypt, 567
OSPITEL, Peru, 561
OTC, Australia, 577
OY Radiolinja AB, Finland, 529

Pacific Bell, 472
Pacific Cable Board, 309
Pacific Century Cyber Works, 570
Pacific Telegraph Co., 100
Pacific Telesis, 554
Pacifictel SA, Ecuador, 562
Packet Radio Net, U.S., 584
Palapa, Indonesia, 432
Panaftel, 385, 466
PanAmSat, 429, 441
Pasific Satellite Nusantara, 432
PAV Data Systems, 390
Philips, 227, 493
Pirelli, 460, 463
Postal Telegraph Co. U.S., 135, 301
Poulsen Wireless Telephone and Telegraph Co., 275
PowerNets Japan Communications Inc., 572
Press Association, U.K., 304
Prof. Braun's Telegraphie GmbH, 215
PT Indosat, Indonesia, 574
PT Telkom, Indonesia, 574
Pyramid Research, 389

Qualcomm Inc., 435, 437, 440, 533, 540

Racal Millcom Ltd., 524
Rand Corporation, 583
RASCOM, 429, 466, 565
Raynet Corp., 474, 475
Raytheon, 364
RBOC, 548, 554
RCA, 289, 433
Red Sea and India Telegraph Co., 125
Reliance Telecom Ltd, India, 569
River Plate Telegraph Co., 138
Rochester (New York) Telephone Corp., 554
Rostelecom, Russia, 576
Rotterdam Municipal Telephone Administration, 261
Royal Electric Telegraph Administration, Sweden, 110
Russian Rocket Research Institute NII-I, 411

S.B. Submarine Systems Company Ltd., 465
SA TSF, France, 273, 380

Samsung, 534
SAPO, South Africa, 566
SAPT, South Africa, 566
Satelindo, Indonesia, 574
Satmex, Mexico, 433
SATRA, South Africa, 566
Scherbiusschen Chiffriermaschinen AG, 352
Sea Launch Co., 424, 439
SEL, Germany, 376, 380, 381, 491, 492
SES, Luxembourg, 429, 441
SFR, France, 279, 347, 380, 383, 529
SGT, France, 169
Shanghai Bell Telephone Equipment Manufacturing Company, 569
Shinawatra Satellite Public Co., Thailand, 432
Shyam Telelink, India, 569
SIA, Singapore, 531
Siemens & Halske, 70, 76, 78, 94, 104, 125, 134, 170, 175, 186, 226, 227, 241–243, 248, 266, 296, 306, 319, 333, 354, 380
Siemens AG, 378–381, 455, 458, 460, 476, 480, 491, 492
Siemens Brothers, 125, 134, 135, 315
Siemens Edison Swan Ltd,, 490
Singapore Cable Vision, 573
SingTel, Singapore, 573
SIP, Italy, 474
SIT, France, 169
SkyBridge, 440
Skyphone Mobil Connect, 532
SLE, France, 490
Smithsonian Institution, 155, 160, 390
SOCOTEL, France, 491
Sony, 572
SOTELEC, France, 491
Southwest Bell, 557
Southwestern Bell, 554
Space Systems/Loral, 421
Spacecom, China, 570
Spaceway, U.S., 435, 440
Springfield, Albany, and Buffalo Telegraph Co., 63
Sprint, 549, 555, 556
SR Telecom, 386
SRI, U.S., 584
StarHub, Singapore, 573
STC, U.K., 248, 259, 343, 349, 380, 381, 459, 467, 490
STC, Australia, 349
Stentor, Canada, 556
STET, Italy, 561, 563
STL, U.K., 446
Stockholm Allmänna Telefonaktiebolag, 176
Stockholm Telephone Co., 174
Strategis Group, 389
Stromberg-Carlson Co., 489
Strowger Automatic Telephone Exchange Co., 194

Submarcom, 406, 460, 467
Subtel, Chile, 560
Svyazinvest, Russia, 576
Swedish Telecom, 472
Swiss PTT, 475
Swisscom, 575

TAS, Singapore, 573
Tata Teleservices, India, 569
TCNZ, New Zealand, 577
TdP, Peru, 561
Teishinsho, Japan, 279
Telautograph Co., U.S., 152
Telcel, Switzerland, 567
Telco Norte, Argentina, 561
Telco Sur, Argentina, 561
Tele Danmark, 529, 575
Telebras, Brasil, 548, 559
Telecom:
    Argentina, 561
    Finland, 529
    Italia, 563, 575
    Portugal, 474
    Colombia, 561
Telecomm, Telecomunicaciones de Mexico, 557
Teledesic, 438, 440
Telefonbau und Normalzeit, Germany, 491, 493
Telefónica de Argentina, 561
Telefónica de España, 475, 560, 561
Teléfonica Nacional S.A., Mexico, 556
Telefunken, 226, 271, 273, 278, 279, 283, 290, 295,
    296, 343–345, 354, 380, 446
Teleglobe Inc., 436, 461, 556
Telegraph Cable Co., U.K., 324
Telegraph Construction and Maintenance Co.,
    U.K., 132, 136
Telekom Malaysia, 566, 573
Telekomunikacja Polska, 576
Telenor, Norway, 575
Telephone Corporation of Hull, 235
Telesat Canada, 431, 556
Telettra, Italy, 381, 384
Teletype Corporation, U.S., 302
Televerket, Sweden, 176, 261, 265, 489, 493, 523
TeleWay Japan Corp., 571
Telgua, Guatemala, 557
Telia, Sweden, 575
Telkom SA, South Africa, 566
Telmex, Mexico, 556, 557, 588
Telstra, Australia, 383, 577
TelstraSaturn, New Zealand, 577
Telus, Canada, 556
TeraBeam Networks, U.S., 90
Tethers Unlimited Inc., U.S., 441
Texas Instruments, 365
Thaicom, 432
Thintana Communications, South Africa, 566

Thomson CSF, 375, 376, 378
Thomson-Houston, 260, 380
Thuraya Satellite Communications Co., 439
TIM, Italy, 525
Tine Telekom, Malaysia, 573
T-Mobil, Germany, 525, 529
Todd Corporation, New Zealand, 577
Tong Guang-Nortel, China, 569
Tongasat, Tonga, 431
Toronto, Hamilton, and Niagara Electro Magnetic
    Telegraph, 66
TOT, Thailand, 573
Toyota, 572
TRAI, India, 569
Transandine Telegraph Co., 139
TransTelecom, Russia, 456
TRESS, Australia, 510
TTCL, Tanzania, 567
Tu-Ka Cellular Group, Japan, 534
Türksat, 431
TV New Zealand, 577

UCAID, U.S., 589
UCLA, U.S., 584, 585
UCSB, U.S., 584
United Kingdom Telegraph Co., 104, 106
United Tel, U.S., 554
United Telephone Co., U.K., 168
United Wireless, U.S., 273, 275
Universal Private Telegraph Co., U.K., 106
University of the World, 592
Unix, 503
US West, 554
USC/ISI, U.S., 585
USITA, U.S., 231
USTSA, U.S., 533
UTL, Uganda, 567

Verein für Raumschiffahrt, Germany, 409
Verizon, U.S., 555
Vodafone, U.K., 524, 529
Voies Navigables de France, 458
VSNL, India, 568

Washington and New Orleans Telegraph Co., 63
West Coast of America Telegraph Co., 139
West European Telegraph Union, 218
West India and Panama Telegraph Co., 138
Western and Brazilian Telegraph Co., 137
Western Electric, 156, 180, 189, 192, 194, 226, 234,
    238, 248, 252, 254, 255, 259, 280, 302, 314,
    364, 555
Western Telegraph Co., 137, 138
Western Union Telegraph Co., 65, 98–105, 131,
    135, 138, 141, 145, 164–169, 176, 178, 273,
    301–303, 314, 376, 433, 510, 515
Westinghouse, 290

Wharf Holdings New T&T (Hong Kong) Ltd., 570
Wireless Telegraphs and Telephones, Russia, 273
WISeKey S.A., Switzerland, 594
Wolfs Telegraphenbureau, Germany, 88
WorldCom, 455, 549, 555
WTC, World Trade Centre, 594
WTO, World Trade Organization, 557, 572

Xerox, 515, 503

Yukon Telegraph Co., 103

Zhongxing Telecom, 458
ZTE, China, 534

PART III   SUBJECTS

@ symbol, 584
7*-unit code, 301

ABC code, 351
ABC pointer telegraph, 70
ACT-1108, 494
ADFOC, 459
ADPCM code, 536
AKE 11, 489
All-Relay System, 252
Aloha protocol, 503, 584
AM, 347, 371
AMPS, 520–528, 531–534
Annales télégraphiques, 221
Annunciator system, 188
Anzcan cable, 405
ARABVISTA, 588
ARCOS, 463
Aristotle, 491
ARPANET, 368, 503, 583–586
ASCII, 512
ASR, 508
AT&T Telephone Statistics of the World, 229
Atlantic Crossing-1, 465
ATM, 507
Audion, 226, 277
Autographic telegraph, 151
Automanual system, 251
AXE 10, 500

B2B, 593
B2C, 593
Barretter, 277
Baudot code, 107, 512
Baudot telegraph, 139, 357
Belinogram, 296
Belinograph, 296
BERKOM, 474

BIGFON, 474
B-ISDN, 474, 505–506
Bletchley Park, 354, 355
Bomba, 353
Bombe, 354
BORSCHT functions, 497
Breguet telegraph, 73, 107, 111, 115, 139
Breguet-Bétancourt telegraph, 37
Breguet-Foy telegraph, 73
Bumber-7 rocket, 411
Butterstamp telephone, 163, 171, 185

C 450, 521, 524
Cable ships, 97, 310–311, 403–404, 460, 465
    Agamemnon, 131
    Dacia, 311
    Faraday, 97, 134
    Goliath, 97
    Gomer, 137
    Great Eastern, 97, 132, 133, 136
    Niagara, 131
    Stephan, 310
Callpoint, U.K., 536
Callzone, Singapore, 537
Cantat:
    -1, 404
    -2, 405
    -3, 461
Carbon granule telephone transmitter, 168, 176
Carbon telephone transmitter, 168, 173
Carrier frequency, 9, 325, 326, 327
Carterfone, 552
CB, 192, 238, 240
CDMA, 525, 530, 533, 540–542, 570
CDPD, 533
Chatley Heath semaphore tower, 45
Chemical telegraph, 147
Chinese telegraph codebook, 116
cHTML, 535
CLOAX, coaxial cable, 398
Coaxial cable, 9, 314, 327, 331–337, 376, 383, 449, 456
Coherer, 205, 208, 214, 215, 225, 279
Colossus, 355, 496
Columbus-2, 466
Communications Act of 1934, U.S., 552
Communications and Multimedia Act of 1999, Malaysia, 573
Communications Satellite Act of 1962, U.S., 427
Compac, Commonwealth-Pacific cable, 405
Cooke and Wheatstone telegraph, 60, 66–76, 87, 107, 111, 139
Cosmonautic paradoxon, 409
Courier I-B, 413
CP-400, 491
Creed teleprinter, 303–305, 510
Cross-polarization interference canceling, 374

Cryptograph, 352, 355
CSNET, U.S., 585

d'Arsonval telephone, 315
D-AMPS, 525, 527, 531–533
DAV, 373
DAVID, 373
DCME, 460
DCS, 524–538
DDM, 390
DEC-Digital Network Architecture, 503
DECnet, 503
De-coherer, 209
DECT, 389, 536–539
Desk-Fax, 515
Dial telephone, 197
Dialing disc, 196, 247, 256
Dietl calling device, 246
Digital divide, 551, 590, 594
Digital watermark, 587
Diode, 9, 225
Director system, 238, 245
DM quad, 320
DNS, 585, 587
Dollis Hill, 490, 496
Dollond telescope, 16
DPLMRS, U.S., 286
DSL, 389, 477–478, 506
Duplex telegraphy, 97
DUV, 373
DWDM, 390, 450, 457, 465

Early Bird, 367, 428
Earth station, 413–419, 428, 431, 433
    Goonhilly, U.K., 416, 584
    Pleumeur-Bodou, France, 416
    Raisting, Germany, 417
    Tanum, Sweden, 585
EC-DC, 594
Echo I and II, 410, 413
ECO, 489
e-commerce, 594
EDFA, 450–455, 461, 470
EDGE, 531
Edison-Effect, 225
e-entertainment, 594
e-government, 594
e-health care, 594
e-learning, 594
Electrochemical telegraph, 30, 54
Electrodynamics, 31
Electromagnetic field, 199
Electromagnetic waves, 90, 199, 203–206, 225, 269, 277, 279
Electronic signatures in global and national commerce act, 594

Electronic valve (vacuum tube), 9, 226, 274, 280, 283, 308, 314, 324, 331, 338, 355, 364, 490
Electrostatic telegraph, 49
e-mail, 10, 17, 147, 512, 542, 566, 584–585
EMAX, 489
EMD, 481
Enchanted lyre, 153
Energia, 421
Enigma, 352–355
Enlaces, Chile, 592
EPABX, 489
EPIRB, 434
e-service, 594
e-signature, 593
ESM matrix, 492
E-TACS, 524
ETHERNET, 503
Eunicid worm, 133
Eutelsat, 429
EVA, 491
EWS, 492, 500
Explorer, 411–412
eZiText, China, 525
EZWeb, 535

FDD, 541
FDM, 373
FEAX-302A, 494
Female telephone operator, 191
FET, 370
First:
    commercial radio-relay network, 348
    electrical telegraph line, 67
    e-mail, 584
    radio emergency call, 211
    submarine coaxial cable, 334
    submarine telephone cable, 315
    telefax transmission, 1847, 148
    transatlantic telegraph cable, 1866, 10
Five-unit binary code, 107
Flag Atlantic-1, 10, 461, 464, 466
Flag Euro-Asia, 464
FM, 344–348
Fonepoint, 536
FPLMTS, 435
Frame relay, 504
Frictional electricity, 49
FSO, 390

Gallows telephone, 162
Galvanic electricity, 30
Galvanometer, 32, 96, 200, 205, 225, 295, 312
Gemini cable, 461
GERAN, 531
GII, 9, 10, 581–582, 587
Global Village, 595

GMDSS, 142, 434
GMPCS, 368, 435
Gooseneck telephone, 168
Gorizont, 428
Gower-Bell telephone, 315
GPRS, 531
GPS, 434, 439, 537
Graham Act of 1921, U.S., 585, 231
Graphophone, 179
Great Northern Telegraph Cable, 113
Greenpoint, The Netherlands, 536
GSM 1900, 533
GSM Global Roaming Forum, 533
GSM, 439, 523, 525, 528–534, 539, 540, 589
GSMPro, 531
GSM-R(ail), 531
Gutta-percha, 7, 94, 129, 131, 133, 159, 209, 314–315, 324
Gutta-percha press, 94

Handy Walkie, 286
Handy-Phone, 523
Harmonic telegraph, 160, 161, 176
Hasler Hs 31 system, 251
Haut-Bar semaphore station, 45
HDW switch, 243, 480
Heptode, 227
Herkon relay, 493
Hertzian waves, 205, 207–208, 277
Heterodyne principle, 278
Hexode, 227
HITEX-1/3, 493, 494
Hohmann ellipse, 421
HSCSD, 531
HTML, 531, 586
HTTP, 586
Hughes telegraph, 126, 139, 169, 221, 306, 357
Hydraulic microphone, 275
Hydrofoil HD-4, 180
Hypernet Internet backbone, Malaysia, 595

iBop, 536
Icecan, 405
IDD, 248, 484
iDEN, 533
iDNS, 588
IES 2000, 594
Image telegraphy, 7
i-mode, 537
IMP, 584
Imperial Chain, 291
IMT-2000, 435, 540–544
IMTS, 519
IN, 487
Indian Telegraph Act, 1854, 113
INS, Japan, 505

Integrated circuit, 365
Intelsat satellites 1 to 4A, 367, 428, 484, 584
International Polar Year, 412
International Property Rights, 540
International radio cartel, 290
Internet, 10, 368, 453, 503, 512, 532, 535, 542, 580, 583–590, 592, 594
Internetting, 585
Interplanetary Internet, 589
IP, 6, 487, 585
IPP, 517
IPv4, 588
IPv6, 589
Iridium system, 13, 368, 442
ISDN, 487, 505–506
ISP, 587
ISS, 488
ITA1, 108
ITU Telecom Surplus Fund, 600
IVR, 508

Janus switch, 267
Journal télégraphique, 221
J-Sky, Japan, 535
J-TACS, 523
Jupiter C rocket, 412

Karolus-cell, 296
Kennelly-Heaviside layers, 280
Kingsbury:
    Agreement, 231
    Commitment, 552
Klystron, 339, 343, 345
Knockenhauer spirals, 202
Kopiertelegraph, 294
Köln-Flittard semaphore tower, 45
Krarub cable, 319, 312–322, 324

Laser, 9, 445, 454, 515
Law on telegraph lines, No. 6801, France, 39, 169
Leyden jar, 49, 199, 200, 202
Lorenz-Korn picture telegraph, 295
LSIC, 366

Magnetron, 339, 343
Marconi radio, 207–212
Marconi station, Australia, 123
Marshall Plan Aid, 363
Maxwell's theory, 201, 202, 204
McBerthy selector, 259
MCS, micro cabling system, 476
MCS, mobile control station, Japan, 520
MCVD, 448
Merchants Code, 351
Microphonic effect, 168
Microprocessor, 366

Microsoft Explorer, 586
MILNET, 585
Minitel, 580–581
Mirror galvanometer, 50, 96
Missing Link, 546, 551
Model 7 teleprinter, Creed, 304
Molniya, 428
Monotelephone, 324
Morelos, 433
Morse code, 10, 76, 141, 143, 206, 217, 303
Morse telegraph, 55–65, 72, 74, 76, 81, 87, 90,
    111, 113, 126, 139, 141, 142, 169, 214, 217,
    220, 306, 350, 357, 510
MSC, Malaysia, 594
MTI, 343
Multi-carrier transmission, 374
Multimedia, 10, 580–596
Multiplicator, 32
Murray multiplex system, 303
Musical telegraph, 159, 160, 164

Nahuel I, 433
N-AMPS, 523
NAP, 594
National Telegraph Review, 221
NCP, 584
Negative feedback, 326
NII, 582
NMT, 521, 523, 525, 527
No. 1 ESS, 487
No. 101 ESS, 488
No. 5 ESS digital switch, 506
Non-zero-dispersion fiber, 450
NSFNET, 585, 586, 587
N-TACS, 523

o/w line, 91, 184, 228, 314, 315, 317, 319, 323,
    349, 350
    Austria, 217
    Central America, 128
    Germany, 173
    Japan, 327
    multiplex, 325, 326, 331
Octode, 227
Ohm's law, 32, 93
OPPW, 459
Optical Ethernet, 503
Optical fiber cable, 9, 374, 378, 445–479
    Australia, 124, 383
    Japan, 327
OPWG, 458
Orbiter, 412
OVD, 448

PABX, 386, 491, 506
Packet Radio Net, 585

Packet-switching, 6, 500, 583
PACS, 536
PAM, 489, 493, 496
Parametron, 493, 494
PCM, 12, 331, 366, 446, 489, 496, 497
PCN, 525, 527, 530
PCS, 533, 535
PDC, 525, 527, 534
PDH, 366
PDM, 455
Pegasus rocket, 440
Pentode, 227
Permalloy cable, 308, 314, 322
Perminvar cable, 324
PFGI-POF, 477
Phonautograph, 160
PhoneNet, 585
Phonepoint, 536
Photo telegraphy, 7
Photoelectric cell, 296, 298
Photophone, 179, 445
PIT, 566
Playfair cipher, 352
PM, 343, 348, 380, 381
P-MP, 386–389, 426
PMR, 531
PN-PN elements, 493
POF, plastic optical fiber, 477
Pointel, France, 536
Polyethylene, 314, 403
Pony express:
    Australia, 123
    United States, 98–99
POTS, 486
Powerline telecommunications, 478
Prepaid service, 525
PSDN, 500, 512
PSTN, 386, 500, 505, 520, 537, 538, 570, 588
PTAT-1 and 2, 461
Public telephone box, 386
Pupin coil, 12, 13, 90, 228, 231, 323

R6 system, 260
RACE, 474
Radar, 343, 348
Radio emergency call, 282, 283, 358, 434
RadioCom 2000, 521, 524
Radiotelegraphy, 7, 270, 350
Raduga, 428
Raman amplification, 452, 470
Ramses, 491
RCR, Japan, 534
Readmail, 584
Reed relay, 480, 492
Relay, 94, 252, 364, 485, 492
Rheinland cable, 319, 323

RITL, 388
Rolls Royce, 10
RTMS, 521, 524

SAT-1, 392, 405
SAT-2, 466
Satellite launching center, 422–424
    Baikonur Cosmodrome, Russia, 422, 423
    Cape Canaveral, U.S., 411, 416, 424, 437
    Kapistin, Russia, 441
    Kourou, French Guiana, 422, 430
    Khrunichev Space Center, 423, 424, 437
    Pletsesk, Russia, 441
    Taiyuan, China, 422, 424
    Tanegashima Island, Japan, 423
Satellite launching vehicle, 421–426
    Atlas, U.S., 422, 426
    Ariane, France, 422, 426, 430, 432
    Cosmos, Russia, 441
    Delta, U.S., 416, 426, 436
    Energia, Russia, 421
    H-2, Japan, 422
    Long March, China, 422, 424, 426, 436
    OP Polyot, Russia, 441
    Proton, Russia, 422, 424–426
    Space shuttle, U.S., 421
    Titan, U.S., 422
Satnet, 584, 585
SchoolNet, South Africa, 592
Score, 413
Scotice, 405
SCTP, 588
SDH, 367, 374, 378, 506
Seacom 1, 405
SEA-ME-WE 1, 406, 462
SEA-ME-WE 2, 461, 466
SEA-ME-WE 3, 462
Secret teleprinter, 353, 355, 356
Selenium cell, 180, 294, 296, 312
Semaphore, 14, 29, 39, 55
SESAT, 429
SIM card, 524, 534
Singapore One, 594
Single-sideband transmission, 325
Siphon galvanometer, 206
Siphon recorder, 96, 308, 311, 314
SkyBridge, 442
SLA, semi-conductor laser amplifiers, 450
Slaby-Arco radio system, 214
SLIC, 497
SMS, 525, 531, 532, 537, 594
SNA, 503
Socrate, 491
SOHO, 506
Solidaridad 1, 433
Soliton transmission, 9, 451–455

SONET, 367, 572
SOS, 142, 434
SPC, 485
Spread spectrum frequency hopping, 367
Sputnik, 412, 423, 583
SS 7, 487, 588
SSB-AM, 371
Star quad, 320
Static electricity, 49
Submarine cable, 10, 87, 88, 91, 142, 308
    Australia, 119, 120, 123, 124, 382
    China, 115
    India, 113, 124
    Japan, 113, 114
    Scandinavia, 128
    U.K.–Germany, 126
Subspace flying base station, 10
Substitution coding, 350
Superheterodyne circuit, 278, 280
Symmetrical cable, 327, 332
SYNCOM I and II, 419, 420
Syntony, 206
System 12, 13, 500
System E 10, 500

Tachygraphe, 19
TACS, 521, 524, 527
TAE, 458
TASI, 402, 406
Tasman cable, 405
TCP, 564, 585
TCP/IP, 585, 587
TDD, 541
TDMA, 533, 541
Telautograph, 152, 295
Telebanking, 590
Telebooking, 590
Telecom exhibition, 599
Telecommunications Act of 1989, Australia,
    577
Telecommunications gap, 546
Teleconsulting, 590
Telecopier, 515
Teledeltos, 298
Tele-entertaining, 590
Telegraph Act, U.K., 106, 168, 210
Telegraph operator, 351
Telegraphic plateau, 131
Telegraphoscope, 295
Telelearning, 590, 592
Telemedicine, 590, 592
Telenet, U.S., 585
Telephone operator, 247, 251
Telephoto, 298
Telepix, 297
Telepoint, 536

Teleprinter, 7, 353
Teleshopping, 590
Télétel, France, 580
Teletex, 512
Teletext, 580
Teletypewriter, 142, 301
Teletypewriter Exchange Service, 302
Teleworking, 590
Telewriter, 152
Telstar, 413, 414, 416, 417, 551
Teredo worm, 133
Terminator Tether, 441
Tesla transformer, 205
Tetrode, 227
Timed wire service, 302
Titanic disaster, 282–283, 358
TLD, 585
Trans-Andean telephone line, 324
Transatlantic telegraph cable, 103, 129–135, 164, 308, 351
Transatlantic telephone cable, 10, 324, 334, 399, 551
  TAT-1, 10, 401, 403, 460, 484
  TAT-2, 402, 403
  TAT-3, 402, 404
  TAT-4, 403, 405
  TAT-5, 392, 403
  TAT-6, 403, 460
  TAT-7, 403
  TAT-8–12, 460
  TAT-13, 461
  TAT-14, 425, 461
Transcontinental telephony, 231
Transistor, 13, 227, 364–366, 485, 489, 494, 515
Transmultiplexer, 373
Transpac 1/2/3, 405
Transposition coding, 350
Trans-Siberian telegraph line, 128
Traveling wave tube, 339, 370, 380
Triode, 9, 226, 231, 279, 322
Trunking, 519
Type-bar page printer, 301

Type-wheel printing telegraph, 301
Typotelegraph, 75

UMTS, 540, 541
URL, 586
UWC-136, 541

V-2, 408, 410–412
Van Allen belts, 413
Vandenbergh Air Force base, 436
Vanguard rocket, 412
Vapor axial deposition, 448
Velocity-modulated tube, 339
Victorian Internet, 107
Videotex, 580, 581
Vigigraphe, 39
Virtual classroom, 592
Vodas, 287
VoIP, 504, 570, 573, 588
Voltaic pile, 30
VSAT, 429

Wac Corporal, 411
WACS, 539
WAP, 531
WDM, 9, 10, 453–455, 474
WESTAR I, 432
Wheatstone bridge, 67
Windows 95, 586
Wireless Girdle Round the World, 273, 289
Wireless Telegraph and Signal Co. Ltd., 210
WLL, 9, 388, 561
WML, 531
Worldphone 1800, Thailand, 530
World Wide Web, 586

Yagi antenna, 281

Zeitschrift des Deutsch-Österreichischen Telegraphenvereins, 221
Zero water peak fiber, 450
Zonepoint, 536